HANDBOOK
of
RECEPTORS
and
CHANNELS

LIGAND-
and
VOLTAGE-GATED
ION CHANNELS

Edited by

R. Alan North

CRC Press
Boca Raton Ann Arbor London Tokyo

Library of Congress Cataloging-in-Publication Data

Ligand- and voltage-gated ion channels / edited by R. Alan North
p. cm. — (Handbook of receptors and channels)
Includes bibliographical references and index.
ISBN 0-8493-8322-6
1. Ion channels. 2. Neurotransmitter receptors. I. North, R. Alan. II. Series.
QH603.I54L53 1994
574.87′5—dc20

94-11520
CIP

Direct all inquiries to CRC Press, Inc., 2000 Corporate Blvd., N.W., Boca Raton, Florida 33431.

International Standard Book Number 0-8493-8322-6
Library of Congress Card Number 94-11520
Printed in the United States of America 1 2 3 4 5 6 7 8 9 0
Printed on acid-free paper

HANDBOOK of RECEPTORS and CHANNELS Series

Stephen J. Peroutka, Series Editor

President and Chief Executive Officer
Spectra Biomedical, Inc.
4040 Campbell Avenue
Menlo Park, California 94025

Published Titles

G Protein Coupled Receptors
Edited by Stephen J. Peroutka

Ligand- and Voltage-Gated Ion Channels
Edited by R. Alan North

Topics covered in future volumes

Transporter/Uptake Receptors

Tyrosine Kinase Receptors

The Editor

R. Alan North is Principal Scientist at the Glaxo Institute for Molecular Biology in Geneva, Switzerland. Dr. North graduated B.Sc. (Physiology, 1969), M.B., Ch.B. (Medicine, 1969), and Ph.D. (Pharmacology, 1973) from the University of Aberdeen. He served as House Officer and Registrar in the Aberdeen Hospitals and is a registered medical practitioner in the United Kingdom. During 18 years in the U.S., Dr. North held appointments as Associate Processor of Pharmacology at Loyola University Stritch School of Medicine, Chicago; Professor of Neuropharmacology at the Massachusetts Institute of Technology, Cambridge; and Senior Scientist at the Vollum Institute of Oregon Health Sciences University, Portland. He has also held Fellowships or Visiting Professorships at the Max Planck Institute for Psychiatry in Munich; the John Curtin School of Medical Research at the Australian National University, Flinders University in Adelaide; the Bogomoletz Institute of Physiology in Kiev; the Johann Wolfgang Goethe Universitaet in Frankfurt; and the University of Melbourne. Dr. North has received several prizes and awards including the Gaddum lectureship of the British Pharmacological Society.

Dr. North's professional interests are reflected in his membership of the Physiological Society, The Society of General Physiologists, The British Pharmacological Society, The Society for Neuroscience, and the American Society for Pharmacology and Experimental Therapeutics. His work has been centered around a quantitative understanding of drug and transmitter action at the level of single cells and single molecules, primarily by electrical measurements. His extensive publications deal with the drug and neurotransmitter receptors, structure and function of potassium channels, drug abuse and drug dependence, the physiology of the autonomic and enteric nervous systems, pain mechanisms, psychoactive drugs, and mental illness.

Abstract

The last 10 years have witnessed an explosive increase in our understanding of the molecular and cellular diversity of membrane ion channels. This volume presents a current view of this information in a readily available format. It is an authoritative and comprehensive picture of the structure and function of each of the known classes of these ion channels. Each chapter provides up-to-date amino acid sequences (with data base accession numbers), current views of secondary, tertiary, and quaternary channel structure, and evolutionary relationships among channels. The relationship between primary structure and channel properties such as ligand binding, permeation, and gating have been elucidated by several approaches — from mutagenesis to modelling; these are clearly presented. Three chapters deal with voltage-gated channels (Potassium, K. G. Chandy and G. A. Gutman; Sodium, A. L. Goldin; Calcium, A. Stea, T. W. Soong, and T. P. Snutch), five describe the main families of ligand-gated channels (Nicotinic Acetylcholine, J. Lindstrom; 5-Hydroxytryptamine, J. J. Lambert, J. A. Peters, and A. G. Hope; Excitatory Amino Acids, R. Sprengel and P. H. Seeburg; γ-Aminobutyric Acid, R. F. Tyndale, R. W. Olsen, and A. J. Tobin; Glycine, D. Langosch), and one chapter covers cyclic nucleotide-gated cation channels (K.-W. Yau and T.-Y. Chen). The volume is an important source of information for researchers and students in molecular and cell biology, whether their primary interest be the physiology of membrane currents or the structure of membrane proteins.

Contributors

K. George Chandy, Ph.D., M.D.
Department of Physiology and Biophysics
College of Medicine
University of California, Irvine
Irvine, California

Tsung-Yu Chen, Ph.D.
Deparatment of Neuroscience
The Johns Hopkins University School of Medicine
Baltimore, Maryland

Alan L. Goldin, M.D., Ph.D.
Department of Microbiology and Molecular Genetics
University of California, Irvine
Irvine, California

George A. Gutman, Ph.D.
Department of Microbiology and Molecular Genetics
University of California, Irvine
Irvine, California

Anthony G. Hope, Ph.D.
Department of Pharmacology and Clinical Pharmacology
Ninewells Hospital and Medical School
University of Dundee
Dundee, Scotland

Jeremy J. Lambert, Ph.D.
Department of Pharmacology and Clinical Pharmacology
Ninewells Hospital and Medical School
University of Dundee
Dundee, Scotland

Dieter Langosch, Ph.D.
Department of Neurochemistry
Max-Planck Institute for Brain Research
Frankfort, Germany

Jon Lindstrom, Ph.D.
Department of Neuroscience
University of Pennsylvania School of Medicine
Philadelphia, Pennsylvania

R. Alan North, M.B., Ch.B., Ph.D.
Glaxo Institute for Molecular Biology
Geneva, Switzerland

Richard W. Olsen, Ph.D.
Department of Pharmacology
Center for the Health Sciences
University of California, Los Angeles School of Medicine
Los Angeles, California

John A. Peters, Ph.D.
Department of Pharmacology and Clinical Pharmacology
Ninewells Hospital and Medical School
University of Dundee
Dundee, Scotland

Peter H. Seeburg, Ph.D.
Department of Molecular Endocrinology
Center for Molecular Biology
University of Heidelburg
Heidelburg, Germany

Terry P. Snutch, Ph.D.
Biotechnology Laboratory
University of British Columbia
Vancouver, Canada

Tuck Wah Soong, Ph.D.
Biotechnology Laboratory
University of British Columbia
Vancouver, Canada

Rolf Sprengel, Ph.D.
Department of Molecular Endocrinology
Center for Molecular Biology
University of Heidelberg
Heidelburg, Germany

Anthony Stea, Ph.D.
Biotechnology Laboratory
University of British Columbia
Vancouver, Canada

Allan J. Tobin, Ph.D.
Department of Biology
University of California, Los Angeles
Los Angeles, California

Rachel F. Tyndale, Ph.D.
Department of Pharmacology and Biology
University of California, Los Angeles
Los Angeles, California

King-Wai Yau, Ph.D.
Department of Neuroscience
The John Hopkins University
Baltimore, Maryland

Table of Contents

1

Voltage-Gated Potassium Channel Genes

K. George Chandy and George A. Gutman

2.1.0 Introduction

2.1.0.1 Potassium Channels

Potassium channels play a vital role in the functioning of diverse cell types (Rudy, 1988; Hille, 1993). They regulate membrane potential in electrically excitable cells such as nerves and muscle, as well as in nonexcitable cells such as lymphocytes (Hille, 1993). Although many vertebrate K^+ channel genes have been isolated by molecular cloning (see Gutman and Chandy, 1993), and a variety of experimental strategies have defined functional domains within the channel proteins (see Miller, 1990, 1992; Jan and Jan, 1992), little progress has been made in the biochemical characterization of these proteins. In addition, little effort has been made to define the mechanisms that control transcription and functional expression of these important molecules. Several excellent reviews have dealt with the structural features that underlie particular biophysical and pharmacological properties of K^+ channels (Miller, 1990, 1992; MacKinnon, 1991a; Jan and Jan, 1992; Pongs, 1992, 1993), and others discuss the *Shaker* and related K^+ channel genes in flies (Salkoff et al., 1992; Pongs, 1993), the mammalian Isk K^+ channel (Swanson et al., 1993), and the calcium-activated K^+ channel (Garcia et al., 1993). This chapter will focus on the mammalian voltage-gated K^+ (Kv) channel gene family. We discuss their amino acid sequence alignments and evolutionary relationships, compare their electrophysiological and pharmacological properties, and discuss possible mechanisms that may underlie their tissue-specific expression.

2.1.0.2 The Extended Kv Channel Gene Family

Early in 1987, three groups, using genetic techniques in combination with molecular strategies, isolated the gene encoded by the *Shaker* (Sh) locus in

0-8493-8322-6/95/$0.00+$.50

Drosophila melanogaster (Kamb et al., 1987; Papazian et al., 1987, Schwarz et al., 1988; Pongs et al., 1988). Three related fly genes, *Shab, Shaw,* and *Shal,* were soon isolated, and each of these produced functionally distinct channels (Butler et al., 1989, 1990; Wei et al., 1990; Covarrubias et al., 1991). A topology has been proposed for this family of channels (shown in Figure 1) on the basis of hydropathy plots and structure–function analyses (Miller, 1990, 1992; Jan and Jan, 1992; Pongs, 1992; Durell and Guy, 1992). The channel is thought to have six membrane-spanning segments, termed S1 through S6, with both the amino- and carboxy-termini located intracellularly. The region between the S5 and S6 segments (the "P-region," or "pore") is thought to participate in forming the ion conduction pathway, while the S4 segment forms a major part of the voltage sensor (Miller, 1990; Papazian et al., 1991; Jan and Jan, 1992; Pongs, 1992; Durell and Guy, 1992). The loop linking S1 and S2 has recently been experimentally confirmed to be located extracellularly (Chua et al., 1992; Shen et al., 1993). A more detailed description is provided below.

Using *Shaker* probes, Tempel et al. (1988) and Baumann and colleagues (1988) isolated the first mammalian homologues of the *Shaker* gene. In the next four years a total of 17 vertebrate genes encoding voltage-gated K^+ channels were isolated from mammals and frogs. The current status of this

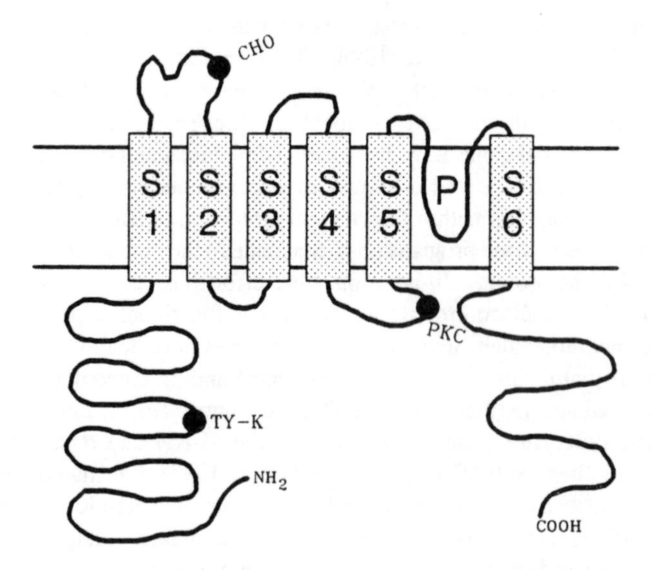

FIGURE 1. Schematic diagram of the presumed structure of voltage-gated K^+ channels. The six putative membrane-spanning domains are labeled S1 through S6, and "P" indicates the P-region, which is thought to form part of the ion conducting pore. Three sites of possible posttranslational modification are indicated, an N-glycosylation ("CHO") and a tyrosine kinase ("TY-K") site both present in mammalian *Shaker*-subfamily channels, and a protein kinase C ("PKC") site present in all known voltage-gated K^+ channels (see text).

continuously growing list is shown in Table 1, which includes the assigned gene names conforming with the recently adopted standardized nomenclature (Chandy et al., 1991). Most of these genes segregate into four clearly defined gene subfamilies (Chandy et al., 1991; Gutman and Chandy, 1993; see references in Table 1), each of which (Kv1.1-1.8, Kv2.1-2.2, Kv3.1-3.4, Kv4.1-4.3) is structurally and evolutionarily related to one of the four fly genes *(Shaker, Shab, Shaw,* and *Shal).* Two additional mammalian genes have been isolated (IK8/Kv5.1 and K13/Kv6.1) for which fly homologues have not yet been identified (Drewe et al., 1992); their inclusion in this scheme of nomenclature should be treated with caution, as their functionality and voltage dependence have yet to be confirmed by expression of the cloned sequences. In addition, two *Shaker* homologues from *Xenopus* (XSha1, XSha2) have been described and have been incorporated into this standardized nomenclature. *Shaker* homologues have also been described in *Aplysia* (Pfaffinger et al., 1991), in *C. elegans* (Wei et al., 1991), in leeches (Johansen, 1990), and in schistosomes (E. Kim, personal communication). The incorporation of the *Aplysia* channel (Shen et al., 1993) into the standardized nomenclature is currently questionable, as their divergence appears to predate the gene duplications that gave rise to the multimembered Kv gene subfamilies in mammals and frogs (discussed below).

2.1.1 *Shaker*-Related Gene Subfamily

2.1.1.1 Protein Sequence Comparisons

Figure 2 shows the alignment of the amino acid sequences of all known *Shaker* homologues, including those of humans, rats, mice, dogs, cattle, frogs, and the mollusc *Aplysia,* together with the *Drosophila Shaker* sequence.* An obvious feature of this alignment is the striking sequence conservation within certain regions, including most of the hydrophobic core

* In discussing this and other sequence alignments, we would emphasize several cautionary notes:

1. Sequences from less than full-length mRNA can obscure authentic start sites (e.g., DRK1), or suggest spurious polyadenylation sites (e.g., MBK1).
2. In the case of protein-coding sequences determined from genomic DNA alone, lacking mRNA sequences with which to compare them, the possible presence of unidentified introns may obscure the sites of translation initiation or termination.
3. The presence of unspliced or only partially spliced mRNAs, which appears to be not uncommon among K^+ channels, may mask the presence of introns (e.g., MBK1) and also obscure translational start and stop sites (e.g., *Shaker).*
4. Not surprisingly, a variety of unresolved sequencing errors exists in published K^+ channel sequences. While many may be of no major consequence, several cause local shifts in the reading frame (e.g., mKv3.3 vs rKShIIID, and RCK4 vs RHK1), or alter either translational start sites (mKv3.3 vs rKShIIID) or termination sites (RK5 vs Shal1).

In our discussion, we have attempted to avoid drawing conclusions from sequence differences that are likely to be the result of technical errors, and to rely largely on features that can be confirmed by comparison between independently determined sequences of the same gene, or of homologous genes between different species.

Table 1a. Nomenclature of Mammalian Voltage-Dependent Potassium Channel Genes[a]

	Names in use[c]			
Standard names[b]	**Mouse**	**Rat**	**Human**	**(Chrom.)**
Shaker-related subfamily 1				
Kv1.1 (KCNA1)	MBK1[1]	RCK1[2]	HuK(I)[3,4,41]	(12p13[35,45,62,80])
	MK1[5,17]	RBK1[7]	HKC-1[71]	
		RK1[17,27]		
Kv1.2 (KCNA2)	MK2[5]	BK2[6]	HuK(IV)[3,4,41]	
	MK5[17]	RCK5[9]		
		NGK1[10]		
		RK2[27]		
		RAK[40]		
		RH1[72]		
Kv1.3 (KCNA3)	MK3[5,11,17]	RCK3[9]	HuK(III)[3,4]	(1p13.3[31])
		RGK5[13]	hPCN3[12]	(1p21[59])
		KV3[14]	HLK3[31]	
			HGK5[32]	
Kv1.4 (KCNA4)	MK4[17]	RCK4[9]	HuK(II)[3,4,41]	(11q13-14[57])
	mKv1.4[80]	RHK1[15]	hPCN2[12a]	(11p14.1[58,80])
		RK4[17,27]	HK1[30]	
		RK8[17]		
Kv1.5 (KCNA5)[d]	—	KV1[14]	HuK(VI)[3,4]	(12p[33,45])
		RK3[27]	hPCN1[12]	
		RMK2[36]	HK2[30]	
			HCK1[33]	
			fHK[73]	
Kv1.6 (KCNA6)	MK2[17]	RCK2[16]	HBK2[16]	
	MK6[56]	KV2[14]	HuK(V)[3,4]	
Kv1.7 (KCNA7)	MK6[17]	RK6[17]	—	(19q13.3[39,45])
	MK4[18]			
Shab-related subfamily 2				
Kv2.1 (KCNB1)	mShab[19]	DRK1[20]	hDRK1[67,68]	
Kv2.2 (KCNB2)	—	CDRK[34]	—	
Shaw-related subfamily 3				
Kv3.1 (KCNC1)[e]	NGK2[10]	Kv4[21]	hKv3.1[35]	(11p15[35,38])
	mShaw22[22]	Raw2[29]	NGK2-KV4[37,38]	(11p14.3-15.2[80])
Kv3.2 (KCNC2)[e]	mShaw12[22]	Rshaw12[22]	HKShIIIA[37]	(12[35,45])
		RKShIIIA[23]		(19q13.3-13.4[84])
		Raw1[29]		
		rKv3.2b,c[60]		
Kv3.3 (KCNC3)[e]	mKv3.3[24]	RKShIIID[37]	HKShIIID[37]	(19q13.3-4[24,39,44,45])
	mShaw19[22]		hKv3.3[44]	
Kv3.4 (KCNC4)	mKv3.4[24]	Raw3[25]	HKShIIIC[28]	(1p21[24,28,45])
Shal-related subfamily 4				
Kv4.1 (KCND1)	mShal[26]	—	—	
Kv4.2 (KCND2)	—	RK5[27]	—	
		Shal1[46]		
Kv4.3 (KCND3)	—	RKShIVB[54,55]	—	
Subfamily 5				
Kv5.1 (KCNF1)	—	IK8[43]	—	
Subfamily 6				
Kv6.1 (KCNG1)	—	K13[43]	—	

Table 1a. (Continued)

[a] See Chandy et al., *Nature* 352, 26, 1991; Gutman and Chandy, *Sem. Neurosci.* 5, 101–106, 1993.
[b] Human locus names assigned by the Human Gene Mapping Workshop (HGMW) are given in parentheses.
[c] Homologues of Kv1.1, Kv1.5, Kv1.6 and Kv1.7 have been isolated from hamster,[17] two *Shaker*-related genes *(Xsha1* and *Xsha2)* from *Xenopus,*[8,61] homologues of Kv1.2 (BGK5)[69] and Kv1.4 (BAK4)[63] from cattle, Kv1.2 (dKv1.2)[70] from dog, and several clones showing similarity with Kv1.1–1.4 from rabbit.[42] Homologues of the *Drosophila eag* gene have been isolated from mouse (meag), rat (reag), and human (heag), representing at least two distinct members of a cyclic nucleotide-binding K+ channel family.[65,66] Genomic and cDNA sequences encoding the IsK/minK channel, a putative single-membrane-spanning domain channel, have been isolated from rats, mice and humans;[47–49] the human homologue has been assigned the HGMW name KCNE1, and the gene has been localized to human chromosome 21q22.[35,45,64]
[d] Termed "KCNA1" in Reference 33.
[e] Alternately spliced forms are known.

Table 1b. K+ Channel Sequences in GenBank/EMBL Databases[a]

Name[b]	Comments	Access. no.	Ref.
mKv1.1	MBK1	Y00305	1
	MK1	M30439[c]	5
rKv1.1	RCK1	X12589	2
	RBK1	M26161	6,7
hKv1.1	HuK(I)	L02750	3,4,41
xKv1.1	XSha1	M94258[c]	61
mKv1.2	MK2	M30440[c]	5
rKv1.2	BK2	J04731	6
	RCK5	X16003	9
	RAK	M74449	40
hKv1.2	HuK(IV)	L02752	3,4,41
bKv1.2	BGK5[e]	L23170	69
xKv1.2	XSha2	M35664[c]	8
mKv1.3	MK3	M30441[c]	5,11
dKv1.2	dKv1.2	L19740	70
rKv1.3	RCK3	X16001	9
	RGK5	M30312[c]	13
	KV3	M31744[c]	14
hKv1.3	hPCN3	M55515[c]	12
	HLK3	M85217	31
	HGK5	M38217[c]	32
rKv1.4	RCK4	X16002	9
	RHK1	M32867	15
hKv1.4	HuK(II)	L02751	3,4,41
	hPCN2	M55514	12
	HK1	M60450	30
hKv1.4		U037322-3[c]	80
bKv1.4	BAK4	X57033	63
rKv1.5	KV1	M27158[c]	14
hKv1.5	hPCN1	M55513	12
	HK2	M60451	30
	HCK1	M83254[c,d]	33
mKv1.6	MK1.6	M96688	56
rKv1.6	RCK2	X17621	16
	KV2	M27159[c,e,f]	14
hKv1.6	HBK2	X17622	16

Table 1b. (Continued)

Name[b]	Comments	Access. no.	Ref.
mKv2.1	mShab	M64228	19
rKv2.1	DRK1	X16476[g]	20
hKv2.1	hDRK1	L02840[g]	67
	hDRK1	X68302	68
rKv2.2	CDRK	M77482	34
mKv3.1	NGK2	Y07521	10
rKv3.1	KV4	M68880	21
	Raw2	X62840	29
hKv3.1	hKv3.1	M96749[e]	35
rKv3.2[h]	RKShIIIA	M34052	23
	Raw1	X62839	29
	rKv3.2b	M59211	60
	rKv3.2c	M59313	60
mKv3.3	mKv3.3	X60796-7[c]	18,24
rKv3.3[h]	RKShIIID	M84210-1	37
hKv3.3	hKv3.3	Z11585[c,e]	44
mKv3.4	mKv3.4	M81253[c,e]	18,24
rKv3.4	Raw3	X62841	25
hKv3.4	HKShIIIC	M64676	28
mKv4.1	mShal	M64226	26
rKv4.2	RK5	M59980[i]	27
	Shal1	S64320	46
rKv5.1	IK8	M81783	43
rKv6.1	K13	M81784	43
Other Kv channels			
Shaker	*Drosophila*	M17211	50
Shab	*Drosophila*	M32659	51
Shal	*Drosophila*	M32660	51
Shaw	*Drosophila*	M32661	51
KC2...	Rabbit	M81350-4[e]	42
APLK	*Aplysia*	M95914	83
IsK (minK) channels			
mIsK	Mouse heart	S57779	49
rIsK	Rat kidney	M22412	47
hIsK	Human	M26685[c]	48
Other K^+ channels			
Eag	*Drosophila*	M61157	52
Slo[h]	*Drosophila*	M69053[e]	53
	Drosophila	M96840	82
mSlo	Mouse	L16912	74
KAT1	*A. thaliana*	M86990	75
AKT1	*A. thaliana*	X62907	76
ROMK1	Rat	X72341	77
IRK1	Rat	X73052	78
GIRK1	Rat	L25264, U09243	85,86
ECOKCH	*E. coli*	L12044	79

[a] cDNA sequences, unless otherwise indicated.
[b] Prefixes to standard names: m, mouse; r, rat; h, human; b, bovine; x, *Xenopus*.
[c] Genomic sequence.
[d] 1 bp insertion causes early termination (compare with HK2, HPCN1).
[e] Partial coding sequence only.

Table 1b. (Continued)

[f] Full published sequence not incorporated.
[g] Published correction in start site (see Reference 43) not incorporated.
[h] Alternately spliced cDNAs.
[i] 2 bp deletion results in early termination (compare with Shal1).

Table 1 references:

1. Tempel, B. L., Jan, Y. N., Jan, L. Y. (1988). *Nature* 332, 837–839.
2. Baumann, A., Grupe, A., Ackermann, A., and Pongs O. (1988). *EMBO J.* 7, 2457–2463.
3. Kamb, A., Weir, M., Rudy, B., Varmus, H., and Kenyon, C. (1989). *Proc. Natl. Acad. Sci. U.S.A.* 86, 4372–4376.
4. Mathew, M. K., Ramashwami, M., Gautam, M., Kamb, A., Rudy, B., and Tanouye, M. A. (1989). *Soc. Neurosci. Abstr.* 15, 540.
5. Chandy, K. G., Williams, C. B., Spencer, R. H., Aguilar, B. A., Ghanshani, S., Tempel, B. L., and Gutman, G. A. (1990). *Science* 247, 973–975.
6. McKinnon, D. (1989). *J. Biol. Chem.* 264, 8230–8236.
7. Christie, M. J., Adelman, J. P., Douglass, J., and North, R. A. (1989). *Science* 244, 221–224.
8. Ribera, A. B. (1990). *Neuron* 5, 691–701.
9. Stuhmer, W., Ruppersberg, J. P., Schroter, K. H., Sakmann, B., Stocker, M., Giese, K. P., Perschke, A., Baumann, A., and Pongs, O. (1989). *EMBO J.* 8, 3235–3244.
10. Yokoyama, S., Imoto, K., Kawamura, T., Higashida, H., Iwabe, N., Miyata, T., and Numa, S. (1989). *FEBS Lett.* 259, 37–42.
11. Grissmer, S., Dethlefs, B., Wasmuth, J. J., Goldin, A. L., Gutman, G. A., Cahalan, M. D., and Chandy, K. G. (1990). *Proc. Natl. Acad. Sci. U.S.A.* 87, 9411–9415.
12. Philipson, L. H., Hice, R. E., Schaefer, K., LaMendola, J., Bell, G. I., Nelson, D. J., and Steiner, D. F. (1991). *Proc. Natl. Acad. Sci. U.S.A.* 88, 53–57.
13. Douglass, J., Osborne, P. B., Cai, Y.-C., Wilkinson, M., Christie, M. J., and Adelman, J. P. (1990). *J. Immunol.* 144, 4841–4850.
14. Swanson, R., Marshall, J., Smith, J. S., Williams, J. B., Boyle, M. B., Folander, K., Luneau, C. J., Antanavage, J., Oliva, C., Buhrow, S. A., Bennett, C., Stein, R. B., and Kaczmarek, L. K. (1990). *Neuron* 4, 929–939.
15. Tseng-Crank, J. C., Tseng, G.-N., Schwartz, A., and Tanouye, M. A. (1990). *FEBS Lett.* 268, 63–68.
16. Grupe, A., Schroter, K. H., Ruppersberg, J. P., Stocker, M., Drewes, T., Beckh, S., and Pongs, O. (1990). *EMBO J.* 9, 1749–1756.
17. Betsholtz, C., Baumann, A., Kenna, S., Ashcroft, F. M., Ashcroft, S. J., Berggren, P. O., Grupe, A., Pongs, O., Rorsman, P., Sandblom, J., and Welsh, M. (1990). *FEBS Lett.* 263, 121–126.
18. Chandy, K. G., Williams, C. B., Spencer, R. H., Aguilar, B. A., Ghanshani, S., Chandy, G., Tempel, B., and Gutman, G. A. (1990). *Biophys. J.* 57, 110a.
19. Pak, M. D., Covarrubias, M., Ratcliffe, A., and Salkoff, L. (1991). *J. Neurosci.* 11, 869–880.
20. Frech, G. C., VanDongen, A. M., Schuster, G., Brown, A. M. and Joho, R. H. (1989). *Nature (London)* 340, 642–645.
21. Luneau, C. J., Williams, J. B., Marshall, J., Levitan, E. S., Oliva, C., Smith, J. S., Antanavage, J., Folander, K., Stein, R. B., and Swanson R. (1991). *Proc. Natl. Acad. Sci. U.S.A.* 88, 3932–3936.
22. Pak, M. and Salkoff, L. A. Unpublished data.
23. McCormack, T., Vega-Saenz de Miera, E. C., and Rudy, B. (1991). *Proc. Natl. Acad. Sci. U.S.A.* 87, 5227–5231 (1990).
24. Ghanshani, S., Pak, M., McPherson, J. D., Strong, M., Dethlefs, B., Wasmuth, J. J., Salkoff, L., Gutman, G. A., and Chandy, K. G. (1992). *Genomics* 12, 190–6.
25. Schroter, K.-H., Ruppersberg, J. P., Wunder, F., Rettig, J., Stocker, M., and Pongs, O. (1991). *FEBS. Lett.* 278, 211–216.
26. Pak, M. D., Baker, K., Covarrubias, M., Butler, A., Ratcliffe, A., and Salkoff, L. (1991). *Proc. Natl. Acad. Sci. U.S.A.* 88, 4386–4390.
27. Roberds, S. L. and Tamkun, M. M. (1991). *Proc. Natl. Acad. Sci. U.S.A.* 88, 1798–1802.

Table 1b. (Continued)

28. Rudy, B., Sen, K., Vega-Saenz de Miera, E., Lau, D., Ried, T., and Ward, D. C. (1991). *J. Neurosci. Res.* 29, 401–412.
29. Rettig, J., Wunder, F., Stocker, M., Lichtinghagen, R., Mastiaux, F., Beckh, S., Kues, W., Pedarzani, P., Schroter, K. H., Ruppersberg, J. P., Veh, R., and Pongs, O. (1992). *EMBO J.* 11, 2473–2486.
30. Tamkun, M. M., Knoth, K. M., Walbridge, J. A., Kroemer, H., Roden, D. M., and Glover, D. M. (1991). *FASEB J.* 5, 331–337.
31. Attali, B., Romey, G., Honore, E., Schmid-Alliana, A., Mattei, M. G., Lesage, F., Ricard, P., Barhanin, J., and Lazdunski, M. (1992). *J. Biol. Chem.* 267, 8650–8657.
32. Cai, Y.-C., Osborne, P. B., North, R. A., Dooley, D. C., and Douglass, J. (1992). *DNA Cell Biol.* 11, 163–172.
33. Curran, M. E., Landes, G. M., and Keating, M. T. (1992). *Genomics* 12, 729–737.
34. Hwang, P. M., Glatt, C. E., Bredt, D. S., Yellen, G., and Snyder, S. H. (1992). *Neuron* 8, 473–481.
35. Grissmer, S., Ghanshani, S., Dethlefs, B., McPherson, J. D., Wasmuth, J., Gutman, G. A., Cahalan, M. D., and Chandy, K. G. (1992). *J. Biol. Chem.* 267, 20971–20979.
36. Matsubara, H., Liman, E. R., Hess, P., and Koren, G. (1991). *J. Biol. Chem.* 266, 13324–13328.
37. Vega-Saenz de Miera, E., Moreno, H., Fruhling, D., Kentros, C., and Rudy, B. (1992). *Proc. R. Soc. London Ser. B* 248, 9–18.
38. Reid, T., Rudy, B., Vega-Saenz de Miera, E., Lau, D., Ward, D. C., and Sen, K. (1992). *Genomics* 15, 405–411.
39. Mohrenweiser, H. and Chandy, K. G. (1992). Unpublished data.
40. Paulmichl, M., Nasmith, P., Hellmiss, R., Reed, K., Boyle, W. A., Nerbonne, J. M., Peralta, E. G., and Clapham, D. E. (1991). *Proc. Natl. Acad. Sci. U.S.A.* 88, 7892–7895.
41. Ramashwami, M., Gautam, M., Kamb, A., Rudy, B., Tanouye, M. A. and Mathew, M. K. (1990). *Mol. Cell. Neurosci.* 1, 214–223.
42. Desir, G. V., Hamlin, H. A., Puente, E., Reilly, R. F., Hildebrandt, F., and Igarashi, P. (1992). *Amer. J. Physiol.* 262, F151–F157.
43. Drewe, J. A., Verma, S., Frech, G., and Joho, R. H. (1992). *J. Neurosci.* 12, 538–548.
44. Lee, J. E., Garbutt, J. H., Phillips, K. L., and Roses, A. D. Unpublished data.
45. McPherson, J. D., Wasmuth, J. J., Chandy, K. G., Swanson, R., Dethlefs, B., Chandy, G., Wymore, R., and Ghanshani, S. (1991). In *Eleventh International Workshop on Human Gene Mapping.* Solomon, E. and Rawlings, C., Eds., Karger, New York.
46. Baldwin, T. J., Tsaur, M. L., Lopez, G. A., Jan, Y. N., and Jan, L. Y. (1991). *Neuron* 7, 471–483.
47. Takumi, T., Ohkubo, H., and Nakanishi S. (1988). *Science* 242, 1042–1045.
48. Murai, T., Kakizuka, A., Takumi, T., Ohkubo, H., and Nakanishi, S. (1989). *Biochem. Biophys. Res. Commun.* 161, 176–181.
49. Honore, E., Attali, B., Romey, G., Heurteaux, C., Ricard, P., Lesage, F., Lazdunski, M., and Barhanin, J. (1991). *EMBO J.* 10, 2805–2811.
50. Papazian, D. M., Schwarz, T. L., Tempel, B. L., Jan, Y. N., and Jan, L. Y. (1987). *Science* 237, 749–770.
51. Wei, A., Covarrubias, M., Butler, A., Baker, K., Pak, M., and Salkoff, L. (1990). *Science* 248, 599–603.
52. Warmke, J., Drysdale, R., and Ganetzky, B. (1991). *Science* 252, 1560–1562.
53. Atkinson, N. S., Robertson, G. A., and Ganetzky, B. (1991). *Science* 253, 551–555.
54. Vega-Saenz de Miera, E., Chiu, N., Sen, K., Lau, D., Lin, J. W., and Rudy, B. (1991). *Biophys. J.* 59, 197a.
55. Rudy, B., Kentros, C., and Vega-Saenz de Miera, E. (1991). *Mol. Cell. Neurosci.* 2, 89–102.
56. Migeon, M. B., Street, V. A., Demas, V. P., and Tempel, B. L. (1992). *Epilepsy Res.* 6(suppl), 173–180.
57. Philipson, L. H., Eddy, R. L., Shows, T. B., and Bell, G. I. (1993). *Genomics* 15, 463–464.
58. Gessler, M., Grupe, A., Grzeschik, K. H., and Pongs, O. (1992). *Human Genet.* 90, 319–321.
59. Rudy, B. (1992). Unpublished data, cited in Reference 38.
60. Luneau, C., Wiedmann, R., Smith, J. S., and Williams, J. B. (1991). *FEBS Lett.* 288, 163–167.
61. Ribera, A. and Nguyen, D.-A. (1993). *J. Neurosci.* 13, 4988–4996.

Table 1b. (Continued)

62. Tempel, B. L. (1993). Personal communication.
63. Garcia-Guzman, M., Calvo, S., Cena, V., and Criado, M. (1992). *FEBS Lett.* 308, 283–289.
64. Chevillard, C., Attali, B., Lesage, F., Fontes, M., Barhanin, J., Lazdunski, M., and Mattei, M. G. (1993). *Genomics* 15, 243–245.
65. Warmke, J. and Ganetzky, B. (1993). *Biophys. J.* 64, A340.
66. Robertson, G. A., Warmke, J., and Ganetzky, B. (1993). *Biophys. J.* 64, A340.
67. Ikeda, S. R., Soler, F., Zuhlke, R. D., Joho, R. H., and Lewis, D. L. (1992). *Eur. J. Physiol.* 422, 201–203.
68. Albrecht, B., Lorra., C., Stocker, K., and Pongs, O. (1993). Unpublished data.
69. Reid, P. F., Pongs, O., and Dolly, J. O. (1992). *FEBS Lett.* 302, 31–34.
70. Hart, P. J., Overturf, K. E., Russell, S. N., Carl, A., Hume, J. R., Sanders, K. M., and Horowitz, B. (1993). *Proc. Natl. Acad. Sci. U.S.A.* 90, 9659–9663.
71. Freeman, S. N., Conley, E. C., Brennand, J. C., Russell, N. J. W., and Brammar, W. J. (1990). *Biochem. Soc. Trans.* 18, 891.
72. Ishii, K., Nunoki, K., Murakoshi, H., and Taira, N. (1992). *Biochem. Biophys. Res. Commun.* 184, 1484–1489.
73. Fedida, D., Wible, B., Wang, Z., Fermini, B., Faust, F., Nattel, S., and Brown, A. M. (1993). *Circ. Res.* 73, 210–216.
74. Butler, A., Tsunoda, S., McCobb, D. P., Wei, A., and Salkoff, L. (1993). *Science* 261, 221–224.
75. Anderson, J. A., Huprikar, S. S., Kochian, L. V., Lucas, W. J., and Gaber, R. F. (1992). *Proc. Natl. Acad. Sci. U.S.A.* 89, 3736–3740.
76. Sentenac, H., Bonneaud, N., Minet, M., Lacroute, F., Salmon J. M., Gaymard, F., and Grignon, C. (1992). *Science* 256, 663.
77. Ho, K., Nichols, C. G., Lederer, W. J., Lytton, J., Vassilev, P. M., Kanazirska, M. V., and Hebert, S. C. (1993). *Nature (London)* 362, 31–38.
78. Kubo, Y., Baldwin, T. J., Jan, Y. N., and Jan, L. Y. (1993). *Nature* 362, 127–132.
79. Milkman, R. (1993). *Proc. Natl. Acad. Sci. U.S.A.* 91, 3510–3514.
80. Wymore, R., Korenberg, J. R., Coyne, C., Chen, X.-N., Hustad, C. M., Copeland, N. G., Gutman, G. A., Jenkins, N. A., and Chandy, K. G. (1993). Submitted for publication.
81. Butler, A., Wei, A., and Salkoff, L. (1990). *Nucleic Acids Res.* 18, 2173–2174.
82. Adelman, J. P., Shen, K.-Z., Kavanaugh, M.P., Warren, R. A., Wu, Y.-N., Lagrutta, A., Bond, C. T., and North, R. A. (1992). *Neuron* 9, 209–216.
83. Pfaffinger, P. J., Furukawa, Y., Zhao, B., Dugan, D., and Kandel, E. R. (1991). *J. Neurosci.* 11, 918–927.
84. Haas, M., Ward, D. C., Lee, J., Roses, A. D., Clarke, V., D'Eustachio, P., Lau, D., Vega-Saenz de Miera, E., and Rudy, B. (1993). *Mammalian Genome* 4, 711–7115.
85. Kubo, Y., Reuveny, E., Slesinger, P.A., Jan, Y. N., and Jan, L. Y. (1993). *Nature* 364, 802–806.
86. Depaoli, A. M., Bell, G. I., and Stoffel, M. (1994). Unpublished observations.

of the protein (the region that includes the six membrane-spanning domains S1 through S6, and the P or pore-forming region), as well as a substantial portion of sequence on the N-terminal side of S1. Equally striking is the substantial *divergence* of sequences at both ends of the protein, as well as within two regions that are thought to form extracellular loops (between S1/S2 and S3/S4).

2.1.1.1.1 Divergence of N- and C-Terminal Regions; Repetitive Sequences

The extensive sequence divergence at both ends of these proteins suggests either that these regions have very different functions in the different channel isoforms, or that whatever functions they do have depend relatively little on

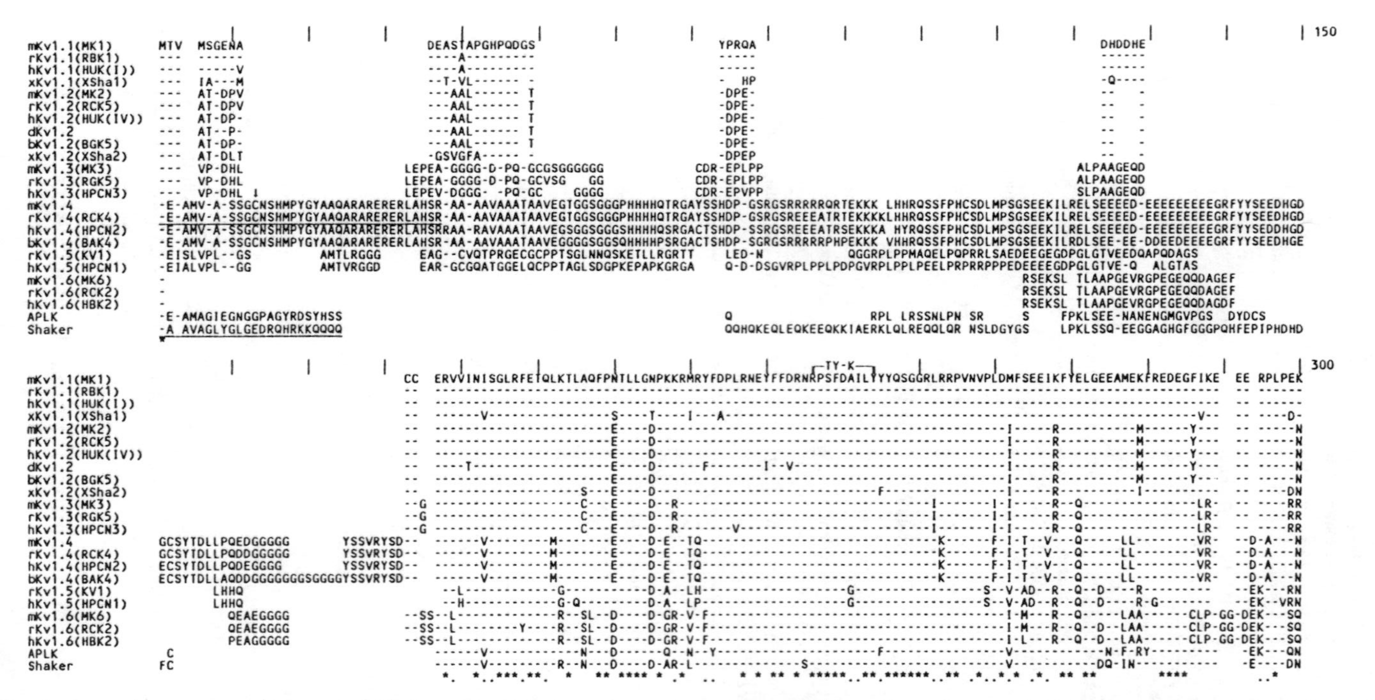

FIGURE 2. Amino acid sequence alignment of *Shaker*-related K⁺ channels, made by manual editing of an alignment initially generated with the help of the program CLUSTALV (Higgins et al., 1992). Dashes indicate identity with the sequence at the top of the alignment (mKv1.1), and hash marks above the top sequence mark every tenth residue. Below the alignment, an asterisk indicates positions at which all sequences are identical, and a period indicates positions that show only conservative substitutions. Shown by brackets above the alignment are the six putative membrane-spanning regions (S1–S6), the "pore" motif (P), a potential tyrosine kinase recognition sequence (TY-K), which is present in all *Shaker*-related sequences, and two potential protein kinase C sites (PKC), at least one of which is present in *all* voltage-gated K⁺ channels. Asterisks above the alignment in the region between S1 and S2 mark the positions of potential N-glycosylation sites in one or more of the sequences below, each of which is underlined in the sequence itself. The underlined sequences at the N-terminus of rKv1.4 and *Shaker* represent peptides that have been shown to mediate fast inactivation in these channels; the arrow marks the cysteine residue within this peptide in rKv1.4 whose oxidation removes fast inactivation (see text).

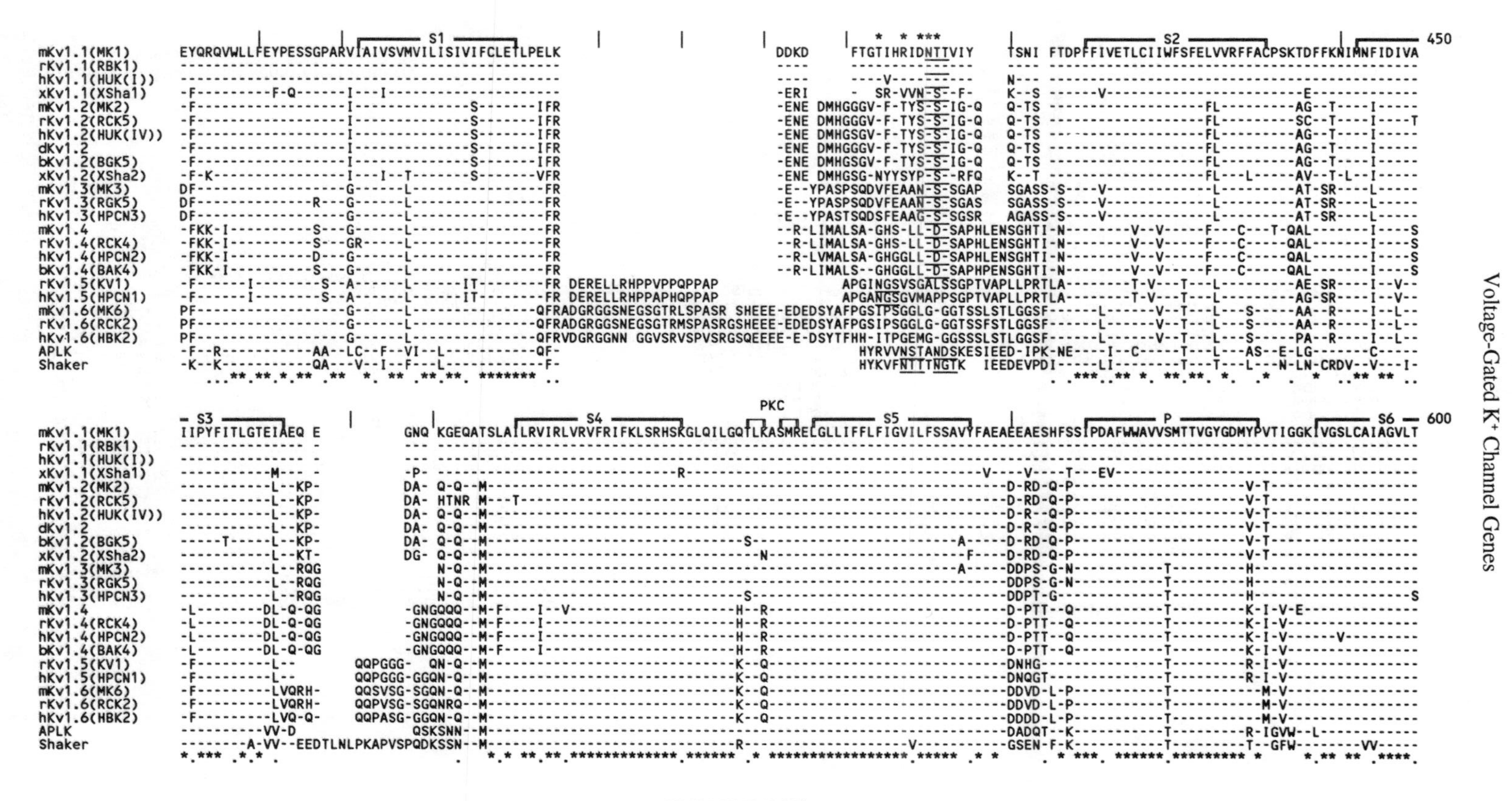

FIGURE 2(2).

```
                                                                                                                                                                  748
mKv1.1(MK1)      IALPVPVIVSNFNYFYHRETEGEE QAQLLH VSS PNLASDSD                   LSR RSSSTIS   KSEY                  MEIEEDMNNSIAHYRQANIRTGN    CTTA DQ  NCVNKSKLLTDV
rKv1.1(RBK1)     ------------------------ ------ --- --------                   --- -------   ----                  -----------------------    --AT --  ------------
hKv1.1(HUK(I))   ------------------------ ------ --- --------                   --- -----M-   -Y--                  ----------------V----A-    ---- N-  ------------
xKv1.1(XSha1)    ------------------------ ------ --- -----N--                   --- ----AM-   ----                  ------L----DNF-E-------    --I- N-  ------------
mKv1.2(MK2)      ------------------------ ---Y-Q -T-C-KIP-SP-                   -KKS--A----   --D-                  ---Q-GV---NEDF-EE-LK-A-    --L- NT  -Y--IT-M----
rKv1.2(RCK5)     ------------------------ ---Y-Q -T-C-KIP-SP-                   -KKS--A----   --D-                  ---Q-GV---NEDF-EE-LK-A-    --L- NT  -Y--IT-M----
hKv1.2(HUK(IV))  ------------------------ ---Y-Q -T-C-KIP-SP-                   -KKS--A----   --D-                  ---Q-GV---NEDF-EE-LK-A-    --L- NT  -Y--IT-M----
dKv1.2           ------------------------ ---Y-Q -T-C-KIP-SP-                   -KKS--A----   --D-                  ---Q-GV---NEDF-EE-LK-A-    --L- NT  -Y--IT-M----
bKv1.2(BGK5)     ------------------------ ---Y-Q -T-C-KIP-SP-                   -KKS--A----   --D-                  ---Q-GV---NEDF-EE-
xKv1.2(XSha2)    ------------------------ ---Y-Q -T-C-KIP-SP-                   -QKS--A--L-   --D-                  ---Q-GV-H-NEDF-EK-LK-A-    --LG NT  -Y--IT-M----
mKv1.3(MK3)      ------------------------ ---YM- -G-CQH-S-SAE                  E-RKA--N--L-   ----                  -V---GGM- QSAFP-TPFK---STAT---N NNPNS---IK-IF---
rKv1.3(RGK5)     --------L--------------- ---YM- -G-CQH-S-SAE                  E-RKA--N--L-   ----                  -V---GGM- HSAFP-TPFK---STAT---N NNPNS---IK-IF---
hKv1.3(HPCN3)    ------------------------ -S-YM- -G-CQH-S-SAE                  E-RKA--N--L-   ----                  -V---GGM- HTAFP-TPFK---STAT---N NNPNS---IK-IF---
mKv1.4           ---------------------NV- -T--TQNAV-C-Y-P-NL                    -KKF---TSS-LGD----                  L-M--GVKE-LCGKEEKCQGK-DESET      DKN--S-AKAVE---
rKv1.4(RCK4)     ---------------------N-- -T--TQNAV-C-Y-P-NL                    -KKF---TSS-LGD----                  L-M--GVKE-LCGKEEKCQGK-DDSET      DKN--S-AKAVE---
hKv1.4(HPCN2)    ---------------------N-- -T--TQNAV-C-Y-P-NL                    -KKF---TSS-LGD----                  L-M--GVKE-LCAKEEKCQAK-DDSET      DKN--S-AKAVE---
bKv1.4(BAK4)     ---------------------N-- -T--TQNAV-C-Y-P-NL                    -KKF---TSS-LGD----                  L-M--GVKE-LCAKE KCQGK-DDSET      DKN-VS-AKAVE---
rKv1.5(KV1)      --------------------DH-- --A-KEEQGNQRRESGLDT                  GGQ-KV-C-KA-FCKTGGS                  L-SSDSIRRGSCPLEKCHLKAKSNVDL  RR SLYAL-LDT-R E--L
hKv1.5(HPCN1)    --------------------DH-- P-V-KEEQGTQSQGPGLDR                  GVQ-KV-G-RG-FCKAGGT                  L-NADSARRGSCPLEKC-VKAKSNVDL  RR SLYAL-LDT-R E--L
mKv1.6(MK6)      ---------------------Q-- -G-YT-  VTCGQPTP                     D-K    ATDNGLG -PDF                     A-ASRERRPS-LPTPH-AYAE                 KRM--E-
rKv1.6(RCK2)     ---------------------Q-- -G-YT-  VTCGQPTP                     D-K    ATDNGLG -PDF                     A-ASRERRSS-LPTPH-AYAE                 KRM--E-
hKv1.6(HBK2)     ---------------------Q-- -G-YT-  VTCGQP-P                     D-R    ATDNGLG -PDF                     P-ANRERRPS-LPTPH-AYAE                 KRM--E-
APLK             -------------------G-STDKG  QYKH-Q-C--YPE      KKDSLDSECGSDIMEMEEGNH-T                                      PLTEKVKE-HAIKANNPGSDYGL            E---
Shaker           -------------------ADR--M-S-NFNH-T-CSY-PGALGQHLKKSSL SESSSDIMD-DDGIDAT-PGLTDHTGRHMVPFLRTQQSFEKQQLQLQLQLQQQSQSPHGQQMTQQQQL-QNGLRS-NSLQLRH-NAMAVSIE---
                 *.******.***.******  .  .
```

FIGURE 2(3).

their structure (since a region that is completely functionless would be expected to be progressively lost over evolutionary time). A clue to the nature of such functions may be provided by the fact that the Kv1.4 proteins have the longest N-terminal region of the mammalian *Shaker* subfamily, and also contain some strikingly repetitive sequences (e.g., runs of 11/13 alanines, 14/15 glutamines, as well as shorter runs of glycines, histidines and lysines); the presence of glutamine repeats in the N-terminal portion of *Shaker* has already been noted (Schwarz et al., 1988).

Simple repetitive sequences are widely dispersed throughout the genomes of eukaryotes, recent highly publicized examples including the trinucleotide repeats associated with mutations in the gene responsible for Huntington's disease (Huntington's Disease Collaborative Research Group, 1993), and the dinucleotide repeats reported to be associated with a predisposition to colorectal cancer (Aaltonen et al., 1993; Thibodeau et al., 1993). One mechanism by which simple repetitive sequences may be expanded is slipped-strand mispairing. It has been suggested that in nucleotide sequences constrained only by the requirement for their presence (and not for a particular structure) one might expect a progressive replacement by simple repeats (Levinson and Gutman, 1987), and the presence of such simple repeats in *Shaker* and *Shaker*-related genes might be the result of such a process. The required conditions could be fulfilled if, for example, the nonconserved region were required only to provide a suitable spacer arm ("chain") for the N-terminal inactivating "ball" known to be responsible for fast inactivation in *Shaker* (Jan and Jan, 1992). Kv1.4, in fact, is the only mammalian *Shaker* homologue that shows a *Shaker*-like rapid inactivation, and it also contains the longest N-terminal region as well as the most striking amino acid repeats within this family; these repeats begin immediately after a stretch of charged residues that are thought to form the fast-inactivation ball of this channel (indicated in Figure 2; see also Section 2.1.5). While other members of this family are delayed rectifiers and do not exhibit rapid "ball and chain" type inactivation, analogous "spacer" regions might be required for structures involved in subunit assembly or other functions, and the presence of simple amino acid repeats might be associated with such spacers.

2.1.1.1.2 Posttranslational Modification

Three potential sites of posttranslational modification are indicated in Figure 2 (as well as Figure 1). First, an N-glycosylation site (Marshall, 1972) is present in the extracellular loop between S1 and S2 in all the mammalian *Shaker*-related sequences shown in Figure 2 other than Kv1.6; the absence of a glycosylation site in this region is shared by both mouse and human Kv1.6. Two such sites are evident in the corresponding region of *Aplysia* and *Shaker* K^+ channels. The presence of such a site in almost all of this subfamily of channels is all the more striking given the considerable divergence in both the sequence and length of the S1-S2 loop. Using an *in vitro* translation system, Shen et al. (1993) demonstrated glycosylation at this site in the *Aplysia* K^+ channel, N-glycosylation of the *Shaker* channel has also been shown

(Rosenberg and East, 1992; Santacruz-Toloza et al., 1994); it has not yet been determined, however, whether this site is glycosylated in native channels. Other N-glycosylation consensus sequences exist in these and other channels, including several in the C-terminal region, but as these are thought to be located intracellularly they are unlikely to be utilized and we do not indicate them in any of our alignments.

Second, all *Shaker*-related proteins have a potential tyrosine kinase site in the N-terminal region (RPSFDAILY), indicated in Figure 2; while its conservation suggests it is functional, this site has not been experimentally shown to be utilized, and deletion of this region from the mKv1.3 channel had no apparent affect on channel function (Aiyar et al., 1993a). Third, all of the *Shaker*-subfamily proteins show one or two protein kinase C (PKC) consensus sites (Ser/Thr-X-Arg/Lys) in the cytoplasmic loop between S4 and S5 (Kemp and Pearson, 1990), a region that has been modeled to form part of the internal surface of the ion conduction pathway (Isacoff et al., 1991; Jan and Jan, 1992; Durell and Guy, 1992). In fact, a PKC site in this region is present in *all* known voltage-gated K^+ channels (further discussed below), and there is evidence that phosphorylation of this site can affect channel function. Phorbol esters, which activate PKC, suppress Kv1.3 currents in oocytes and in Jurkat T cells, probably by phosphorylating these sites, and the effect is reversibly inhibited by PKC inhibitors (Attali et al., 1992a; Payet and Dupuis, 1992; Aiyar et al., 1993b). Rat Kv1.3, translated *in vitro,* has been reported to be phosphorylated by PKC, presumably at one or both of these sites (Cai and Douglass, 1993). Similar results have been reported for the rKv3.1 (Critz et al., 1993), hKv1.4 (Fahrig et al., 1992), rKv4.2 (Blair et al., 1992), and *Shaker* channels (Moran et al., 1991), and removal of both PKC sites by site-directed mutation (K $\rightarrow$ D and R $\rightarrow$ D) in the *Shaker* protein produced nonfunctional channels (Isacoff et al., 1992).

2.1.1.1.3 Species Conservation

The mammalian *Shaker*-related genes generally show a high level of interspecies conservation. For example, comparison of the Kv1.2 protein sequences of human, rat, mouse, dog, and cattle reveals greater than 98% identity over the entire protein (Figure 3A); even the more distantly related *Xenopus* sequence is greater than 91% identical to the mouse. This is not universally true, however, and one exception is illustrated in Figure 3B, which shows that rat and human Kv1.5 are only 87% identical overall (ignoring one 11-residue size difference), and less than 64% identical in the amino-terminal 100 residues. The more typical Kv1.2, on the other hand, shows 99.4% overall identity between its human and rodent homologues. These apparent differences in rates of divergence are not the consequence of sequencing artifacts, since the nucleotide sequences of the five human Kv1.5 cDNAs that have been independently isolated are identical in this region (Ramaswami et al., 1990; Philipson et al., 1991; Tamkun et al., 1991; Curran et al., 1992; Fedida et al., 1993), as are the sequences of three independent rat Kv1.5 clones (Swanson et al., 1990).

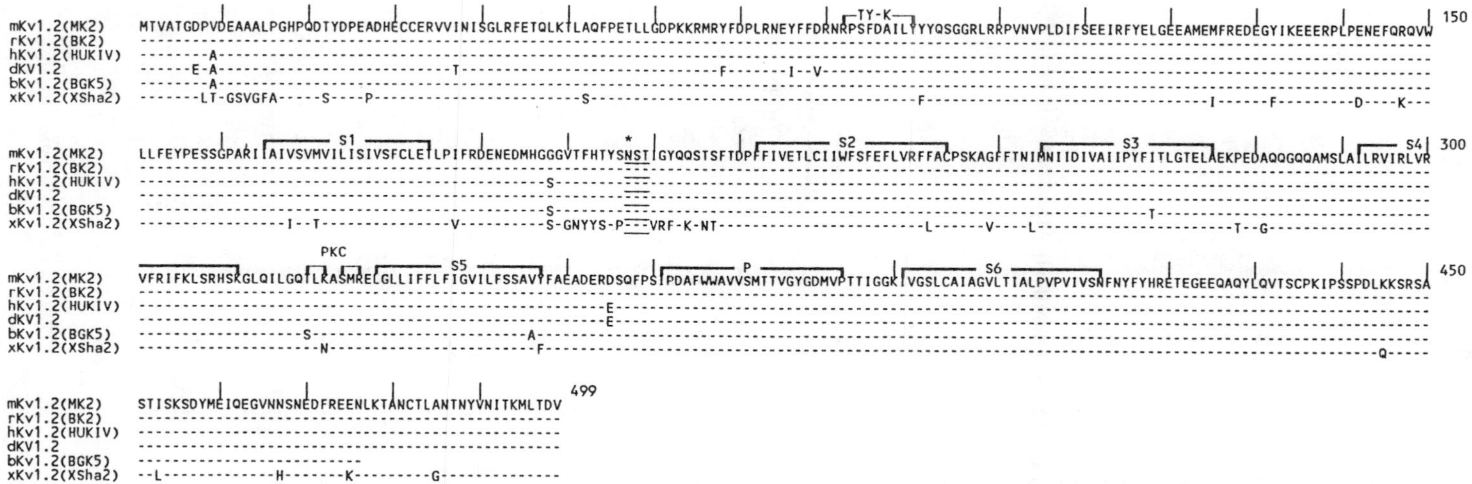

```
mKv1.2(MK2)     MTVATGDPVDEAAALPGHPQDTYDPEADHECCERVVINISGLRFETQLKTLAQFPETLLGDPKKRMRYFDPLRNEYFFDRNRPSFDAILYYYQSGGRLRRPVNVPLDIFSEEIRFYELGEEAMEMFREDEGYIKEEERPLPENEFQRQVW 150
rKv1.2(BK2)     ------------------------------------------------------------------------------------------------------------------------------------------------------
hKv1.2(HUKIV)   --------A-----------------------------------------------------------------------------------------------------------------------------------------
dKV1.2          ------E-A---------------------------T------------------------------F-------I--V-----------------------------------------------------------------------
bKv1.2(BGK5)    --------A-----------------------------------------------------------------------------------------------------------------------------------------
xKv1.2(XSha2)   -------LT-GSVGFA-----S----P------------------------S--------------------------------------F---------------------------------I------F---------D----K---

mKv1.2(MK2)     LLFEYPESSGPARIIAIVSVMVILISIVSFCLETLPIFRDENEDMHGGGVTFHTYSNSTIGYQQSTSFTDPFFIVETLCIIWFSFEFLVRFFACPSKAGFFTNIMNIIDIVAIIPYFITLGTELAEKPEDAQQGQQAMSLAILRVIRLVR 300
rKv1.2(BK2)     ------------------------------------------------------------------------------------------------------------------------------------------------------
hKv1.2(HUKIV)   -----------------------------------------------S------------------------------------------------------------------------------------------------------
dKV1.2          ------------------------------------------------------------------------------------------------------------------------------------------------------
bKv1.2(BGK5)    -----------------------------------------------S---------------------------------------------------------------------T--------------------------------
xKv1.2(XSha2)   -----------------I--T---------------V----------S-GNYYS-P---VRF-K-NT------------------------L------V----L-----------------------T--G-------------------

                PKC
mKv1.2(MK2)     VFRIFKLSRHSKGLQILGQTLKASMRELGLLIFFLFIGVILFSSAVYFAEADERDSQFPSIPDAFWWAVVSMTTVGYGDMVPTTIGGKIVGSLCAIAGVLTIALPVPVIVSNFNYFYHRETEGEEQAQYLQVTSCPKIPSSPDLKKSRSA 450
rKv1.2(BK2)     ------------------------------------------------------------------------------------------------------------------------------------------------------
hKv1.2(HUKIV)   ------------------------------------------------------E-----------------------------------------------------------------------------------------------
dKV1.2          ------------------------------------------------------E-----------------------------------------------------------------------------------------------
bKv1.2(BGK5)    -------------------S-------------------------A--------------------------------------------------------------------------------------------------------
xKv1.2(XSha2)   ---------------------N------------------------F------------------------------------------------------------------------------------------------Q-----

mKv1.2(MK2)     STISKSDYMEIQEGVNNSNEDFREENLKTANCTLANTNYVNITKMLTDV 499
rKv1.2(BK2)     -------------------------------------------------
hKv1.2(HUKIV)   -------------------------------------------------
dKV1.2          -------------------------------------------------
bKv1.2(BGK5)    --------------------------
xKv1.2(XSha2)   --L-------------H-------K---------G--------------
```

FIGURE 3. (A) Amino acid sequence alignment of rat, mouse, human, dog, bovine (partial sequence), and frog homologues of Kv1.2, illustrating their high degree of similarity. Putative functional regions are indicated as in Figure 2. (B) Amino acid sequence alignment of rat and human homologues of Kv1.5, showing their unusually extensive sequence divergence in the N-terminal region. Putative functional regions are indicated as in Figure 2.

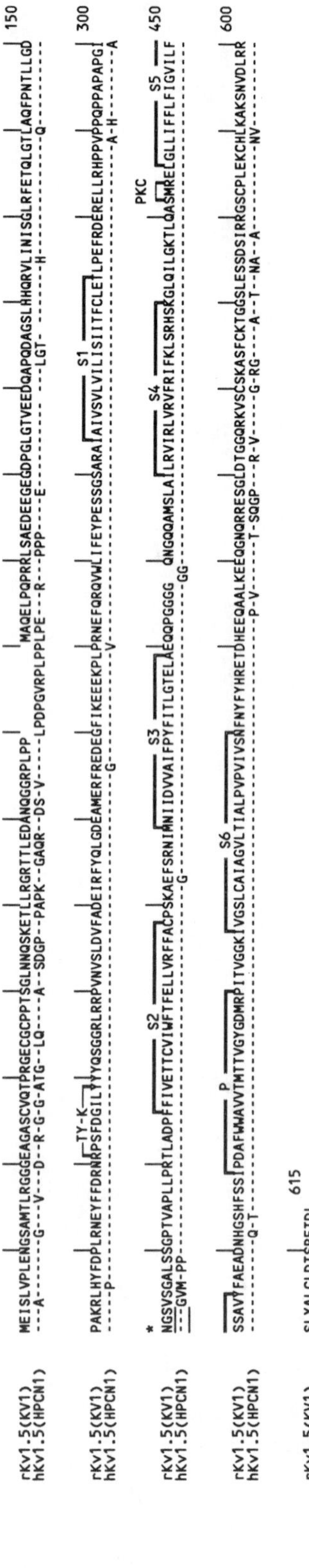

```
rKv1.5(KV1)     MEISLVPLENGSAMTLRGGGEAGASCVQTPRGECGCPPTSGLNNQSKETLLRGRTTLEDANQGGRPLPP          MAQELPQPRRLSAEDEEGEGDPGLGTVEEDQAPQDAGSLHHQRVLINISGLRFETQLGTLAQFPNTLLGD  150
hKv1.5(HPCN1)   ---A-------G---V---D--R-G-G-ATG--LQ----A--SDGP--PAPK--GAQR--DS-V-----LPDPGVRPLPPLPE---R---PPP----E---------- ---LGT- -------H-------------Q---------

                                  TY-K                                                                                                            S1
rKv1.5(KV1)     PAKRLHYFDPLRNEYFFDRNRPSFDGILYYYQSGGRLRRPVNVSLDVFADEIRFYQLGDEAMERFREDEGFIKEEEKPLPRNEFQRQVWLIFEYPESSGSARAIAIVSVLVILISIITFCLETLPEFRDERELLRHPPVPPQPPAPAPGI  300
hKv1.5(HPCN1)   -----P---------------------------------------------------G------------V----------------------------------------------------------A-H--------A

                *                         S2                         S3                                  S4                        PKC        S5
rKv1.5(KV1)     NGSVSGALSSGPTVAPLLPRTLADPFFIVETTCVIWFTFELLVRFFACPSKAEFSRNIMNIIDVVAIFPYFITLGTELAEQQPGGGG  QNGQQAMSLAILRVIRLVRVFRIFKLSRHSKGLQILGKTLQASMRELGLLIFFLFIGVILF  450
hKv1.5(HPCN1)   ---GVM-PP---------------------------------G------------------------------------GG-----------------------------------------------------------------

                                          P                          S6
rKv1.5(KV1)     SSAVYFAEADNHGSHFSSIPDAFWWAVVTMTTVGYGDMRPITVGGKIVGSLCAIAGVLTIALPVPVIVSNFNYFYHRETDHEEQAALKEEQGNQRRESGLDTGGQRKVSCSKASFCKTGGSLESSDSIRRGSCPLEKCHLKAKSNVDLRR  600
hKv1.5(HPCN1)   ----------Q-T------------------------------------------------------------P-V------T-SQGP---R-V-----G-RG----A--T--NA--A----------NV----------

                               615
rKv1.5(KV1)     SLYALCLDTSRETDL
hKv1.5(HPCN1)   ---------------
```

FIGURE 3B.

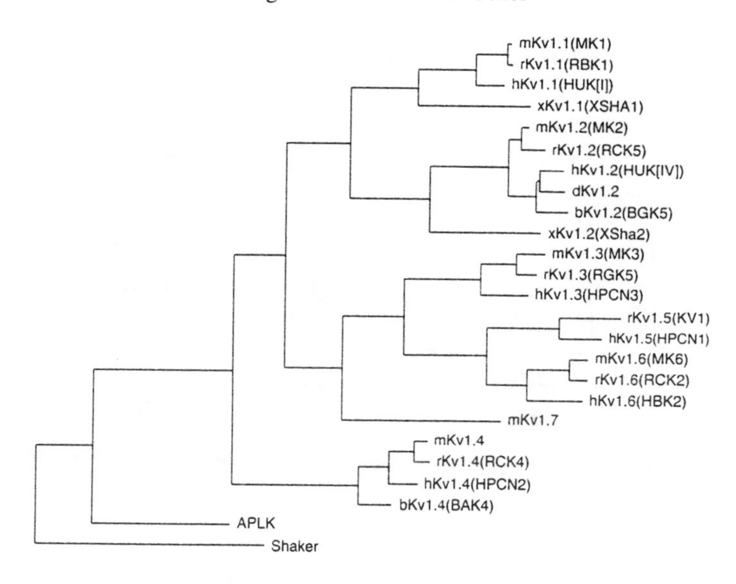

FIGURE 4. Proposed phylogenetic tree of *Shaker*-related K^+ channel genes, based on parsimony analysis of nucleotide sequence alignments using the program PAUP (Swofford, 1993). The placement of mKv1.7 is poorly supported (see text).

Two additional *Shaker*-related mammalian genes are known to exist, but sequences for these genes either exist only as short published fragments [mouse and rat Kv1.7 (Betsholtz et al., 1990)] or have not yet been published [mouse Kv1.8 (B. Tempel, personal communication); mouse Kv1.7, K. Kalman et al., unpublished data].

2.1.1.1.4 *Shaker*-Subfamily Phylogenetic Tree

A proposed phylogeny for the *Shaker*-related genes is shown in Figure 4. While our earlier analyses placed the *Xenopus* Xsha2 gene outside the mammalian group (Strong et al., 1993), the current larger dataset places each of the two *Xenopus* genes within one of the mammalian groups of homologues, Xsha1 with Kv1.1 and Xsha2 with Kv1.2. This larger dataset also positions the Kv1.2 group differently than previously, making it a sister group of the Kv1.1 cluster. The remaining features of this tree are the same as we have already proposed (Strong et al., 1993), including our inability to determine the position the mouse Kv1.7 sequence in this tree with any degree of confidence. It should be noted that the terminal patterns of branching in the Kv1.2 and Kv1.4 clusters, the two that include bovine homologues as well as rodent and human, are inconsistent with each other. Neither of these patterns is compellingly supported by bootstrap analysis, and this inconsistency reflects the current uncertainty regarding the true phylogenetic relationships between many mammalian orders (Novacek, 1992; Graur, 1993).

If this phylogeny is correct, then the duplications that gave rise to the mammalian *Shaker*-related subfamily of genes must have occurred before the divergence of mammals and amphibians, some 350 million years ago (MYA; for discussion of times of divergence see Nei, 1987). Reptiles and birds, which diverged from mammals more recently (about 300 MYA), would therefore be expected to also show the presence of multiple members of the known mammalian homologues, and the nomenclature proposed for mammalian K^+ channel genes (Chandy et al., 1991; Gutman and Chandy, 1993) should also be applicable to these other vertebrates when cloned genes become available. The placement of the *Aplysia* sequence outside the mammalian group implies that the vertebrate duplications occurred after the divergence of molluscs as well as insects from the vertebrate lineage (~600 MYA). Whether these duplications predate the divergence of older vertebrate lineages, i.e., fish and cyclostomes (400–450 MYA), remains to be determined.

2.1.1.2 Genomic Organization

The *Shaker* gene in *Drosophila* contains multiple exons spread over 120 kb and can be alternatively spliced to generate at least five distinct functional transcripts (Pongs et al., 1988; Schwarz et al., 1988). The various resulting *Drosophila Shaker* channels inactivate with different time courses (Timpe et al., 1988; Stocker et al., 1990; Iverson et al., 1990; Wittka et al., 1991); ShB1, for example, inactivates much more rapidly than ShA1, although it recovers equally rapidly from inactivation (Stocker et al., 1990; Iverson et al., 1988; Witta et al., 1991).

In dramatic contrast, the coding regions of all but one of the known mammalian *Shaker* homologues, Kv1.1-Kv1.6, and Kv1.8, are uninterrupted in the genome (Chandy et al., 1990b; Swanson et al., 1990; Douglass et al., 1990; Wymore et al., 1994; B. Tempel, personal communication). The Kv1.7 gene is the only known exception, having at least one intron in its coding sequence in the loop linking S1–S2 (Chandy et al., 1990a; K. Kalman et al., unpublished data). While the existence of intronless coding regions would seem to preclude the generation of diverse channel proteins through alternate exon utilization, different forms of protein could nevertheless be generated by alternative use of undiscovered splice sites, either sites defining new introns, or potential sites within known coding regions (see cautions regarding sequence interpretation in footnote). The *Xenopus* Kv1.1 and Kv1.2 genes also lack introns in their coding regions (Ribera, 1990; Ribera and Nguyen, 1993) suggesting that this genomic organization may be common to all vertebrates. The evolutionary significance of this difference in the structure of fly and vertebrate channel genes and the mechanisms by which this family of intronless genes have evolved remain obscure.

2.1.1.2.1 5′-Noncoding Introns in Kv1.1 and Kv1.2; Posttranscriptional Regulation?

Kv1.1–Kv1.3 genes express several large transcripts (8–9.5 kb) along with several smaller ones (Tempel et al., 1988; Stuhmer et al., 1989; McKinnon,

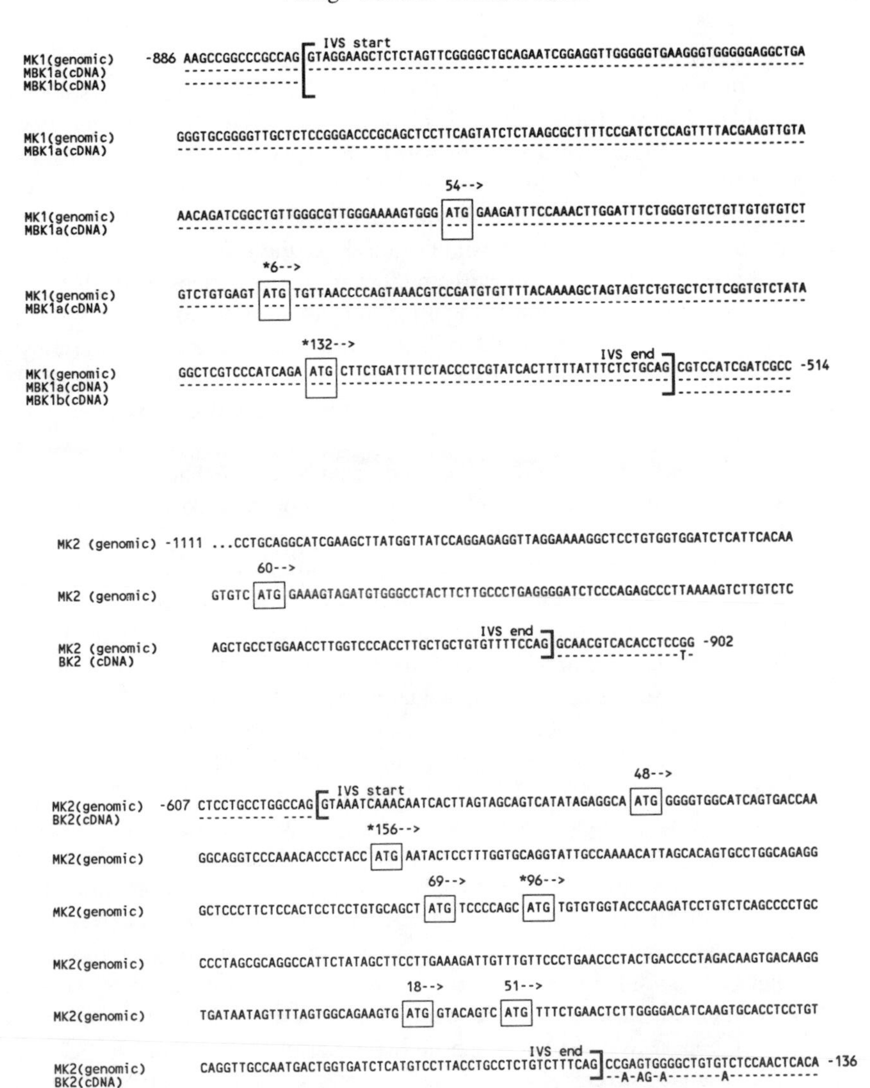

FIGURE 5. (A) Nucleotide sequence alignment of a portion of the genomic 5′ untranslated sequence (5′-UTS) of mouse mKv1.1 (from position –886 to –514 relative to the beginning of the MK1 protein-coding region), with two distinct mKv1.1 cDNAs (MBK1a and b). The borders of an intervening sequence present both in the genomic sequence and in MBK1a are indicated by brackets. Three potential translation start sites (PNNATG, where P indicates either A or G, and N indicates any nucleotide) are boxed, and the lengths of their downstream open reading frames are indicated above. Two sites that display the preferred initiation sequence, ANNATG, are indicated by asterisks. (B) Alignment of a portion of the genomic 5′-UTS of mouse Kv1.2 (from position –1111 to –902) with that of a rat Kv1.2 cDNA. The 3′ end of a putative intervening sequence is indicated by a bracket, and a potential translational start site is indicated as in A. (C) Alignment of another region of the genomic 5′-UTS of mouse Kv1.2 (from position –607 to –136) with that of a rat Kv1.2 cDNA, showing the presence of an intron whose ends are indicated by brackets. Six potential translation start sites are indicated as in A and B.

1989; Douglass et al., 1990; Spencer et al., 1993). In the case of mKv1.1, alternative splicing of a short intron within its 5′-NCR has been shown to yield both spliced and unspliced polyadenylated RNA (Tempel et al., 1988). As seen in Figure 5A, this 5′ intron (position –886 to –514) contains three potential initiation codons (consensus sequence PNNATG, where P indicates either A or G, and N is any nucleotide), each generating an open reading frame whose length is indicated above the ATG codon; two of these (indicated by asterisks) show the preferred consensus sequence ANNATG. According to the scanning model for translation initiation (Kozak, 1991), these upstream ATGs could inhibit translation of the Kv1.1 protein by delaying the arrival of scanning ribosomes at the authentic start site; Kv1.1 protein might therefore be inefficiently produced from transcripts which contain this intron.

Comparison of the mKv1.2 genomic sequence with that of rKv1.2 cDNA also reveals the presence of at least two introns in the 5′-NCR (Figure 5B,C). We have identified only the splice acceptor site of the upstream intron, at position –920 (Figure 5B), and this putative intron contains at least one ATG initiation codon. The downstream intron (–607 to –136) contains 6 possible translational start sites producing open reading frames ranging from 18 to 156 bp (Figure 5C), two of which match the preferred Kozak consensus sequence.

The existence of 5′ introns containing potential initiation codons raises the intriguing possibility of post-transcriptional regulation of K^+ channel expression by alternative splicing in the 5′-UTS. Kozak (1991) has reviewed examples of messenger RNAs of differing translational efficiency being produced by alternative sites of transcription initiation, and of the demonstrated deleterious effects of upstream ATG codons on translational efficiency. Her cautionary notes regarding the difficulty of unambiguously identifying the structure of functional mRNAs are particularly relevant to our discussion; in the case of Kv1.1, for example, neither end of the transcription unit has been identified, and the intron-containing form of mRNA has not been demonstrated to be functional. Nevertheless, mRNAs containing this intron appear to constitute a substantial minority of Kv1.1 messenger RNA in mouse brain (Tempel et al., 1988), and alternative splicing of such introns could potentially play a role in both the tissue specificity of expression of Kv proteins and in their up- and down-regulation.

2.1.1.2.2 Mapping Kv1.4 and Kv1.5 mRNAs

The transcripts of Kv1.4 and Kv1.5 are generally shorter (2.4–4.5 kb) than those of Kv1.1–Kv1.3 (Tamkun et al., 1991; Swanson et al., 1990; Philipson et al., 1990). A schematic diagram of the mKv1.4 gene and its transcripts is shown in Figure 6. mKv1.4 appears to have a single, large intron (3.4 kb) in its 5′-NCR (Wymore et al., 1994). Like most other members of the *Shaker*-related family, however, its protein coding sequence, as well as an additional 2 kb of 5′- and 3′-UTS, is contiguous in the genome. Although neither the transcription initiation site nor the 3′ polyadenylation site of this mRNA has been definitively identified, they are unlikely to be far from the limits indicated in the figure, defined by the 5′ end of hPCN2 and the 3′ end of RHK1.

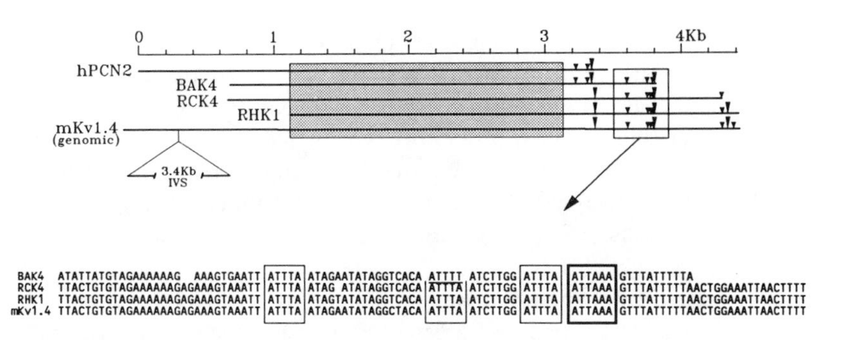

FIGURE 6. Schematic diagram (above) of Kv1.4 transcripts from human (hPCN2), cattle (BAK4), rat (RCK4 and RHK1), and genomic DNA from mouse Kv1.4. The shaded box encloses the protein coding region. Within the 3′-UTS are indicated the positions of potential polyadenylation signals (tall arrowheads) and ATTTA motifs (short arrowheads) implicated in destabilization of mRNA (see text). The aligned nucleotide sequences for the boxed region in the 3′-UTS are shown below, with the polyadenylation signals and ATTA sequences indicated. The ruler above indicates nucleotides (in kb) beginning with the 5′ end of the hPCN2 transcript.

A total of more than 4.4 kb of the Kv1.4 transcript is thus accounted for in the mouse genomic sequence, leaving no more than about 100 bp possibly remaining at both ends of the longest message.

Unusual features of the 3′-UTS of Kv1.4 suggest the possibility of this region playing a role in posttranscriptional regulation of expression. The rat, mouse, and bovine Kv1.4 genes have at least three polyadenylation recognition signals in the 3′-NCR, which are indicated in Figure 6. Differential utilization of these signals may be partially responsible for the generation of the three Kv1.4 mRNAs (2.4, 3.5, and 4.5 kb), although omission of a large portion of the 5′-NCR (through an alternative transcriptional start site, or splicing of an unidentified intron) would also need to be invoked to account for a functional message as small as 2.4 kb. Many conserved ATTTA sequences are also present in the 3′-NCRs of the rat, mouse, and bovine Kv1.4 genes (also shown in Figure 6), motifs that have been reported to reduce mRNA stability when present in the 3′-NCRs of cytokine genes (Shaw and Kamen, 1986). If the presence of these sequences has a destabilizing effect on Kv1.4 transcripts, then one might expect that the shortest Kv1.4 mRNAs, those containing the fewest ATTTA sequences, would be more stable than longer transcripts and may therefore be the most efficiently translated. Such AT-rich sequences have also been reported to interfere with translation, independently of their effects on message stability (Kruys et al., 1989). It is worth noting that the shortest Kv1.4 transcripts, 2.4 kb in length, can have no more than 400 nt of total UTS; therefore, if they have intact protein coding regions, they must lack *all* the ATTTA motifs indicated in Figure 6.

The 5′-NCR of rKv1.5 as well as its protein coding region appear to be intronless, and its transcription start site has been reported to be located 774 bp upstream of the initiating ATG (Mori et al., 1993). A cAMP response element located in the 5′-NCR at position +636 (relative to the transcription start site) confers cAMP responsiveness and binds CREB and CREM DNA-binding proteins (Mori et al., 1993). Recently, Shelton et al. (1993) have used the β-globin locus-control region (LCR) to direct transcription from the hKv1.5 promoter region, located just upstream of the coding region, resulting in the expression of hKv1.5 currents in mouse erythroleukemic cells.

2.1.1.2.3 Chromosomal Locations

The Kv1.1, Kv1.5, and Kv1.6 genes have been mapped to mouse chromosome 6 (Lock et al., 1994) near the *ophisthotonus* and *deaf waddler* loci, and have been recovered together on a single yeast artificial chromosome (YAC) about 275 kb in length (Migeon et al., 1992). No recombination was detected between these three genes in 113 backcrosses studied, confirming their close proximity (Lock et al., 1994). Human Kv1.1 has been localized to human 12p13, a region homologous to mouse chromosome 6 (Grissmer et al., 1992; Wymore et al., 1994); hKv1.5 and hKv1.6 would therefore also be expected to map to this region. In fact, hKv1.5 has been mapped to human 12p13 (Curran et al., 1992; Attali et al., 1993). Mutations in hKv1.1, 1.5, or 1.6 might therefore result in human diseases linked to human 12p13, possibly resembling *ophisthotonus* or *deaf waddler,* although none have yet been described. Trisomy of the 12p region is associated with an epileptiform disorder (Guerrini et al., 1990), and it is tempting to implicate one of these three K^+ channel genes in this disorder. One other *Shaker*-subfamily gene has been reported to be on human chromosome 12, namely Kv1.2 (Grissmer et al., 1990); however, its mouse homologue maps to mouse chromosome 3, which is homologous to human 1p (Lock et al., 1994), and this conflict needs to be resolved.

Human Kv1.3 has been mapped to human chromosome 1p13/1p21 (Table 1; Attali et al., 1992b); its mouse homologue mKv1.3 as well as mKv1.8 have also been mapped to mouse chromosome 3 (Lock et al., 1994). Kv1.7 lies on human 19q13.3 (K. Kalman et al., unpublished), and on mouse chromosome 7 in close proximity to mKv3.3 (K. Kalman et al., unpublished). The recessive mutant mouse gene *quivering* is also located in this region.

The defective gene for some forms of the long QT (LQT) syndrome, an autosomal dominant cardiac disease, lies within 10 cM of H-*ras* at human chromosome 11p15.5 (Keating et al., 1991; Wymore et al., 1994). The QT electrocardiogram interval, which is abnormally long in this disease, is a measure of cardiac repolarization. Since K^+ channel opening during the action potential is responsible for cardiac repolarization, a defect in a cardiac Kv gene such as Kv1.4 could lead to the prolongation of the QT interval, and consequently to the LQT syndrome. In the mouse, Kv1.4 is located close to the gene encoding FSHB (follicle stimulating hormone B) on chromosome 2 (Wymore et al., 1994), and it occupies a homologous position at human chromosome 11p14.3, approximately 25 Mb from H-*ras* (Gessler et al., 1993; Wymore et al., 1994). Thus, Kv1.4 appears to be too far from H-*ras* to be considered a candidate gene at least for this form of the LQT syndrome. However, the *anx* gene that causes anorexia, body tremors, and uncoordinated gait in mice also maps to mouse chromosome 2 in the same region as mKv1.4 (Maltais et al., 1984).

2.1.2 *Shab-, Shaw-,* and *Shal*-Subfamily Genes

2.1.2.1 *Shab*-Subfamily

Only two mammalian *Shab*-related isoforms are known, Kv2.1 and Kv2.2, represented by five cDNA sequences, one from mouse and two each from rat and human (Table 1). An amino acid sequence alignment of the rat homologues of Kv2.1 and Kv2.2, together with the fly *Shab* gene, is shown in Figure 7. Some of the features discussed for the *Shaker* alignment hold for this comparison as well, namely the striking conservation of the hydrophobic core of the proteins and the presence of a PKC site in the S4–S5 intracellular loop. However, the S1/S2 loop, which in the *Shaker* channels is highly variable in length and in sequence, is represented by a short conserved segment in the *Shab*-related proteins; while the *Drosophila Shab* protein has an N-glycosylation site in this segment (as do most *Shaker*-related channels), this site is absent from both of its mammalian homologues. The S3–S4 extracellular loop is also shorter and more highly conserved in *Shab*-subfamily genes, and this loop contains an N-glycosylation site in all three proteins. A region of 200 N-terminal residues adjacent to S1 is very highly conserved between all three proteins, while the *Shab* protein has about 240 additional residues at its N-terminus, marked by the presence of some strikingly repetitive segments (e.g., poly-Glu, poly-GlyAla). On the other hand, the long C-terminal region beginning about 100 residues past the end of S6 (the longest among all mammalian Kv proteins) shows very little sequence conservation, a feature similar to the *Shaker* comparisons. It remains possible, however, that alternate splicing at the 3′ end of the coding regions of the *Shab* genes may contribute to this diversity, as it does for *Shaw*- and *Shal*-related channels (see below), since nothing is yet known about the intron/exon structure of the *Shab*-subfamily genes. Kv2.1 and Kv2.2 are located on mouse chromosomes 2 (some distance from mKv1.4) and 1, respectively (Lock et al., 1994).

2.1.2.2 *Shaw*-Subfamily

2.1.2.2.1 Amino Acid Sequence Alignment

Four mammalian *Shaw* homologues have been described (Kv3.1–Kv3.4; Table 1), and an alignment of four representative protein sequences from rats together with the *Shaw* sequence is shown in Figure 8. The hydrophobic core is highly conserved, as well as a ~70-residue region within the N-terminal region, but here separated from the hydrophobic core by a highly variable segment of 50–70 residues. Unlike the mammalian *Shaker* homologues that have seven positive charges as part of the repeated motif in S4 (a positively charged residue at every third position), the *Shaw* channel has only four positive charges in the region, which has been aligned under the mammalian S4, lacking the first two arginines through one deletion and one substitution.

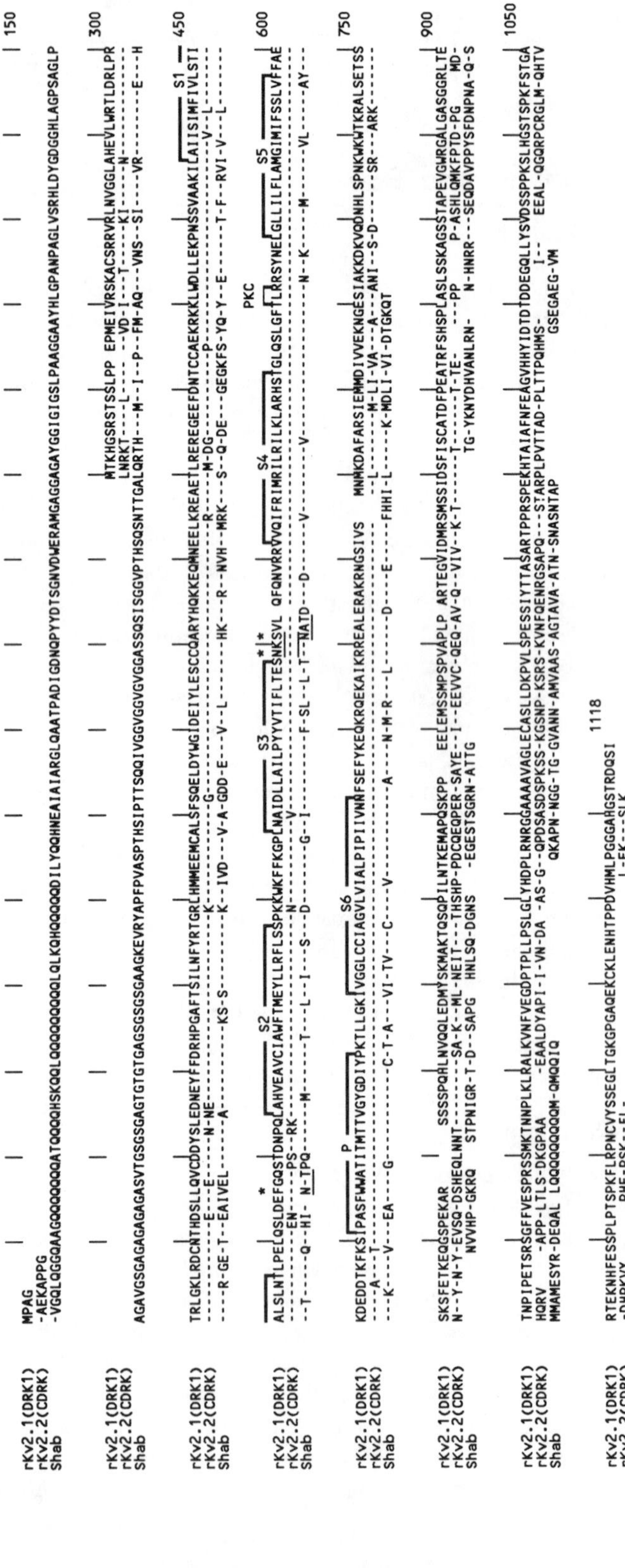

```
                                                                                                                                        150
rKv2.1(DRK1)  MPAG
rKv2.2(CDRK)  -AEKAPPG
Shab          -VGQLQGGQAAGQQQQQQQATQQQQHSKQQLQQQQQQQQQLQLKQHQQQQQDILYQQHNEAIAIARGLQAATPADIGDNQPYYDTSGNVDWERAMGAGGAGAYGGIGIGSLPAAGGAAYHLGPANPAGLVSRHLDYGDGGHLAGPSAGLP

                                                                                                                                        300
rKv2.1(DRK1)                                                                                                  MTKHGSRSTSSLPP EPMEIVRSKACSRRVRLNVGGLAHEVLWRTLDRLPR
rKv2.2(CDRK)                                                                                                  LNRKT----L---- --VD-I---T-----KI-----N------------
Shab          AGAVGSGAGAGAGAGASVTGSGSGAGTGTGTGAGSGSGSGAAGKEVRYAPFPVASPTHSIPTTSQQIVGGVGGVGVGGASSQSISGGVPTHSQSNTTGALQRTH---M--I--P--FM-AQ---VNS--SI----VR--------E---H

                                                                                                                              S1 —  450
rKv2.1(DRK1)  TRLGKLRDCNTHDSLLQVCDDYSLEDNEYFFDRHPGAFTSILNFYRTGRLHMMEEMCALSFSQELDYWGIDEIYLESCCQARYHQKKEQMNEELKREAETLREREGEEFDNTCCAEKRKKLWDLLEKPNSSVAAKILAIISIMFIVLSTI
rKv2.2(CDRK)  -----------E---E----N-NE--------------------K-----------G------------------------------R-----M-DG--------P----------------------V--L------
Shab          ----R-GE-T--EAIVEL------A----------KS-S---------K--IVD---V-A-GDD-E---V--L-------HK---R--NVH--MRK---S--Q-DE---GEGKFS-YQ-Y--E-----T-F--RVI-V---L------

                                                                                                     PKC
                    *                   S2                       S3         *|*          S4                        S5                  600
rKv2.1(DRK1)  ALSLNTLPELQSLDEFGQSTDNPQLAHVEAVCIAWFTMEYLLRFLSSPKKWKFFKGPLNAIDLLAILPYYVTIFLTESNKSVL QFQNVRRVVQIFRIMRILRILKLARHSTGLQSLGFTLRRSYNELGLLILFLAMGIMIFSSLVFFAE
rKv2.2(CDRK)  -----------EN-----PS--RK-------------------------N---------V-----------------------------------------------------------------------------------
Shab          --T-----Q--HI- N-TPQ------M------T---L---I---S---D-------G--I---------F-SL--L-T--NATD---D------V-------V------------------N--K-----M------VL-----AY---

                             P                    S6                                                                                    750
rKv2.1(DRK1)  KDEDDTKFKSIPASFWWATITMTTVGYGDIYPKTLLGKIVGGLCCIAGVLVIALPIPIIVNNFSEFYKEQKRQEKAIKRREALERAKRNGSIVS   MNMKDAFARSIEMMDIVVEKNGESIAKKDKVQDNHLSPNKWKWTKRALSETSS
rKv2.2(CDRK)  ----A---T-----------------------------------------------------------------------------------   --L-------M-LI-VA---A---ANI--S-D------SR---ARK-------
Shab          ---K----V---EA----G----------C-T-A---VI-TV---C----V-----------A----N-M-R---L------D----E-----FHHI-L-----K-MDLI-VI-DTGKQT

                                                                                                                                        900
rKv2.1(DRK1)  SKSFETKEQGSPEKAR        SSSSPQHLNVQQLEDMYSKMAKTQSQPILNTKEMAPQSKPP    EELEMSSMPSPVAPLP ARTEGVIDMRSMSSIDSFISCATDFPEATRFSHSPLASLSSKAGSSTAPEVGWRGALGASGGRLTE
rKv2.2(CDRK)  N--Y-N-Y-EVSQ-DSHEQLNNT--------SA-K--ML-NEIT---THSHP-PDCQEQPER-SAYE--I--EEVVC-QEQ-AV-Q--VIV--K-T------T------T-TE-    ---PP      P-ASHLQMKFPTD-PG      MD-
Shab                  NVVHP-GKRQ   STPNIGR-T-D--SAPG  HNLSQ-DGNS    -EGESTSGRN-ATTG                                        TG-YKNYDHVANLRN-    N-HNRR---SEQDAVPPYSFDNPNA-Q-S

                                                                                                                                        1050
rKv2.1(DRK1)  TNPIPETSRSGFFVESPRSSMKTNNPLKLRALKVNFVEGDPTPLLPSLGLYHDPLRNRGGAAAAVAGLECASLLDKPVLSPESSIYTTASARTPPRSPEKHTAIAFNFEAGVHHYIDTDTDDEGQLLYSVDSSPPKSLHGSTSPKFSTGA
rKv2.2(CDRK)  HQRV     -APP-LTLS-DKGPAA      -EAALDYAPI-I-VN-DA -AS-G--QPDSASDSPKSS-KGSNP-KSRS-KVNFQENRGSAPQ---STARPLPVTTAD-PLTTPQHMS-        I--    EEAL-QGQRPCRGLM-QHTV
Shab          MMAMESYR-DEQAL LQQQQQQQQM-QMQQIQ                      QKAPN-NGG-TG-GVANN-AMVAAS-AGTAVA-ATN-SNASNTAP                   GSEGAEG-VM

                                                                                1118
rKv2.1(DRK1)  RTEKNHFESSPLPTSPKFLRPNCVYSSEGLTGKGPGAQEKCKLENHTPPDVHMLPGGGAHGSTRDQSI
rKv2.2(CDRK)  -DHPKVY          PHE-PSK--FL-                           L-EK---SLK
Shab                 -MGAVSMTTT-P--R             D-                              P-R
```

FIGURE 7. *Shab*-family amino acid sequence alignment, including the rat homologs of Kv2.1 and Kv2.2 and the *Drosophila Shab* protein. Putative functional regions are indicated as in Figure 2.

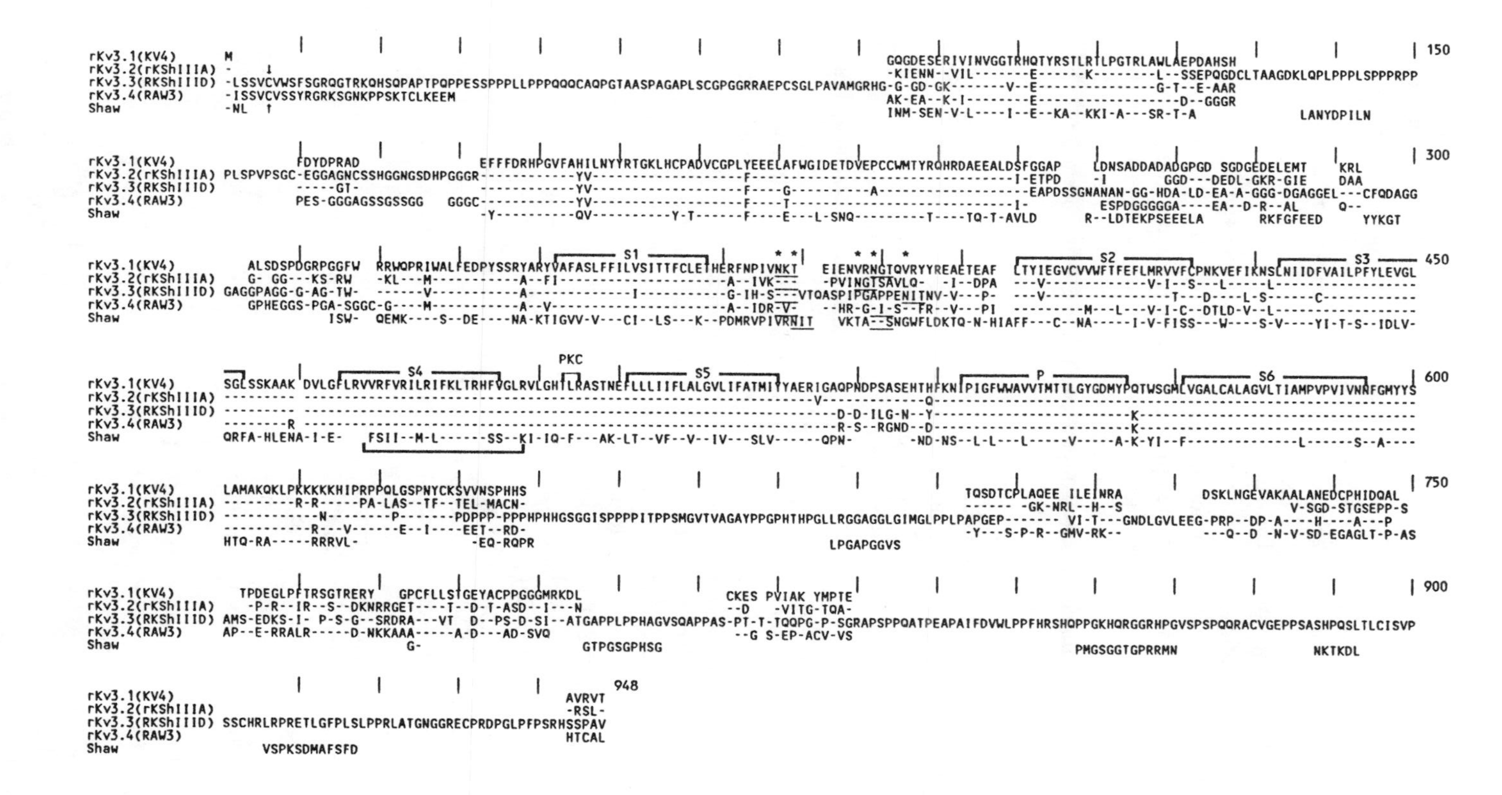

FIGURE 8. *Shaw*-family amino acid sequence alignment, including the rat homologs of Kv3.1–Kv3.4 and the *Drosophila Shaw* protein. Putative functional regions are indicated as in Figure 2. The proposed S4 region of the *Shaw* protein is indicated by a bracket below, in consideration of the size difference evident between this protein and its mammalian homologues (see text). Arrows mark the position of a cysteine residue at position 6 of rKv3.3 and rKv3.4, suggested to correspond to the cysteine involved in fast inactivation in Kv1.4 (see Figure 2, and text).

However, an additional arginine is present six residues downstream, which is in phase with the last arginine in the repeated motif. Given the three-residue deletion in *Shaw* relative to its mammalian homologues, its S4 region should probably be extended to this arginine (as indicated by the bracket below the alignment), thus bringing to five the total number of positive charges in S4, which might contribute to voltage sensing (discussed further below).

2.1.2.2.2 Posttranslational Modification

Two N-glycosylation sites are present in the poorly conserved S1/S2 loop of all of these *Shaw*-related proteins, and the universal PKC site is present between S4 and S5. There is a highly variable C-terminal region, although alternate mRNA splicing (discussed below) is known to contribute to this diversity. Regions containing short amino acid repeats are evident, particularly in the N-terminal region of Kv3.2 and in the C-terminal insertion just outside the conserved core in Kv3.3.

2.1.2.2.3 Introns and Alternative Splicing

The transcripts of all four *Shaw* genes are heterogeneous and include large forms, ranging from 4.5–13 kb, and none has yet been mapped. While complete genomic sequence is not available for any *Shaw*-related protein, one intron is known to be present at the 5′ margin of the S1 transmembrane segment in Kv3.1, 3.3, and 3.4 (Ghanshani et al., 1992; Luneau et al., 1991a,b). In addition, alternative splicing is known to contribute to the diversity of the C-termini of Kv3.1 and Kv3.2; the Kv3.1b channel differs from its alternately spliced counterpart, Kv3.1a, in that the last 10 amino acids of Kv3.1a are replaced by an unrelated sequence of 84 amino acids in Kv3.1b. In the case of Kv3.2, alternative splicing at their 3′ ends can yield four distinct Kv3.2 mRNAs, as shown schematically in Figure 9. Two of these transcripts (RAW1/rKShIIIA2 and rKShIIIA) share a common 3′ segment, indicated by cross-hatching; this shared region does not contribute to the translated product of the RAW1/rKShIIIA2 transcript, however, due to the presence of a termination codon (indicated by a heavy bar) in the shorter exon, which is absent from the rKShIIIA transcript (Luneau et al., 1991b; Rettig et al., 1992). While these two exons are shown as being separated by an intervening sequence, the existence of such an intron is unproven, since the genomic sequence is not known; the two exons might be contiguous in the genome, as they are in the RAW1/rKShIIIA2 transcript, and the rKShIIIA transcript would then result from the use of a splice acceptor site within this exon. The same caution holds true for many other Kv genes for which multiple transcripts are known but no genomic sequence is available.

2.1.2.2.4 *Shaw*-Subfamily Phylogenetic Tree

Figure 10 shows a proposed phylogenetic tree of all known *Shaw*-related genes, resolving the ambiguities in our earlier analyses, which resulted from a paucity of data (Strong et al., 1993). This phylogeny implies that three separate gene duplications of a *Shaw*-related ancestor gave rise sequentially to the four known mammalian *Shaw* homologues. No nonmammalian *Shaw*-

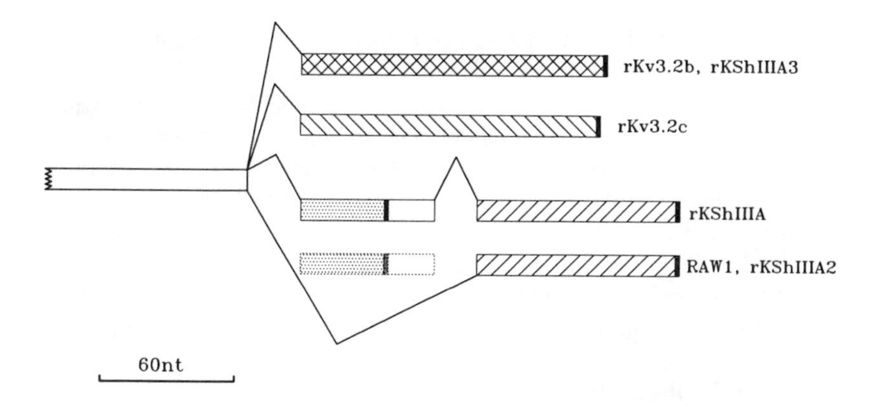

FIGURE 9. Four alternatively spliced variants at the 3′ end of rat Kv3.2 mRNAs, showing the hypothetical splicing patterns. Four distinct exon-containing regions are indicated by different fill patterns, and the positions of translation termination codons are indicated by solid vertical bars. The scale bar indicates a length of 60 nucleotides.

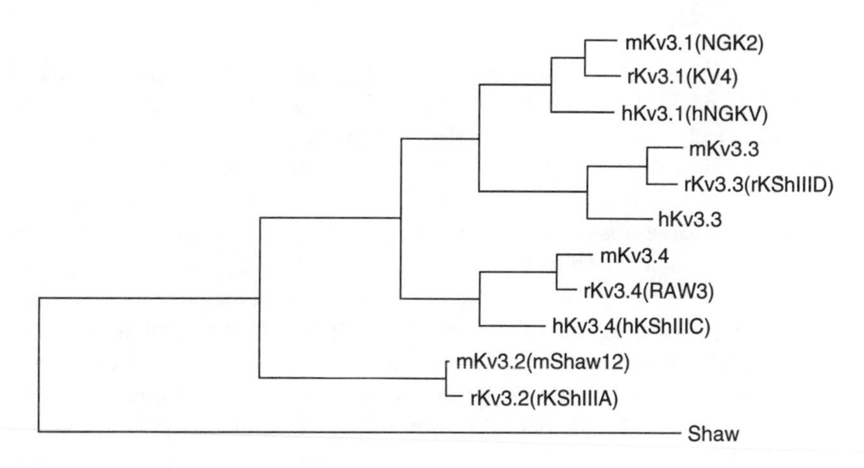

FIGURE 10. Proposed phylogenetic tree of *Shaw*-related K^+ channel genes, based on parsimony analysis of nucleotide sequence alignments using the program PAUP (Swofford, 1993).

related sequences are yet available to determine at what period in the vertebrate radiation these duplications occurred.

2.1.2.2.5 Chromosomal Locations

The four *Shaw*-related genes (Kv3.1, Kv3.2, Kv3.3, Kv3.4) have been reported to be located on human chromosomes 11, 12/19, 19, and 1, respectively (see references in Table 1). Kv3.1 has been sublocalized to human chromosome 11p14.3–15.2 (Reid et al., 1993; Wymore et al., 1994); it appears to be about >25 Mb away from H-*ras,* and therefore is unlikely to be a candidate gene for the H-*ras*-linked form of the LQT syndrome (Wymore et al., 1994). In the mouse, Kv3.1 is within 1 cM of myoD1 on chromosome 7 (Wymore et al., 1994), a region that is homologous to human 11p15.1/15.2 (Reid et al., 1993; Wymore et al., 1994). While Kv3.2 maps to mouse chromosome 10 in a region homologous to human 12q (Lock et al., 1994), conflicting reports have placed it on human 12 (Curran et al., 1992; McPherson et al., 1991) and on 19q13.3-13.4 (Haas et al., 1993). Kv3.3 lies close to Kv1.7 both in the mouse (on chromosome 7) and in humans (on human 19q13.3; K. Kalman et al., unpublished data). Kv3.4 has been mapped to mouse chromosome 3 and the homologous region of human 1p21 (Rudy et al., 1991a; Ghanshani et al., 1992; B. Tempel, personal communication).

2.1.2.3 *Shal*-Subfamily

An alignment of proteins representing two mammalian isoforms of *Shal*-related channels is shown in Figure 11, together with that of the *Drosophila Shal* protein; one additional *Shal*-related mammalian gene, Kv4.3, is known to exist, but its sequence has not yet been published (Rudy et al., 1991b). The hydrophobic core is highly conserved, and the universal PKC site is present between S4 and S5. A potential tyrosine kinase site is present in the N-terminal region of Kv4.2, but not Kv4.1, and the only potential external N-glycosylation site is in the extracellular loop linking S5 to the P-region of Kv4.1 and *Shal* (but not Kv4.2). Interestingly, sodium channel proteins are known to be heavily N-glycosylated, at a position thought to be between S5 and the pore region (Miller et al., 1983; James and Agnew, 1987; Ukomadu et al., 1992). The S4 segment of the *Shal*-subfamily genes contains only six positive charges that are part of the repeated motif, in contrast to the 7 positive charges present in *Shaker*-subfamily genes. Like the *Shab* subfamily (but unlike the *Shaker* and *Shaw* families), the N-terminal region of *Shal* subfamily genes is almost perfectly conserved in its length, and substantially conserved in its sequence. While the C-terminal regions show extensive sequence divergence, nothing is known of possible alternative 3′ splicing within these genes. Kv4.1 is located on the mouse X chromosome (Lock et al., 1994).

2.1.3 The Extended Voltage-Gated K^+ Channel Gene Family: Amino Acid Sequence Comparisons

Figure 12 shows an alignment of the amino acid sequences of 16 mammalian K^+ channel genes, one representative of every known mammalian Kv gene, together with the four known fly homologues. There is clearly substantial sequence conservation in a large portion of the hydrophobic core of these proteins, as well as a more limited region in the N-terminal portion, about 100 residues away from S1; these relatively conserved regions are indicated by brackets below the alignment. Within these regions, comprising some 262 residues altogether, 44 positions show complete identity among all 20 sequences, while at an additional 65 positions only conservative substitutions

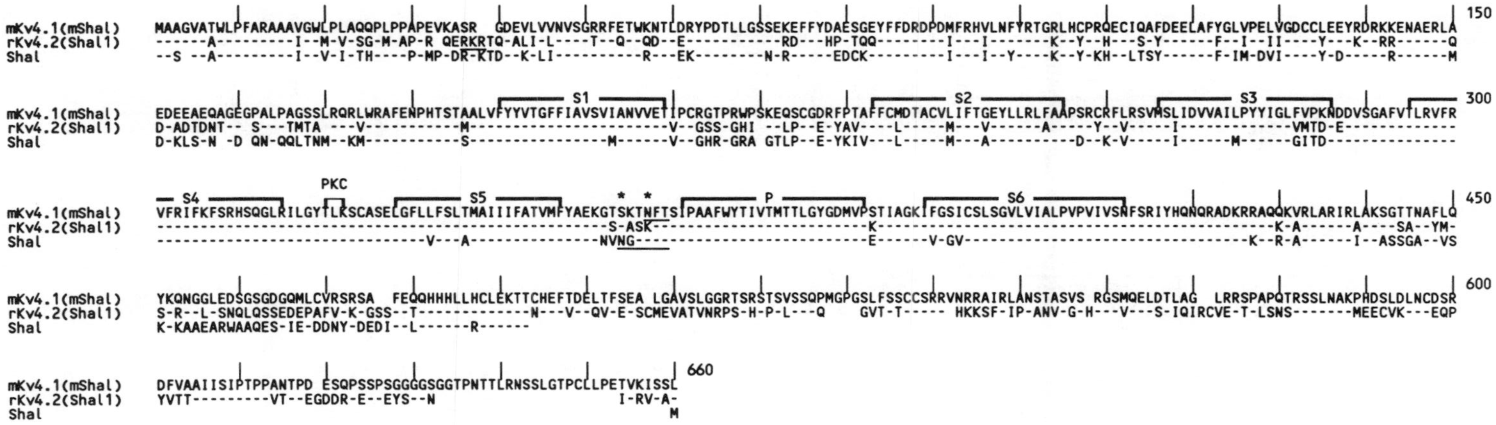

```
                                                                                                                                   150
mKv4.1(mShal)  MAAGVATWLPFARAAAVGWLPLAQQPLPPAPEVKASR  GDEVLVVNVSGRRFETWKNTLDRYPDTLLGSSEKEFFYDAESGEYFFDRDPDMFRHVLNFYRTGRLHCPRQECIQAFDEELAFYGLVPELVGDCCLEEYRDRKKENAERLA
rKv4.2(Shal1)  ------A---------I--M-V-SG-M-AP-R QERKRTQ-ALI-L----T--Q--QD--E-----------RD---HP-TQQ--------I---I-------K--Y--H---S-Y------F--I--II----Y---K--RR------Q
Shal           --S --A---------I--V-I-TH----P-MP-DR-KTD--K-LI----------R---EK--------N-R-----EDCK---------I---I--Y----K--Y-KH--LTSY------F-IM-DVI----Y-D-----R------M

                                                       S1                                   S2                                S3                   300
mKv4.1(mShal)  EDEEAEQAGEGPALPAGSSLRQRLWRAFENPHTSTAALVFYYVTGFFIAVSVIANVVETIPCRGTPRWPSKEQSCGDRFPTAFFCMDTACVLIFTGEYLLRLFAAPSRCRFLRSVMSLIDVVAILPYYIGLFVPKNDDVSGAFVTLRVFR
rKv4.2(Shal1)  D-ADTDNT-- S---TMTA ---V-----------M-----------------------V--GSS-GHI --LP--E-YAV----L-----M---V------A-----Y--V-----I-------------VMTD-E-------------
Shal           D-KLS-N -D QN-QQLTNM--KM-----------S----------------M------V--GHR-GRA GTLP--E-YKIV---L-----M---A-----------D--K-V-----I------M------GITD--------------

                        PKC
               — S4                     S5                   *  *      P                      S6                                                  450
mKv4.1(mShal)  VFRIFKFSRHSQGLRILGYTLKSCASELGFLLFSLTMAIIIFATVMFYAEKGTSKTNFTSIPAAFWYTIVTMTTLGYGDMVPSTIAGKIFGSICSLSGVLVIALPVPVIVSNFSRIYHQNQRADKRRAQQKVRLARIRLAKSGTTNAFLQ
rKv4.2(Shal1)  ------------------------------------------------S-ASK-------------------------K--------------------------------------------K-A------A----SA--YM-
Shal           -----------------------------V---A--------------NVNG----------------------------E------V-GV------------------------------------K--R-A------I--ASSGA--VS

                                                                                                                                                   600
mKv4.1(mShal)  YKQNGGLEDSGSGDGQMLCVRSRSA  FEQQHHHLLHCLEKTTCHEFTDELTFSEA LGAVSLGGRTSRSTSVSSQPMGPGSLFSSCCSRRVNRRAIRLANSTASVS RGSMQELDTLAG  LRRSPAPQTRSSLNAKPHDSLDLNCDSR
rKv4.2(Shal1)  S-R--L-SNQLQSSEDEPAFV-K-GSS--T-------------N---V--QV-E-SCMEVATVNRPS-H-P-L---Q    GVT-T-----  HKKSF-IP-ANV-G-H---V---S-IQIRCVE-T-LSNS-------MEECVK---EQP
Shal           K-KAAEARWAAQES-IE-DDNY-DEDI--L------R------

                                                                 660
mKv4.1(mShal)  DFVAAIISIPTPPANTPD ESQPSSPSGGGGSGGTPNTTLRNSSLGTPCLLPETVKISSL
rKv4.2(Shal1)  YVTT---------VT--EGDDR-E--EYS--N                     I-RV-A-
Shal                                                                      M
```

FIGURE 11. *Shal*-family amino acid sequence alignment, including the rat homologs of Kv4.1 and Kv4.2 and the *Drosophila Shal* protein. Putative functional regions are indicated as in Figure 2. The "RKR" sequence at position 36 of rKv4.2, known to be important in fast inactivation (see text), is underlined.

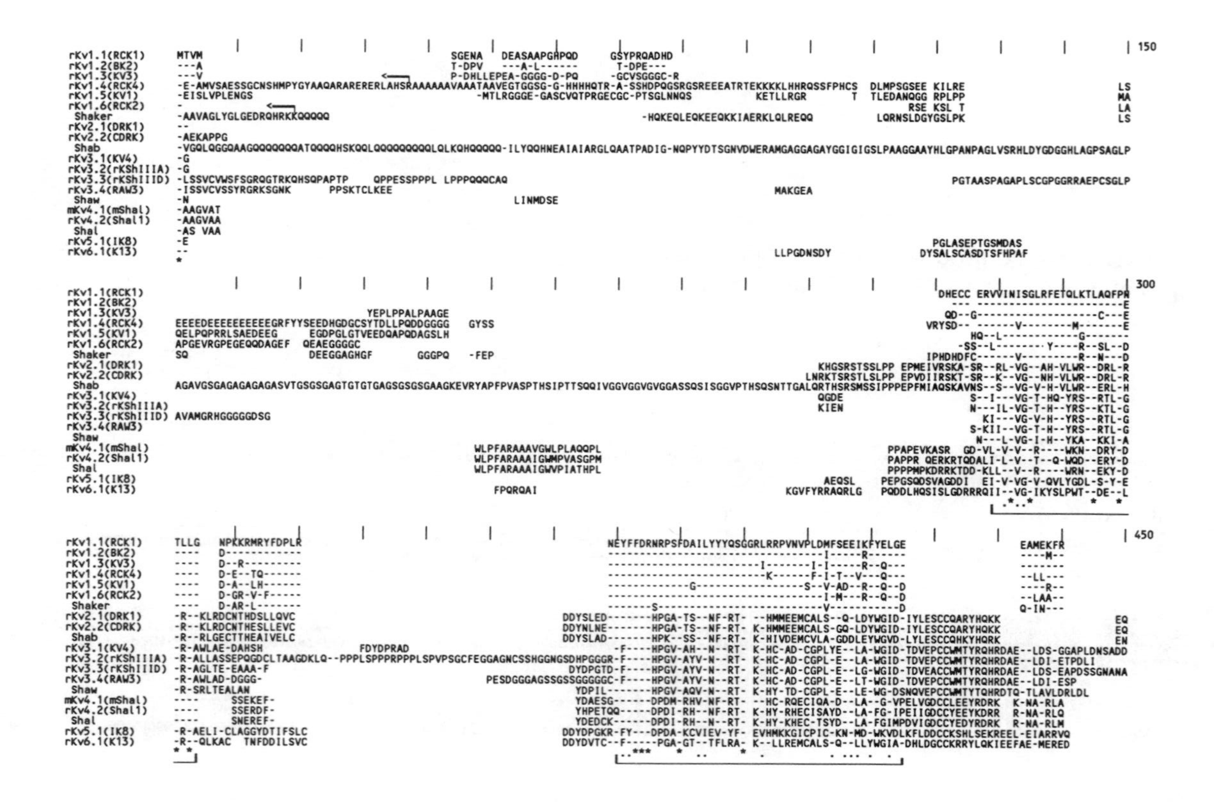

FIGURE 12. Amino acid sequence alignment of rat homologues of each of the known mammalian Kv proteins (with the exception of Kv4.1, which is represented by its mouse homologue). Putative functional regions are indicated by double brackets, and positions of identity or similarity in all sequences are indicated below by asterisks or periods, as in Figure 2. Five single brackets below the alignment indicate regions that show substantial conservation among all members of this family (see text). Arrows near the N-terminus of rKv1.4 and *Shaker* indicate the margins of peptides that have been shown to be involved in fast inactivation of these channels (see text).

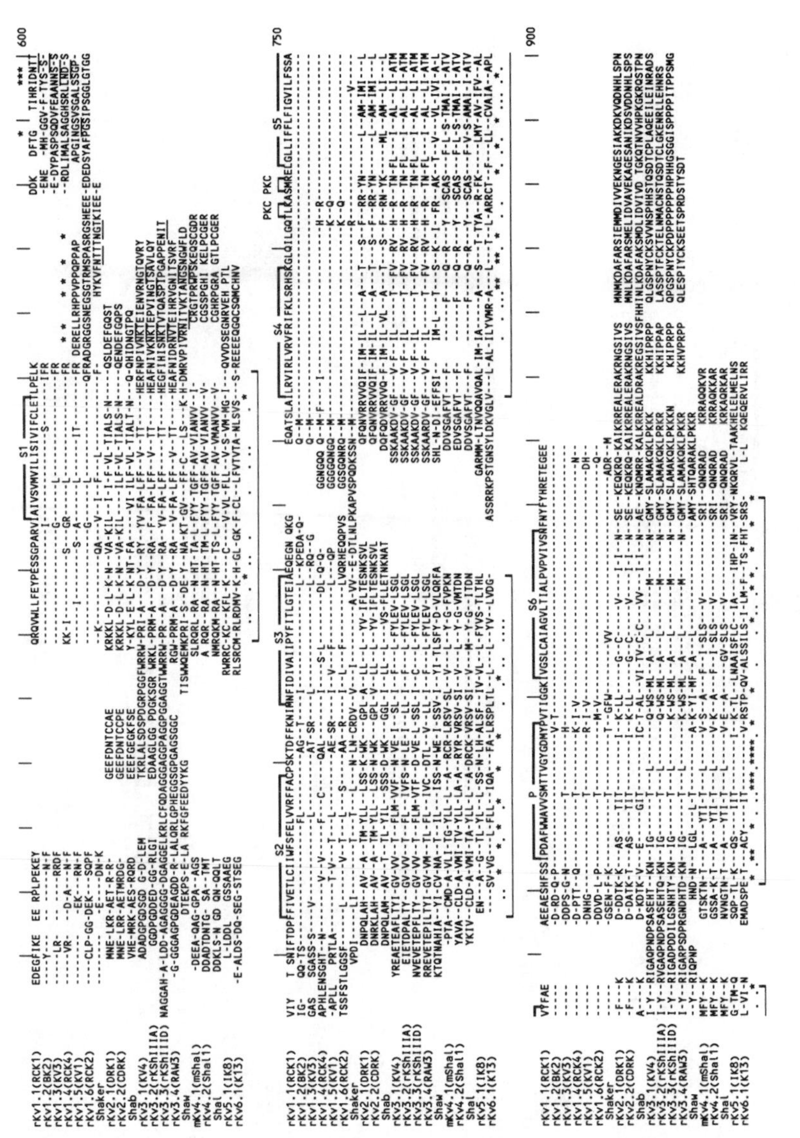

FIGURE 12(2).

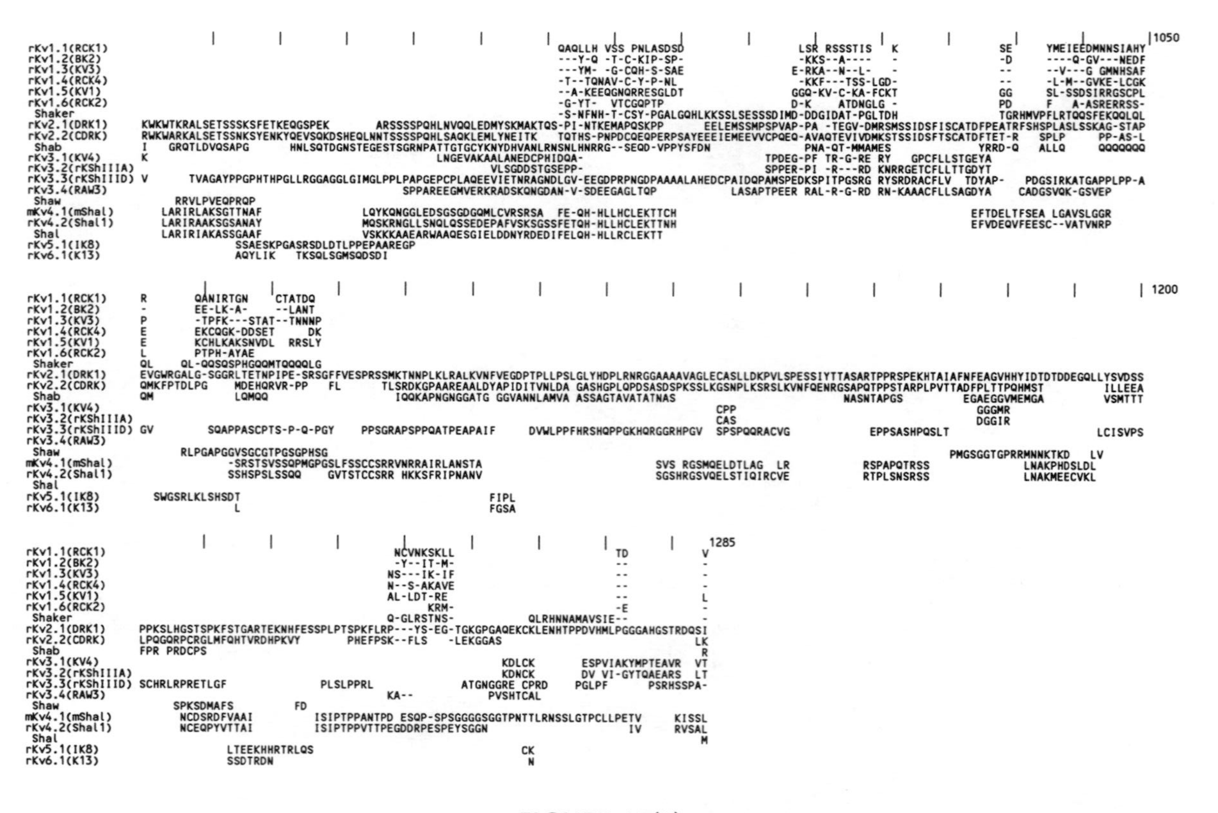

FIGURE 12(3).

are seen. In current models of K^+ channels, the six membrane-spanning domains define the overall architecture of the channels and determine its gating properties, and the P-region participates in forming the ion-conducting pore (Durell and Guy, 1992); substantial conservation of these sequences is therefore not difficult to understand.

2.1.3.1 Amino and Carboxy-Terminal Regions

While size differences within the hydrophobic core of Kv channels is limited to the regions between the putative membrane-spanning segments, there is considerable diversity in the length of their C- and N-terminal regions, as illustrated schematically in Figure 13. The *Shaker*-related channels all have longer N-terminal regions than C-terminal; the *Shab*-related proteins are the longest overall, and its mammalian homologues have the longest known C-termini, while the *Shab* protein has the longest N-terminus among all known Kv channels. Alternative RNA splicing can contribute substantially to diversity in the lengths of these regions, the C-terminus of Kv3.3 (for example) varying more than twofold in length in alternate forms.

While this diversity of structure suggests a corresponding diversity of function, relatively little is known about the specific roles of the cytoplasmic "tails" of these proteins. In some channels (e.g., *Shaker* and Kv1.4), an N-terminal peptide has been shown to participate in rapid inactivation, although

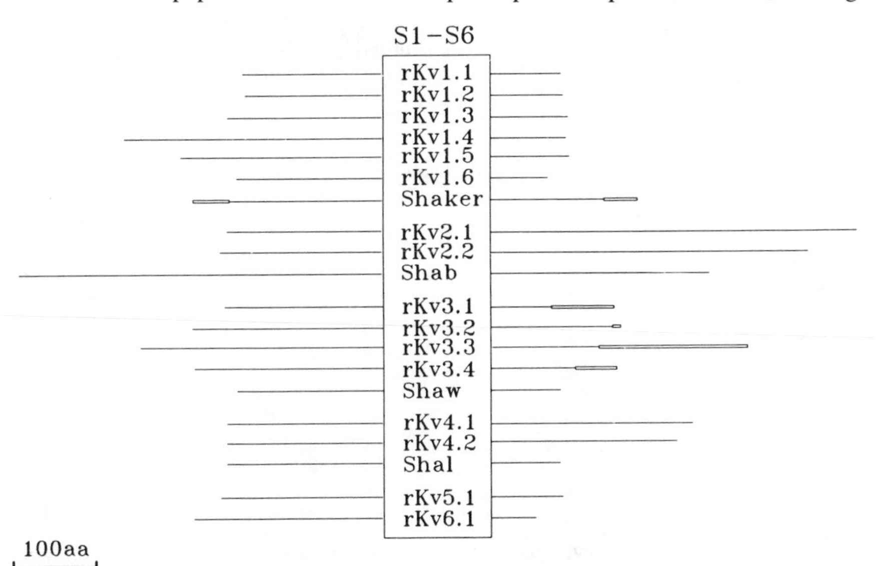

FIGURE 13. Schematic diagram showing the relative lengths of the N-terminal and C-terminal cytoplasmic "tails" of Kv proteins. The boxed region represents the hydrophobic core of the proteins from S1 through S6, and size differences within this region have been ignored. Those segments known to vary in length through alternative mRNA splicing are indicated by open boxes. The scale bar indicates a length of 100 residues.

there is relatively little sequence conservation of such peptides. The strongly conserved region in the N-terminal cytoplasmic region of these proteins (the region including the YFFDR motif) could be important in establishing the specificity of subunit interactions in its native tetrameric form (Li et al., 1992; Shen et al., 1993). Another (short) conserved motif is represented by the two adjacent cysteines present in the N-terminal region of all Kv genes other than the *Shaker* subfamily (IYLES*CC*QARY in rKv2.1, at position 382 in Figure 12), in a region with no homologue in the *Shaker* subfamily; a conserved cysteine is also located about 25 residues upstream in this same group of proteins (HMMEEM*C*ALS in rKv2.1, at position 361 in Figure 12), although no function has been ascribed to any of these cysteines. Two additional conserved cysteines, one in S2 and the other in S6 (see Figure 12), when mutated in Kv2.1, did not affect channel function (Zuhlke et al., 1993).

2.1.3.2 The P-Region

Portions of the P-region (spanning some 22 residues) are among the most highly conserved sequences within this family of proteins. In fact, as illustrated in the alignment in Figure 14, this region can be identified in *all* known K⁺-selective channels, including all Kv channels (which we have discussed here), calcium-activated K⁺ channels of fly and mouse (Atkinson et al., 1991; Butler et al., 1993), inward rectifier K⁺ channels in plants and mammals (Anderson et al., 1992; Sentenac et al., 1992; Ho et al., 1993; Kubo et al., 1993a,b) including the G-protein-coupled rat GIRK channel (Kubo et al., 1993b), the fly *eag* K⁺ channel (Warmke et al., 1991, 1993), and a putative K⁺ channel isolated from *E. coli* (Milkman, 1994; also see Table 1).

```
Shaker          NSFFKSIPDAFWWAVVTMTTVGYGDMTPVGFWGKIV
rKv1.1(RCK1)    E-H-S-----------=----------Y--TIG----
Shab            D=K-V---=-----G=----------=C-TTAL--==
rKv2.1(DRK1)    D=K-----AS----T=----------=Y-KTLL----
Shaw            HND-N---LGL---=------------A-KT=I-MF-
rKv3.1(KV4)     H=H--N--IG----------=-----Y-QT=S-M=-
Shal            A-K-T---A---=T=-----=-----V-KTIA---F
mKv4.1(mShal)   K=N-T---A---=T=-----=-----V-STIA---F
rKv5.1(IK8)     E=L-----QS---==----------=Y-KTTL--=N
rKv6.1(K13)     SPE-T---AC=----=----------V-RSTP-Q=-
                   F sIP  fWwa vtMTTvGYGDm P    G        Kv consensus
ECOKCH          -PRIE-=MT--==S=E--=-------=V--SESA==F
slo             -AHRL-YWTCV==L=---=-------=YCETVL-=TF
mSlo            -NQAL=YW=CV=LL=---=-------=YAKTTL-==F
KAT1            A-L=NRYVT-L=-S=T-=--T----FHAENPREM=F
AKT1            E-L=MRYVTSM=-S=T-=--------=H--NTKEM-F
ROMK1           VENING=TS--L=S=E-QV-=---FRFVTEQCATA=
IRK1            V-EVN-FTA--L=S=E-Q--=---FRCVTDECPIA-
GIRK1           VANVYNF-S--L=F=E-EA-=---YRYITDKCPEG=
eag             P-RKSMYVT-L==T=TC--=--=-N=AAETDNE-=F
                                tmttvGyGd                all K+ consensus
```

FIGURE 14. Amino acid sequence alignment of the putative pore region of K⁺ channel proteins (see text). The symbol "–" indicates identity with the topmost sequence, while "=" indicates a conservative substitution. The consensus sequences for either the Kv family alone (below rKv1.6), or *all* K⁺ channels (bottom), represent positions that either are completely conserved (upper case) or at which a single amino acid predominates (lower case).

The region aligned in Figure 14 contains 13 residues that are identical in all Kv channels, and an additional 5 that show only conservative changes within this group. Among the larger group of all K^+-selective channels, only two residues are perfectly conserved *(G[Y/F]G)*, and an additional three positions show exclusively conservative changes. In fact, the *eag* channel is the only known K^+-selective channel that lacks the tyrosine in the otherwise universally conserved *GYG* motif, replacing it with the very similar phenylalanine; the recent demonstration that this channel is also permeable to Ca^{2+} (Bruggemann et al., 1993) could be relevant in this regard, although the Ca^{2+} permeability of other K^+ channels has not been systematically examined. It should be noted that Isk/Mink channel lacks *any* sequence motifs resembling the K^+ channel pore; however, a recent report by Attali et al. (1993) suggests that Isk may not be a channel at all, but rather an activator of endogenous K^+ or Cl^- channels present in *Xenopus* oocytes.

The presence of this pore motif in both eubacteria and vertebrates suggests that its evolution predated the early branching of the "tree of life" more than three billion years ago. The fact that other cation channels (e.g., Na^+ and Ca^{2+}) do not share this motif suggests that the P-region plays a key role in defining the K^+ selectivity of these channels, and that acquisition of K^+ selectivity was an ancient evolutionary development.

The loop linking S5 to the P-region is thought to form part of the outer vestibule of the ion conduction pathway (Durell and Guy, 1992), since residues in this region have been reported to be involved in toxin binding (MacKinnon et al., 1989, 1990; Hurst et al., 1991; Goldstein et al., 1992). This loop is more highly conserved in its length than in its sequence, being 13–15 amino acids long in all but the mammalian *Shaw* homologues which have 22 residues (Figure 12).

2.1.3.3 The S4 Segment and Leucine Zipper

The S4 segment of K^+ channels is thought to form a major element of the voltage sensor, and mutagenesis of its positive charges (and some hydrophobic residues) has been shown to alter the gating properties of *Shaker* (Papazian et al., 1991; Liman et al., 1991; Koren et al., 1990; Logothetis et al., 1992, 1993; Lopez et al., 1991; also see review by Jan and Jan, 1992). The S4 motif with its characteristic pattern of positive charges every third residue is also present in Ca^{2+} and Na^+ channels (Catterall, 1988), supporting its importance for voltage sensing. The presence of a similar motif in cGMP-gated channels seems to be at variance with their ligand-gated properties, but may account for their moderate voltage dependence (Kaupp et al., 1989; Ludwig et al., 1990).

The S4 segment, as indicated in Figure 12, is generally assumed to be about 20 amino acids in length, consistent with the presumed length of α-helix required to span the cell membrane. Given this disposition, the S4 segments of the *Shaker*-subfamily channels contain seven conserved positive charges (five arginines, two lysines) in the repeated motif, while the *Shaw*-subfamily proteins have six, and *Shab*-subfamily channels as well as Kv5.1 (IK8) and Kv6.1 (K13) each has five. Alignment of the *Shal*-subfamily channels with the other Kv channels reveals a three amino acid deletion in the S4 segment, which eliminates the second arginine in the repeated motif, and in place of the last lysine they contain a glutamine, leaving them with only five positive charges. However, the *Shal* proteins contain an additional downstream arginine in phase with the repeated motif (at a position at which *Shaker* and *Shab* channels have a glutamine), and inclusion of this residue in the S4 segment would bring the total number of positive charges to six. If the S4 segment is of the same length in these different channels, then this sixth positively charged residue would be included in the voltage sensor of the mammalian *Shal*-subfamily proteins, as well as in the fly *Shaw* protein. Nor has the possibility been excluded that the arginine at this position in the mammalian *Shaw* proteins also participates in voltage sensing.

Voltage sensing in these channels certainly depends on structural features outside the S4 region itself. While specific residues outside S4 influencing this process have not yet been identified, charged amino acids at highly conserved positions may be attractive candidates. It is tempting, for example, to suggest that the conserved arginine present in the S2 segment of all Kv protein may influence the voltage dependence of activation of these channels. Several negatively charged residues are also conserved in all Kv channels (the FF*D*R motif in the N-terminal region already noted, a glutamic acid in S2, an aspartic acid in S3, and a glutamic acid at each end of S5); these and other charged residues could influence voltage sensing in the various Kv isoforms.

A leucine heptad motif is repeated five times at the end of the S4 segment and S4/S5 loop of all the *Shaker*- and *Shab*-subfamily channels (Figure 12), forming a "leucine zipper" motif thought to be important in many protein–protein interactions (McCormack et al., 1989; Alber, 1992). Leucines 2 and 5 (L2 and L5) are present in all Kv channels. L1 is also found in all the channels, with the exception of the *Shal*-subfamily channels, which have a phenylalanine at that position. Similarly, phenylalanine replaces L4 in the *Shaw* channels and Kv6.1, while phenylalanine and alanine are present in *Shaw* and Kv6.1 in the place of L3. Substitutions of these leucines in the *Shaker* channel has been shown to alter the voltage-dependent steps during channel gating (McCormack et al., 1991; Schopa et al., 1992). While the significance of this leucine zipper-like structure is still not understood, it clearly can influence channel gating, perhaps through modification of interdomain interactions.

2.1.3.4 Kv-Channel Phylogenetic Tree

Figure 15 shows a proposed phylogenetic tree for the extended Kv family. This topology was determined by first examining trees including only members of the Kv1, Kv2, Kv3, and Kv4 subfamilies, and constraining the relationships within the *Shaker* and *Shaw* groups to be those shown in the trees in Figures 4 and 10; the most parsimonious tree found in this way makes neighbors of the *Shaker* and *Shal* groups on the one hand, and the *Shaw* and *Shab* groups on the other, as in our earlier analyses (Strong et al., 1993).

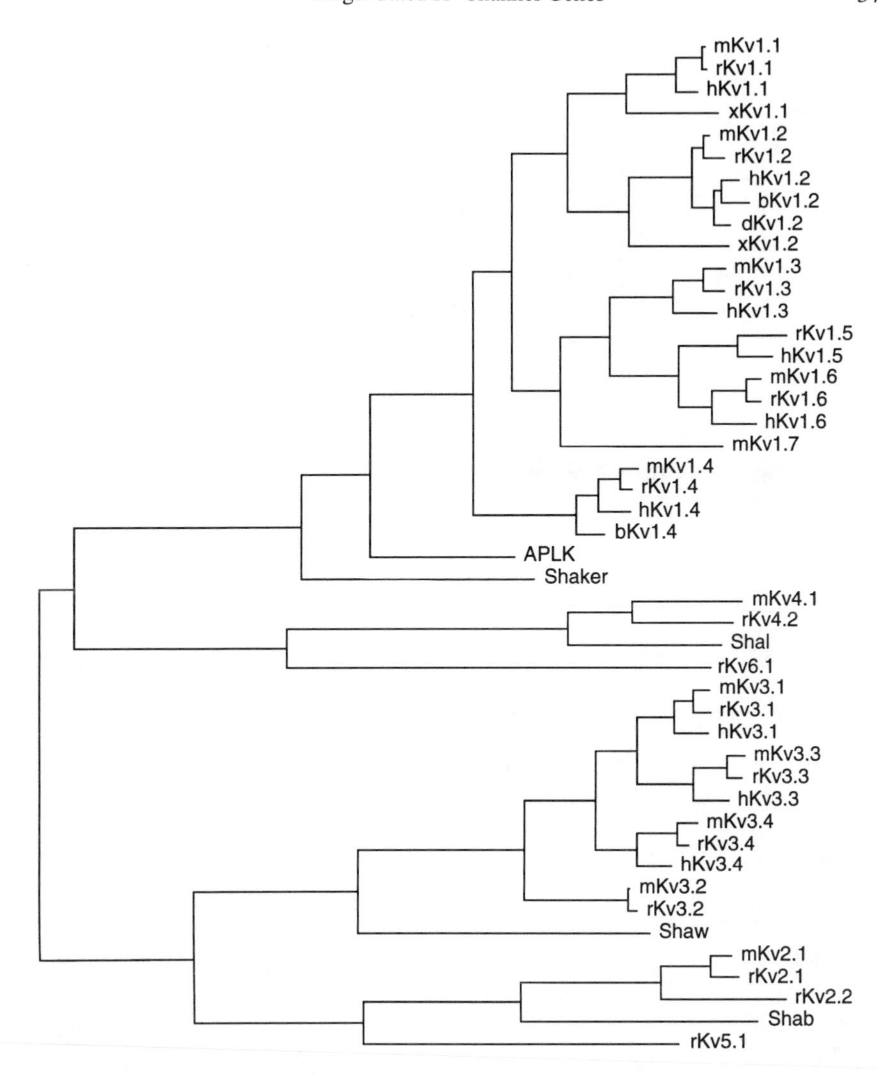

FIGURE 15. Proposed phylogenetic tree of voltage-gated K^+ channel genes, based on parsimony analysis of nucleotide sequence alignments using the program PAUP (Swofford, 1993; see text).

Adding rKv5.1 (IK8) and rKv6.1 (K13) to such a tree places the former on the branch leading to *Shab,* and the latter on that leading to *Shal,* as shown in the figure. However, neither the deep branching between the Kv1 and Kv4 subfamilies in this tree *(Shak/Shal* versus Shab/Shaw), nor the placement on this tree of Kv5.1 and Kv1.6 is compellingly supported by bootstrapping. The deep divisions within this group of genes, together with their as yet unresolved rooting (Strong et al., 1993), may therefore have to await the availability of additional sequence data, or more powerful analytic methods, for their satisfactory resolution.

Table 2a. Tissue Distribution of Voltage-Gated Potassium Channel Genes

Gene	Expression in tissue[a]															Comments
	Brain	Atr	Vent	Kid	Retina	Lung	Liver	Skm	Islet	Thy	Spl	Lym	C_2C_{12}	GH3	Aorta	
Kv1.1	+[1,2]	+[3]	–[3]		+[4]		–[3]	+[5]	+[6]				–[7]		–[3]	Olfactory bulb +[1]
Kv1.2	+[1,2]	+[3,8]	+[3,8]				–[3]	–[3]	+[6]				–[7]		–[3]	Atrium > vent,[3] NG108-15 +[9]
Kv1.3	+[1]	–[10,11]	–[10,11]	–[10,11]		+[11]	–[10,11]	–[11]	+[6]	+[12,13]	+[10,11]	+[10–15]	–[7]			Fibroblasts +,[10] pre-B +[16]
Kv1.4	+[1,17]	+[18]	+[18]	–[19]			–[3]	+[7]	+[6]				+[7]	+[20]	–[19]	Atrium = vent,[18] 2.3 kb mRNA in skm;[7] adrenal medulla +, stomach –[9]
Kv1.5	+[22]	+[3,11]	+[3,11]	+[11]		+[11]	–[11]	+[7]	+[23]				–[7]	+[24]		Ant. pituitary +,[24] hypothalamus +[24] 3 kb mRNA skm[7]
Kv1.6	+[25]	–[22]	–[22]	–[22]		–[22]	–[22]	–[22]	+[6]							
Kv2.1	+[25,26]	+[25,26]	+[25,26]	+[25,26]	+[25,26]	–[22]	–[22]	+[25,26]					+[7]			Retina[26]>kid[26]>vent[26] >skm[25]>atrium[25] >olfactory epithelium[25] >tongue[25]
Kv2.2	+[26]	–[26]	+[26]		+[26]		–[26]	–[26]								Tongue[26]>olfactory epithelium[26] >ventricle[26]>retina[26]
Kv3.1	+[27]	–[28]	–[28]	–[28]		–[28]	–[28]	+[7]			+[29]	+[29]	+[7]			Human Louckes B-cells +[29] NG108-15 +,[9] AtT20 +[30]
Kv3.2	+[31]	–[22]	–[22]	–[22]				–[22]								
Kv3.3	+[32,33]						+[33]									
Kv3.4	+[34]							+[34]								
Kv4.1	+[7,35]							–[7]					+[7]			
Kv4.2	+[17]	+[3]	+[3]				–[3]	–[3]							+[3]	Atrium = vent[3]
Kv5.1	+[22]	–[22]	–[22]	–[22]		+[22]	+[22]	–[22]								
Kv6.1	+[22]	–[22]	–[22]	–[22]	–[22]		–[22]	–[22]								

[a] Atr, heart atrium; Vent, heart ventricle; Kid, kidney; Skm, skeletal muscle; Islet, pancreatic islet; Thy, thymus; Spl, spleen; Lym, lymphocytes; C_2C_{12}, C_2C_{12} cells in culture; GH3, GH3 cells in culture.

Table 2b. Distribution of Voltage-Gated Potassium Channel Genes in Mammalian Brain

Gene	Comments
Kv1.1	Pons/medulla[1]>mid brain,[1] superior and inferior colliculus,[1] cerebellum[1] >hippocampus [CA3>dentate gyrus>CA1],[1,2] thalamus,[2] cerebral cortex[1,2] >corpus striatum,[1] olfactory bulb[1]
Kv1.2	Pons/medulla[1] > cerebellum,[1,2] inferior colliculus[1] >hippocampus [CA3>dentate gyrus = CA1],[1,2] thalamus,[2] cerebral cortex,[1,2] superior colliculus[1] >mid brain,[1] corpus striatum,[1] olfactory bulb[1]
Kv1.3	Inferior colliculus[1] >olfactory bulb,[1] pons/medulla[1] >mid brain,[1] superior colliculus,[1] corpus striatum,[1] hippocampus,[1] cerebral cortex[1]
Kv1.4	Olfactory bulb,[1] corpus striatum,[1] > hippocampus,[1,17] superior and inferior colliculus[1] >cerebral cortex,[1] mid brain,[1] basal ganglia[17] > pons/medulla[1]
Kv1.5	Cerebellum [purkinje and granular cells],[22] hypothalamus[24]
Kv1.6	Medulla/pons[25] > inferior colliculus[25] > corpus striatum[25]
Kv2.1	Cerebral cortex[26] > hippocampus[22] > cerebellum[22] > olfactory bulb[22]
Kv2.2	Olfactory bulb [granule cell layer>olfactory tubercle][26] >cortex[26] >hippocampus [CA1–CA4 region, dentate gyrus][26] >cerebellum[granule cell layer][26]
Kv3.1	Cerebellum[27] >globus pallidus,[27] subthalamus,[27] substantia nigra[27] > reticular thalamic nuclei,[27] cortical and hippocampal interneurons[27] >inferior colliculi,[27] cochlear and vestibular nuclei;[27] while Kv3.1a is expressed mainly in the adult brain Kv3.1b is the major transcript in embryonic and perinatal neurons; their distribution within the brain is identical[27]
Kv3.2	Thalamus[31] >neocortex [layers 4–6],[31] piriform nucleus,[31] red nucleus,[31] hippocampus [CA3][31]
Kv3.3	Cerebellum [purkinje cells][32,33]
Kv3.4	Brain[34]
Kv4.1	Brain[18,35]
Kv4.2	Cerebellum [granular cells][17] >hippocampus[granule cells, CA3/CA1 pyramidal cells],[17] thalamus,[17] medial habenular nucleus[17] >cerebral cortex[17]
Kv5.1	Brain[22]
Kv6.1	Brain[22]

Table 2 references:

1. Beckh, S. and Pongs, O. (1990). *EMBO J.* 9, 777–782.
2. Tsaur, M.-L., Sheng, M., Lowenstein, D. H., Jan, Y. N., and Jan, L. Y. (1992). *Neuron* 8, 1055–1067.
3. Roberds, S. L. and Tamkun, M. M. (1991). *Proc. Natl. Acad. Sci. U.S.A.* 88, 1798–1802.
4. Klumpp, D. J., Farber, D., Bowes, C., Song, E.-J., and Pinto, L. H. (1991). *Cell. Mol. Neurobiol.* 11, 611–622.
5. Matsubara, H., Liman, E. R., Hess, P., and Koren, G. (1991). *J. Biol. Chem.* 266, 13324–13328.
6. Betsholtz, C., Baumann, A., Kenna, S., Ashcroft, F. M., Ashcroft, S. J. H., Berggren, P.-O., Gruoe, A., Pongs, O., Rorsman, P., Sandblom, J., and Welsh, M. (1990). *FEBS Lett.* 263, 121–126.
7. Lesage, F., Attali, B., Lazdunski, M., and Barhanin, J. (1992). *FEBS Lett.* 310, 162–166.
8. Paulmichl, M., Nasmith, P., Hellmiss, R., Reed, K., Boyle, W. A., Nerbonne, J. M., Perlata, E. G., and Clapham, D. E. (1991). *Proc. Natl. Acad. Sci. U.S.A.* 88, 7892–7895.
9. Yokoyama, S., Imoto, K., Kawamura, T., Higashida, H., Iwabe, N., Miyata, T., and Numa, S. (1989). *FEBS Lett.* 259, 37–42.
10. Grissmer, S., Dethlefs, B., Wasmuth, J. J., Goldin, A. L., Gutman, G. A., Cahalan, M. D., and Chandy, K. G. (1990). *Proc. Natl. Acad. Sci. U.S.A.* 87, 9411–9415.
11. Swanson, R. A., Marshall, J., Smith, J. S., Williams, J. B., Boyle, M. B., Folander, K., Luneau, C. J., Antanavage, J., Oliva, C., Buhrow, S. A., Bennet, C., Stein, R. B., and Kaczmarek, L. K. (1990). *Neuron* 4, 929–939.
12. Douglass, J., Osborne, P. B., Cai, Y.-C., Wilkinson, M., Christie, M. J., and Adelman, J. P. (1990). *J. Immunol.* 144, 4841–4850.
13. Cai, Y.-C., Osborne, P. B., North, R. A., Dooley, D. C., and Douglass, J. (1992). *DNA Cell Biol.* 11, 163–172.
14. Attali, B., Romey, G., Honore, E., Schmid-Alliana, A., Matte, G., Lesage, F., Ricard, P., Barhanin, J., and Lazdunski, M. (1992). *J. Biol. Chem.* 267, 8650–8657.

Table 2b. (Continued)

15. Honore, E., Attali, B., Romey, G., Lesage, F., Barhanin, J., and Lazdunski, M. (1992). *EMBO J.* 11, 2465–2471.
16. Spencer, R. H., Chandy, K. G., and Gutman, G. A. (1993). *Biochem. Biophys. Res. Commun.* 191, 201.
17. Sheng, M., Tsaur, M.-L., Jan, Y.-N., and Jan, L. Y. (1992). *Neuron* 9, 271–284.
18. Tamkun, M. M., Knoth, K. M., Walbridge, J. A., Kroemer, H., Roden, D. M., and Glover, D. M. (1991). *FASEB J.* 5, 331–337.
19. Tseng-Crank, J. C. L., Tseng, G.-N., Schwartz, A., and Tanouye, M. (1990). *FEBS Lett.* 268, 63–68.
20. Meyerhoff, W., Schwarz, J. R., Bauer, C. K., Hubel, A., and Richter, D. (1992). *J. Neuroendocrinol.* 4, 245–253.
21. Garcia-Guzman, M., Calvo, S., Cena, V., and Criado, M. (1992). *FEBS Lett.* 308, 283–289.
22. Drewe, J. A., Verma, S., Frech, G., and Joho, R. (1992). *J. Neuroscience* 12, 538–548.
23. Philipson, L. H., Hice, R. E., Schaefer, K., LaMendola, J., Bell, G. I., Nelson, D. J., and Steiner, D. F. (1991). *Proc. Natl. Acad. Sci. U.S.A.* 88, 53–57.
24. Levitan, E. S., Hemmick, L. M., Birnberg, N. C., and Kaczmarek, L. K. (1991). *Mol. Endocrinol.* 5, 1903–1908.
25. Grupe, A., Schroter, K. H., Ruppersberg, J. P., Stocker, M., Drewes, T., Beckh, S., and Pongs, O. (1990). *EMBO J.* 9, 1749–1756.
26. Hwang, P. M., Glatt, C. E., Bredt, D. S., Yellen, G., and Snyder, S. H. (1992). *Neuron* 8, 473–481.
27. Perney, T. M., Marshall, J., Martin, K. A., Hockfield, S., and Kaczmarek, L. K. (1992). *J. Neurophysiol.* 68, 756–766.
28. Luneau, C. J., Williams, J. B., Marshall, J., Levitan, E. S., Oliva, C., Smith, J. S., Antanavage, J., Folander, K., Stein, R. B., Swanson, R., Kaczmarek, L. K., and Buhrow, S. A. (1991). *Proc. Natl. Acad. Sci. U.S.A.* 88, 3932–3936.
29. Grissmer, S., Ghanshani, S., Dethlefs, B., McPherson, J. D., Wasmuth, J. J., Gutman, G. A., Cahalan, M. D., and Chandy, K. G. J. (1992). *J. Biol. Chem.* 267, 20971–20979.
30. Hemmick, L. M., Perney, T. M., Flamm, R. E., Kaczmarek, L. K., and Birnberg, N. C. J. (1992). *Neuroscience* 12, 2007–2014.
31. Rudy, B., Kentros, C., Weiser, M., Fruhling, D., Serodio, P., deMiera, V.-S. E., Ellisman, M. H., Pollock, J. A., and Baker, H. (1992). *Proc. Natl. Acad. Sci. U.S.A.* 89, 4603–4607.
32. Goldman-Wohl, E. C. and Heintz, N. (1992). *Soc. Neurosci. Abstr.* 18, 41.
33. Ghanshani, S., Pak, M., McPherson, J. D., Strong, M., Dethlefs, B., Wasmuth, J. J., Salkoff, L., Gutman, G. A., and Chandy, K. G. (1992). *Genomics* 12, 190–196.
34. Rettig, J., Wunder, F., Stocker, M., Lichtinghagen, R., Mastiaux, F., Beckh, S., Kues, W., Pedarzani, P., Schroter, K. H., Ruppersberg, J. P., Veh, R., and Pongs, O. (1992). *EMBO J.* 11, 2473–2486.
35. Pak, M. D., Baker, K., Covarrubias, M., Butler, A., Ratcliffe, A., and Salkoff, L. (1991). *Proc. Natl. Acad. Sci. U.S.A.* 88, 4386–4390.

2.1.4 Tissue Distribution of Kv Channels

2.1.4.1 Tissue-Specific Expression of Kv Genes

Although all known Kv genes are expressed in the brain, their distribution in other tissues is restricted (Table 2a). Even within the brain the localization of each of these genes is highly restricted and variable (Table 2b; e.g., Rudy et al., 1992a). The molecular mechanisms underlying these tissue-specific expression patterns are not understood. For the sodium channel genes, particular silencer and enhancer elements govern the tissue-specific expression of these genes in brain and/or skeletal muscle (Maue et al., 1990; Mori et al., 1990;

Z.-H. Sheng et al., 1992). Future studies may determine whether similar mechanisms are responsible for the restricted tissue-distribution patterns of the Kv genes.

Kv channels can also show very specific patterns of subcellular localization. Although Kv1.4 and Kv4.2 mRNAs are both expressed in the same neuronal cells, Kv1.4 protein is found in axons and possibly nerve terminals, while Kv4.2 is mainly concentrated in dendrites and soma (M. Sheng et al., 1992). Kv2.1 also segregates to the cell body and dendrites of rat brain neurons (Trimmer, 1991). The processes that determine subcellular distribution patterns of channel proteins remain to be defined.

2.1.4.2 Developmental Regulation and Induction

In addition to their unique tissue distribution, Kv channel expression is tightly regulated during differentiation and in mature cells, several examples of which are listed below. Kv3.1 mRNA expression is first evident in the cerebellum of rats, 11 days postnatal, and continues to increase throughout adulthood (Drewe et al., 1992). In the mouse pituitary cell line AtT20, transfection of the human H-*ras* oncogene results in a six-fold enhancement of a TEA-sensitive voltage-gated K^+ current (presumably Kv3.1), and a parallel induction of Kv3.1 mRNA expression (Hemmick et al., 1992; Perney et al., 1992). Kv3.1 mRNA expression can also be triggered in these cells by a short depolarizing pulse (via an influx of extracellular calcium), as well as by basic fibroblast growth factor (Perney and Kaczmarek, 1993). Kv1.5 transcripts are down-regulated during the development of the rat ventricle, completely disappearing by 6 months, whereas atrial expression of this gene remains unaltered (Matsubara et al., 1991); dexamethasone induces Kv1.5 mRNA expression in clonal pituitary cell lines (Levitan et al., 1991; Takimoto et al., 1993), whereas cAMP reduces expression of these mRNAs in the same cells (Mori et al., 1993). The dexamethasone-induced increase in Kv1.5 mRNA is a consequence of enhanced transcription rather than a reduction in mRNA turnover (Takimoto et al., 1993). Pentylenetetrazole-induced seizures in rats are accompanied by a reduction of Kv1.2, Kv4.2, and Kv3.1 mRNA (Tsaur et al., 1992; Perney and Kaczmarek, 1993). Of the three Kv2.1 transcripts (4.3 kb, 9 kb and 10 kb), only the 10-kb mRNA is induced by nerve growth factor in the rat pheochromocytoma, PC12 (Rudy et al., 1992b). Lastly, rKv1.5 mRNA expression increases in cAMP-dependent manner in neonatal atrial cells, and a canonical cAMP-response element (TGACGTCA) located 435 bp downstream to the transcription start site may be responsible for this modulation (Matsubara et al., 1992; Mori et al., 1993).

Many of the Kv genes have multiple transcripts that may be selectively expressed in a particular tissue, providing an additional level of complexity to the system. For example, the Kv1.4 gene generates three mRNAs (4.5, 3.5, and 2.4 kb), and only the 2.4-kb transcript appears to be expressed in skeletal muscle and in the mouse myoblast cell line, C_2C_{12} (Lesage et al., 1992). The molecular mechanisms underlying the tissue-specific expression of different mRNAs have not yet been defined.

We have already discussed posttranslational modifications (phosphorylation and glycosylation) that may modulate Kv function, and such processes may be important in the induction and developmental regulation of Kv channel expression. Diverse cell lines can begin expressing functional voltage-gated K^+ currents within minutes after being exposed to low-dose radiation, heat-shock, or free radicals (Kuo et al., 1992). This process is not blocked by DNA or protein synthesis inhibitors, suggesting that nonfunctioning pools of Kv protein exist in these cells, and become functional in response to environmental stimuli. The recent generation of specific antibodies against several Kv channels may facilitate studies to delineate the nature of those posttranslational events required for expression and regulation of Kv channels (Barbas et al., 1989; Trimmer, 1991; Schwarz et al., 1990; M. Sheng et al., 1992; Spencer et al., 1993; Perney et al., 1993).

2.1.5 Structural and Electrophysiological Analyses of Kv Channels

2.1.5.1 Tetrameric Structure of Functional Kv Channels

Homomultimers of the single domain structures described above form functional channels, and some of these (Kv1.3, Kv1.4, Kv1.5, Kv3.1) resemble specific native channels found in mammalian cells (discussed above). Coexpression of two kinetically or pharmacologically distinct K^+ channels results in ionic currents with hybrid behavior (Christie et al., 1990; Isacoff et al., 1990; Ruppersberg et al., 1990; K. McCormack et al., 1990) indicating that heteromultimeric channels may contribute to functional diversity. In fact, the existence of functional heteromultimers of *Shaker*-related peptides has recently been demonstrated in brain and myelinated nerves of rats and mice (Sheng et al., 1993; Wang et al., 1993). However, Kv proteins appear to be capable of associating only with other Kv members of the same subfamily (Covarrubias et al., 1991).

The α subunits of voltage-gated sodium and calcium channel contain four homologous domains that are thought to assemble around a central pore to form a functional channel. Each of these domains is clearly homologous to the product of a Kv gene, and the functional K^+ channel has therefore been proposed to be a tetramer. This proposed model has been experimentally supported by analysis of heteropolymers formed from K-channel domains with differing drug sensitivities (MacKinnon, 1991b; Liman et al., 1992). Constructs containing four K-channel domains linked in tandem can produce functional channels (Liman et al., 1992), also supporting this model. A 115 amino acid region in the amino-terminus of the *Shaker* channel (residues 82–196) has been reported to contain determinants responsible for tetramer formation, and deletion of this region from the *Shaker* channel abolishes expression of the K^+ current (Li et al., 1992; Shen et al., 1993). However, deletion of the corresponding segment from mKv1.3 (Aiyar et al., 1993b) or rKv2.1 (VanDongen et al., 1990) does not affect channel function, implying

Table 3. Electrophysiological and Pharmacological Properties of Cloned Kv Channels

			Pharmacology, 50% inhibitory concentrations[a]					
Channel	Activation V_n (mV)	Single channel currents (pS)	TEA (m*M*)	CTX (n*M*)	DTX (n*M*)	MCDP (n*M*)	4-AP (m*M*)	Quinine (μ*M*)
mKv1.1	–27,[1]–34[2]	10[2]	0.4,[1,12]0.3[2]	>1500[2]	21[2]	490[2]	1.1,[1]0.29[2]	41[2]
rKv1.1	–30[3,4]	9,[3,5]14.2[6]	0.6,[3,5]0.8[4]	22[3]	12[3,5]	45[3,5]	1.0,[3,5]0.16[4]	>100[4]
hKv1.1	–30[7]	?	20[7]	>100[7]	?	?	1.1[7]	?
rKv1.2	+3,[2]–34,[3]+4.8[8]	18[2,9]	>560,[2]129,[3]10[9]	17,[2]6,[3]1.7[8]	24,[2]4,[3]2.8[8]	440,[2]180[3,8]	0.6,[2]0.8,[3]0.2[9]	
800,[9]600[10]								
hKv1.2	–5[7]	?	>50[7]	10[7]	?	?	0.8[7]	?
mKv1.3	–23,[2]–35[11]	14,[2]13[11]	11[2,11]	2.6,[2]0.5–2.0[11]	250[2]	>2000[2]	0.20,[2]0.4[11]	40[11]
rKv1.3	–25,[3]–14[12]	9.6[3]	50,[3]11,[12]>40[13]	1[3]	>500[3,12]	>1000[3]	1.5[3]	80[12]
hKv1.3	–13,[14]–20[15]	13[16]	14,[14]30[15]	0.8[15]	?	?	0.19,[14]0.3[15]	55[14]
mKv1.4	–27[38]	?	>1600[38]	>1000[38]	?	?	?	22[38]
rKv1.4	–22[3]	4.7[3]	>100[3]	>40[3]	>200[3]	>2000[3]	13,[3]1.2[17]	?
hKv1.4	–5,[7]–34[18]	?	>50[7]	?	?	?	0.8,[7]0.7[18]	?
rKv1.5	–13,[9]–3[13]	7.9[19]	>40[23]	>200[23]	>200[23]	>600[23]	0.4[13]	?
hKv1.5	–25[2]	8[2]	330,[2]>40[20]	>1000[2]	>1000[2]	>10000[2]	0.27,[2]<0.1[20]	?
rKv1.6	–13,[13]–17[21]	8.7,[21]8.2[22]	4,[13]7,[21]1.7[22]	1,[21]>3000[22]	20,[21]25[22]	10,[21]200[22]	1.5,[21]0.3[22]	?
hKv1.6	–21[21]	9.1[21]	7[21]	1[21]	20[21]	10[21]	1.5[21]	?
mKv2.1	–14[23]	?	5[23]	?	?	?	>100[23]	?
rKv2.1	–9.2[24]	8.1[24]	10,[24]5.6[25]	>1000[24]	?	?	0.5[24]	?
rKv2.2	?	?	7.9[26]	?	?	?	?	?
mKv3.1a	+16[27]	26[27]	0.1[27]	?	?	?	0.6[27]	1000[27]
mKv3.1b	+21,[2]–1[28]	27[2,28]	0.15[2,28]	>1000[2]	>1000[2]	>2000[2]	0.021,[2]0.18[28]	4–20[2,28]
rKv3.1b	+19,[29]+15[30]	22,[24]11[29]	0.1,[29]0.15[30]	>1000[30]	>100[29]	>1000[30]	0.25,[29]0.15[30]	1000[29]
rKv3.2	+5.6,[29]+13[31]	11[29]	0.1,[29]<1[31]	?	>100[29]	>200[29]	0.9,[29]<1[31]	130[31]
rKv3.3	+7[32]	?	0.14[32]	?	?	?	1.2[32]	?
rKv3.4	–3.4,[29]+14[33]	14[33]	0.3[29,33]	?	>100[29,33]	>200[29]	0.5[29,33]	500[29]
hKv3.4	+19[34]	?	0.088[34]	?	?	?	0.6[34]	?
mKv4.1	0[35]	?	>10[35]	?	?	?	9[35]	?

Table 3. (Continued)

[a] TEA, tetraethylammonium; CTX, charybdotoxin; DTX, dendrotoxin; MCDP, mast cell degranulating peptide; 4-AP, 4-aminopyridine. Reported values are rounded to two significant figures, and comparable values differing by less than 10% are shown as identical.

Table 3 references:

1. Klumpp, D. J., Farber, D., Bowes, C., Song, E.-J., and Pinto, L. H. (1991). *Cell. Mol. Neurobiol.* 11, 611–622.
2. Grissmer, S., Nguyen, A. N., Aiyar, J., Hanson, D. C., Mather, R. J., Gutman, G. A., Karmilowicz, M.J., Auperin, D. D., and Chandy, K. G. (1994). *Mol. Pharmacol.* 45, 1227–34.
3. Stuhmer, W., Ruppersberg, J. P., Schroter, K. H., Sakmann, B., Stocker, M., GIese, K. P., Perschke, A., Baumann, A., and Pongs, O. (1989). *EMBO J.* 8, 3235–3244.
4. Christie, M. J., Adelman, J. P., Douglass, J., and North, R. A (1989). *Science* 244, 221–224.
5. Stuhmer, W., Stocker, M., Sakmann, B., Seeburg, P., Baumann, A., Grupe, A., and Pongs, O. (1988). *FEBS Lett.* 242, 199–206.
6. Koren, G., Liman, E. R., Logothetis, D. E., Nadal-Ginard, B., and Hess, P. (1990). *Neuron* 2, 39–51.
7. Ramaswami, M., Gautam, M., Kamb, A., Rudy, B., Tanouye, M., and Mathew, M. K. (1990). *Mol. Cell. Neurosci.* 1, 214–223.
8. Werkman, T. R., Kawamura, T., Yokoyama, S., Higashida, H., and Rogawski, M. A. (1992). *Neuroscience* 50, 935–946.
9. Yokoyama, S., Imoto, K., Kawamura, T., Higashida, H., Iwabe, N., Miyata, T., and Numa, S. (1989). *FEBS Lett.* 259, 37–42.
10. Paulmichl, M., Nasmith, P., Hellmiss, R., Reed, K., Boyie, W. A., Nerbonne, J. M., Periata, E. G., and Clapham, D. E. (1991). *Proc. Natl. Acad. Sci. U.S.A.* 88, 7892–7895.
11. Grissmer, S., Dethlefs, B., Wasmuth, J. J., Goldin, A. L., Gutman, G. A., Cahalan, M. D., and Chandy, K. G. (1990). *Proc. Natl. Acad. Sci. U.S.A.* 87, 9411–9415.
12. Douglass, J., Osborne, P. B., Cai, Y-C., Wilkinson, M., Christie, M. J., and Adelman, J. P. (1990). *J. Immunol.* 144, 4841–4850.
13. Swanson, R. A., Marshall, J., Smith, J. S., Williams, J. B., Boyle, M. B., Folander, K., Luneau, C. J., Antanavage, J., Oliva, C., Buhrow, S. A., Bennet, C., Stein, R. B., and Kaczmarek, L. K. (1990). *Neuron* 4, 929–939.
14. Cai, Y.-C., Osborne, P. B., North, R. A., Dooley, D. C., and Douglass, J. (1992). *DNA Cell Biol.* 11, 163–172.
15. Attali, B., Romey, G., Honore, E., Schmid-Alliana, A., Matte, G., Lesage, F., Ricard, P., Barhanin, J., and Lazdunski, M. (1992). *J. Biol. Chem.* 267, 8650–8657.
16. Honore, E., Attali, B., Romey, G., Lesage, F., Barhanin, J., and Lazdunski, M. (1992). *EMBO J.* 11, 2465–2471.
17. Tseng-Crank, J. C. L., Tseng, G.-N., Schwartz, A., and Tanouye, M. A. (1990). *FEBS Lett.* 268, 63–68.
18. Po, S., Snyders, D. J., Baker, R., Tamkun, M. M., and Bennett, P. B. (1992). *Circ. Res.* 71, 732–736.
19. Matsubara, H., Liman, E. R., Hess, P., and Koren, G. (1991). *J. Biol. Chem.* 266, 13324–13328.
20. Philipson, L. H., Hice, R. E., Schaefer, K., LaMendola, J., Bell, G. I., Nelson, D. J., and Steiner, D. F. (1991). *Proc. Natl. Acad. Sci. U.S.A.* 88, 53–57.
21. Grupe, A., Schroter, K. H., Ruppersberg, J. P., Stocker, M., Drewes, T., Beckh, S., and Pongs, O. (1990). *EMBO J.* 9, 1749–1756.

Table 3. (Continued)

22. Kirsch, G. E., Drewe, J. A., Verma, S., Brown, A. M., and Joho, R. (1991). *FEBS Lett.* 278, 55–60.
23. Pak, M. D., Covarrubias, M., Ratciiffe, A., and Salkoff, L. (1991). *J. Neurosci.* 11, 869–880.
24. Hartmann, H. A., Kirsch, G. E., Drewe, J. A., Taglialatela, M., Joho, R., and Brown, A. M. (1991). *Science* 251, 942–944.
25. Kirsch, G. E., Drewe, J. A., Hartmann, H. A., Taglialatela, M., deBiasi, M., Brown, A. M., and Joho, R. H. (1992). *Neuron* 8, 499–505.
26. Hwang, P. M., Glatt, C. E., Bredt, D. S., Yellen, G., and Snyder, S. H. (1992). *Neuron* 8, 473–481.
27. Yokoyama, S. *et al.* (1989). *FEBS Lett.* 259, 37–42.
28. Grissmer, S., Ghanshani, S., Dethlefs, B., McPherson, J. D., Wasmuth, J. J., Gutman, G. A., Cahalan, M. D., and Chandy, K. G. (1992). *J. Biol. Chem.* 267, 20971–20979.
29. Rettig, J., Wunder, F., Stocker, M., Lichtinghagen, R., Mastiaux, F., Beckh, S., Kues, W., Pedarzani, P., Schroter, K. H., Ruppersberg, J. P., Veh, R., and Pongs, O. (1992). *EMBO J.* 11, 2473–2486.
30. Luneau, C. J., Williams, J. B., Marshall, J., Levitan, E. S., Oliva, C., Smith, J. S., Antanavage, J., Folander, K., Stein, R. B., Swanson, R., Kaczmarek, L. K., and Buhrow, S. A. (1991). *Proc. Natl. Acad. Sci. U.S.A.* 88, 3932–3936.
31. McCormack, T., Vega-Saenz de Miera, E. C., and Rudy, B. (1990). *Proc. Natl. Acad. Sci. U.S.A.* 87, 5227–5231.
32. deMiera, E. V.-S., Moreno, H., Fruhling, D., Kentros, C., and Rudy, B. (1992). *Proc. R. Soc. London Ser. B* 248, 9–18.
33. Schroter, K.-H., Ruppersberg, J. P., Wunder, F., Rettig, J., Stocker, M., and Pongs, O. (1991). *FEBS Lett.* 278, 211–216.
34. Rudy, B., Sen, K., Vega-Saenz de Miera, E. C., Ried, T., and Ward, D. C. (1991). *J. Neurosci. Res.* 29, 401–412.
35. Pak, M. D., Baker, K., Covarrubias, M., Butler, A., Ratcliffe, A., and Salkoff, L. (1991). *J. Neurosci.* 88, 4386–4390.
36. Blair, T. A., Roberds, S. L., Tamkun, M. M., and Hartshorne, R. P. (1991). *FEBS Lett.* 295, 211–213.
37. Baldwin, T. J., Tsaur, M., Lopez, G., Jan, Y. N., and Jan, L. Y. (1991). *Neuron* 7, 471–483.
38. Grissmer, S., unpublished observations on pulled patches from *Xenopus* oocytes.

that other residues, presumably in the hydrophobic core, are responsible for assembly of mammalian K^+ channel multimers. Recent electron microscopic analysis of purified K^+ channel protein has begun to provide provide direct confirmation of its tetrameric structure (Lie et al., 1994).

2.1.5.2 Biophysical Properties of Cloned Kv Channels

Table 3 summarizes some of the electrophysiological and pharmacological properties of the channels encoded by each of the cloned Kv genes. Of the 16 mammalian Kv channels that have been biophysically characterized, only five (Kv1.4, Kv3.3, Kv3.4, Kv4.1, and Kv4.2) display rapid (A-type) inactivation, and the remainder are delayed rectifiers (Table 3). Many of these measurements have been made on whole oocytes with the two-voltage clamp or the macropatch methods. With the availability of a variety of cell lines stably expressing transfected channel genes, which are listed in Table 4, it has now become possible to study the biophysical properties of the cloned channels with traditional patch clamp recording techniques. It is also possible to perform patch clamp measurements on membrane patches isolated from oocytes expressing channel genes.

One must exercise considerable caution in interpreting the results of such studies, however, as the results obtained with these different methods are not always in agreement, even for identical channels. For example, the single K^+ conductance for rKv1.1 was 9 pS when measured in oocytes with the macropatch method, whereas studies on a fibroblast cell line stably expressing this gene revealed a single K^+ conductance of 14.2 pS (see Table 3). Another difference between oocyte and patch clamp experiments is in the resulting measurements of sensitivity to TEA, quinine, and verapamil. For example, the TEA sensitivity of Kv1.3 is lower in intact oocytes (K_d = 50 m*M*) than in patches (K_d = 11 m*M*). In addition, the time course of inactivation measured in whole oocytes is much slower than in oocyte patches or in mammalian cells (Grissmer et al., 1992). The larger single K^+ conductance measured in mammalian cells may in part be due to the higher external K^+ concentration in mammalian Ringer (4.5 m*M*) versus *Xenopus* Ringer (2 m*M*). The differences in pharmacological sensitivity and time course of inactivation may result from involvement of cytoplasmic factors (Marom et al., 1993). One must also consider the overall level of channel expression when interpreting oocyte data, since qualitatively different K^+ currents may be generated by different levels of a single cRNA (Honore et al., 1992). Lastly, grossly conflicting results (such as the CTX sensitivity of rKv1.6 shown in Table 3) can result from contaminated toxin preparations (MacKinnon and Miller, 1989; Garcia-Calvo et al., 1992); this is discussed further below.

2.1.5.3 Relating Cloned Kv Channel Genes to Native K^+ Channels

One goal of molecular biological studies on K^+ channels is to identify those Kv genes that encode known native channels found in normal cells. A commonly used approach has been to compare the biophysical and pharma-

Table 4. Stable Expression of Transfected Kv Channels in Cultured Cells

Channel	Cell line
mKv1.1	CHO cells,[1] L929[2]
rKv1.1	Sol-8 myoblasts,[3] mouse L cells,[4] MEL (mouse erythroleukemia) cells[5]
rKv1.2	B82 mouse fibroblasts[6]
mKv1.3	A4[2]
hKv1.3	IM-9 B cells[7]
bKv1.4	Neuro-2A[8]
hKv1.4	MEL cells[5]
hKv1.5	CHO cells,[9] MEL cells[5]
mKv3.1	Human embryonic kidney cells (293),[10] L929 cells[2]

Table 4 references:

1. Robertson, B. and Owen, D. G. (1993). *Br. J. Pharmacol.* 109, 725–735.
2. Grissmer, S., unpublished observations.
3. Koren, G., Liman, E. R., Logothetis, D. E., Nadal-Ginard, B., and Hess, P. (1990). *Neuron* 2, 39–51.
4. Deal, K. K., Lovinger, D. M., and Tamkun, M. M. (1993). *Biophys. J.* 64: A196.
5. Shelton, P. A., Davies, N. W., Antoniou, M., Grosveld, F., Needham, M., Hollis, M., Brammar, W. J., and Conley, E. C. (1993). *Receptors Channels* 1, 25–37.
6. Werkman, T. R., Kawamura, T., Yokoyama, S., Higashida, H., and Rogawski, M. A. (1992). *Neuroscience* 50, 935–946.
7. Honore, E., Attali, B., Romey, G., Lesage, F., Barhanin, J., and Lazdunski, M. (1992). *EMBO J.* 11, 2465–2471.
8. Garcia-Guzman, M., Calvo, S., Cena, V., and Criado, M. (1992). *FEBS Lett.* 308, 283–289.
9. Malayev, A., Philipson, L. H., Kuznetsov, A., Chang, C., and Nelson, D. J. (1993). *Biophys. J.* 64, A196.
10. Critz, S. D., Wible, B. A., Lopez, H. S., and Brown, A. M. (1993). *J. Neurochem.* 60, 1175–1178.

cological properties of each of the cloned channels with particular native channels. Unfortunately, the quantitative differences in the measurements made by the two-electrode recording technique in whole oocytes, compared with those made by the patch clamp on native mammalian cells, has complicated comparisons of this type. Patch clamp experiments on mammalian cell lines that stably express Kv genes (Table 4), or that have been injected with Kv-cRNA (Ikeda et al., 1992), may be helpful in this regard. Another factor that complicates such comparisons is that native channels may be composed of products of more than one Kv gene. The formation of functional heteromultimeric channels within the mammalian *Shaker*-related subfamily had been shown in *Xenopus* expression systems (Ruppersberg et al., 1990; Isacoff et al., 1990), and their existence has recently been demonstrated in rat brain (Sheng et al., 1993; Wang et al., 1993).

2.1.5.3.1 Mammalian Kv Channels

Only five cloned Kv genes have thus far been convincingly related to native mammalian channels. Kv1.3 and Kv3.1 encode the well-characterized types *n* and *l* K+ channels in lymphocytes and the Kv1.5 gene encodes a rapidly activating delayed rectifier in human heart, both discussed further below. The cloned Kv1.4 channel has properties indistinguishable from a K^+ channel in GH3 pituitary cells, and Kv1.4 cDNA has been isolated from these cells, suggesting that the native channel is a homomultimer formed from Kv1.4 subunits (Meyerhoff et al., 1992). Similar considerations suggest that Kv1.1 encodes a delayed rectifier in glial cells (Wang et al., 1992).

Lymphocyte Channels. Since several reviews (e.g., Cahalan et al., 1991; Chandy et al., 1993) have described the properties and roles of three distinct voltage-gated K^+ channels, types *n, n′,* and *l,* expressed in T and B lymphocytes, this subject is only briefly discussed here. The Kv1.3 gene encodes the *n*-type channel while the *l* channel is a product of the Kv3.1 gene (see Chandy et al., 1993). Both native channels are probably homomultimers, since their properties are nearly identical to that of their cloned counterparts (Grissmer et al., 1990, 1992; Douglass et al., 1990; Attali et al., 1992b). The gene encoding the type *n′* channel has not been identified. Blockade of Kv1.3 channels in T cells by chemically disparate pharmacological agents (see Table 3), including highly selective toxins (margatoxin, noxiustoxin, and CTX), results in membrane depolarization, diminution of the calcium signal, and suppression of the lymphocyte activation cascade (reviewed in Chandy et al., 1993). Drugs that selectively block Kv1.3 could therefore be developed into immunosuppressants for use in prevention of graft rejection and inflammation. The restricted tissue distribution of Kv1.3 (Table 2) adds to its attractiveness as a target for immunosuppression. These features have prompted several pharmaceutical groups to actively search for novel compounds that block Kv1.3 channels with a high degree of potency and selectivity. Interestingly, human platelets (Mahaut-Smith et al., 1990) and the melanoma cell line IRG1 (Nilius et al., 1990) express a CTX-sensitive voltage-gated K^+ channel, which is remarkably similar to the Kv1.3 channel.

Cardiac Channels. Several delayed rectifier Kv channels (Kv1.1, Kv1.2, Kv1.5, Kv1.6, Kv2.1, Kv2.2,) are present in the heart (see Table 2). As mentioned above, a rapidly activating delayed rectifier in the heart is most likely formed from four Kv1.5 subunits (Fedida et al., 1993). The remaining genes probably encode subunits of other types of cardiac I_k delayed rectifier channels that are responsible for repolarization during the cardiac action potential. Direct comparisons of the properties of these cloned channels and cardiac delayed rectifiers do not reveal any close similarities, suggesting that the native channels may be heteromultimeric.

A rapidly inactivating K^+ current, I_{to}, modulates the early plateau phase of the cardiac action potential. At least two A-type (rapidly inactivating) Kv channels, Kv1.4 and Kv4.2, are expressed in the heart (Table 2), while the expression in heart of the other three A-type channels, Kv3.3, Kv3.4, and Kv4.1, has not been examined. Like the I_{to} channel, Kv1.4 has small conductance (4.7 pS), is blocked by 4-AP, and is resistant to CTX, DTX, and external TEA (Po et al., 1992); in addition, the level of Kv1.4 transcript is elevated in ventricular hypertrophy (Matsubara et al., 1993), a condition associated with an increased I_{to} current. However, rat and human Kv1.4 recover from inactivation about 10-fold more slowly than the native I_{to} channel, suggesting that the native I_{to} channel may be heteromultimeric (Po et al., 1993). In fact, heteromultimeric channels formed by coinjection into oocytes of Kv1.4:Kv1.2 or Kv1.4:Kv1.5 cRNA show the rapid recovery from inactivation characteristic of the I_{to} channel (Po et al., 1993).

Drugs that modulate cardiac Kv channels have been successfully utilized in the treatment of ventricular arrhythmias. These class III antiarrhythmic agents (e.g., UK-68798 or Dofetilide, E-4031, Almokalant, and Sematilide) prolong myocardial refractoriness, and suppress reentrant atrial and ventricular arrhythmias in dog and pig models (see reviews by Lynch et al., 1992; Colatsky et al., 1993). The recent increase in our understanding of the molecular structure of cardiac Kv channels may lead to the identification of more selective and/or potent class III antiarrhythmics.

2.1.5.3.2 Drosophila Kv Channels

Shaker channels are expressed in diverse excitable cells in the fly, including muscle and photoreceptors (Solc et al., 1987; Hardie et al., 1991). The products of the different *Shaker* transcripts inactivate with different time courses, ShB1, for example, inactivating more rapidly than ShA1 (Timpe et al., 1988; Stocker et al., 1990; Iverson et al., 1988; Wittka et al., 1991). The biophysical properties of the ShB1 channel, expressed in oocytes, are similar to the channels found in pupal muscle (Timpe et al., 1988). However, Pongs (1992), in a recent review, points out the existence of subtle differences in properties (mean open-time duration, kinetics of inactivation, toxin sensitivity) between cloned *Shaker* channels and their native counterparts; he suggests that the unique properties of native channels may arise as a consequence of interactions between *Shaker* subunits and other proteins [e.g., unidentified β subunits or *ether-a-go-go (eag)* proteins].

2.1.5.3.3 *Aplysia* Kv Channels

The *Shaker*-related *Aplysia* K+ gene AK01a (Pfaffinger et al., 1991) encodes a protein of 515 amino acids containing the conserved glycosylation (in the S1-S2 loop) and PKC (in the S4–S5 loop) consensus sequences (Figure 2). Its ~15-kb mRNAs are present in the central nervous system, heart, gill muscle, and buccal muscle. The channel demonstrates A-type inactivation, half-activates at +5 mV, is 4-AP sensitive, and resistant to 10 m*M* TEA [presumably because of the arginine in the TEA binding site (GYGDM*R*P); see discussion below], 200 n*M* CTX and 1 μ*M* MCDP. The AK01a channel most closely resembles the IA_{depol} current detected in R15 neuronal cells of the abdominal ganglion of *Aplysia* (Pfaffinger et al., 1991).

2.1.5.3.4 *Xenopus* Kv Channels

Two Kv genes have been isolated from *Xenopus*. The xKv1.2 gene (XSha2), a homologue of the mammalian Kv1.2 genes, is expressed in the spinal cord and brain, but not in skeletal muscle, heart, liver, or eye (Ribera, 1990). Like rKv1.2 channels, the xKv1.2 channel is a delayed rectifier that activates positive to –20 mV, is ~70% blocked by 1 m*M* 4-AP and is resistant to TEA (Ribera, 1990). K^+ channels with similar properties are found in *Xenopus* spinal neurons (Ribera, 1990). Although the single K^+ conductance of xKv1.2 has not been determined (and therefore cannot be compared with channels in spinal neurons), the mammalian homologues of xKv1.2 (see Table 3) have single K^+ conductances of 15 to 30 pS, which resemble those of the delayed rectifiers in spinal neurons. The second channel, xKv1.1, shares 88% amino acid sequence identity with mKv1.1 and rKv1.1, and is expressed in primary spinal neurons (Rohon-Beard cells) and ganglia, trigeminal ganglia, and gill arches (Ribera and Nguyen, 1993). Like mammalian Kv1.1, the xKv1.1 gene (xSha1) encodes a delayed rectifier that activates positive to –40 mV and is ~60% blocked by 1.5 m*M* external TEA. The physiological and developmental roles of these channels remain to be elucidated.

2.1.5.4 Identifying the Residues That Underlie Channel Inactivation

At least three distinct parts of the protein (N-terminus, P-region, and S6 segment) contribute to channel inactivation. Rapid inactivation involving the N-terminus has been called "N-type" inactivation, alternative modes being referred to as "C-type" and "P-type" inactivation (Iverson and Rudy, 1990; DeBiasi et al., 1993a; Hoshi et al., 1991; Lopez-Barneo et al., 1993).

2.1.5.4.1 Rapid Inactivation Associated with the Amino-Terminus

Deletion of the NH_2-terminus from the *Shaker* A2 channel removes rapid inactivation (Hoshi et al., 1990), while addition of an N-terminal peptide (MAAVAGLYGLGED*R*QH*RKK*Q) to the inner surface of this channel restores inactivation (Zagotta et al., 1990). The four basic residues indicated within this segment are thought to form a tethered ball that interacts with a receptor elsewhere in the channel, possibly the S4–S5 linker (Isacoff et al., 1991), and substitution of glutamine for two or more of these basic residues slows down inactivation (Hoshi et al., 1991). The N-termini of the five mammalian A-type channels (Kv1.4, Kv3.3, Kv3.4, Kv4.1, Kv4.2) contain charged residues that may constitute a similar inactivation ball (see Figures 2, 8, and 11). Peptides corresponding to this region in Kv1.4 (RARERERLAHSR) and Kv3.4 (RGKKSGNKPPSKTCLK) have been shown to accelerate inactivation of these channels (Ruppersberg et al., 1991b); similar studies have not yet been carried out on Kv3.3, Kv4.1, and Kv4.2. Deletion of residues 3–25 from rKv1.4 abolishes N-type inactivation indicating that the ball might lie in this segment (Tseng and Tseng-Crank, 1992). A similar deletion (residues 2–32, including one arginine; Figure 11) from mKv4.1 slows down, but does not eliminate, rapid inactivation (Pak et al., 1991a). Deletion of three basic residues (PAPPRQE*RKR*TQ) further downstream in the N-terminus of rKv4.2 (Figure 11) also reduces N-type inactivation of this channel (Baldwin et al., 1991).

A modulatory site has been identified near the amino-terminus of mammalian A-type K^+ channels (Ruppersberg et al., 1991a). Rapid inactivation of the A-current channels, rKv1.4 and rKv3.4, completely disappears when inside-out patches are excised from oocytes expressing these genes, and can be restored by exposing the inner surface of these channels to the reducing agent glutathione (Ruppersberg et al., 1991a). This loss of inactivation is due to the oxidation of a cysteine located upstream to the charged ball in both rKv1.4 (SSG*C*NS) and Kv3.4 (SSV*C*VS) (Ruppersberg et al., 1991a). The *Shaker* channel lacks the regulatory cysteine and inactivation of this channel does not disappear in excised patches (Ruppersberg et al., 1991a). Since Kv3.3 also has a cysteine at this position (see Figure 8), inactivation of this channel might also be sensitive to oxidation.

2.1.5.4.2 The Receptor for the N-Terminal-Inactivating Ball

Mutations in the intracellular loop linking S4 and S5 of the *Shaker* channel (LQILGRTLKASMREL) alter inactivation kinetics suggesting that this region may be the receptor for the N-terminal ball (Isacoff et al., 1991; McCormack et al., 1991). For example, replacement of the conserved glutamate (MR*E*L) with either glutamine or aspartic acid dramatically slows down inactivation, whereas substitution of a conserved leucine (QI*L*GR) with either alanine or valine (amino acids with smaller hydrophobic side chains) accelerates inactivation (Isacoff et al., 1991). The leucine–alanine mutation also induces rapid inactivation of the normally noninactivating rKv2.1 (DRK1) channel (Isacoff et al., 1991).

The N-terminal ball is thought to cause inactivation by physically occluding the pore. Internal application of TEA dramatically slows N-type inactivation in the *Shaker* channel, probably by competing for a common binding site (Choi et al., 1991), and mutations of the conserved leucine and the glutamic acid in this region alter the TEA_i sensitivity of the *Shaker* channel (Slesinger et al., 1993). Collectively these data suggest that the receptor for the fast-inactivation ball must lie at, or close to, the cytoplasmic mouth of the ion conduction pathway. Once this site is occupied by the ball, the gating charges within the voltage sensor (presumably including S4 segment) may be immobilized and prevented from returning to its original position (Bezanilla et al., 1991).

Delayed rectifier and calcium-activated K^+ channels, despite the fact that they do not normally show rapid inactivation, appear to contain a receptor site for the *Shaker* inactivation ball, since the *Shaker* N-terminal peptide has been shown to induce rapid N-type inactivation of several of these channels (Zagotta et al., 1990; Dubinsky et al., 1992; Isacoff et al., 1991; Foster et al., 1992).

2.1.5.4.3 Inactivation Linked to the P-Region

In the rKv2.1 channel, removal of either the N- or C-terminus retards a slow form of inactivation called C-type inactivation, whereas rKv2.1 channels with both termini deleted behave like the wild-type channel (VanDongen et al., 1990). These studies suggest that residues within the hydrophobic core are responsible for slow inactivation. Recent studies point to residues within the P-region and S6 segment as being important for this type of inactivation

(DeBiasi et al., 1993a; Lopez-Barneo et al., 1993). Many of these experiments have been performed on *Shaker* channels with deleted termini (to remove N-type inactivation). In these truncated *Shaker* channels, replacement of threonine at the C-terminal end of the P-region (GYGDM*T*P) with glutamic acid or lysine (lysine occupies the position equivalent to threonine in Kv1.4), speeds up inactivation ~100-fold (Lopez-Barneo et al., 1993). Substitution of the threonine with tyrosine or valine (residues with large hydrophobic side chains) accelerates inactivation ~8-fold, while replacement with histidine (the homologous residue in Kv1.3) slows down inactivation ~5-fold (Lopez-Barneo et al., 1993). TEA, a K^+ channel blocker that interacts with residues at this site, also slows inactivation (Choi et al., 1991).

Similar experiments have been conducted on the rKv1.3 channel which has a histidine in place of the threonine present in the *Shaker* channel. Substitution of the histidine with a tyrosine reduces C-type inactivation roughly 5-fold, while replacement of the homologous tyrosine in rKv1.1 with histidine enhances inactivation ~2-fold (Busch et al., 1991).

Another site deeper within the P-region also influences inactivation. Replacing an isoleucine in the P-region of the rKv2.1 channel with leucine speeded up inactivation ~50-fold, and reduced mean open time at 0 mV, without affecting the single channel conductance (DeBiasi et al., 1993a). This type of inactivation has been termed P-type inactivation to distinguish it from C-type inactivation, which is slowed by external TEA (DeBiasi et al., 1993a).

2.1.5.4.4 Inactivation Linked to the S6 Segment

Two exons (exons 1 and 2), at the 3′-end of the *Shaker* coding sequence can be alternatively utilized to yield two channels, A1 and A2, which have distinct S6 segments and C-terminal regions. The two alternative S6 segments differ at only a single position, a valine in *Shaker* A1 (VGSLC*V*IAG) being replaced with an alanine in *Shaker* A2 (VGSLC*A*IAG). *Shaker* A1 currents show only rapid (N-type) inactivation, while *Shaker* A2 channels exhibit both N-type and C-type forms of inactivation (Wittka et al., 1991). Exchange of the residues at this position resulted in a transfer of inactivation properties (Wittka et al., 1991). All the mammalian *Shaker*-subfamily channels have alanine at this position (Figure 2) and could therefore be expected to exhibit slow inactivation as well.

2.1.5.4.5 β-Subunits and Inactivation

Two distinct β-subunits accelerate inactivation of voltage-gated sodium channels (Isom et al., 1992; also see review by Goldin in this volume). Similar Kv-associated small subunit proteins have been identified in rats (Rehm and Lazdunski, 1988), chickens (Schmidt and Betz, 1989), and cattle (Parcej et al., 1992). Low mol wt mRNA has been shown to encode a protein that can accelerate A-type inactivation of the mKv4.1 channel (Chabala and Covarrubias, 1993), and genes for these putative β-subunits have recently been cloned (Rettig et al., 1994; Scott et al., 1994).

2.1.5.5 Delineating the Binding Sites of Pharmacological Agents

Many of the pharmacological blockers discussed below are thought to bind within the ion conduction pathway and have therefore been used to probe the structure of the channel pore.

2.1.5.5.1 External TEA- and K^+-Binding Site

Kavanaugh et al. (1991, 1992) and MacKinnon and Yellen (1990) identified a tyrosine at the C-terminal end of the P-region (GYGDX*Y*PXX) of Kv channels that confers sensitivity to external TEA (TEA_e). Replacement of this tyrosine in rKv1.1 with valine (the equivalent residue in the TEA_e-insensitive rKv1.2) makes the channel resistant to TEA_e, and the reverse mutation in rKv1.2 makes the channel TEA_e sensitive (Kavanaugh et al., 1991). Similarly, introducing a tyrosine or phenylalanine in place of threonine in the *Shaker* channel greatly enhances the TEA_e sensitivity of the channel, and there is a direct relationship between the number of subunits containing tyrosine and degree of TEA_e block, suggesting that the four Kv subunits interact simultaneously with TEA to form a high-affinity binding site (Heginbotham and MacKinnon, 1992; Kavanaugh et al., 1992). Nine mammalian K^+ channels (Kv1.1, Kv1.6, Kv2.1-2.2, Kv3.1–Kv3.4, Kv5.1) contain this tyrosine. Five of these (Kv1.1, Kv3.1–Kv3.4) are extremely sensitive to TEA_e (K_d = ~0.1–1 m*M*), three (Kv1.6, Kv2.1–Kv2.2) are moderately sensitive (K_d = ~1–10 m*M*), and one (Kv5.1) has not been expressed (see Table 3). Clearly, the presence of tyrosine alone is not enough to make a channel highly TEA_e sensitive. The presence of hydrophobic residues (valine in Kv1.2, Kv4.1, Kv4.2, *Shal,* Kv6.1; alanine in *Shaw*) at the homologous position reduces the TEA sensitivity of channels, while positively charged residues (lysine in Kv1.4, arginine in Kv1.5) (Table 3) renders the channels TEA insensitive.

Channels with a positively charged residue at this position have an absolute requirement of external K^+ for channel opening (Pardo et al., 1992). For example, the rKv1.4 channel does not open in the absence of external K^+, but replacement of the lysine (GYGDM*K*PIT) with tyrosine allows the channel to open under such conditions (Pardo et al., 1992). The opening of Kv1.3 is similarly dependent on external K^+, presumably because the corresponding histidine is weakly positive under physiological conditions (Pardo et al., 1992). Kv1.5, the only other channel with a positively charged residue at this site (GYGDM*R*P; see Figure 12), has not been tested for its dependence on external K^+ for channel opening.

2.1.5.5.2 Internal TEA-Binding Site

Internal TEA (TEA_i) blocks Kv channels by interacting with a second site on the inner surface of the pore. Kv1.1, Kv1.3, Kv1.6, Kv2.1, and the *Shaker* channel are half-blocked by ~0.3 m*M* TEA_i (Taglialatela et al., 1991). This block is voltage dependent, becoming less potent at depolarized potentials.

The site of interaction of TEA_i is thought to lie ~20% within the cytoplasmic boundary of the membrane's electric field (Taglialatela et al., 1991; Kirsch et al., 1991b).

Mutagenesis studies have been used to localize the TEA_i binding site. Replacement of threonine (TM*T*TVGYG) with serine in the P-region of the *Shaker* channel markedly reduces TEA_i sensitivity (Yellen et al., 1991). The valine two residues downstream is also thought to participate in TEA binding (Kirsch et al., 1992a,b; Aiyar et al., 1993a). All the *Shaker*- and *Shab*-subfamily Kv channels have a valine in this position in place of the leucine present in mammalian *Shaw*- and *Shal*-subfamily channels (Figure 12). A Kv2.1/Kv3.1 chimera containing the Kv3.1 P-region (and therefore the leucine), is resistant to TEA_i, whereas a double reversion mutant (V3691 + L374V) restores the sensitivity to TEA_i (Kirsch et al., 1992a). In the Kv3.1 channel, replacement of this leucine with valine makes the channel more sensitive to TEA_i (Aiyar et al., 1993b). These studies indicate that the P-region must extend at least partially through the cell membrane, and that residues within this segment must form part of the TEA_i binding site.

Residues in S6 also appear to interact with TEA_i. Mutation of a threonine in the S6 segment of *Shaker* (GVL*T*IAL) to serine reduced the sensitivity to TEA_i by about six-fold, and this residue also appears to participate in binding long chain alkyl derivatives of TEA (Choi et al., 1993).

2.1.5.5.3 Dendrotoxin-Binding Site

Peptide toxins [including dendrotoxin (DTX), charybdotoxin (CTX), and mast cell degranulation peptide (MCDP)] have been used to block channels with a high degree of selectivity and potency, and could be used as molecular calipers to probe the channel pore. DTX, a 59-amino acid peptide component of venom from the green mamba snake, *Dendroaspis angusticeps*, blocks Kv1.1 (K_d = 12–50 nM), Kv1.2 (K_d 3–4 nM), and Kv1.6 (K_d = 20–25 nM) with high affinity (Table 3). All the other cloned channels that have been studied (see Table 1) are resistant or only weakly sensitive to this toxin. Using site-specific mutational strategies, Hurst et al. (1991) have shown that the presence of an alanine and a glutamic acid (FAEAEE*AE*SH) in the loop linking S5 with the P-region in rKv1.1 confer sensitivity to DTX. The tyrosine associated with TEA binding also appears to affect the interaction of DTX with rKv1.1 (Hurst et al., 1991). It is important to determine whether the residues at the corresponding positions in Kv1.2 (FAEADE*RD*SQ), and Kv1.6 (FAEADD*VD*SL) underlie the DTX sensitivity of these channels.

2.1.5.5.4 Charybdotoxin-Binding Site

CTX, a toxin from the scorpion, *Leiurus quinquestriatus,* was originally described by Miller et al. (1985) as a specific blocker of calcium-activated K^+ channels, but is now recognized to also block voltage-gated K^+ channels (see Table 3). In early experiments CTX was reported to block the *Shaker* K^+ channel with high affinity, but more recent studies (Garcia-Calvo et al., 1992;

Garcia et al., 1994) have shown that the inhibitory activity was due to contaminants present in some CTX preparations, and recombinant CTX does not block the *Shaker* channel. Similarly, studies with contaminated preparations initially suggested that Kv1.6 was CTX sensitive (Grupe et al., 1990), but subsequent experiments with recombinant CTX (Kirsch et al., 1991a) have not confirmed this finding (Table 3). The toxins that contaminated the earlier CTX preparations (agitoxins) have been purified to homogeneity, and shown to block *Shaker* channels with high affinity (Garcia-Calvo et al., 1992; Garcia et al., 1994). The channels most sensitive to recombinant CTX are Kv1.3 (K_d = 0.5–2 n*M*) and Kv1.2 (K_d = 1.7–22 n*M*), while all the other Kv channels that have been characterized are CTX insensitive (see Table 3).

Replacement of a glycine (residue volume ~60 Å^3) at the mouth of the CTX-sensitive Kv1.3 channel (PSS*G*FNSIPD) with larger residues such as glutamine or tyrosine (with volumes ranging from 145 to 195 Å^3) rendered the channel resistant to recombinant CTX, presumably by sterically hindering CTX entry into the outer vestibule; substitution with intermediate sized residues (e.g., aspartic acid or valine, with volumes ranging from 110 to 140 Å^3) reduced CTX sensitivity only moderately (R. Swanson, personal communication). Conversely, replacing phenylalanine at the homologous position in the CTX-resistant *Shaker* channel with the smaller glycine facilitated high-affinity binding between recombinant CTX and the channel, suggesting that the *Shaker* channel's resistance to CTX blockade is due to the narrow entry to its outer vestibule, rather than to a lack of CTX-binding sites (Goldstein and Miller, 1992). The CTX insensitivity of other channels (Table 3) may also be due to the presence of large residues at the entrance to the outer vestibule (Figure 12). Surprisingly, the identical position in the CTX-sensitive Kv1.2 channel (see Figure 2) is occupied by glutamine, which conferred resistance in the mutated Kv1.3 channel; this implies that compensatory differences at neighboring sites may make the opening to the outer vestibule wide enough for CTX to enter.

Studies with the contaminated CTX preparation had suggested that an aspartic acid at the N-terminal end of the pore region of the CTX-resistant *Shaker* protein (KSIP*D*AFWW) is involved in CTX binding (MacKinnon and Miller, 1989; MacKinnon et al., 1990). Surprisingly, neutralization of this aspartic acid in mKv1.3 (D to N, the identical mutation used in the *Shaker* channel) did not alter its sensitivity to recombinant CTX, implying that the toxin may interact differently with different channels (J. Aiyar, S. Grissmer, and K. G. Chandy, unpublished data). Similar studies with recombinant CTX have not yet been conducted on the CTX-sensitive Kv1.2 channel.

2.1.5.5.5 Mast Cell Degranulation Peptide-Binding Site

MCDP, a 22-amino acid peptide isolated from the honey bee, *Apis mellifera*, has been reported to affect both fast-inactivating (A-type) and slow-inactivating (delayed rectifier) channels (Stansfield et al., 1987; Brau et al., 1990). The channels most sensitive to MCDP are Kv1.6 (K_d = 10–200 n*M*), Kv1.1 (K_d = 45–490 n*M*), and Kv1.2 (K_d = 180 n*M*), all three of which are delayed rectifiers (Table 3). The only cloned A-type channel that has been examined, Kv1.4, is MCDP resistant (Table 3). It is not yet known whether the other A-type Kv channels, Kv3.3, Kv3.4, and Kv4.2, are MCDP sensitive, nor have the residues that determine sensitivity to MCDP been delineated.

2.1.5.5.6 4-Aminopyridine-Binding Site

The classical K^+ channel blocker, 4-aminopyridine, blocks all known Kv channels with moderate potency (K_d ranging from 0.2–9 m*M*), Kv3.1 being most potently blocked. Recent experiments with Kv2.1/Kv3.1 chimeras suggests that the 4-AP-binding site is formed from the association of the N-terminal end of S5 and C-terminal-end of S6, which are both thought to lie in the inner vestibule of the channel pore (Kirsch et al., 1993; Yao and Tseng, 1993).

2.1.5.5.7 Ethanol and Halothane

Both these agents block the noninactivating fly *Shaw* channel at concentrations ranging from 70 to 170 m*M,* whereas other cloned channels are affected only at concentrations greater than 200 m*M* (Covarrubias and Rubin, 1993). While these experiments support the protein hypothesis for anesthetic action, the molecular sites of action of these drugs remain to be determined.

2.1.5.6 Experimentally Defining the Ion Conduction Pore

Since drugs that occlude the channel pore (TEA_e, TEA_i, DTX, and CTX) interact with residues in the loop linking the S5 and S6 segments, and in the P-region (see discussion above), it was initially suggested that the P-region contributed to the ion conduction pathway. To test this idea further, Hartman and colleagues (1991) exchanged the P-region of the low conductance rKv2.1 channel (8 pS) for that of the high conductance rKv3.1 channel (26–27 pS). The resulting Kv2.1/Kv3.1 chimera had all of the pore properties of the donor Kv3.1 channel. Additional mutational experiments identified residues that were purported to determine ion selectivity (Kirsch et al., 1992a,b; Yool and Schwarz, 1991; DeBiasi et al., 1993a,b). Collectively, these studies indicated that all or most of the pore determinants were contained in the P-region. This conclusion is further supported by the fact that the P-region is the only recognizably conserved segment in all K^+-selective channels (see Figure 15).

Recent studies suggest that the S6 segment and the loop linking the S4–S5 segments also contribute to the formation of the ion conduction pathway (Kirsch et al., 1993; Lopez et al., 1993). Choi et al. (1991) showed that mutations in the S6 segment of the *Shaker* channel altered binding to alkyl-triethylammonium blockers, which implied that this region lined part of the inner surface of the channel pore. More recently, Lopez et al. (1993) were able to transfer most of the pore properties of the rKv3.1 channel to the *Shaker* channel by transplantation of the rKv3.1-S6 segment, but not the P-region; only sensitivity to external TEA was transferred by the P-region transplant. Lastly, we have recently found that a mutation in the S6 segment of mKv3.1 (M430L) altered single channel properties (i.e., induced a channel flicker that was dependent on the permeant ion) without affecting sensitivity to internal TEA, although the converse mutation (L427M) in mKv1.3 did not produce any change (J. Aiyar, S. Grissmer, and K. G. Chandy, unpublished data). Mutations in the S5–S5 linker of the *Shaker* channel have also been reported to change the channel's sensitivity to TEA_i and to alter its single channel properties (Isacoff et al., 1991; Slesinger et al., 1993). Thus, the P-region along with the S6 segment and the S4–S5 linker appear to contain most of the pore determinants. In fact, the model for K^+ channels proposed by Durell and Guy (1992) predicts an ion conduction pathway comprised of these three regions.

We would emphasize, however, that considerable caution must be exercised in interpreting all such mutational experiments. For one thing, it is clear that an amino acid substitution in any protein may have distant as well as local consequences. While many of the results of P-region mutagenesis have been dramatic and compelling, the fact that a particular substitution affects ion selectivity, for example, is not positive proof that this residue interacts directly with the permeant ion. As a result of such distant interactions, the functional properties of a particular residue in one Kv channel may be quite different from its properties at the same position in chimeras or other channels. For example, a valine/leucine exchange in the P-region of rKv2.1, or that of a Kv2.1/Kv3.1 chimera containing the Kv3.1 P-region, have different consequences from the identical mutation in native Kv3.1. Replacing a valine in the P-region of rKv2.1 (ITMTT*V*GYG) with leucine alters single channel properties, and so does the converse leucine to valine mutation in the Kv2.1/Kv3.1 chimera (Kirsch et al., 1992a,b). The leucine to valine mutation in native Kv3.1, however, does not change its single channel properties (Aiyar et al., 1993b). These conflicting results are most likely due to differences in the interactions between the leucine (or valine) and residues outside the P-region.

2.1.6 Conclusion

Potassium channels are the largest and most diverse known family of ion channels, and the voltage-gated K^+ channels constitute the largest group among these. The explosion of cloned genes and cDNAs over the past few years has led to an extraordinary accumulation of information on their functional diversity, gene organization, tissue-specific expression, and structure–function relationships, information that we have attempted to organize in this review.

Over the next few years we expect to witness major advances in our knowledge of K^+ channels, advances of both basic biological and practical significance. The availability of high-level expression systems for Kv channels should facilitate the biochemical characterization of these proteins, identifying the processes involved in their synthesis and assembly, and the variety of posttranslational modifications that they undergo. Such systems should also be capable of providing purified protein of sufficient quality and quantity for direct structural analysis, and the simple schematic diagrams that now constitute our best images of Kv channels may finally be replaced by empirically determined structures.

Identification of those genetic elements and mechanisms that regulate Kv gene transcription will provide a rationale for understanding their complex tissue-specific expression, at the same time explaining the surprising heterogeneity of Kv transcripts and their peculiarly long 5′ and 3′ untranslated regions. Resolution of these issues will also help in understanding the selective forces that have resulted in the expansion of the Kv subfamilies in vertebrates, and the striking lack of introns in the protein-coding regions of most of the *Shaker* subfamily genes.

While therapeutic drugs affecting other ion channels have long been known, Kv channels are just beginning to be developed as drug targets. The great diversity of these channels, and their involvement in myriad cellular functions, should provide the possibility of developing highly selective agents for the treatment of diseases affecting both electrically excitable and nonexcitable tissues. Finally, establishing associations between genetically determined diseases and mutations in specific Kv genes will forge additional links in the chain connecting basic biological knowledge with the treatment of human disease.

Acknowledgments

We thank Stephan Grissmer, Gary Yellen, and Tim Jegla for helpful suggestions. This work was supported in part by grants from Pfizer Inc. and from the NIH (AI-24783).

REFERENCES

Aaltonen, L. A., Peltomaki, P., Leach, F. S., Sistonen, P., Pylkkanen, L., Mecklin, J. P., Jarvinen, H., Powell, S. M., Jen, J., Hamilton, S. R., Petersen, G. M., Kinzler, K. W., Vogelstein, B., and de la Chapelle, A. (1993), Clues to the pathogenesis of familial colorectal cancer. *Science* 260, 812–816.

Adelman, J. P., Shen, K. Z., Kavanaugh, M. P., Warren, R., A., Wu, Y.-N., Lagrutta, A., Bond, C. T., and North, R. A. (1992). Calcium-activated potassium channels expressed from cloned complementary cDNAs. *Neuron* 9, 209-216.

Aiyar, J., Grissmer, S., and Chandy, K. G. (1993a). The L401V mutation in Kv3.1 does not alter single channel conductance or K^+/Rb^+ selectivity. *Biophys. J.* 64, A197.

Aiyar, J. Grissmer, S., and Chandy, K. G. (1993b). Full-length and truncated Kv1.3 K^+ channels are modulated by 5-HT_{1c} receptor activation, and independently, by PKC. *Am. J. Physiol.* 265, C1571–C1578.

Alber, T. (1992). Structure of the leucine zipper. *Curr. Opin. Genet. Dev.* 2, 205–210.

Albrect, B., Lorra, C., Stocker, K., and Pongs, O. (1993). Cloning, expression and chromosomal localization of the delayed rectifier K^+ channel 2.1 (h-DRK1). Unpublished (Gen Bank accession number X68302).

Anderson, J. A., Huprikar, S. S., Kochian, L. V., Lucas, W. J., and Gaber, R. F. (1992). Functional expression of a probable *Arabidopsis thaliana* potassium channel in *Saccharomyces cerevisiae. Proc. Natl. Acad. Sci. U.S.A.* 89, 3736–3740.

Atkinson, N., Robertson, G. A., and Ganetzky. (1991). A component of the calcium-activated potassium channels encoded by the Drosophila *slo* locus. *Science* 253, 551–55.

Attali, B., Honore, E., Lazdunski, M., and Barhanin, J. (1992a). Regulation of a major cloned voltage-gated K+ channel from human T lymphocytes. *FEBS Lett.* 303, 229–232.

Attali, B., Romey, G., Honore, E., Schmid-Alliana, A., Mattei, M G., Lesage, F., Ricard, P., Barhanin, J., and Ladzunski, M. (1992b). Cloning, functional expression, and regulation of two K+ channels in human T lymphocytes. *J. Biol. Chem.* 267, 8650–8657.

Attali, B., Guillemare, E., Lesage, F., Honore, E., Romey, G., Lazdunski, M., and Barhanin, J. (1993). The protein IsK is a dual activator of K+ and Cl- channels. *Nature (London)* 365, 850–852.

Attali, B., Lesage, F., Ziliani, P., guillemare, E., Honore, E., Waldman, R., Hugnot, J. P., Mattei, M. G., Lazdunski, M., and Barhanin, J. (1993). Multiple mRNA isoforms encoding the mouse cardiac Kv1-5 delayed rectifier K+ channel. *J. Biol. Chem.* 268, 24283–24289.

Baldwin, T. J., Tsaur, M-L., Lopez, L. A., Jan, Y. N., and Jan, L. Y. (1991). Characterization of a mammalian cDNA for an inactivation voltage-sensitive K+ channel. *Neuron* 7, 471–483.

Barbas, J. A., Rubio, N., Pedrosa, E., Pongs, O., and Ferrus, A. (1989). Antibodies against *Drosophila* potassium channels identify membrane protein across species. *Mol. Brain Res.* 5, 171–176.

Baumann, A., Grupe, A., Ackermann, A., and Pongs, O. (1988). Structure of the voltage-dependent potassium channel is highly conserved from Drosophila to vertebrate central nervous systems. *EMBO J.* 7, 2457–2463.

Beckh, S. and Pongs, O. (1990). Members of the RCK family are differentially expressed in the rat nervous system. *EMBO J.* 9, 777–782.

Betsholtz, C., Baumann, A., Kenna, S., Ashcroft, F. M., Ashcroft, S. J. H., Berggren, P-O., Grupe, A., Pongs, O., Rorsman, P., Sandblom, J., and Welsh, M. (1990). Expression of voltage-gated K+ channels in insulin-producing cells. Analysis by polymerase chain reaction. *FEBS Lett.* 263, 121–126.

Bezanilla, F., Perozo, E., Papazian, D. M., and Stefani, E. (1991). Molecular basis of gating charge immobilization in Shaker potassium channels. *Science* 254, 679–683.

Blair, T. A., Roberds, S. L., Tamkun, M. M., and Hartshorne, R. P. (1991). Functional characterization of RK5, a voltage-gated K+ channel cloned from the rat cardiovascular system. *FEBS Lett.* 295, 211–213.

Blair, T. A., Roberds, S. L., Tamkun, M. M., and Hartshorne, R. P. (1992). Protein kinase C modulates the RK5 K+ current. *Biophys J.* 61, A382.

Brau, M. E., Dreyer, F., Jonas, P., Repp, H., and Vogel, W. (1990). A K+ channel in Xenopus nerve fibers selectively blocked by bee and snake toxins: Binding and voltage-clamp experiments. *J. Physiol. (London)* 420, 365–385.

Bruggemann, A., Pardo, L. A., Stuhmer, W., and Pongs, O. (1993). *Ether-a-go-go* encodes a voltage-gated channel permeable to K+ and Ca^{2+} and modulated by cAMP. *Nature* 365, 445–448.

Busch, A. E., Hurst, R. S., North, R. A., Adelman, J. P., and Kavanaugh, M. P. (1991). Current inactivation involves a histidine residue in the pore of the rat lymphocyte potassium channel RGK5. *Biochem. Biophys. Res. Commun.* 179, 1384–1390.

Butler, A., Wei, A., Baker, K., and Salkoff, L. (1989). A family of K+ channel genes in *Drosophila. Science* 243, 943.

Butler, A., Wei, A., and Salkoff, L. (1990). *Shal, Shab,* and *Shaw:* Three genes encoding potassium channels in *Drosophila. Nucl. Acid. Res.* 18, 2173–2174.

Butler, A., Tsunoda, S., McCobb, D. P., Wei, A., and Salkoff, L. (1993). mSlo, a complex mouse gene encoding "maxi" calcium-activated potassium channels. *Science* 261, 221–224.

Cahalan, M. D., Chandy, K. G., and Grissmer, S. (1991). Potassium channels in development, activation, and disease in T lymphocytes. *Cur. Topics Membranes: Dev. Biol. Membrane Transport Syst.* 39, 358–394.

Cai, Y. C. and Douglass, J. (1993). *In vivo* and *in vitro* phosphorylation of the T lymphocyte type *n* (Kv1.3) potassium channel. *J. Biol. Chem.* 268, 23720–23727.

Cai, Y. C., Osborne, P. B., North, R. A., Dooley, D. C., and Douglass, J. (1992). Characterization and functional expression of genomic DNA encoding the human lymphocyte type *n* potassium channel. *DNA Cell Biol.* 11, 163–172.

Catterall, W. A. (1988). Structure and function of voltage-sensitive ion channels. *Science* 242, 50–61.

Chabala, L. D. and Covarrubias, M. (1993). Low-molecular weight poly-(A^+) mRNA encodes factors that modulate non-*Shaker* A-type K^+ currents. *Biophys. J.* 64, A227.

Chandy, K. G., Williams, C. B., Spencer, R. H., Aguilar, B. A., Ghanshani, S., Chandy, G., Tempel, B. L., and Gutman, G. A. (1990a). Multiple genes contribute to K^+ channel diversity in the mouse. *Biophys. J.* 57, 110a.

Chandy, K. G., Williams, C. B., Spencer, R. H., Aguilar, B. A., Ghanshani, S., Tempel, B. L., and Gutman, G. A. (1990b). A family of three mouse potassium channel genes with intronless coding region. *Science* 247, 973–975.

Chandy, K. G., Douglass, J., Gutman, G. A., Jan, L., Joho, R., Kaczmarek, L., McKinnon D., North, R. A., Numa, S., Philipson, L., Ribera, A. B., Rudy, B., Salkoff, L., Swanson, R. A., Steiner, D., Tanouye, M., and Temple, B. L. (1991). A simplified gene nomenclature. *Nature* 352, 26.

Chandy, K. G., Gutman, G. A., and Grissmer, S. (1993). Physiological role, molecular structure and evolutionary relationships of voltage-gated potassium channels in T lymphocytes. *Sem. Neurosci.* 5, 125–134.

Chevillard, C., Attali, B., Lesage, F., Fontes, M., Barhanin, J., Lazdunski, M., and Mattei, M. G. (1993). Localization of a potassium channel gene (KCNE1) to 21q22.1-q22 by in-situ hybridization and somatic cell hybridization. *Genomics* 15, 243–245.

Choi, K. L., Aldrich, R. W., and Yellen, G. (1991). Tetraethylammonium blockade distinguishes two inactivation mechanisms in voltage-activated K^+ channels. *Proc. Natl. Acad. Sci. U.S.A.* 88, 5092–5095.

Choi, K. L., Mossman, C., Aube, J., and Yellen, G. (1993). The internal quartenary ammonium receptor site of *Shaker* potassium channels. *Neuron* 10, 533–541.

Christie, M. J., Adelman, J. P., Douglass, J., and North, R. A. (1989). Expression of a cloned rat brain potassium channel in *Xenopus* oocytes. *Science* 244, 221–224.

Christie, M. J., North, R. A., Osborne, P. B., Douglass, J., and Adelman, J. P. (1990). Heteropolymeric potassium channels expressed in *Xenopus* oocytes from cloned subunits. *Neuron* 4, 405.

Chua, K., Tytgat, J., Liman, E., and Hess, P. (1992). Membrane topology of RCK1 K-channels. *Biophys. J.* 61, A289.

Colatsky, T. J., Follmer, C. H., and Starmer, C. F. (1993). Channel specificity in antiarrhythmic drug action: Mechanism of potassium channel block and its role in suppressing and aggravating cardiac arrhythmias. *Circulation* 82, 2235–2242.

Covarrubias, M. and Rubin, E. (1993). Ethanol selectively blocks a noninactivating K^+ current expressed in *Xenopus* oocytes. *Proc. Natl. Acad. Sci. U.S.A.* 90, 6957–6960.

Covarrubias, M., Wei., A. and Salkoff, L. (1991). *Shaker, Shal, Shab,* and *Shaw* express independent K^+ current systems. *Neuron* 7, 763–773.

Critz, S. D., Wible, B. A., Lopez, H. S., and Brown, A. M. (1993). Stable expression and regulation of a rat brain K^+ channel. *J. Neurochem.* 60, 1175–1178.

Curran, M. E., Landes, G. M., and Keating, M. T. (1992). Molecular cloning, characterization, and genomic localization of a human potassium channel gene. *Genomics* 12, 729–737.

Deal, K. K., Lovinger, D. M., and Tamkun, M. M. (1993). Examination of K^+ channel biosynthesis: Evidence for glycosylation and cotranslational subunit assembly. *Biophys. J.* 64, A196.

DeBiasi, M., Hartmann, H. A., Drewe, J. A., Taglialatela, M., Brown, A. M., and Kirsch, G. E. (1993a). Inactivation determined by a single site in K^+ pores. *Pflügers Arch.* 422, 354–363.

DeBiasi, M., Drewe, J. A., Kirsch, G. E., and Brown, A. M. (1993b). Histidine substitution identifies a surface position and confers Cs^+ selectivity on a K^+ pore. *Biophys. J.* 65, 1235–1242.

Desir, G. V., Hamlin, H. A., Puente, E., Reilly, R. F., Hildebrandt, F., and Igarishi, P. (1992). Isolation of putative voltage-gated epithelial and K-channel isoforms from rabbit kidney and LLC-PK_1 cells. *Am. J. Physiol.* 262, F151–F157.

Douglass, J., Osborne, P. B., Cai, Y. C., Wilkinson, M., Christie, M. J., and Adelman, J. P. (1990). Characterization of RGK5, a genomic clone encoding a lymphocyte channel. *J. Immunol.* 144, 4841–4850.

Drewe, J., Verma, S., Frech, G., and Joho, R. (1992). Distinct spatial and temporal expression patterns of K^+ channel mRNAs from different subfamilies. *J. Neurosci.* 12, 538–548.

Dubinsky, W. P., Mayorga-Wark, O., and Schultz, S. G. (1992). A peptide from the *Drosophila Shaker* K^+ channel inhibits a voltage-gated K^+ channel in basolateral membranes of *Necturus* enterocytes. *Proc. Natl. Acad. Sci. U.S.A.* 89, 1770–1774.

Durell, S. R. and Guy, H. R. (1992). Atomic scale structure and functional models of voltage-gated potassium channels. *Biophys. J.* 62, 238–247.

Fahrig, S. A., Snyders, D. J., Knoth, K. M., Tamkun, M. M., and Murray, K. T. (1992). Depression of a human cardiac K^+ current by stimulation of protein kinase C. *Biophys. J.* 61, A383.

Fedida, D., Wible, B., Wang, Z., Fermini, B., Faust, F., Nattel, S., and Brown, A. M. (1993). Identity of a novel delayed rectifier current from human heart with a cloned K^+ channel current. *Circ. Res.* 73, 210–216.

Foster, C. D., Chung, S., Zagotta, W. N., Aldrich, R. W., and Levitan, I. B. (1992). A peptide derived from the *Shaker* B K^+ channel produces short and long blocks of reconstituted Ca(2+)-dependent K^+ channels. *Neuron* 9, 229–236.

Frech, G. C., VanDongen, A. M. J., Schuster, G., Brown, A. M., and Joho, R. H. (1989). A novel potassium channel with delayed rectifier properties isolated from rat brain by expression cloning. *Nature (London)* 340, 642–645.

Freeman, S. N., Conley, E. C., Brennand, J. C., Russell, N. J. W., and Brammar, W. J. (1990). Cloning and characterization of a cDNA encoding a human brain potassium channel. *Biochem. Soc. Trans.* 18, 891–892.

Garcia, M. L., Garcia-Calvo, M., Hidalgo, P., Lee, A., and MacKinnon, R. (1994). Purification and characterization of thre inhibitors of voltage-dependent K^+ channels from *Leirus quinquestriatus* var, hebraeus venom. Biochemistry 33, 6834–6839.

Garcia, M. L., Garcia-Calvo, M., Knaus, H-G., and Kaczrowoski. (1993). High conductance Ca^{2+}-activated K^+ channels: Pharmacology and molecular characterization. *Sem. Neurosci.* 5, 107–115.

Garcia-Calvo, M., Leonard, R., Giangiacomo, K., McManus, O. B., Kaczrowski, G. J., and Garcia, M. L. (1992). Identification of a toxin from Leurius quinquestriatus venom that blocks a *Shaker* K^+ channel. *Biophys. J.* 61, A377.

Garcia-Guzman, M., Calvo, S., Cena, V., and Criado, M. (1992). Molecular cloning and permanent expression in a neuroblastoma cell line of a fast inactivating potassium channel from bovine adrenal medulla. *FEBS. Lett.* 308, 283–289.

Gessler, M., Grupe, A., Grzeschik, K. H., and Pongs, O. (1993). The potassium channel HK1 maps to human chromosome 11p14.1 close to the FSHB gene. *Human Genet.* 90, 319–321.

Ghanshani, S., Pak, M., McPherson, J. D., Strong, M., Dethlefs, B., Wasmuth, J. J., Salkoff, L., Gutman, G. A., and Chandy, K. G. (1992). Genomic organization, nucleotide sequence, cellular distribution of a *Shaw*-related potassium channel gene, Kv3.3, and mapping of Kv3.3 and Kv3.4 on human chromosomes 19 and 1. *Genomics* 12, 190–196.

Goldman-Wohl, E. C. and Heintz, N. (1992). A novel potassium channel cDNA clone is developmentally regulated and is expressed in specific cell types in the mouse cerebellum. *Soc. Neurosci. Abstr.* 18, 41.

Goldstein, S. A. and Miller, C. (1992). A point mutation in a *Shaker* K^+ channel changes its charybdotoxin binding site from low to high affinity. *Biophys. J.* 62, 5.

Graur, D. (1993). Towards a resolution of the ordinal phylogeny of the eutherian mammals. *FEBS Lett.* 325, 152–159.

Grissmer, S., Dethlefs, B., Wasmuth, J. J., Goldin, A. L., Gutman, G. A., Cahalan, M. D., and Chandy, K. G. (1990). Expression and chromosomal localization of a lymphocyte K^+ channel gene. *Proc. Natl. Acad. Sci. U.S.A.* 87, 9411–9415.

Grissmer, S., Ghanshani, S., Dethlefs, B., McPherson, J. D., Wasmuth, J. J., Gutman, G. A., Cahalan, M. D., and Chandy, K. G. (1992). The *Shaw*-related K^+ channel gene on human chromosome 11 encodes the type *l* K^+ channel in T cells. *J. Biol. Chem.* 267, 20971–20979.

Grissmer, S., Nguyen, A. N., Aiyar, J., Hanson, D. C., Mather, R. J., Gutman, G. A., Karmilowicz, M. J., Auperin, D. D., and Chandy, K. G. (1994). Pharmacological characterization of five cloned voltage-gated K^+ channels, types Kv1.1, 1.2, 1.3, 1.5, and 3.1, stably expressed in mammalian cell lines. *Mol. Pharmacol.* 45, 1227–1234.

Grupe, A., Schroter, K. H., Ruppersberg, J. P., Stocker, M., Drewes, T., Beckh, S., and Pongs, O. (1990). Cloning and expression of a human voltage-gated potassium channel. A novel member of the RCK potassium channel family. *EMBO J.* 9, 1749–1756.

Guerrini, R., Bureau, M., Mattei, M. G., Battaglia, A., and Galland, M. C. (1990). Trisomy 12p syndrome: A chromosomal disorder associated with generalized 3-Hs spike and wave discharges. *Epilepsia* 31, 557–566.

Gutman, G. A. and Chandy, K. G. (1993). Nomenclature for vertebrate voltage-gated K^+ channels. *Sem. Neurosci.* 5, 101–106.

Haas, M., Ward, D. C., Lee, J., Roses, A. D., Clarke, V., D'Eustachio, P., Lau, D., Vega-Saenz de Miera, E., and Rudy, B. (1993). Localization of *Shaw*-related K^+ channel genes on mouse and human chromosomes. *Mamm. Genome* 4, 711–715.

Hardie, R. C., Voss, D., Pongs, O., and Laughlin, S. B. (1991). Novel potassium channels encoded by the *Shaker* locus in *Drosophila* photoreceptors. *Neuron* 6, 477–486.

Hart, P. J., Overturf, K. E., Russell, S. N., Carl, A., Hume, J. R., Sanders, K. M., and Horowitz, B. (1993). Cloning and expression of a Kv1.2 class delayed rectifier K^+ channel from canine colonic smooth muscle. *Proc. Natl. Acad. Sci. U.S.A.* 90, 9659–9663.

Hartmann, H. A., Kirsch, G. E., Drewe, J. A., Tagliatela, M., Joho, R. H., and Brown, A. M. (1991). Exchange of conduction pathways between two related K^+ channels. *Science* 251, 942–944.

Heginbotham, L. and MacKinnon, R. (1992). The aromatic binding site for tetraethylammonium ion on potassium channels. *Neuron* 8, 483–491.

Hemmick, L. M., Perney, T. M., Flamm, R. E., Kaczmarek, L. K., and Birnberg, N. C. (1992). Expression of the H *ras* oncogene induces potassium conductance and neuron-specific potassium channel mRNAs in the AtT20 cell line. *J. Neurosci.* 12, 2007–2014.

Higgins, D. G., Bleasby, A. J., and Fuchs, R. (1992). CLUSTALV: Improved software for multiple sequence alignment. *CABIOS* 8, 189–191.

Hille, B. (1993). *Ionic Channels of Excitable Membranes,* 2nd ed. Sinauer, Sunderland, MA.

Ho, K., Nichols, C. G., Lederer, W. J., Lytton, J., Vassilev, P. M., Kanazirska, M. V., and Hebert, S. C. (1993). Cloning and expression of an inwardly rectifying ATP-regulated potassium channel. *Nature* 362, 31–37.

Honore, E., Attali, B., Romey, G., Heurteaux, C., Ricard, P., Lesage, F., Lazdunski, M., and Barhanin, J. (1991). Cloning, expression, pharmacology and regulation of a delayed rectifier in mouse heart. *EMBO J.* 10, 2805–2811.

Honore, E., Attali, B., Romey, G., Lesage, F., Barhanin, J., and Lazdunski, M. (1992). Different kinds of K^+ current are generated by different levels of a single mRNA. *EMBO J.* 11, 2465–2471.

Hoshi, T., Zagotta, W. N., and Aldrich, R. W. (1990). Biophysical and molecular mechanisms of Shaker potassium channel inactivation. *Science* 250, 533–538.

Hoshi, T., Zagotta, W. N., and Aldrich, R. W. (1991). Two types of inactivation in *Shaker* K^+ channels: Effects of alterations in the carboxy terminal region. *Neuron* 7, 547–556.

Huntington's Disease Collaborative Research Group. (1993). A novel gene containing a trinucleotide repeat that is expanded and unstable in Hungtington's disease chromosomes. *Cell* 72, 971–983.

Hurst, R. S., Busch, A. E., Kavanaugh, M. P., Osborne, P. B., North, R. A., and Adelman, J. P. (1991). Identification of amino acid residues involved in dendrotoxin block of rat voltage-dependent potassium channels. *Mol. Pharmacol.* 40, 572.

Hwang, P. M., Glatt, C. E., Bredt, D. S., Yellen, G., and Snyder, S. H. (1992). A novel K^+ channel with unique localizations in mammalian brain: Molecular cloning and characterization. *Neuron* 8, 473–481.

Ikeda, S. R., Soler, F., Zuhlke, R. D., Joho, R. H., and Lewis, D. L. (1992). Heterologous expression of the human potassium channel Kv2.1 in clonal mammalian cells by direct cytoplasmic microinjection of cRNA. *Pflügers Archiv.* 422, 210–203.

Isacoff, E. Y., Jan, Y. N., and Jan, L. Y. (1990). Evidence for formation of heteromultimeric potassium channels in *Xenopus* oocytes. *Nature (London)* 345, 530–534.

Isacoff, E. Y., Jan, Y. N., and Jan, L. Y. (1991). Putative receptor for the cytoplasmic inactivation gate in the *Shaker* K^+ channel. *Nature (London)* 353, 86–90.

Isacoff, E. Y., Kimmerly, W., Jan, Y. N., and Jan, L. Y. (1992). Phosphorylation and modulation of the *Shaker* K^+ channel by protein kinase C (PKC). *Biophys. J.* 61, A13.

Ishii, K., Nunoki, K., Murakoshi, H., and Taira, N. (1992). Cloning and modulation by endothelin-1 of rat cardiac K channel. *Biochem. Biophys. Res. Commun.* 184, 1484.

Iverson, L. E. and Rudy, B. (1990). Role of the divergent amino and carboxyl domains on the inactivation properties of potassium channels derived from the *Shaker* gene of *Drosophila. J. Neurosci.* 10, 2903–2916.

Iverson, L. E., Tanouye, M. A., Lester, H. A., Davidson, N., and Rudy, B. (1988). A-type potassium channels expressed from *Shaker* locus cDNA. *Proc. Natl. Acad. Sci. U.S.A.* 85, 5723–5727.

James, W. M. and Agnew, W. S. (1987). Multiple oligosaccharide chains in the voltage-sensitive Na channel from *Electrophorus electricus:* Evidence for alpha-2,8-linked polysialic acid. *Biochem. Biophys. Res. Commun.* 148, 817–826.

Jan, L. Y. and Jan, Y. N. (1992). Structural elements involved in specific K^+ channel functions. *Annu. Rev. Physiol.* 54, 537–555.

Johansen, K., Wei, A. G., Salkoff, L., and Johansen, J. (1990). Leech gene sequence homologous to *Drosophila* and mammalian *Shaker*-K^+ channels. *J. Cell Biol.* 111, 60a.

Kamb, A., Iverson, L. E., and Tanouye, M. A. (1987). Molecular characterization of *Shaker,* a *Drosophila* gene that encodes a potassium channel. *Cell* 50, 405–413.

Kamb, A., Weir, M., Rudy, B., Varmus, H., and Kenyon, C. (1989). Identification of genes from pattern formation, tyrosine kinase, and potassium channel families by DNA amplification. *Proc. Natl. Acad. Sci. U.S.A.* 86, 4372–4376.

Kaupp, U. B., Niidome, T., Tanabe, T., Terada, S., Bonigk, W., Stuhmer, W., Cook, N. J., Kangawa, K., Matsuo, H., and Hirose, T. (1989). Primary structure and functional expression from complementary DNA of the rod photoreceptor cyclic-GMP-gated channel. *Nature (London)* 342, 762–766.

Kavanaugh, M. P., Varnum, M. D., Osborne, P. B., Christie, M. J., Busch, A. E., Adelman, J. P., and North, R. A. (1991). Interaction between tertaethylammonium and amino acid residues in the pore of cloned voltage dependent potassium channels. *J. Biol. Chem.* 266, 7583–7587.

Kavanaugh, M. P., Hurst, R. S., Yakel, J., Varnum, M. D., Adelman, J. P., and North, R. A. (1992). Multiple subunits of a voltage-dependent potassium channel contribute to the binding site for tetraethylammonium. *Neuron* 8, 493–497.

Keating, M., Atkinson, D., Dunn, C., Timothy, K., Vincent, G. M., and Leppert, M. (1991). Linkage of a cardiac arrhythmia, the long QT syndrome, and the Harvey ras-1 gene. *Science* 252, 704–706.

Kemp, B. E. and Pearson, R. B. (1990). Protein kinase recognition sequence motifs. *Trends Biochem.* 15, 342–346.

Kirsch, G. E., Drewe, J. A., Verma, S., Brown, A. M., and Joho, R. H. (1991a). Electrophysiological characterization of a new member of the RCK family of rat brain K^+ channels. *FEBS Lett.* 278, 55–60.

Kirsch, G. E., Taglialatela, and M., Brown, A. M. (1991b). Internal and external TEA block in single cloned K^+ channels. *Am. J. Physiol.* 261, C583–C590.

Kirsch, G. E., Drewe, J. A., Taglialatela, M., Joho, R. H., DeBiasi, M., Hartmann, H. A., and Brown, A. M. (1992a). A single nonpolar residue in the deep pore of related K^+ channels acts as a K^+:Rb^+ conductance switch. *Biophys. J.* 62, 136–143.

Kirsch, G. E., Drewe, J. A., Hartmann, H. A., Taglialatela, M., DeBiasi, M., Brown, A. M., and Joho, R. H. (1992b). Differences between the deep pores of K^+ channels determined by an interacting pair of nonpolar amino acids. *Neuron* 8, 499–505.

Kirsch, G. E., Shieh, C. C., Drewe, J. A., Vener, D. F., and Brown, A. M. (1993). Segmental exchanges define 4-aminopyridine binding and the inner mouth of K^+ pores. *Neuron* 11, 1–20.

Klumpp, D. J., Farber, D. B., Bowes, C., Song, E-J., and Pinto, L. H. (1991). The potassium channel MBK1 (Kv1.1) is expressed in the mouse retina. *Cell. Mol. Neurobiol.* 11, 611–622.

Koren, G., Liman, E. R., Logothetis, D. E., Nadal-Ginard, B., and Hess, P. (1990). Gating mechanism of a cloned potassium channel expressed in frog oocytes and mammalian cells. *Neuron* 2, 39–51.

Kozak, M. (1991). An analysis of vertebrate mRNA sequences: Intimations of translational control. *J. Cell. Biol.* 115, 887–903.

Kruys, V., Marinx, O., Shaw, G., Deschampes, J., and Huez, G. (1989). Translational blockade imposed by cytokine-derived UA-rich sequences. *Science* 245, 852–855.

Kubo, Y., Reuveny, E., Slesinger, P. A., Jan, Y. N., and Jan, L. Y. (1993). Primary structure and functional expression of a rat G-protein-coupled muscarinic potassium channel. *Nature* 364, 802–806.

Kubo, Y., Baldwin, T. J., Jan, Y. N., and Jan, L. Y. (1993a). Primary structure and functional expression of a mouse inward rectifier potassium channel. *Nature (London)* 362, 127–133.

Kubo, Y., Reuveny, E., Slesinger, P. A., Jan, Y. N., and Jan, L. Y. (1993b). Primary structure and functional expression of a rat G-protein-coupled muscarinic potassium channel. *Nature (London)* 364, 802–806.

Kuo, S. S., Saad, A. H., Koong, A. C., Hahn, G. M., and Giaccia, A. J. (1992). Potassium-channel activation in response to low doses of gamma-irradiation involves reactive oxygen intermediates in nonexcitatory cells. *Proc. Natl. Acad. Sci. U.S.A.* 90, 908–912.

Lesage, F., Attali, B., Lazdunski, M., and Barhanin, J. (1992). Developmental expression of voltage-sensitive K^+ channels in mouse skeletal muscle and C_2C_{12} cells. *FEBS. Lett.* 310, 162–166.

Levinson, G. and Gutman, G. A. (1987). Slipped-strand mispairing: A major mechanism of DNA sequence evolution. *Mol. Biol. Evol.* 4, 203–221.

Levitan, E. S., Hemmick, L. M., Birnberg, N. C., and Kaczmarek, L. K. (1991). Dexamethasone increases potassium channel messenger RNA and activity in clonal pituitary cells. *Mol. Endocrinol.* 5, 1903–1908.

Li, M., Jan, Y. N., and Jan, L. Y. (1992). Specification of subunit assembly by the hydrophilic amino-terminal domain of the Shaker potassium channel. *Science* 257, 1225–1230.

Li, M., Unwin, N., Stauffer, K. A., Jan, Y. N., and Jan, L. Y. (1993). Images of purified Shaker potassium channels. *Curr. Biol.* 4, 110–115..

Liman, E. R., Hess, P., Weaver, F., and Koren, G. (1991). Voltage-sensing residues in the S4 region of a mammalian K^+ channel. *Nature (London)* 353, 752–756.

Liman, E. R., Tytgat, J., and Hess, P. (1992). Subunit stochiometry of a mammalian K^+ channel determined by construction of multimeric cDNAs. *Neuron* 9, 861–871.

Lock, L. F., Gilbert, D. J., Street, V. A., Migeon, M. B., Jenkins, N. A., Copeland, N. G., and Tempel, B. L. (1994). Voltage-gated potassium channel genes are clustered in paralogous regions of the mouse genome. *Genomics* 20, 354–362.

Logothetis, D. E., Movehedi, S., Satler, C., Lind-Paintner, K., and Nadal-Ginard, B. (1992). Incremental reductions of positive charge within the S4 region of voltage-gated K+ channel result in corresponding decreases in gating charge. *Neuron* 8, 531–540.

Logothetis, D. E., Kammen, B. F., Lindpaintner, K., Bisbas, D., and Nadal-Ginard, B. (1993). Gating charge differences between two voltage-gated K+ channels are due to the specific charge content of their respective S4 regions. *Neuron* 10, 1121–1129.

Lopez, G. A., Jan, Y. N., and Jan, L. Y. (1991). Hydrophobic substitution mutations in the S4 sequence after voltage-dependent gating in *Shaker* K+ channels. *Neuron* 7, 327–336.

Lopez, G. A., Jan, Y. N., and Jan, L. Y. (1993). Mutations in the S6 domain affecting conductance, TEA, and barium blockade in the permeation pathway of the *Shaker* potassium channel. *Biophys. J.* 64, A113.

Lopez-Barneo, J., Hoshi, T., Heinemann, S. H., and Aldrich, R. W. (1993). Effects of external cations and mutations in the pore region on C-type inactivation of *Shaker* potassium channels. *Receptors Channels* 1, 66–71.

Ludwig, J., Margalit, T., Eisman, E., Lancet, D., and Kaupp, U. B. (1990). Primary structure of cAMP-gated channel from bovine olfactory epithelium. *FEBS Lett.* 270, 24–29.

Luneau, C. J., Williams, J. B., Marshall, J., Levitan, E. S., Oliva, C., Smith, J. S., Anatanavage, J., Folander, K., Stein, R. B., Swanson, R., Kaczmarek, L. K., and Buhrow, S. A. (1991a). Alternative splicing contributes to K+ channel diversity in the mammalian central nervous system. *Proc. Natl. Acad. Sci. U.S.A.* 88, 3932–3936.

Luneau, C., Wiedmann, R., Smith, J. S., and Williams, J. B. (1991b). *Shaw*-like rat brain potassium channel cDNAs with divergent 3′ ends. *FEBS. Lett.* 288, 163–167.

Lynch, J. J., Sanguinetti, Kimura, S., and Bassett, A. L. (1992). Therapeutic potential of modulating potassium currents in the diseased myocardium. *FASEB J.* 6, 2952–2960.

MacKinnon, R. (1991a). New insights into the structure and function of potassium channels. *Curr. Opinion Neurobiol.* 1, 14–19.

MacKinnon, R. (1991b). Determination of the subunit stochiometry of a voltage-activated potassium channel. *Nature (London)* 350, 232–235.

MacKinnon, R. and Miller, C. (1989). Mutant potassium channels with altered binding of charybdotoxin, a pore-blocking peptide inhibitor. *Science* 245, 1382–1385.

MacKinnon, R. and Yellen, G. (1990). Mutations affecting TEA blockade and ion permeation in voltage-activated K+ channels. *Science* 250, 276–279.

MacKinnon, R., Heginbotham, L., and Abramson, T. (1990). Mapping the receptor site for charybdotoxin, a pore-blocking potassium channel inhibitor. *Neuron* 5, 767–771.

Mahaut-Smith, M. P., Rink, T. J., Collins, S. C., and Sage, S. O. (1990). Voltage-gated potassium channels and the control of membrane potential in human platelets. *J. Physiol.* 428, 723–735.

Malayev, A., Philipson, L. H., Kuznetsov, A., Chang, C., and Nelson, D. J. (1993). High-level expression of wild type and C-terminus chimeric human K+ channel genes (HPCN-Kv1.5) in stable mammalian cell lines: Characterization of channel function, detection of the gene product and fluorescence-activated cell sorting using two-site directed antibodies. *Biophys. J.* 64, A196.

Maltais, L. J., Lane, P. W., and Beamer, W. G. (1984). Anorexia, a recessive mutation causing starvation in preweanling mice. *J. Hered.* 75, 468–472.

Marom, S., Goldstein, S. A. N., Kupper, J., and Levitan, I. B. (1993). Mechanism and modulation of inactivation of the Kv3 potassium channel. *Receptors Channels* 1, 81–88.

Marshall, R. D. (1972). Glycoproteins. *Annu. Rev. Biochem.* 41, 673–702.

Mathew, M. K., Ramashwami, M., Gautam, M., Kamb, C. A., Rudy, B., and Tanouye, M. A. (1989). Cloning and characterization of human potassium channel genes. *Soc. Neurosci. Abstr.* 15, 540.

Matsubara, H., Liman, E. R., Hess, P., and Koren, G. (1991). Pretranslational mechanisms determine the type of potassium channels expressed in the rat skeletal and cardiac muscles. *J. Biol. Chem.* 266, 13324–13328.

Matsubara, H., Mori, Y., and Koren, G. (1992). cAMP modulates the expression of a delayed rectifier at the pretranslational level. *Biophys. J.* 61, A146.

Matsubara, H., Suzuki, J., and Inada, M. (1993). Shaker-related potassium channel, Kv1.4, mRNA regulation in cultured rat heart myocytes and differential expression of Kv1.4 and Kv1.5 genes in myocardial development and hypertrophy. *J. Clin. Invest.* 92, 1659–1666.

Maue, R. A., Kraner, S. D., Goodman, R. H., and Mandel G. (1990). Neuron-specific expression of the rat-brain type II sodium channel gene is directed by upstream regulatory elements. *Neuron* 4, 223–231.

McCormack, K., Campanelli, J. T., Ramaswami, M., Mathew, M. K., and Tanouye, M. A. (1989). Leucine zipper motif update. *Nature (London)* 340, 103.

McCormack, K. J., Lin, W., Iverson, L. E., and Rudy, B. (1990). *Shaker* K^+ channel subunits form heteromultimeric channels with novel functional properties. *Biochem. Biophys. Res. Commun.* 171, 1361–1371.

McCormack, K., Tanouye, M. A., Iverson, L. E., Lin, J. W., Ramaswami, M., McCormack, T., Pampanelle, J. T., Mathew, M. K., and Rudy, B. (1991). A role for hydrophobic residues in the voltage-dependent gating of *Shaker* K^+ channels. *Proc. Natl. Acad. Sci. U.S.A.* 88, 2931–2935.

McCormack, T., Vega-Saenz de Miera, E. C., and Rudy, B. (1990). Molecular cloning of a member of a third class of Shaker-family K^+ channel genes in mammals. *Proc. Natl. Acad. Sci. U.S.A.* 87, 5227–5231.

McKinnon, D. (1989). Isolation of a cDNA clone coding for a putative second potassium channel indicates the existence of a gene family. *J. Biol. Chem.* 264, 8230–8236.

McPherson, J. D., Wasmuth, J. J., Chandy, K. G., Swanson, R., Dethlefs, B., Chandy, G., Wymore, R., and Ghanshani, S. (1991). Chromosomal localization of 7 potassium channel genes. In *Eleventh International Workshop on Human Gene Mapping,* Solomon, E. and Rawlings, C. Eds., Karger, New York.

Meyerhoff, W., Schwarz, J. R., Bauer, C. K., Hubel, A., and Richter, D. (1992). A rat pituitary tumor K^+ channel expressed in frog oocytes induces a transient K^+ current indistinguishable from that recorded in native cells. *J. Neuroendocrinol.* 4, 245–253.

Migeon, M. B., Street, V. A., Demas, V. P., and Tempel, B. L. (1992). Cloning, sequence and chromosomal localization of MK6, a murine potassium channel gene. *Epilepsy Res.* 6(Suppl.), 173–181.

Milkman, R., (1994). An Escherichia coli homologue of eukaryotic potassium channel proteins. *Proc. Natl. Acad. Sci. U.S.A.* 91, 3510–3514.

Miller, C. (1990). 1990: Annus Mirabilis of potassium channels. *Science* 252, 1092–1096.

Miller, C. (1992). Hunting for the pore of voltage-gated channels. *Curr. Biol.* 11, 573–575.

Miller, C., Moczydlowski, E., Latorre, R., and Phillips, M. (1985). Charybdotoxin, a protein inhibitor of single Ca2+-activated K^+ channels from mammalian skeletal muscle. *Nature (London)* 313, 316–318.

Miller, J. A., Agnew, W. S., and Levinson, S. R. (1983). Principal glycopeptide of the tetrodotoxin/saxitoxin binding protein from *Electrophorus electricus*: Isolation and partial chemical and physical characterization. *Biochemistry* 22, 462–470.

Moran, O., Dascal, N., and Lotan, I. (1991). Modulation of a *Shaker* potassium A-channel by protein kinase C activation. *FEBS Lett.* 279, 256–260.

Mori, N., Schoenherr, C., Vandenbergh, D. V., and Anderson, D. J. (1990). A common silencer element in the SCG10 and type II Na^+ channel genes binds a factor present in non-neuronal cells but not in neuronal cells. *Neuron* 9, 45–54.

Mori, Y., Matsubara, H., Folco, E., Siegel, A., and Koren, G. (1993). The transcription of a mammalian voltage-gated potassium channel is regulated by cAMP in a cell-specific manner. *J. Biol. Chem.* 268, 26482–26493.

Murai, T., Kakizuka, A., Takumi, T., Ohkubo, H., and Nakanishi, S. (1989). Molecular cloning and sequence analysis of human genomic DNA encoding a slowly activating potassium channel activity. *Biochem. Biophys. Res. Commun.* 161, 176–181.

Nei, M. (1987). *Molecular Evolutionary Genetics.* Columbia University Press, New York.

Nilius, B., Bohm, T., and Wohlrab, W. (1990). Properties of a potassium-selective ion channel in human melanoma cells. *Pflügers Arch.* 417, 269–277.

Novacek, M. J. (1992). Mammalian phylogeny: Shaking the tree. *Nature (London)* 356, 121–125.

Pak, M. D., Baker, K., Covarrubias, M., Butler, A., Ratcliffe, A., and Salkoff, L. (1991a). mShal, a subfamily of A-type K^+ channel cloned from mammalian brain. *Proc. Natl. Acad. Sci. U.S.A.* 88, 4386–4390.

Pak, M. D., Covarrubias, M., Ratcliffe, A., and Salkoff, L. (1991b). A mouse brain homolog of the *Drosophila Shab* K^+ channel with conserved delayed rectifier properties. *J. Neurosci.* 11, 869–880.

Papazian, D. M., Schwarz, T. L., Tempel, B. L., Jan, Y. N., and Jan, L. Y. (1987). Cloning of genomic and complementary DNA from Shaker, a putative potassium channel gene from *Drosophila. Science* 237, 749–753.

Papazian, D. M., Timpe, L. C., Jan, Y. N., and Jan, L. Y. (1991). Alteration of voltage-dependence of *Shaker* potassium channel by mutations in the S4 sequence. *Nature (London)* 349, 305–310.

Parcej, D. N., Scott, V. E., and Dolly, J. O. (1992). Oligomeric properties of alpha-dendrotoxin-sensitive potassium ion channels purified from bovine brain. *Biochemistry* 31, 11084–11088.

Pardo, L. A., Heinemann, S. H., Terlau, H., Ludewig, U., Lorra, C., Pongs, O., and Stuhmer, W. (1992). Extracellular K^+ specifically modulates a rat brain potassium channel. *Proc. Natl. Acad. Sci. U.S.A.* 89, 2466–2470.

Paulmichl, M., Nasmith, P., Hellmiss, R., Reed, K., Boyle, W. A., Nerbonne, J. M., Peralta, E. G., and Clapham, D. E. (1991). Cloning and expression of a rat cardiac delayed rectifier potassium channel. *Proc. Natl. Acad. Sci. U.S.A.* 88, 7892–7895.

Payet, M. D. and Dupuis, G. (1992). Dual regulation of the *n*-type K^+ channel in Jurkat T-lymphocytes by protein kinase-A and kinase-C. *J. Biol. Chem.* 267, 18270–18273.

Perney, T. M. and Kaczmarek, L. (1993). Expression and regulation of mammalian K^+ channel genes. *Sem. Neurosci.* 5, 135–145.

Perney, T. M., Marshall, J., Martin, K. A., Hockfield, S., and Kaczmarek, L. K. (1992). Expression of the mRNAs for the Kv3.1 potassium channel gene in the adult and developing rat brain. *J. Neurophysiol.* 68, 756–766.

Pfaffinger, P. J., Furukawa, Y., Zhao, B., Dugan, D., and Kandel, E. R. (1991). Cloning and expression of an *Aplysia* K^+ channel and comparison with native *Aplysia* K^+ currents. *J. Neurosci.* 11, 918–927.

Philipson, L. H., Schaefer, K., LaMendola, J., Bell, G. I., and Steiner, D. F. (1990). Sequence of a human fetal skeletal muscle potassium channel cDNA related to RCK4. *Nucl. Acid. Res.* 18, 7160.

Philipson, L. H., Hice, R. E., Schaeffer, K., Lamendola, J., Bell, G. I., Nelson, D. J., and Steiner, D. F. (1991). Sequence and functional expression in *Xenopus* oocytes of a human insulinoma and islet cell potassium channel. *Proc. Natl. Acad. Sci. U.S.A.* 88, 53–57.

Philipson, L. H., Eddy, R. L., Shows, T. B., and Bell, G. I. (1993). Assignment of human potassium channel gene, KCNA4(Kv1.4,PCN2) to chromosome 11q13.4-q14.1. *Genomics* 15, 462–464.

Po, S., Snyders, D. J., Baker, R., Tamkun, M. M., and Bennett, P. B. (1992). Functional expression of an inactivating potassium channel cloned from human heart. *Circ. Res.* 71, 732–736.

Po, S., Roberds, S., Snyders, D. J., Tamkun, M. M., and Bennett, P. B. (1993). Heteromultimeric assembly of human potassium channels: Molecular basis of a transient outward current? *Circ. Res.* 72, 1326–1336.

Pongs, O. (1992). Molecular biology of voltage-dependent potassium channels. *Physiol. Rev.* 72, S69–S88.

Pongs, O. (1993). Shaker-related K^+ channels. *Sem. Neurosci.* 5, 93–100.

Pongs, O., Kecskemethy, N., Muller, R., Krah-Jentgens, I., Baumann, A., Kiltz, H. H., Canal, I., Llamazares, S., and Ferrsu, A. (1988). *Shaker* encodes a family of putative potassium channel proteins in the nervous system of *Drosophila. EMBO J.* 7, 1087–1096.

Ramaswami, M., Gautam, M., Kamb, A., Rudy, B., Tanouye, M. A., and Mathew, M. K. (1990). Human potassium channel genes: Molecular cloning and functional expression. *Mol. Cell. Neurosci.* 1, 214–223.

Rehm, H. and Lazdunski, M. (1988). Purification and subunit structure of a putative K^+-channel protein identified by its binding properties for dendrotoxin I. *Proc. Natl. Acad. Sci. U.S.A.* 85, 4919–4923.

Reid, P. F., Pongs, O., and Dolly, J. O. (1992). Cloning of a bovine voltage-gated K^+ channel gene utilizing partial amino acid sequence of a dendrotoxin-binding protein from brain cortex. *FEBS Lett.* 302, 31–34.

Reid, T., Rudy, B., Vega-Saenz de Miera, E., Lau, D., Ward, D. C., and Sen, K. (1993). Localization of a highly conserved human potassium channel gene (NgK2-Kv4; KCNC1) to chromosome 11p15. *Genomics* 15, 405–411.

Rettig, J., Wunder, F., Stocker, M., Lichtinghagen, R., Mastiaux, F., Beckh, S., Kues, W., Pedarzani, P., Schroter, K. H., Ruppersberg, J. P., Veh, R., and Pongs, O. (1992). Characterization of a *Shaw*-related potassium channel family in the brain. *EMBO. J.* 11, 2473–2486.

Rettig, J., Heinemann, S. H., Wunder, F., Lorra, C., Parcej, D. N., Dolly, J. O., and Pongs, O. (1994). Inactivation properties of voltage-gated K^+ channels altered by presence of beta-subunit. *Nature* 369, 289–294.

Ribera, A. B. (1990). A potassium channel gene is expressed at neural induction. *Neuron* 5, 691–701.

Ribera. A. B. and Nguyen, D.-A. (1993). Primary sensory neurons express a *Shaker*-like potassium channel gene. *J. Neurosci.* 13, 4988–4996.

Roberds, S. L. and Tamkun, M. M. (1991). Cloning and tissue-specific expression of five voltage-gated potassium channel cDNAs expressed in rat heart. *Proc. Natl. Acad. Sci. U.S.A.* 88, 1798–1802.

Robertson, B. and Owen, D. G. (1993). Pharmacology of a cloned potassium channel from mouse brain (MK-1) expressed in CHO cells: Effects of blockers and an 'inactivation peptide'. *Br. J. Pharmacol.* 109, 725–735.

Robertson, G. A., Warmke, J., and Ganetzky, B. (1993). Functional expression of the Drosophila EAG K^+ channel gene. *Biophys. J.* 64, A340.

Rosenberg, R. L. and East, J. E. (1992). Cell-free expression of functional Shaker potassium channels. *Science* 360, 166–168.

Rudy, B. (1988). Diversity and ubiquity of K^+ channels. *Neuroscience* 25, 729–749.

Rudy, B., Sen, K., Vega-Saenz de Miera, Lau, E., Ried, T., and Ward, D. (1991a). Cloning of a human cDNA expressing a high voltage activating TEA sensitive, type A K^+ channel which maps to human chromosome 1 band p21. *J. Neurosci. Res.* 29, 401–412.

Rudy, B., Kentros, C., and Vega-Saenz de Miera, E. (1991b). Families of potassium channel genes in mammals: Toward an understanding of the molecular basis of potassium channel diversity. *Mol. Cell Neurosci.* 2, 89–102.

Rudy, B., Kentros, C., Weiser, M., Fruhling, D., Serodio, P., Vega-Saenz, de Miera, Ellisman, M. H., Polock, J. A., and Baker, H. (1992a). Region-specific expression of K^+ channel gene in brain. *Proc. Natl. Acad. Sci. U.S.A.* 89, 4603–4607.

Rudy, B., Lau, D., Lin, J. W., Pollack, J., and Kentros, C. (1992b). DRK1 mRNA is induced by NGF in rat pheochromocytoma PC12 cells. *Soc. Neurosci. Abstr.* 18, 77.

Ruppersberg, J. P., Schroter, K. H., Sakmann, B., Stocker, M., Sewing, S., and Pongs, O., (1990). Heteromultimeric channels formed by rat brain potassium-channel proteins. *Nature (London)* 345, 535–537.

Ruppersberg, J. P., Stocker, M., Pongs, O., Heinemann, S. H., Frank, R., and Koenen, M. (1991a). Regulation of fast inactivation of cloned mammalian I_K (A) channels by cysteine oxidation. *Nature (London)* 352, 711–714.

Ruppersberg, J. P., Frank, R., Pongs, O., and Stocker, M. (1991b). Cloned neuronal I_K(A) channels reopen during recovery from inactivation. *Nature (London)* 353, 657–660.

Salkoff, L., Baker, K., Butler, A., Covarrubias, M., Pak, M. D., and Wei, A. (1992). An essential set of K^+ channels conserved in flies, mice and humans. *Trends Neurosci.* 15, 161–166.

Santacruz-Toloza, L., Huang, Y., John, S. A., and Papazian, D. M. (1994). Glycosylation of *shaker* potassium channel protein in insect cell culture and in Xenopus oocytes. *Biochemistry* 33, 5607–5613.

Schopa, N., McCormack, K., Tanouye, M. A., and Sigworth, F. J. (1992). The size of gating charge in wild-type and mutant *Shaker* potassium channels. *Science* 255, 1712–1715.

Schroter, K-H., Ruppersberg, J. P., Wunder, F., Rettig, J., Stocker, M., and Pongs, O. (1991). Cloning and functional expression of a TEA-sensitive A-type potassium channel from rat brain. *FEBS. Lett.* 278, 211–216.

Schmidt, R. R. and Betz, H. (1989). Cross-linking of beta-bungarotoxin to chick brain membranes. Identification of subunits of a putative voltage-gated K^+ channel. *Biochemistry* 28, 8346–8350.

Schwarz, T. L., Tempel, B. L., Papazian, D. M., Jan, Y. N., and Jan, L. Y. (1988). Multiple potassium-channel components are produced by alternative splicing at the *Shaker* locus in *Drosophila. Nature (London)* 331, 137–142.

Schwarz, T. L., Papazian, D. M., Caretto, R. C., Jan, Y. N., and Jan, L. Y. (1990). Immunological characterization of K^+ channel components from the *Shaker* locus and differential distribution of splicing variants in *Drosophila. Neuron* 2, 119–27.

Scott, V. E. S., Parcej, D. N., Keen, J. N., Findlay, J. B. C., and Dolly, J. O. (1990). α-dendrotoxin acceptor from bovine brain is a K^+ channel protein: Evidence from the N-terminal sequence of its larger subunit. *J. Biol. Chem.* 265, 20094–20097.

Scott, V. E., Rettig, J., Parcej, D. N., Keen, J. N., Findlay, J. B., Pongs, O., and Dolly, J. O. (1994). Primary structure of a beta subunit of alpha-dendrotoxin-sensitive K^+ channels from bovine brain. *Proc. Natl. Acad. Sci U.S.A.* 91, 1637–1641.

Sentenac, H., Bonneaud, N., Minet, M., Lacroute, F., Salmon, J.-M., Gaymard, F., and Grignon, C. (1992). Cloning and expression in yeast of a plant potassium ion transport system. *Science* 256, 663–665.

Shaw, G. and Kamen, R. (1986). A conserved AU sequence from the 3′ untranslated region of GM-CSF mRNA mediates selective mRNA degradation. *Cell* 46, 659–667.

Shelton, P. A., Davies, N. W., Antoniou, M., Grosveld, F., Needham, M., Hollis, M., Brammar, W. J., and Conley, E. C. (1993). Regulated expression of K^+ channel genes in electrically silent mammalian cells by linkage to beta-globin gene-activation elements. *Receptors Ion Channels* 1, 25–37.

Shen, N. V., Chen, X., Boyer, M. M., and Pfaffinger, P. J. (1993). Deletion analysis of K^+ channel assembly. *Neuron* 11, 67–76.

Sheng, M., Tsaur, M.-L., Jan, Y. N., and Jan, L. Y. (1992). Subcellular segregation of two A-type K^+ channel proteins in rat central neurons. *Neuron* 9, 271–284.

Sheng, M., Liao, Y. J., Jan, Y. N., and Jan, L. Y. (1993). Presynaptic A-current based on heteromultimeric K^+ channels detected *in vivo. Nature (London)* 365, 72–75.

Sheng, Z.-H., Barchi, R. L., and Kallen, R. G. (1992). Transcriptional regulation of the tetrodotoxin-insensitive rat skeletal muscle sodium channel subtype 2 (rSkM2): Positive and negative cis-acting promoter elements. *Biophys. J.* 61, A108.

Solc, C. K., Zagotta, W. N., and Aldrich, R. W. (1987). Single channel and genetic analyses reveal two distinct A-type potassium channels in *Drosophila. Science* 236, 1094–1098.

Spencer, R. H., Chandy K. G., and Gutman, G. A. (1993). Immunological and molecular characterization of the Kv1.3 K^+ channel protein in lymphocytes and a related protein in yeast. *Biochem. Biophys. Res. Commun.* 191, 201–206.

Stansfield, C. E., Marsh, S. J., Parcej, D. N., Dolly, J. O., and Brown, D. A. (1987). Mast cell degranulation peptide and dendrotoxin selectively inhibit a fast-activating potassium current and bind to common neuronal proteins. *Neuroscience* 23, 893–902.

Stocker, M., Stuhmer, W., Wittka, R., Wang, S., Muller, R., Ferrus, A., and Pongs, O. (1990). Alternate *Shaker* transcripts express either rapidly inactivating or noninactivating K^+ channels. *Proc. Natl. Acad. Sci. U.S.A.* 87, 8903–8907.

Strong, M., Chandy, K. G., and Gutman, G. A. (1993). Molecular evolution of voltage-sensitive ion channel genes: On the origins of electrical excitability. *J. Mol. Biol. Evol.* 10, 221–242.

Stuhmer, W., Stocker, M., Sakmann, B., Seeburg, P., Baumann, A., Grupe, A., and Pongs, O. (1988). Potassium channel expressed from rat brain cDNA have delayed rectifier properties. *FEBS Lett.* 242, 199–206.

Stuhmer, W., Ruppersberg, J. P., Schroter, K. H., Sakmann, B., Stocker, M., Glese, K. P., Perschke, A., Baumann, A., and Pongs, O. (1989). Molecular basis of functional diversity of voltage-gated potassium channels in mammalian brain. *EMBO. J.* 8, 3235–3244.

Swanson, R. A., Marshall, J., Smith, J. S., Williams, J. B., Boyle, M. B., Folander, K., Luneau, C. J., Antanavage, J., Oliva, C., Buhrow, S. A., Bennet, C., Stein, R. B., and Kaczmarek, L. K. (1990). Cloning and expression of cDNA and genomic clones encoding three delayed rectifier potassium channels in brain. *Neuron* 4, 929–939.

Swanson R., Hice, R. E., Folander, K., and Sanguinetti, M. C. (1993). The I_{sK} protein, a slowly activating voltage-dependent K^+ channel. *Sem. Neurosci.* 5, 117–124.

Swofford, D. L. (1993). PAUP: Phylogenetic analysis using parsimony, Version 3.1. Computer program distributed by the Illinois Natural History Survey, Champaign, Illinois.

Taglialatela, M., VanDongen, A. M. J., Drewe, J. A., Joho, R. H., Brown, A. M., and Kirsch, G. E. (1991). Patterns of internal and external tetraethylammonium block in four homologous K^+ channels. *Mol. Pharmacol.* 40, 299–307.

Takimoto, K., Fomina, A. F., Gealy, R., Trimmer, J. S., and Levitan, E. S. (1993). Dexamethasone rapidly induces Kv1.5 K^+ channel gene transcription and expression in clonal pituitary cells. *Neuron* 11, 359–369.

Takumi, T., Ohkubo, H., and Nakanishi, S. (1988). Cloning of a membrane protein that induces a slow voltage-gated potassium current. *Science* 242, 1042–1045.

Tamkun, M. M., Knoth, K. M., Walbridge, J. A., Kroemer, H., Roden, D. M., and Glover, D. M. (1991). Molecular cloning and characterization of two voltage-gated K^+ channel cDNAs from human ventricle. *FASEB J.* 5, 331–337.

Tempel, B. L., Jan, Y. N., and Jan, L. Y. (1988). Cloning of a probable potassium channel gene from mouse brain. *Nature (London)* 332, 837–839.

Thibodeau, S. N., Bren, G., and Schaid, D. (1993). Microsatellite instability in cancer of the proximal colon. *Science* 260, 816–9.

Timpe, L. C., Jan, Y. C., and Jan, L. Y. (1988). Four cDNA clones from the *Shaker* locus of Drosophila induce kinetically distinct A-type potassium currents in Xenopus oocytes. *Neuron* 1, 659–667.

Trimmer, J. S. (1991). Immunological identification and characterization of a delayed rectifier K^+ channel polypeptide in rat brain. *Proc. Natl. Acad. Sci. U.S.A.* 88, 10764–10768.

Tsaur, M.-L., Sheng, M., Lowenstein, D. H., Jan, Y. N., and Jan, L. Y. (1992). Differential expression of K^+ channel mRNAs in the rat brain and down regulation in the hippocampus following seizures. *Neuron* 8, 1055–1067.

Tseng, G.-N. and Tseng-Crank, J. (1992). Differential effects of elevating $[K]_o$ on three transient outward potassium channels. Dependence on channel inactivation mechanisms. *Circ. Res.* 71, 657–672.

Tseng-Crank, J. C. L., Tseng, G.-N., Schwartz, A., and Tanouye, M. A. (1990). Molecular cloning and functional expression of a potassium channel cDNA isolated from a rat cardiac library. *FEBS. Lett.* 268, 63–68.

Ukomadu, C., Zhou, J., Sigworth, F. J., and Agnew, W. S. (1992) ξI Na^+ channels expressed transiently in human embryonic kidney cells: Biochemical and biophysical properties. *Neuron* 8, 663–676.

VanDongen, A. M. J., Frech, G. C., Drewe, J. A., Joho, R. H., and Brown, A. M. (1990). Alteration and deletion of K^+ channel function by deletions at the N- and C-termini. *Neuron* 5, 433–443.

Vega-Saenz de Miera, E., Chiu, N., Sen, K., Lau, D., Lin, J. W., and Rudy, B. (1991). Towards an understanding of the molecular composition of K^+ channels: Products of at least nine distinct Shaker family K^+ channel genes are expressed in a single cell. *Biophys. J.* 59, 197a.

Vega-Saenz de Miera, E., Moreno, H., Fruhling, D., Kentros, C., and Rudy, B. (1992). Cloning of ShIII (Shaw-like) cDNAs encoding a novel high-voltage, TEA-sensitive, type A K+ channel. *Proc. R. Soc. London Ser. B.* 248, 9–18.

Wang, H., Kunkel, D. D., Martin, T. M., Schwartzcroin, P. A., and Tempel, B. L. (1993). *Nature* 365, 75–79.

Wang S.-Y., Castle, N. A., and Wang, G. K. (1992). Identification of RBK1 potassium channels in C6 astrocytoma cells. *Glia* 5, 146–153.

Warmke, J. and Ganetzky, B. (1993). A novel potassium channel gene family: *Eag* homologs in *Drosophila,* mouse and humans. *Biophys. J.* 64, A341.

Warmke, J., Drysdale, R., and Ganetzky, B. (1991). A distinct potassium channel polypeptide encoded by the *Drosophila eag* locus. *Science* 252, 1560–1562.

Wei, A. M., Covarrubias, M., Butler, A., Baker, K., Pak, M., and Salkoff, L. (1990). K+ current diversity is produced by an extended gene family conserved in *Drosophila* and mouse. *Science* 248, 599–603.

Wei, A., Jegla, T., and Salkoff, L. (1991). A *C. elegans* potassium channel gene with homology to *Drosophila. Soc. Neurosci. Abstr.* 17, 1281.

Werkman, T. R., Kawamura, T., Yokoyama, S., Higashida, H., and Rogawski, M. A. (1992). Charybdotoxin, dendrotoxin and mast cell degranulating peptide block the voltage-activated K+ current of fibroblast cells stably transfected with NgK1 (Kv1.2) K+ channel complementary DNA. *Neuroscience* 50, 935–946.

Wittka, R., Stocker, M., Boheim, G., and Pongs, O. (1991). Molecular basis for different rates of recovery from inactivation in the *Shaker* potassium channel family. *FEBS Lett.* 286, 193–200.

Wymore, R., Korenberg, J. R., Coyne, C., Chen, X.-N., Hustad, C., Copeland, N. G., Gutman, G. A., Jenkins, N. A., and Chandy, K. G. (1994) Genomic organization, nucleotide sequence and localization of the voltage-gated K+ channel gene, KCNA4/Kv1.4 to mouse chromosome 2/human 11p14, and mapping of KCNC1/Kv3.1 and KCNA1/Kv1.1 to mouse 7/human 11p14.3-15.2 and human 12p13, respectively. *Genomics,* in press.

Yao, J.-A. and Tseng, G.-N. (1993). Removing inactivation of an A-type channel (RHK1) potentiates block by 4-aminopyridine (4-AP). *Biophys. J.* 64, A313.

Yellen, G., Jurman, M. E., Abramson, T., and MacKinnon, R. (1991). Mutations affecting internal TEA blockade identify the probable pore-forming region of a K+ channel. *Science* 251, 939–942.

Yokoyama, S., Imoto, K., Kawamura, T., Higashida, H., Iwabe, N., Miyata, T., and Numa, S. (1989). Potassium channels from NG108-15 neuroblastoma glioma hybrid cells: Primary structure and functional expression from cDNAs. *FEBS. Lett.* 259, 37–43.

Yool, A. J. and Schwarz, T. L. (1991). Alteration of ionic selectivity of a K+ channel by mutation of the H5 region. *Nature (London)* 349, 700–704.

Zagotta, W. N., Hoshi, T., and Aldrich, R. W. (1990). Restoration of inactivation in mutants of *Shaker* potassium channels by a peptide derived from ShB. *Science* 250, 568–571.

Zuhlke, R. D., Zhang, J.-J., and Joho, R. H. (1993). Conserved cysteine residues in S2 and S6 are not essential for expression of Kv2.1 (DRK1) in Xenopus oocytes. *Biophys. J.* 64, A113.

2

Voltage-Gated Sodium Channels

Alan L. Goldin

2.2.0 Introduction

Voltage-gated sodium channels are responsible for the initial depolarization event during impulse conduction in most electrically excitable cells. Membrane depolarization activates the channel, causing a voltage-dependent conformational change that increases the permeability to sodium ions. This is followed by inactivation, in which the channel closes and the permeability to sodium ions returns to the resting level. When the channel opens, it is highly permeable to sodium and lithium ions, and much less so to all other monovalent cations. These three properties define the essential functional features of the voltage-sensitive sodium channel: voltage-dependent activation, inactivation, and selective ion permeability.

The proteins that comprise the sodium channel have been purified from a variety of tissues and species, and cDNA clones encoding the proteins have since been isolated. All voltage-gated sodium channels contain a highly processed α-subunit that is approximately 260 kDa. In addition to the α-subunit, sodium channels from some tissues, such as rat brain and muscle, contain accessory β-subunits (Hartshorne and Catterall, 1981, 1984; Casadei et al., 1986). These subunits are not essential for function, as demonstrated by the synthesis of functional sodium channels following injection into *Xenopus* oocytes of only the α-subunit (Noda et al., 1986b; Auld et al., 1988; Joho et al., 1990; Suzuki et al., 1988; Trimmer et al., 1989). Based on the predicted amino acid sequence, the α-subunit of the sodium channel consists of four homologous domains termed I–IV, as depicted in Figure 1. Each of these domains is 40 to 60% identical to each of the others at the amino acid level. From hydrophobicity profiles there are five hydrophobic segments of approximately 20 amino acids each at equivalent positions in the four domains (S1, S2, S3, S5, and S6). In addition, there is an amphipathic segment of

0-8493-8322-6/95/$0.00+$.50

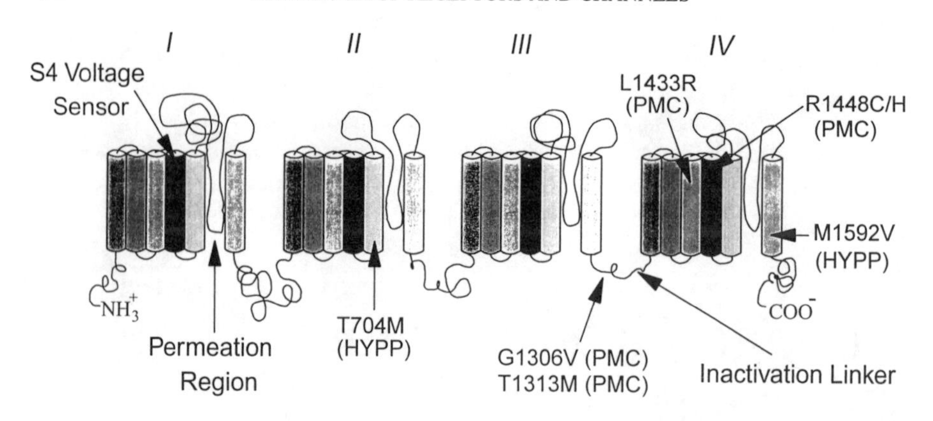

FIGURE 1. Diagrammatic representation of the sodium channel α-subunit. The α-subunit consists of four homologous domains (I–IV), each of which contains 8 putative transmembrane spanning segments. The six segments that are predicted to be α-helical are termed S1–S6, and the region between S5 and S6 is predicted to span the membrane in a more extended conformation as the actual pore of the channel. The S4 segments in each domain have been shown to comprise at least part of the voltage sensor. The permeation region and S4 voltage sensor are indicated only for the first domain, but the comparable regions from all four domains are most likely involved in each property. The cytoplasmic linker connecting domains III and IV is critical for fast sodium channel inactivation. Mutations in the human skeletal muscle sodium channel gene (hNaSk1) that result in either hyperkalemic periodic paralysis (HYPP) or paramyotonia congenita (PMC) are indicated. A more complete description of the mutations, including full citations, is given in section 2.2.5 of the text.

approximately the same size with positively charged residues at every third position (S4). The secondary structure is most often represented with S1 through S6 as membrane-spanning α-helical domains (Catterall, 1988). The region between S5 and S6 most likely spans the membrane as the actual pore of the channel, based on results obtained with the potassium channel (Yool and Schwarz, 1991; Hartmann et al., 1991; Yellen et al., 1991). The linker regions connecting the four homologous domains are all located on the cytoplasmic side of the membrane (Gordon et al., 1987, 1988; Rossie et al., 1987). The voltage-gated calcium channel appears to have a comparable structure to that of the sodium channel, and the voltage-gated potassium channel looks like one of the four sodium channel domains (reviewed by Jan and Jan, 1989). Since the potassium channel functions as a tetramer (MacKinnon, 1991), the three voltage-gated cation selective channels are structurally homologous. A detailed model of the predicted structure of the voltage-gated potassium channel, which is also applicable to the sodium channel, has been proposed by Durell and Guy (1992).

Although voltage-gated sodium channels demonstrate a range of electrophysiological properties, they do not show the extreme diversity of function that is characteristic of calcium and potassium channels. However, sodium channels from different tissues can be distinguished by their sensitivities to various toxins and by some functional differences (Catterall, 1980; Mandel, 1992). For these reasons the voltage-gated sodium channels have been classified on the basis of species and tissue source. In addition, each species or tissue has its unique advantages for experimental manipulation, so that different types of information have been obtained about the channels from different sources. This review will present the various sodium channels categorized by species and tissue distribution, concentrating on the unique information obtained about the sodium channels from each source.

2.2.1 Sodium Channel α-Subunits

2.2.1.1 Electric Eel

One of the first tissues from which sodium channel proteins were purified was the electroplax of the electric eel, *Electrophorus electricus* (Agnew et al., 1978; Nakayama et al., 1982; Miller et al., 1983; Norman et al., 1983). The primary channel component, as assayed by tetrodotoxin binding, is a high-molecular weight peptide of about 260 kDa (Agnew et al., 1980), containing approximately 29% carbohydrate by mass (Miller et al., 1983; James and Agnew, 1987). This size estimate is consistent with the results of irradiation inactivation experiments (Levinson and Ellory, 1973). The high-mol wt peptide was shown to be sufficient for sodium channel activity by reconstitution into planar lipid bilayers (Recio-Pinto et al., 1987) or liposomes (Rosenberg et al., 1984b), in which the channels could be activated by either batrachotoxin treatment (Rosenberg et al., 1984a; Tomiko et al., 1986; Recio-Pinto et al., 1987) or voltage depolarization (Shenkel et al., 1989; Correa et al., 1990). The carbohydrate was shown to affect the electrophysiological properties of the channel, in that removal of sialic acid residues with neuraminidase shifted the voltage required for channel opening in the depolarizing direction (Recio-Pinto et al., 1990). The eel sodium channel is phosphorylated by protein kinase A in the amino- and carboxy-terminal regions, but it is not phosphorylated by protein kinase C (Emerick and Agnew, 1989).

Noda et al. (1984) obtained cDNA clones encoding the electric eel sodium channel α-subunit by combining immunological screening with amino acid sequencing of the purified protein (Table 1). The predicted protein consists of 1,820 amino acids, and could be divided into four homologous domains. The eel α-subunit will be referred to as eNa1 (eel sodium channel 1) in this review (Table 1). The eNa1 cDNA clone was the first one isolated for any voltage-gated cation channel, and the predicted sequence was used to construct a variety of models predicting the channel structure (Noda et al., 1984; Catterall, 1986; Greenblatt et al., 1985; Guy and Seetharamulu, 1986; Kosower, 1985; Sawaryn and Drouin, 1991). Unfortunately, injection into *Xenopus* oocytes of *in vitro* transcribed RNA from eNa1 did not result in functional sodium channels. It has since been shown that the protein is incompletely processed in *Xenopus* oocytes, so that it lacks carbohydrate and

Table 1. Sodium Channel α-Subunit Clones

Name	Original name	Species	Tissue	Size	GenBank number	References
eNa1	Eel Na channel	*Electrophorus electricus*	Electroplax	1820	X01119	Noda et al. (1984)
sNa1	Squid Na channel	*Loligo bleekeri*	Axon	1522	D14525	Sato and Matsumoto (1992)
sNa2	GFLN1	*Loligo opalescens*	Giant fiber lobe	1784	L19979	Rosenthal and Gilly (1993)
jNa1	CYNA1	*Cyanea capillata*	Neuronal	1739	L15445	Anderson et al. (1993)
dNa1	DSC1	*Drosophila melanogaster*	Genomic	Partial	M32078-80	Salkoff et al. (1987)
dNa2	para	*Drosophila melanogaster*	Genomic	1820	X14394-8	Loughney et al. (1989) Ramaswami and Tanouye (1989)
rNaB1	Rat I	*Rattus norvegicus*	Brain	2009	X03638	Noda et al. (1986a)
rNaB2	Rat II	*Rattus norvegicus*	Brain	2005	X03639	Noda et al. (1986a)
rNaB3	Rat III	*Rattus norvegicus*	Brain	1951	Y00766	Kayano et al. (1988) Joho et al. (1990)
rNaB2A	Rat IIA	*Rattus norvegicus*	Brain	2005	X61149	Auld et al. (1988)
rNaSk1	SkM1,μ1	*Rattus norvegicus*	Skeletal muscle	1840	M26643	Trimmer et al. (1989)
rNaSk2 r4NaH1	SkM2 RH1	*Rattus norvegicus*	Denervated skeletal muscle, heart	2018	M27902	Kallen et al. (1990) Rogart et al. (1989)
rNaG1	Na-G	*Rattus norvegicus*	Astrocyte	Partial	M96578	Gautron et al. (1992)
hNaB1	HBSCI	*Homo sapiens*	Brain	Partial	X65362	Lu et al. (1992)
hNaB2	HBSCII HBA	*Homo sapiens*	Brain	2005	X65361 M94055	Lu et al. (1992) Ahmed et al. (1992)
hNaSk1	SkM1	*Homo sapiens*	Skeletal muscle	1836	M81758	George et al. (1992b) Wang et al. (1992)
hNaH1	hH1	*Homo sapiens*	Heart	2016	M77235	Gellens et al. (1992)
hNaH2	$hNa_V2.1$	*Homo sapiens*	Heart	1682	M91556	George et al. (1992a)

hydrophobic domains, which might account for the absence of functional channels (Thornhill and Levinson, 1986, 1987). Because of the lack of functional expression, the eel sodium channel has not been studied as extensively as many of the other sodium channels for which cDNA clones have since been isolated.

2.2.1.2 Squid

The squid giant axon has been the primary focus of electrophysiological studies on ion channel function. Hodgkin and Huxley (1952) used the giant axon of the squid *Loligo* to demonstrate that action potentials are due to increases in membrane permeabilities to sodium and potassium, and many other investigators have since carefully examined the electrophysiological properties of sodium channels in the squid giant axon (reviewed by Armstrong, 1975; Bezanilla, 1985; Patlak, 1991; Catterall, 1992). The proteins comprising the squid channel have not been purified, but cDNA clones encoding α-subunits from two different species of squid have been isolated. Sato and Matsumoto (1992) used the polymerase chain reaction to isolate a cDNA clone encoding the α-subunit from *Loligo bleekeri*. This squid α-subunit will be referred to as sNa1 in this review (Table 1). Rosenthal and Gilly (1993) used the polymerase chain reaction to isolate a cDNA clone encoding the α-subunit from the giant fiber lobes of *Loligo opalescens*. This channel was called GFLN1, and it will be referred to as sNa2 in this review (Table 1). The predicted sNa1 protein is 1522 amino acids in length, and the predicted sNa2 protein is 1784 amino acids. Both proteins are structurally similar to the electric eel α-subunit, but the sNa2 sequence demonstrates greater identity to the vertebrate sodium channels than to the sNa1 channel (Figure 2). The two squid channels share 32% identity, whereas sNa2 is 40% identical to the rat skeletal muscle rNaSk1 channel, 39% identical to the rat brain rNaB3 channel, and 35% identical to the rat brain rNaB1 and rNaB2 channels. Neither squid channel has been expressed in an exogenous system yet, so the properties of the channels encoded by these cDNA clones cannot be compared to those observed *in vivo*. As the most detailed electrophysiological data have been obtained for the squid giant axon sodium channel *in vivo*, it should be informative to make these comparisons once sNa1 or sNa2 has been expressed.

2.2.1.3 Jellyfish

The jellyfish *Cyanea capillata* is a member of the phylum Cnidaria, whose members are the lowest extant organism that possess a nervous system. They are the first organisms that contain cells producing action potentials resulting solely from sodium ion currents, and the sodium channels that have been studied are all relatively resistant to tetrodotoxin. Anderson et al. (1993) used the polymerase chain reaction to isolate a cDNA clone encoding a sodium channel α-subunit from the perirhopalial tissue of *Cyanea.* This sodium channel was termed CYNA1, and it will be referred to as jNa1 in this review (Table 1). The predicted jNa1 protein is 1739 amino acids in length, and it is quite similar to the α-subunits obtained from other species (Figure 2). If only the putative transmembrane segments are considered, the jNa1 channel sequence is 55% identical to the rat brain rNaB2 sequence and 51% identical to the squid Na1 and eel Na1 sequences. However, phylogenetic comparison of the entire sequences demonstrated that jNa1 is most closely related to sNa1, with sNa2 not included in that comparison (Anderson et al., 1993). As with the squid channels, jNa1 has not yet been expressed in an exogenous system, so that the functional significance of sequence differences cannot be evaluated.

ALIGNMENT OF SODIUM CHANNEL α SUBUNIT AMINO ACID SEQUENCES

```
hNaB2   MAQSVL VPPGPDSFRFFTRESLAAIEQRIAEEKAK RPKQER KDEDDENGPKPNSDLEAGKSLPFIYGDI PPEMVSVPLEDLDPYYINKK TFIVLNK GKAISRFSATPALYILTPFNPIRKL
rNaB2   --R--- ------------------------------- ------ ---------------------------- ------E------------- ------- ----------S-------------
hNaB1
rNaB1   -E-T-- --------N-----------R-------- N--PDK -- ---------------N-------- ------E------------- ------- ----F-----S----------L--I
rNaB3   ---AL- -----E---L-----------K-A------ K--K-Q -I----K-----------N-------- ------E---------VS-- --V---- ----F-----S-------L--V--I
hNaSk1  --RPS-CTLARL--ECL-P-------------AV--E-R LQRNKQ MEIE -PER--R-------N--M----P.---VIGI---------S--- ------- ----F---------L-S--SVV-RG
rNaSk1  --S-S-PNL-----HCL-P--P----------AV--E-R LQRNKQ MEIE -PER--R-------N--L----P ---VIGI---------SD-- ------- ----F---------L-S--SIV-RV
hNaH1   --NF LL-R-TS---R-----------K-M--KQ-R G STTLQES- EGLPE-EA-R-QL--Q-S-K--DL--NP -Q-LIGE-------F-STQ- ------- --T-F-----N---V-S--H-V-RA
rNaH1   --NL LL-R-TS---R-----------K-M--KQ-R GGSATSQES- EGLQE-EA-R-QL--Q-S-K--DL--NP -R-LIGE-------F-STQ- ------- --T-F-----N---V-S--H-V-RA
hNaH2   - LAS-E-KGLVP--K--FEL-K-H-- KTH NEDHE-EDL---P---V--K------NL SQG---E----V----YK--N------- NRT-F--N-ASI-CT-S---C--RT
rNaG1
sNa2    --TAKDLEKGDNPKGR--L----T-FE--R---- RRAQAEDADSDE-EN-PEL---PKF----G--PSVE-V -SRYQGR------E--H-N- --C-IGR DRTLY-----K-IFL-S------RF
dNa2    -TEDSD SISEEERSL--P------VQ------A-HE-QKELERKRAEGEVPRYGRKKKQ-EI-YD---EDE--Q-DPT--Q-VPI-VRLQGSF---LA-T----I----S-VL --V-VS- --D-F----SK-MWM-D------RV
eNa1    --R KFSSAR-EM--R--PD--EE-- AFTEL-KSC TL-KK-PEST-RI------P--M----P -EDLLNI-------F-KTQ- ----IS- -NI-N--N-ER----FS------RG
jNa1    -FYH-ARKSHSADAE -GG-ELVD-M--LAEVLTMVDIAERKGVELR VRRE-QADILKMA-TR-EAPSTKRR-KKNYVSTCTL-QS-LGR--K-F-S EQEN--V-V-NRLGKQTVY---TDKS--LMGTN-LA--V
sNa1    - DEKYTAKNRDKTFV -IE-RFKKNI-H----KRS-FLF--R----R-
dNa1    LETTLAVSNWSTAILS QTAPCIKRAFFK -QTVC -VS-RFRKNY-H--TG-KS-FLFY-WS-A-RV
                                                                                                                      | 150

            ----IS1----           ----IS2----                ----IS3----                ----IS4----            ----IS5----
hNaB2   AIKLVHSLFNMLIMCTILTNCVFMTMSNPPDWTKNVEYTFTGIYTFESLIKILARGFCLEDFTFLRDPWNWLDFTVITFAYVTEFVD LGNVSALRTFRVLRALKTISVIPGLKTIVGALIQSVKKLSDVMILTVFCLSVFALI
rNaB2   -----------V----------------------------------------------------N------------------N ----------------------------------------------------
hNaB1
rNaB1   ----------S-----------------------------------I------------------------------------ ---------------------------------------------------
rNaB3   ----------S---------------L-------------------------------------------S--VM-------- ---------------------------------------------------
hNaSk1  ---V-I-A--S-F--I--------------D--P-S-------------------VD--------------S--MM--L----- ---I---------------T--------------------------------V
rNaSk1  ---V-I-A--S-F--I---------------S-S-H--------------M------ID--------------S---M------- ---I---------------T--------------------------------V
hNaH1   -V-----------------------AQHD--P---Y------A-------V-----A---HA---------------S--IM--T----- -------------------------S-----------------A---V---------
rNaH1   -V--------S----------SL--AQHD--P---Y------A-------V---------HA---------------S--VM--T----- -------------------------S-----------------A---V---------
hNaH2   T--V---PF-QLF-LISV-ID----SLT-L-K-RPVL-N-LL------I-V-LF---VWAGS-S--G---------S-TV-EVIIRYSP -DFIPT-Q-A-T--I--I-PLNQ---SL--V--HCL-Q-IG-I---L-F--I-S--
rNaG1
sNa2    --F----PI-SLVVIL--VC-------KEN- LPVS-WI------C-A---LF----I--P--Y---A------V--GL--L--V-- ---L---------------VA-------------IEA-RR-R------G-M--I----
dNa2    --Y----P--SLF-IT---V--IL-I-PTT-T VEST-VI----------AV-VM----I-CP--Y---A------V--AL----MGI- ---LA--------------VAIV----------V-E---N-R--I---M-S------M
eNa1    --RVF-N-A--FF--F--FS--I---I----A-S-I-------------VIV-V-S----IGH--------------S-V-MT-I---I- -R-------------------TIF-------R---E-M-QMG--V-----S-A--T-A
jNa1    FVR-VTNQI-EIF-LL--I------AL-D-- -ES--V-AA-----VIV--S-K--I-HKYSY--NS------F-VIVG---MIPE MNKY-GVK-------------TVK---AM-NT-LK-MRMMT--L---L-FIAI----
sNa1    -VC-ATNVC-DYFL-F--MI----LA-PDIS EFA--I-L----M-MA--LV-G--FIDKY-Y---A--C------MIS-I-LLLQTINDKVISDITG-----------R-L-I------M-N--LRALRM-IS-L--IL---WI-SQA
dNa1    CVY-ATNQF-DYCV-A---F--I-LA-TETV EEA--I-LA--SI-MV---I-K--L-NKY-Y--N-------V---MR-A-IGME V--LAG-------------V-IM------IRP-LH--AQ-AE--T--I--VM---F
                                                                                                                      | 300

                                                                                                              ----ISS1----  SS2
hNaB2   GLQLFMGNLRNKCLQWPPDNSSFEI NITSFFNNSL DGNGTTFNRTVSIFNWDE YIEDKSHFYFLEGQNDALLCGNSSDAGQCPEGYICVKA GRNPNYGYTSFDTFSWAFLSLFRLMTQDFWENLY
rNaB2   ---------------------T--- ---------- -W---A-----NM----- ------------------------------------ ---------------------------------
hNaB1
rNaB1   -------------V----T-A-L-E HSIEK -VTT -Y---LV-E--FE-D-KS --Q-SRYH-----VL-----------------M---- ---------------------------------
rNaB3   -------------S----SD-A--T -T--Y--GTM -S---FV-V-M-T---KD --A-D----V-D--K-P-----G--------------- -----------------------------Y----
hNaSk1  ----------Q--VR---PFNDINTTWYSN DTWYGNDTWYG-EMWYGNDSWYANDTW-SHASWA-NDT-D--A --S-EGN------S-------------H--K--E-I-T ----------Y--------A--------Y----F
rNaSk1  ----------Q--VR---PMNDINTTWYGN DTWYSNDTWYG- DTWYINDTW-SQESWAGN-T-D-EA --N-EGN------S-------------H-----E-I-- ----------Y--------A--------Y----F
hNaH1   ----------H--VR N- TAL---NGSVEADGLV-ESLDL-LS-PENYLLKN-TS-V----------T-----R-L-- -E--DH------S-A----A--------C--R--
rNaH1   ----------H--VR N- TEL---NGSVEADGLV-NSLDV-LN-PANYLLKN-TT-V----------T-----R-L-- -E--DH------S-A----A------H-C--R--
hNaH2   -MG------KH--FR--QE-E -E-LH---GNPY --RETEN--Y---ERY------RT---------V---- -I--DQ-F-N--S-G--LFA-----A--YP-V--
rNaG1
sNa2    -M--YQ-A--S--VR -ND ENMTD-EY QK-VSI-ANWQENFYGSTL V---A-GP-K-G-N-T-LPEIMD---F------N-A--M-CA-------Y-----
dNa2    ---IY--V-TE--IK KFPLDGSW GNLTDENW DYHNRNS-NW-SEDEGISFP----I-G---DDD-V-LQGF-P----------S-G----A----------D--
eNa1    -M--------H--IR--IS-VTLD YES A Y-T-FD-TA ---NEENQ---D-AL-------N---K-----T-M-- ---------NY-N-A-T--C-----L--Y-----
jNa1    -----I-S--SR-VPTDRNI-V-NKAFYT NS QHW YHYNDEEVI---A-T-WN--PN-L-MPGA-S--KN--V-Y-N-LL-M-TSLQVC-L-Y--SV-
sNa1    -V---G-A--H--VLQIHGSPA-GKTYDEFYAEHIENS DNW LAK-NGEYV----ATG--P--TN-T-LPDI-E-------N--SIG-SM-IS-Q-L---Y--DV-
dNa1    A--VY--E-----VRQV PTDWTNVSHTDWQIWV-D- DNW LYDEDELPV----LTG-RH--FE-V-LC V-E---H---N--N-M-SM-TT-Q-I-L-Y---V-
                                                                                                                      | 450
```

FIGURE 2. Alignment of the sodium channel α-subunit amino acid sequences. The initial alignment was generated using the program CLUSTALV (Higgins et al., 1992), after which the alignment was manually edited. Dashes indicate identity with the sequence at the top of the alignment (hNaB2). Above the sequence are indicated the positions of each of the putative membrane-spanning segments in the four domains (S1–S6), as defined by Noda and Numa (1987). Also indicated are the short segments in each domain (SS1 and SS2) that are thought to form part of the channel pore, as defined by Guy and Conti (1990) and modified by Terlau et al. (1991). Vertical bars below the sequences mark the position of every tenth residue.

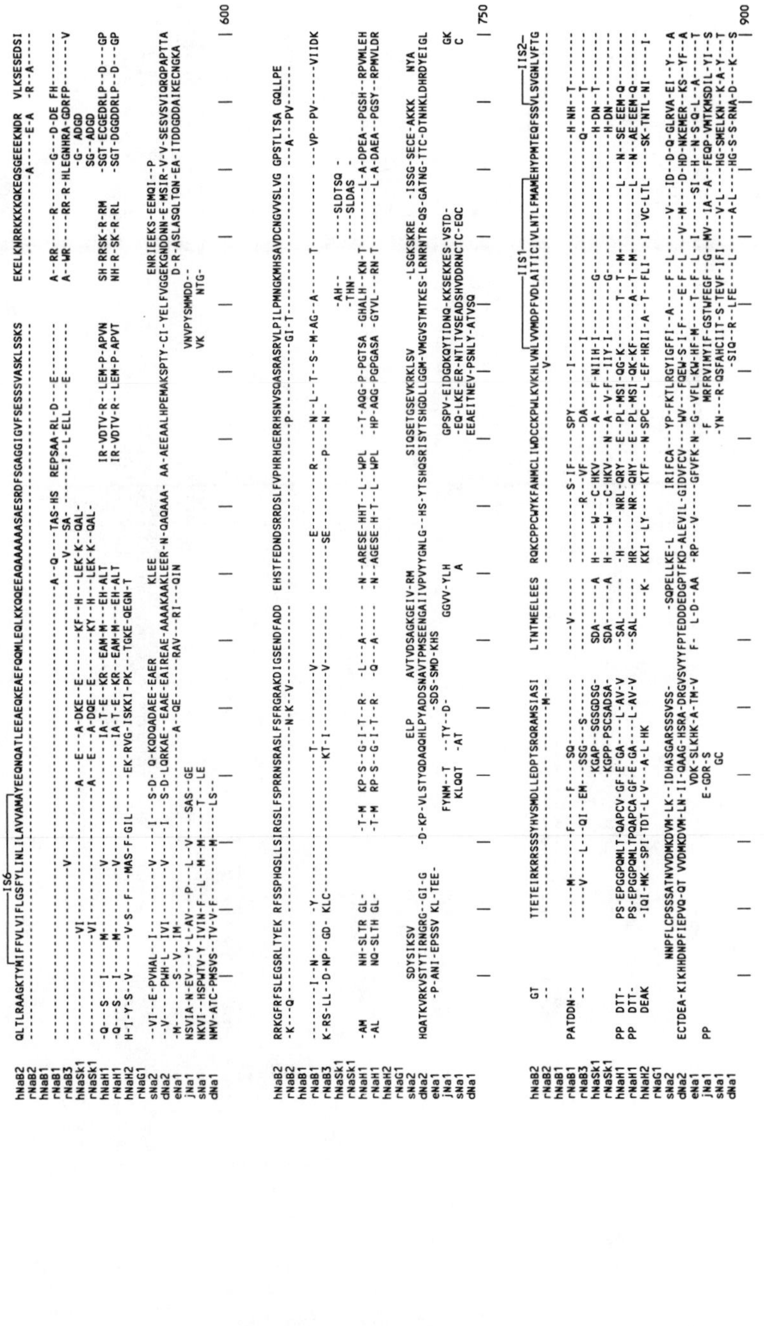

```
                    ┌──────────IS6──────────┐
hNaB2   QLTLRAAGKTYMIFFVLVIFLGSFYLINLILAVVAMAYEEQNQATLEEAEQKEAEFQQMLEQLKKQQEEAQAAAAAASAESRDFSGAGGIGVFSESSSVASKLSSKS      EKELKNRRKKKKQKEQSGEEEKNDR  VLKSESEDSI
rNaB2   ------------------------------------------------------------------------------------------------------------      ----------------A-----E-A  -R--A-----
hNaB1
rNaB1   -------------------------------------------------------------------A--Q----TAS-HS  REPSAA-RL-D---E--------      A--RR-----R-------G---D-DE FH--------
rNaB3   --------------------------V---------------------------------------------V---SA- ------I--L-ELL----E--------      A--WR-----RR-R-HLEGNHRA-GDRFP-------V
hNaSk1  ----------------VI-------------------A---E---A-DKE--E-------KF--H---LEK-K--QAL-                                            -G- ADGD
rNaSk1  ----------------VI-------------------A---E---A-DQE--E-------KY--H---LEK-K--QAL-                                            SG--ADGD
hNaH1   -Q---S---I-----M----------V-------------------IA-T-E--KR--EAM-M---EH-ALT               IR-VDTV-R--LEM-P-APVN      SH-RRSK-R-RM    -SGT-ECGEDRLP--D---GP
rNaH1   -Q---S---I-----M----------V-------------------IA-T-E--KR--EAM-M---EH-ALT               IR-VDTV-R--LEM-P-APVT      NH-R-SK-R-RL    -SGT-DGGDDRLP--D---GP
hNaH2   H-I-Y-S--V------V-S--F---MAS-F-GIL------EK-RVG-ISKKI-PK---TGKE-QEGN-T
rNaG1
sNa2    --VI--E-PVHAL--I----------V-----I---S-D- Q-KQDQADAEE-EAER            KLEE                                         ENRIEEKS-EEMQI--P
dNa2    --V-----PWH-L--IVI--------V-----I---S-D-LQRKAE--EAAE-EAIREAE-AAAAKAAKLEER-N-QAQAAA- AA-AEEAALHPEMAKSPTY-CI-YELFVGGEKGNDDNN-E-MSIR-V-V-SESVSVIQRQPAPTTA
eNa1    -M-------S--V--IM-----------------------------A--QE------RAV---RI---QIN                                          D-R-ASLASQLTQN-EA-ITDDGDDAIKECNGKA
jNa1    NSVIA-N-EV---Y-L-AV---P---L--V----SAS--GE                                                            VNVPYSMMDD--
sNa1    NKVI--HSPWTV-Y-IVIN-F--L--M--M-----T---LE                                                            VK       NTG-
dNa1    NMV-ATC-PMSVS--TV-V-F--------M-----LS--
                 |         |         |         |         |         |         |         |         |         |         |         |         |         |         | 600

hNaB2   RRKGFRFSLEGSRLTYEK RFSSPHQSLLSIRGSLFSPRRNSRASLFSFRGRAKDIGSENDFADD  EHSTFEDNDSRRDSLFVPHRHGERRHSNVSQASRASRVLPILPMNGKMHSAVDCNGVVSLVG GPSTLTSA GQLLPE
rNaB2   -K---Q------------ ----------------------------N-K--V-----------  -------------------------P-----------GI-T-------------------- ---A---PV------
hNaB1
rNaB1   --------I--N------ -Y----------------------T-----------V---------  --------E---------R------N--L--T--S--M-AG--A------T----------- ---VP--PV------VIIDK
rNaB3   K-RS-LL--D-NP--GD- KLC--------------------KT-I---------V---------  -------SE-----------P----N--
hNaSk1                                                                                                            -AH--      ----SLDTSQ -
rNaSk1                                                                                                            -THN-      ----SLDAS  -
hNaH1   -AM        NH-SLTR GL-                -T-M  KP-S--G-I-T--R-  -L---A-----  -N--ARESE-HHT--L--WPL  --T-AQG-P-PGTSA -GHALH--KN-T---------L-A-DPEA--PGSH--RPVMLEH
rNaH1   -AL        NQ-SLTH GL-                -T-M  RP-S--G-I-T--R-  -Q---A-----  -N--AGESE-H-T--L--WPL  -HP-AQG-PGPGASA -GYVL---RN-T---------L-A-DAEA--PGSY--RPMVLDR
hNaH2
rNaG1
sNa2             SDYSIKSV                            ELP       AVTVDSAGKGEIV-RM                SIQSETGSEVKRKLSV            -LSGKSKRE    -ISSG-SECE-AKKK     NYA
dNa2    HQATKVRKVSTYTIRNGRG--GI-G    -D-KP-VLSTYQDAQQHLPYADDSNAVTPMSEENGAIIVPVYYGNLG--HS-YTSHQSRISYTSHGDLLGGM-VMGVSTMTKES-LRNRNTR-QS-GATNG-TTC-DTNHKLDHRDYEIGL
eNa1           -P-ANI-EPSSV KL-TEE-                              -SDS-SMD-KHS
jNa1                                       FYNM--T  --TY--D-                  GGVV-YLH                 GPSPV-EIDGDKQYTIDNQ-KKSEKKES-VSTD-                             GK
sNa1                                        KLQQT  -AT                            A                   -EQ-LKE-ER-NTLTVSEADSHVDDRNCTC-EQC                             C
dNa1                                                                                                  EEAEITNEV-PSNLY-ATVSQ
                 |         |         |         |         |         |         |         |         |         |         |         |         |         |         | 750

                                                                                                                ┌──────────IIS1──────────┐           ┌───IIS2─
hNaB2          GT           TTETEIRKRRSSSYHVSMDLLEDPTSRQRAMSIASI     LTNTMEELEES  RQKCPPCWYKFANMCLIWDCCKPWLKVKHLVNLVVMDPFVDLAITICIVLNTLFMAMEHYPMTEQFSSVLSVGNLVFTG
rNaB2          --           ------------------------- -------M---     -----------  ---------------------------V--------------------------------------------------
hNaB1
rNaB1   PATDDN--            -----M-------F-----F----SQ----------     ----V------  -----------S-IF-----SPY------I--------------------------------H-NH--T--------
rNaB3          --           -----V----L---QI--EM---SSG---S------     -----------  ---------R---VF------DA----------I-----------------------------Q------T--------
hNaSk1                                           -KGAP--SGSGDSG-      SDA------A  H-----W---C-HKV------A----F-NIIH-I--------G----------------------H-DN--T--------
rNaSk1                                           -KGPP-PSCSADSA-      SDA------A  H-----W---C-HKV---N--A--V-F--IIY-I--------G----------------------H-DN-----------
hNaH1   PP  DTT-            PS-EPGGPQMLT-QAPCV-GF-E-GA----L-AV-V     --SAL------  -H------NRL-QRY---E--PL-MSI-QG-K-------T--T--M----------L---N--SE-EEM-Q--------
rNaH1   PP  DTT-            PS-EPGGPQMLTPQAPCA-GF-E-GA----L-AV-V     --SAL------  HR------NR--QHY---E--PL-MSI-QK-KF------A--T--M----------L---N--AE-EEM-Q--------
hNaH2       DEAK            -IQI-MK--SPI-TDT-L-V---A-L-HK                  ----K-  KKI--LY-----KTF---N-SPC---L-EF-HRII-A--T--FLI---I--VC-LTL-----SK-TNTL-NI-----I-
rNaG1
sNa2                NNPFLCPSSSATNVVDMKDVM-LK--IDHASGARSSSVSS-               -SQPELLKE-L       IRIFCA---YP-FKTLRGYIGFFI--A----F--L------V---ID--D-Q-GLRVA-EI--Y---A
dNa2    ECTDEA-KIKHHDNPFIEPVQ-QT VVDMKDVM-LN-II-QAAG-HSRA-DRGVSVYYFPTEDDDEDGPTFKD-ALEVIL-GIDVFCV----WV---FQEW-S-I-F----E-F--L---V--M----D-HD-NKEMER--KS--YF--A
eNa1                                              VDK-SLKHK-A-TM-V     F-  L-D--AA  -RP---V-----GFVFK-N--G--VFL-KW-HF-M----T--F--L--I------SI--H--N-S-Q-L--A-----T
jNa1    PP                                    E-GDR-S                                          -F   MRFRVIMYIF-GSTWFEGF--G--MV--IA--A--FEQP-VMTKMSDIL-YI--S
sNa1                                              GC                                           -YN---R-QSFAHCIIT-S-TEVF-IFI-----V-L----HG-SMELKN--K-A-Y---T
dNa1                                                                                                     -SIQ--R--LFE----L------A-L----HG-S-S-RNA-D---K---S
                 |         |         |         |         |         |         |         |         |         |         |         |         |         |         | 900
```

FIGURE 2(2).

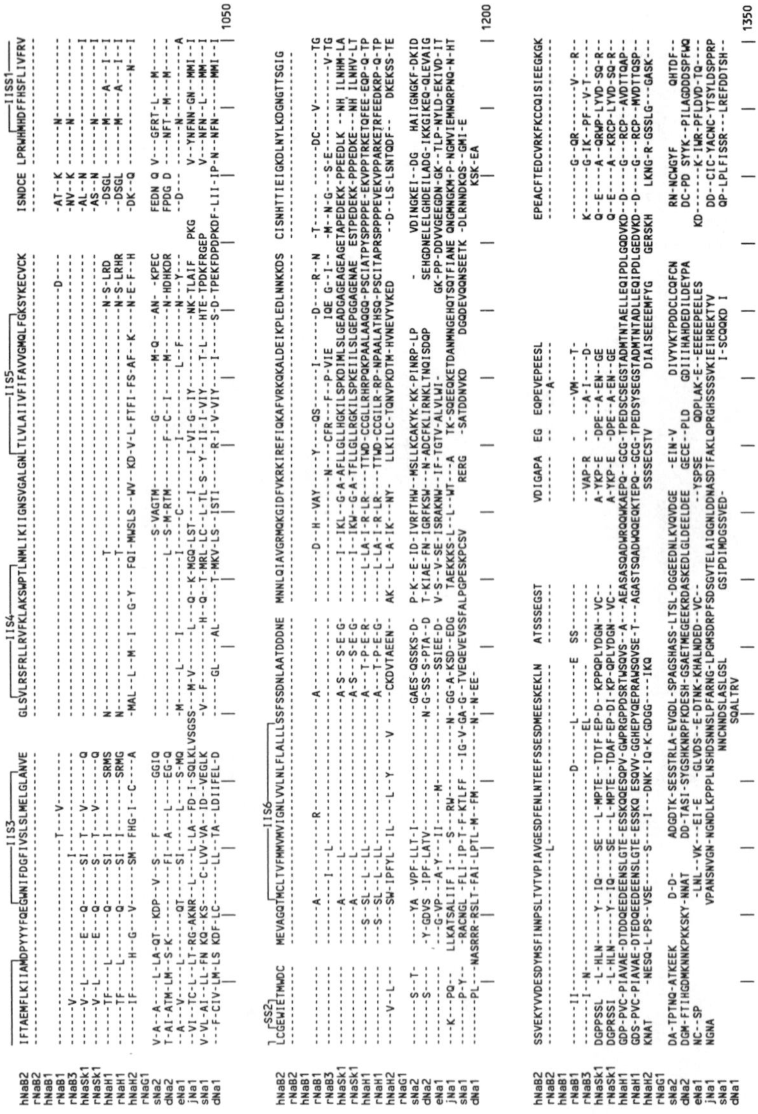

```
                                               IIS3                                  IIS4                                IIS5                                   IISS1
hNaB2   IFTAEMFLKIIAMDPYYYFQEGWNIFDGFIVSLSLMELGLANVE    GLSVLRSFRLLRVFKLAKSWPTLNMLIKIIGNSVGALGNLTLVLAIIVFIFAVVGMQLFGKSYKECVCK      ISNDCE LPRWHMHDFFHSFLIVFRV
rNaB2   --------------------------------------------    -----------------------------------------------------------------------      ------ -------------------
hNaB1
rNaB1   -----------------------------T---V--------      -------------------------------------------------------------------D----      -AT--K ------N------------
rNaB3   ------V---------------------I--------------      -----------------------------------------------------------------------      -NV--K ------N------------
hNaSk1  ------V--L------E---Q------SI--T---V-------Q      -----------------------------------------------------------------------      -AL--N ------------------I
rNaSk1  ------V--L------E---Q------S---T---V-------Q      -----------------------------------------------------------------------      -AS--N ------N-----------I
hNaH1   ------TF----L-------Q------SI--I--------SRMS     N-----------------------T-------------------------------------N-S-LRD         -DSGL ------M----A---I--I
rNaH1   ------TF----L-------Q------SI--I--------SRMG     N-----------------------T-------------------------------------N-S-LRHR       --DSGL ------M----A---I--I
hNaH2   ------IF-----H--G---V------SM--FHG-I--C----A     -MAL--L--M--I---G-Y---FQI-MWSLS--WV--KD-V-L-FTFI-FS-AF--K----N-E-F--H         -DK--Q --------------N---I
rNaG1
sNa2    V-A--A----L-LA-QT--KDP--V--S---F--------GGIQ     --------------------------S-VAGTM----------G--------M-Q----AN- -KPEC        FEDN Q V---GFRT-L---M-----
dNa2    T-AI-ATM-LM--S-K-----------FI--A---L----EG-Q     ----------------------L--S-M-RTM------F--C--I-----M--------N-HDHKDR       FPDG D ----NFT--M---M-----
eNa1    --A---V-----L-------QT-----SI------L----S-MQ     -M-----L----I-----------I-----C------------I-----------L--F-----N---Y---    --D--- -------N-----------A
jNa1    --VI--TC-L--LT-RG-AKNR--L----L-LA--FD-I-SQLKLVSGSS---M-V------L---Q--K-MGQ-LST----I-----I-VI-G--IY-----------NK-TLAIF   PKG     V--YNFNN-GN--MMI--I
sNa1    V-VL-AI--LL-FN KQ--KS----C-LVV-VA--ID--VEGLK    -V--F---------H--Q--T-MRL-LC--L-TL-S--Y--II-I-VIY----T-L---HTE-TPDKFRGEP          V---NFN--L---MM---I
dNa1    ---F-CIV-LM-LS KDF-LC------LL--TA--LDIIFEL-D    ------GL----AL------T-MKV-LS--ISTI------R-I-V-VIY----I-----S-D-TPEKFDPDPKDF-LII-IP-N--NFN------MMI--I
             |        |        |        |        |        |        |        |        |        |        |        |        |        |      | 1050

   SS2                       IIS6
hNaB2   LCGEWIETMWDC   MEVAGQTMCLTVFMMVMVIGNLVVLNLFLALLLSSFSSDNLAATDDDNE   MNNLQIAVGRMQKGIDFVKRKIREFIQKAFVRKQKALDEIKPLEDLNNKKDS   CISNHTTIEIGKDLNYLKDGNGTTSGIG
rNaB2   ------------   ---------------------------------------------------   ---------------------------------------------------   ---------------------------
hNaB1
rNaB1   ------------   ------A------------R-----------------A-----------   --------D--H--VAY-----Y----QS------I-------D----R--N   -T----- ------DC---V--------TG
rNaB3   ------------   ----------I---L-----------------------------------   ------------------N----CFR---F--P-VIE    IQE G--I--   -M--N-G---S-E------------V-TG
hNaSk1  ------------   ------A------L----------------------A-S---S-E-G-   -------I--IKL--G-A-AFLLGLLHGKILSPKDIMLSLGEADGAGEAGEAGETAPEDEKK-PPEEDLK --NH ILNHM-LA
rNaSk1  ------------   ------A------L----------------------A-S---S-E-G-   -------I--IKW--G-A-TFLLGLLRGKILSPKEIILSLGEPGGAGENAE  ESTPEDEKK-PPPEDKE---NH ILNHV-LT
hNaH1   ------------   ---S--SL--L--LL---------------------A---T-P-E-R-   -----L-LA-I-R-LR----TTWD-CCGLLRHRPQKPAALAAQGQ-PSCIATPYSPPPPE-EKVPPTRKETQFEE-EQP-Q-TP
rNaH1   ------------   ---S--SL--L--LL---------------------A---T-P-E-G-   -----L-LA-I-R-LR----TTWD-CCGILR-RP-NPAALATHSQ-PSCITAPRSPPPPEVEKVPPARKETRFEEDKRP-Q-TP
hNaH2   -----V--L---   ------SW-IPFYL--IL----L--Y-----V -----CKDVTAEEN--   AK---L--A-IK---NY-   LLKILC-TQNVPKDTM-HVNEVYVKED          --D--LS-LSNTQDF--    DKEKSS-TE
rNaG1
sNa2    -------S--T-   -----YA -VPF-LLT-I------------------GAES-QSSKS-D-   P-K--E-ID-IVRFTHW--MSLLKCAKYK-KK-PINRP-LP         -        VDINGKEI--DG    HAIIGNGKF-DKID
dNa2    -------S----   -Y-GDVS -IPF-LATV----------------N-G-SS-S-PTA--D   T-KIAE-FN-IGRFKSW---N-ADCFKLIRNKLTNQISDQP          SEHGDNELELGHDEILADG-IKKGIKEQ-QLEVAIG
eNa1    ------------   ---G-VP---A-Y---II-----M------------------SSIEE-D-   V-S--V-SE-ISRAKNW--IF-TGTV-ALVLWI-              GK-PP-DDVVGEEGDN-GK--TLP-NYLD-EKIVD-IT
jNa1    ---K---PQ--   LLKATSALIIF I--  -S---RW-----------N--GG-A-KSD--EDG    TAEKKKS-L--L--WT---A    TK-SQEEQKETDANMNGEHQTSQTFIANE QNGMNGKM-P-NGMVIEMNQRPNQ-N-HT
sNa1    -------P-Y--   -RACNGL -FLI-IP-T-F-KTLFF --IG-V-GA-G--TVEQEVEVSSFALPGPESKPCSV        RERG   -SATDDNVKD      DGQDEVQQNSEETK -DLRNNDKQS--GMI-E
dNa1    -------PL---NASRRR-RSLT-FAI-LPTL-M--FM----------N--N-EE-                                                                  KSK-EA
             |        |        |        |        |        |        |        |        |        |        |        |        |        |      | 1200

hNaB2   SSVEKYVVDESDYMSFINNPSLTVTVPIAVGESDFENLNTEEFSSESDMEESKEKLN   ATSSSEGST          VDIGAPA  EG  EQPEVEPEESL            EPEACFTEDCVRKFKCCQISIEEGKGK
rNaB2   -----------------------------L--------------------------   ---------          -------  --  ----A------            ---------------------------
hNaB1
rNaB1   ------II---------------------------------D------L--------E  SS-------          -------  -   ---VM----T-            --------G--QR-------V---R--
rNaB3   -------I--N-------------------------------EL---------   ---------          --VAP-R  --  --A-I----D-            K-------G-IK--PF--V-T------
hNaSk1  DGPPSSL  -L-HLN-----Y--IQ----SE---L-MPTE--TDTF-EP-D--KPPQPLYDGN--VC--            A-YKP-E  -DPE--A-EN--GE            Q--E----A--QRWP-LYVD-SQ-R--
rNaSk1  DGPRSSI  -L-HLN-----Y--IQ----SE---L-MPTE--TDAF-EP-DI-KP-QPLYDGN--VC--            A-YKP-E  -DPE--A-EN--GE            Q--E----A--KRCP-LYVD-SQ-R--
hNaH1   GDP-PVC-PIAVAE-DTDDQEEDEENSLGTE-ESSKQQESQPV-GWPRGPPDSRTWSQVS--A---AEASASQADWRQQWKAEPQ--GCG-TPEDSCSEGSTADMTNTAELLEQIPDLGQDVKD--D----G---RCP--AVDTTQAP--
rNaH1   GDS-PVC-PIAVAE-DTEDQEEDEENSLGTE-ESSKQ ESQVV-GGHEPYQEPRAWSQVSE-T---AGASTSQADWQQEQKTEPQ--GCG-TPEDSYSEGSTADMTNTADLLEQIPDLGEDVKD--D----G---RCP--MVDTTQSP--
hNaH2   KNAT      -NESQ-L-PS--VSE-----S----I---DNK-IQ-K-GDGG----IKQ                   SSSSECSTV          DIAISEEEEMFYG    GERSKH    LKNG-R-GSSLG---GASK---
rNaG1
sNa2    DA-TPTNQ-ATKEEK       D-D-      ADGDTK-SESSTRLA-EVGDL-SPAGSHASS-LTSL-DGGEEDNLKVQVDGE   -EIN-V        DIVYVKTPDDCLCQFCN          RN-NCWGYF         QHTDF--
dNa2    DGM-FTIHGDMKNNKPKKSKY-NNAT      DD-TASI-SYGSHKNRPFKDESH-GSAETMEGEEKRDASKEDLGLDEELDEE   GECE--PLD    GDIIIHAHDEDILDEYPA          DC-PD SYYK--PILAGDDDSPFWQ
eNa1    NC--SP                -LNL--VK---EI-E   -GLVDS--E-DTNK-KHALNDED--VC--            --YSPSE      QDPLAK-E--EEEEEPEELES      KD-------K-IWR-PFLDVD-TQ----
jNa1    NGNA                VPANSNVGN-NGNDLKPPPLNSHDSNNSLPFARNG-LPGMSDRPFSDSGVTELAIQGNLDDNASDTFAKLQPRGHSSSSVKIEIHREKTYV          DD--CIC-YACNC-YTSYLDSPPRP
sNa1                                   NNCNNDSLASLGSL           GSIPDIMDGSSVED-                  I-SCQQKD I           QP-LPLFISSR---LREFDDTSH--
dNa1                                       SQALTRV
             |        |        |        |        |        |        |        |        |        |        |        |        |        |      | 1350
```

FIGURE 2(3).

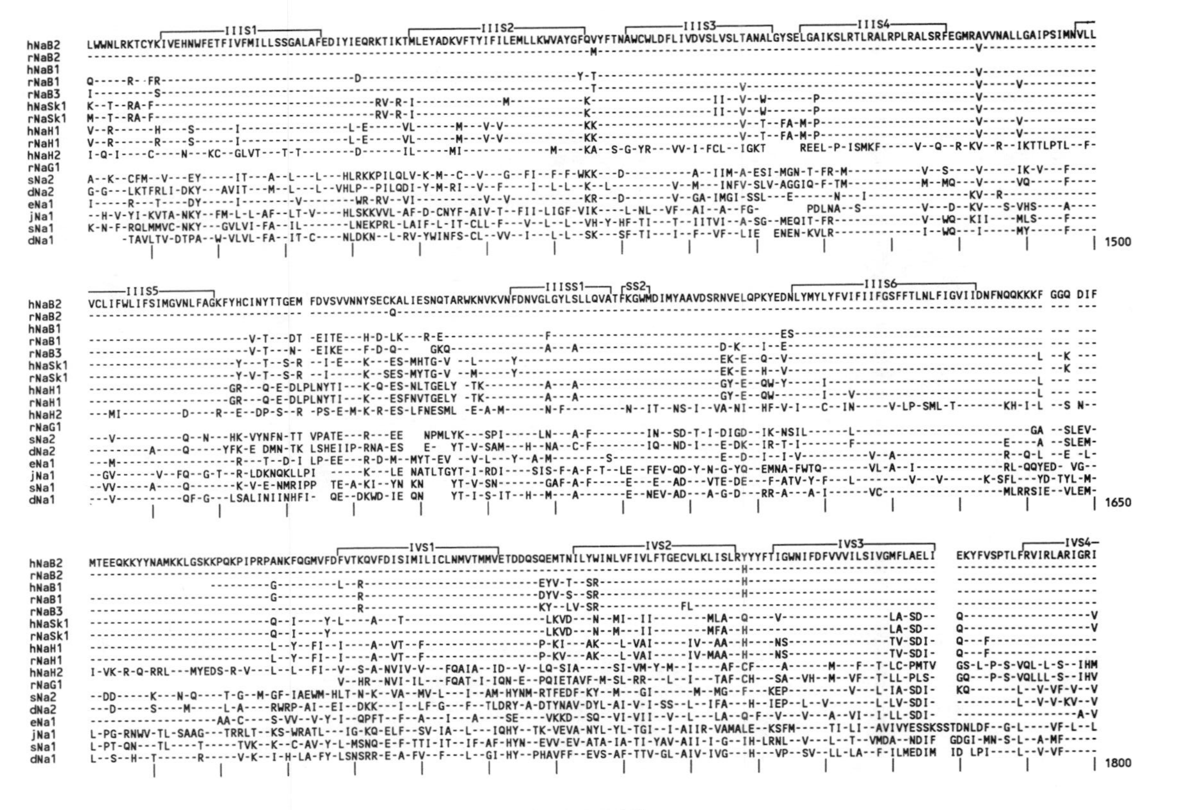

FIGURE 2(4).

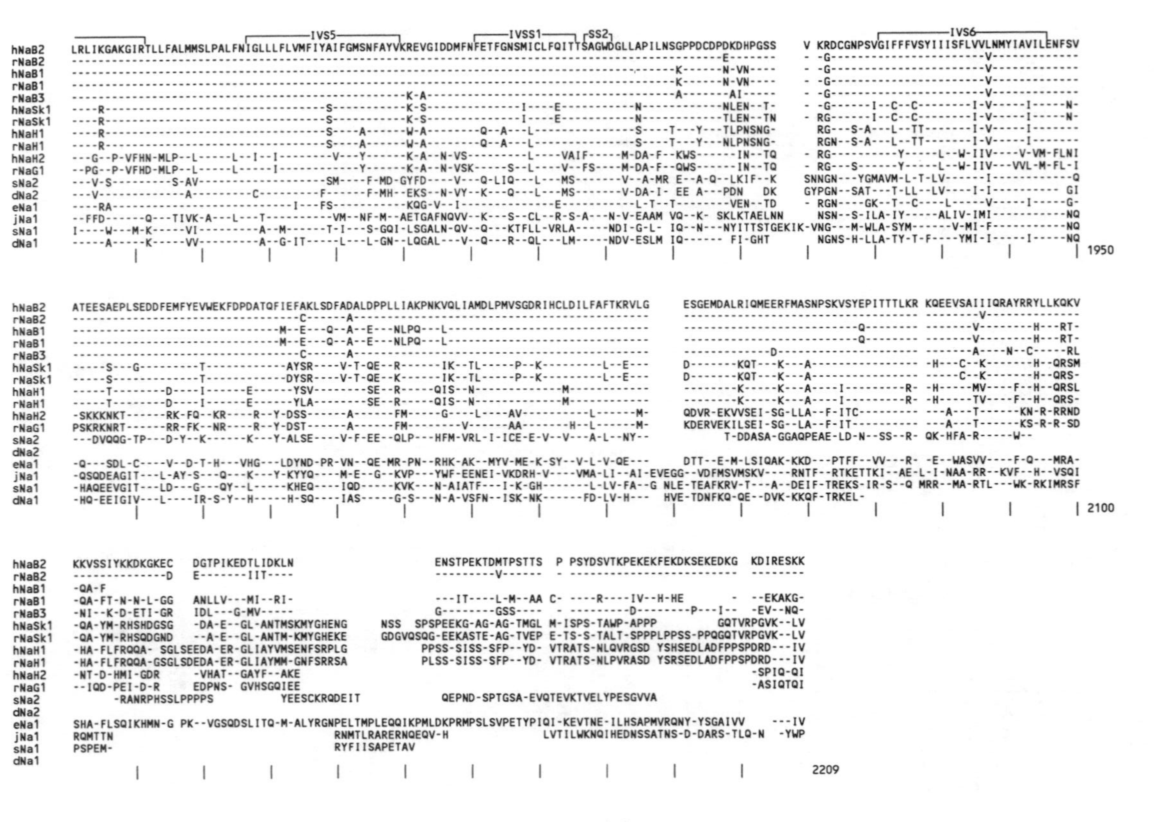

FIGURE 2(5).

2.2.1.4 *Drosophila*

The primary advantage of *Drosophila melanogaster* for the study of sodium channels has been the ability to investigate the genetics of sodium channel function (Ganetzky, 1986). Four loci have been identified that affect sodium channel function: *para, sei, nap,* and *tip.*

Mutations at the paralytic locus *(para^{ts})* result in paralysis at 30°C with recovery at 20°C within seconds (Suzuki et al., 1971). This locus is X-linked (Siddiqi and Benzer, 1976), and it is thought to comprise the structural gene for the sodium channel because of the rapid temperature-sensitive paralysis. Mutations in the *para* gene affect the level of expression of sodium currents in *Drosophila* embryonic neurons, and at least one mutation alters the gating properties of the channels (O'Dowd et al., 1989). A gene with sequence similarity to the sodium channel α-subunit is transcribed from the *para* locus (Loughney et al., 1989; Ramaswami and Tanouye, 1989). This gene extends over 60 kb, with at least 26 separate exons with alternative splicing of small exons (Loughney et al., 1989). The predicted protein is 1820 amino acids in length, with 46 to 47% sequence identity to the rat brain and eel sodium channel α-subunits. The cDNA clone was termed para, but it will be referred to as dNa2 in this review (Table 1).

The second sodium channel locus is seizure *(sei),* in which mutations result in convulsive seizures followed by paralysis at temperature above 38°C (Jackson et al., 1985). This locus is located on the right arm of chromosome 2 (60A-B) (Jackson et al., 1985), and it is also thought to identify a structural gene for the sodium channel because at least one mutation increases the K_D for saxitoxin binding (Jackson et al., 1984). Salkoff and coworkers (1987) have isolated cDNA clones encoding a sodium channel α-subunit on chromosome 2 in a region near but not identical to the *sei* locus (60D-E). This cDNA clone was termed DSC1, but it will be referred to as dNa1 in this review (Table 1). The predicted amino acid sequence is very similar to the eel and rat brain sodium channels in the four homology domains (Figure 2), but the regions between domains I–II and II–III cannot be compared because they were not completely sequenced. This sodium channel gene is highly spliced, with at least 20 introns within the coding region.

The third sodium channel locus in *Drosophila* is no-action potential *(nap^{ts}),* which is also located on chromosome 2 (Wu et al., 1978). Mutations at the *nap* locus result in temperature-sensitive paralysis at 37°C (Wu et al., 1978) and a decreased number of saxitoxin (Jackson et al., 1984) and tetrodotoxin (Kauvar, 1982) binding sites. However, *nap* mutations do not decrease the level of sodium current measured directly in cultured embryonic neurons (O'Dowd and Aldrich, 1988). In fact, *nap* is thought to exert its physiological effects on sodium channels through modulation of *para,* since some temperature-sensitive *para* mutations become unconditionally lethal in a *nap* background (Ganetzky, 1984). Kernan et al. (1991) have since shown that *nap* is an allele of maleless *(mle),* which increases transcription from the X chromosome in males as a means of dosage compensation. The temperature-sensitive phenotype of *nap^{ts}*

may result from a decreased number of sodium channels *(para* is located on the X chromosome), which causes a lowering of the temperature at which action potentials would normally cease (Fakler et al., 1990). However, nap^{ts} is a gain of function allele, so that the mechanism by which *para* gene transcription would be decreased in unclear (Kernan et al., 1991).

The final *Drosophila* sodium channel locus is temperature-induced paralysis *(tip-E),* located on chromosome 3 (Kulkarni and Padhye, 1982). Mutations at this locus also result in temperature-sensitive paralysis, but the nonpermissive temperature is higher than that for mutations at the other three loci (39–40°C). This locus may regulate sodium channel expression, because the sodium current density was reduced in embryonic *Drosophila* neurons from *tip-E* mutant flies (O'Dowd and Aldrich, 1988) and the number of saxitoxin binding sites in adult heads was reduced (Jackson et al., 1986). However, there is as yet no molecular genetic data concerning the function of the *tip-E* locus.

The genetic data concerning *Drosophila* sodium channels suggest that there are at least two structural genes for the channel and at least two regulatory loci. It is of course possible that additional genes are present, but that mutations at those loci have not been identified, perhaps because they are unconditionally lethal. Although genomic DNA clones have been isolated for two distinct sodium channel α-subunits, neither has been expressed in an exogenous system. For this reason a direct comparison between the genetic effects of the different mutations and the electrophysiological properties of the different clones cannot yet be made.

2.2.1.5 Mammalian Brain

2.2.1.5.1 Rat Brain

The rat is one of the primary organisms in which the sodium channel has been analyzed both biochemically and electrophysiologically. The rat brain sodium channel is sensitive to a variety of toxins, including both tetrodotoxin and saxitoxin at nanomolar concentrations (reviewed by Catterall, 1980). The sodium channel purified from rat brain consists of a large subunit of 260 kDa (α) and associated small subunits of 36 kDa (β_1) and 33 kDa (β_2) (Hartshorne and Catterall, 1981, 1984). The β_2-subunit is covalently attached to the α-subunit by disulfide linkage, and the β_1-subunit is noncovalently attached (Hartshorne et al., 1982; Messner and Catterall, 1985). All three sodium channel subunits are highly processed. The α-subunit is glycosylated, sulfated, and palmitylated (Schmidt and Catterall, 1987; Elmer et al., 1985), and glycosylation is essential for maintenance of functional channels in neuroblastoma cells (Waechter et al., 1983). The α-subunit is phosphorylated by protein kinase A (Costa et al., 1982; Costa and Catterall, 1984; Rossie and Catterall, 1987), with most if not all of the phosphorylation occurring on the intracellular loop between domains I and II (Rossie et al., 1987; Rossie and Catterall, 1989). The α-subunit can also be phosphorylated by protein kinase C at a site in the intracellular loop between domains III and IV (Numann et al., 1991; West et al., 1991). The proteins purified from rat brain have been reconstituted into lipid vesicles (Tamkun and Catterall, 1981; Tamkun et al., 1984) and bilayers (Hanke et al., 1984; Hartshorne et al., 1985; Keller et al., 1986), resulting in channels that could be activated by voltage depolarization or batrachotoxin treatment.

Using the cDNA clones that they isolated for the eel α-subunit, Numa and co-workers (Noda et al., 1986a; Kayano et al., 1988) obtained cDNA clones encoding three distinct sodium channel α-subunits from rat brain. These were termed rat I, II, and III, but they will be referred to as rat Na Brain 1 (rNaB1), rNaB2, and rNaB3 in this review (Table 1). Rat NaB1 encodes a predicted protein of 2009 amino acids, rNaB2 encodes a protein of 2005 amino acids, and rNaB3 encodes a protein of 1951 amino acids. The amino acid sequences are very similar, with 85 to 87% similarity between all three channels (Figure 2) (Kayano et al., 1988). RNA transcribed *in vitro* from rNaB1 and expressed in *Xenopus* oocytes resulted in marginal inward currents (about 50 nA) (Noda et al., 1986b), whereas both rNaB2 (Noda et al., 1986b) and rNaB3 (Suzuki et al., 1988; Joho et al., 1990) expressed high levels of sodium current in oocytes. An alternative form of rNaB2, termed rNaB2A, differs from rNaB2 at 6 amino acid positions but is otherwise similar in expression and physiological properties (Auld et al., 1988, 1990).

The genes for the brain sodium channels have not been localized in the rat, but they have been mapped on the mouse chromosomes. All three genes are located on chromosome 2, with rNaB2 and rNaB3 within 600 kb by physical mapping, and rNaB1 and rNaB2 within 0.7 cM by genetic linkage (Malo et al., 1991). Analysis of polymerase chain reaction products containing the largest cytoplasmic loop of the rat sodium channels indicates that some of the genes are expressed in multiple different tissues, and that alternative splicing does occur (Schaller et al., 1992). Alternative splicing has been investigated more specifically for the rNaB2 and rNaB2A gene products (Sarao et al., 1991; Ahmed et al., 1990). These two cDNA clones contain one 90 nucleotide region with 20 differences between them. Expression of each 90 nucleotide segment is mutually exclusive, and most likely represents alternative exon selection (Sarao et al., 1991). The choice of which exon is expressed appears to be regionally and developmentally regulated, so that rNaB2A becomes predominant as development proceeds (Sarao et al., 1991).

The three rat brain sodium channels are differentially expressed both regionally and during development (Mandel, 1992). Rat NaB1 is expressed at high levels in the peripheral nervous system, where rNaB2 and rNaB2A are expressed only at low levels (Beckh, 1990). In central neurons, rNaB2 or rNaB2A is predominant in rostral parts of the brain, such as the forebrain, hippocampus, and cerebellum, whereas rNaB1 is present at higher densities in caudal areas, such as spinal motor neurons, cerebellar Purkinje cells, and hippocampal pyramidal cells (Gordon et al., 1987; Westenbroek et al., 1989; Furuyama et al., 1993). Expression of the rat brain sodium channels is not limited to the brain and peripheral nervous tissues. Rat NaB1 mRNA has been detected in rat myocardium, although rNaB2 mRNA could not be detected under similar conditions (Sills et al., 1989). During development sodium channel α-subunit expression peaks within the first 3 weeks after birth (Scheinman et al., 1989), after which rNaB2A expression increases compared to rNaB2 with increasing age (Yarowsky et al., 1991; Brysch et al., 1991). Rat NaB3 is expressed mainly in the embryonic stages, reaching maximal levels about the time of birth (Brysch et al., 1991; Beckh et al., 1989; Scheinman et al., 1989).

Tissue-specific expression of the rNaB2 sodium channel appears to be the result of at least one negative regulatory element upstream of the rNaB2 gene (Maue et al., 1990). A 28-bp silencer element located in the 5′ flanking region of the rNaB2 gene prevents expression of a minigene in cell lines that do not express the rNaB2 gene, such as fibroblasts, skeletal muscle cells, and some neuronal cell lines (Kraner et al., 1992). A similar 21-bp element was found upstream of another neural-specific gene, SCG10, which is expressed in sympathetic neurons (Mori et al., 1992). A protein binding to these elements was identified, suggesting that these genes are specifically repressed in nonneuronal cells (Kraner et al., 1992; Mori et al., 1992). Regulatory differences between the rNaB1, rNaB2, and rNaB3 genes have not yet been identified.

2.2.1.5.2 Human Brain

The sodium channels from human brain have not been as well characterized as those from rat brain. Complementary DNA clones encoding portions of two distinct human brain sodium channel α-subunit genes have been isolated. The first cDNA clone was termed HBSCI (Lu et al., 1992), and will be referred to as hNaB1 in this review (Table 1). Clones encoding all of domain IV and part of the amino-terminal region of hNaB1 were isolated. In this region hNaB1 is 90% similar to rNaB1, and 85% similar to rNaB2 and rNaB3 (Lu et al., 1992). The second α-subunit was alternatively termed HBSCII (Lu et al., 1992) and HBA (Ahmed et al., 1992), and will be referred to as hNaB2 in this review (Table 1). A full-length cDNA clone for hNaB2 has been obtained, and it encodes a predicted protein of 2005 amino acids (Ahmed et al., 1992). The hNaB2 amino acid sequence is 97% identical to rNaB2, and 88% identical to rNaB1 and rNaB3 (Figure 2) (Ahmed et al., 1992). Both human brain sodium channel genes are located on chromosome 2q23–24.3 (Ahmed et al., 1992; Han et al., 1991; Lu et al., 1992), with tight linkage similar to that observed for the brain channels in the mouse chromosome. Expression of the two sodium channels is distributed in a rostral to caudal manner, similar to the distribution seen for the rat brain sodium channels. Specifically, hNaB2 is expressed at higher levels in the rostral regions, such as the cerebellum, and hNaB1 is expressed at higher levels in the caudal regions, such as the spinal cord (Lu et al., 1992). The hNaB2 cDNA has been expressed by transfection in Chinese hamster ovary cells, and the resulting sodium currents resembled those from human cell lines in both the kinetics and sensitivity to tetrodotoxin (Ahmed et al., 1992). All of these results indicate that the human brain sodium channels are more similar to the rat brain sodium channels than to the other human sodium channels.

2.2.1.6 Mammalian Skeletal Muscle

2.2.1.6.1 Rat Skeletal Muscle

The adult skeletal muscle sodium channel is similar to the brain sodium channel in that it is sensitive to nanomolar concentrations of tetrodotoxin, but it is also sensitive to nanomolar concentrations of μ conotoxin, to which the brain channel is resistant (Cruz et al., 1985; Moczydlowski et al., 1986). A separate muscle sodium channel characteristic of early development and denervated skeletal muscle is relatively resistant to tetrodotoxin. This channel is also expressed in cardiac muscle, and will be discussed under cardiac sodium channels (Section 2.2.1.7). The proteins comprising the adult skeletal muscle sodium channel have been purified from rat sarcolemma (Barchi et al., 1980), and consist of an α-subunit and a single β-subunit of 38 kDa (Casadei et al., 1986). The purified proteins have been reconstituted into phospholipid vesicles (Weigele and Barchi, 1982) and planar lipid bilayers (Furman et al., 1986), resulting in batrachotoxin-activated currents. The proteins are structurally similar to those isolated from rat brain, but they are antigenically distinct (Casadei et al., 1984; Wollner and Catterall, 1985).

A cDNA clone has been isolated encoding the α-subunit of the adult rat skeletal muscle sodium channel (Trimmer et al., 1989). This cDNA clone has been termed μ1 or SkM1, and it will be referred to in this review as rNaSk1 (Table 1). The rNaSk1 clone encodes a predicted protein of 1840 amino acids, which is 78% similar to the eel α-subunit and about 82% similar to each of the rat brain α-subunits. Expression in *Xenopus* oocytes (Trimmer et al., 1989) or human embryonic kidney cells (Ukomadu et al., 1992) results in sodium currents that are inhibited by nanomolar concentrations of tetrodotoxin and conotoxin, comparable to the sensitivities observed for the channels expressed *in vivo.* Expression of the rNaSk1 channel was detected at high levels in adult rat skeletal muscle, at low levels in neonatal skeletal muscle, and not at all in brain or heart (Trimmer et al., 1990).

2.2.1.6.2 Human Skeletal Muscle

A cDNA clone, termed hNaSk1, has been isolated encoding the human adult skeletal muscle sodium channel (George et al., 1992b; Wang et al., 1992) (Table 1). The predicted protein is 1836 amino acids and 92% identical to rNaSk1. The gene encoding hNaSk1 (SCN4A) has been mapped to human chromosome 17q23.1-25.3 (George et al., 1991). The SCN4A gene comprises 32.5 kb (George et al., 1993; Wang et al., 1992), and consists of 24 exons (George et al., 1993). Conservation with the *Drosophila* Na2 *(para)* sodium channel gene is suggested by the fact that 10 of the splice junctions are located at homologous positions in the two genes (George et al., 1993). The SCN4A gene is the only sodium channel gene that has been completely mapped. It has been extensively studied because mutations in it result in the human periodic paralysis (reviewed in Barchi, 1992; Ptacek et al., 1993b), which is discussed in Section 2.2.5.

2.2.1.7 Mammalian Cardiac Muscle

2.2.1.7.1 Rat Cardiac Muscle

Sodium channels expressed in cardiac muscle cells are distinguished by the fact that they are resistant to nanomolar concentrations of tetrodotoxin (Brown et al., 1981), but are more sensitive than neuronal channels to inhibition by lidocaine (Bean et al., 1983). Cardiac sodium channels are generally referred to as tetrodotoxin resistant, although they are blocked by micromolar concentrations of the toxin. Tetrodotoxin-resistant sodium channels are also present in immature skeletal muscle. They disappear during the first weeks of life, and they reappear if adult muscles are denervated (Gonoi et al., 1985; Pappone, 1980). Complementary DNA clones encoding sodium channels have been obtained from both cardiac muscle and denervated skeletal muscle. The cardiac muscle clone is termed rH1 (Rogart et al., 1989), and the denervated skeletal muscle clone is termed SkM2 (Kallen et al., 1990). The two clones are essentially identical, encoding a protein of 2018 amino acids, which is 84% identical to rNaSk1. This sodium channel will be referred to as either rNaH1 or rNaSk2 in this review (Table 1). Both rNaH1 (Cribbs et al., 1990) and rNaSk2 (White et al., 1991) have been expressed in *Xenopus* oocytes, resulting in tetrodotoxin-resistant and μ conotoxin-resistant sodium channels. Messenger RNA encoding rNaH1/rNaSk2 was detected at high levels in rat heart, but not in brain, liver, kidney, or uterus (Rogart et al., 1989; Kallen et al., 1990). Expression could not be detected in adult skeletal muscle, but was detectable in neonatal skeletal muscle and after denervation of adult muscle (Kallen et al., 1990). Both muscle sodium channels (rNaSk1 and rNaSk2) were detected in denervated muscle, but the increase in the level of sodium channel mRNA following denervation (Cooperman et al., 1987) resulted from an induction of rNaSk2 expression (Yang et al., 1991).

2.2.1.7.2 Human Cardiac Muscle

A cDNA clone encoding a human heart sodium channel (hNaH1) has been obtained by hybridization screening with rNaSk2 (Gellens et al., 1992) (Table 1). The human cDNA clone encodes a protein of 2016 amino acids, and is 94% identical to rNaSk2. Messenger RNA encoding hNaH1 was detected in human heart, but not in brain, skeletal muscle, myometrium, liver, or spleen. Expression in *Xenopus* oocytes resulted in sodium current that were resistant to tetrodotoxin and sensitive to lidocaine (Gellens et al., 1992), as expected for the cardiac sodium channel. The hNaH1 currents differed slightly from those observed for rNaSk2, in that the currents were less sensitive to tetrodotoxin and appeared to have a higher single channel conductance (Gellens et al., 1992). It has not yet been determined whether these differences reflect *in vivo* differences of the human heart channels compared to the rat heart channels.

A second and somewhat unusual cDNA clone encoding a sodium channel from human heart has also been isolated (George et al., 1992a). This clone was termed hNa_v-2.1, but it will be referred to in this review as hNaH2 (Table 1). The predicted protein is 1682 amino acids in length, with a comparable secondary structure as the other α-subunits. However, the overall amino acid identity is only about 50% with each of the rat sodium channel cDNA clones. Two major differences are the number of charges in the S4 voltage sensor regions, and a poorly conserved interdomain III–IV linker (Figure 2). Human NaH2 contains fewer charged residues in the S4 regions, which might suggest a weaker voltage sensitivity. The poorly conserved III–IV linker might indicate that hNaH2 inactivates with different kinetics than the other sodium channels analyzed to date, as this linker has been shown to be critical for fast sodium channel inactivation (Patton et al., 1992; West et al., 1992). However, the hNAH2 cDNA clone has not yet been expressed in an *in vitro* system, so the functional significance of these differences cannot be evaluated. Messenger RNA encoding hNAH2 was detected at high levels in human heart, skeletal muscle, and uterus, at low levels in brain, kidney, and spleen, and not at all in liver or smooth muscle.

2.2.1.8 Mammalian Peripheral Nerve

Electrophysiological data obtained using rat dorsal root ganglion cells have indicated that there are at least three distinct types of peripheral nerve sodium channels (Kostyuk et al., 1981; Schwartz et al., 1990; Caffrey et al., 1992). One class of sodium currents is resistant to micromolar concentrations of tetrodotoxin, and demonstrates slow activation and inactivation kinetics. A second class of currents is sensitive to tetrodotoxin, and has rapid activation and inactivation kinetics. The third class is also sensitive to tetrodotoxin, but the kinetics of inactivation are intermediate between the other two types. The three types are differentially expressed, so that large cells express the tetrodotoxin-sensitive intermediate class and small neurons express the other two classes. Although rNaB1 is expressed at high levels in the peripheral nervous system (Beckh, 1990), it has not been demonstrated whether any of these sodium channels result from expression of rNaB1 or any of the other α-subunits that have already been characterized. In addition, it is not known if the different classes of sodium channels represent expression of distinct α-subunits, or if the same α-subunit is modulated differentially in the different cell types. No sodium channel unique to peripheral neurons has yet been isolated or cloned.

2.2.1.9 Mammalian Glial Cells

Voltage-gated sodium channels have been detected in a variety of glial cells. Mammalian Schwann cells exhibit binding sites for both tetrodotoxin and saxitoxin (Villegas et al., 1976), and electrophysiological studies have directly demonstrated voltage-gated sodium channels in Schwann cells from both rabbits and rats (Chiu et al., 1984; Howe and Ritchie, 1990; Shrager et al., 1985). Comparisons between the sodium channels expressed in rat astrocytes and neurons indicated similarities in conductance and sensitivity to saxitoxin, but the glial channels displayed a hyperpolarizing shift in voltage dependence and slower kinetics of inactivation (Barres et al., 1989). The hyperpolarizing shift could make the sodium channels more functional in the glial cells, as astrocytes have a more negative resting potential than the corresponding neuronal cells (Barres et al., 1989). These results indicate that either one of the previously isolated sodium channel α-subunits is functionally modulated in glial cells, or that there exists at least one glial-specific sodium channel gene.

Consistent with the latter alternative, a partial cDNA clone has been isolated encoding the fourth domain and carboxy-terminal region of a sodium channel expressed in cultured rat astrocytes (Gautron et al., 1992). This cDNA clone was termed Na-G, but will be referred to in this review as rNaG1 (Table 1). The rNaG1 channel is significantly different from the other sodium channels, with only about 55% sequence similarity to the rNaB2 channel (Figure 2). Messenger RNA encoding rNaG1 was detected at highest levels in the spinal cord, at intermediate levels in the midbrain, and at low levels in the cerebrum and cerebellum. Expression was also detected in skeletal muscle, spleen, and intestine, but not in liver or kidney. The level of rNaG1 in skeletal muscle was significantly increased following denervation. Since only a partial cDNA clone has been isolated, the functional properties of this sodium channel have not yet been determined.

2.2.1.10 Other Sodium Channels

Sodium channels have also been observed in fibroblasts (Pouyssegur et al., 1980) and epithelial cells (Koefoed-Johnson and Ussing, 1958; Sariban-Sohraby and Benos, 1986). Many of these channels demonstrate tetrodotoxin binding (Munson et al., 1979; Pouyssegur et al., 1980), suggesting that there is structural similarity in the pore region with the sodium channels of electrically excitable tissues. However, the channels are not gated by voltage, but can be activated only by the binding of various ligands, some of which also activate the voltage-gated sodium channels. These channels clearly represent a completely different gene family, as demonstrated by the lack of sequence similarity between the amiloride-sensitive sodium channels (Canessa et al., 1993; Lingueglia et al., 1993) and the voltage-gated sodium channels, and so they will not be discussed in this review.

2.2.2 Sodium Channel β-Subunits

The functional role of the sodium channel β-subunits is unknown. They are clearly not essential for a functional sodium channel in all systems, since many sodium channels contain only an α-subunit. The β-subunits are also not essential for functional expression of the rat brain or muscle sodium channels in *Xenopus* oocytes. Injection into oocytes of size fractionated rat brain RNA that would be unlikely to encode the β-subunits is sufficient for the synthesis of functional channels (Goldin et al., 1986; Hirono et al., 1985; Krafte et al., 1988;

Table 2. Sodium Channel β-Subunit Clones

Name	Original name	Species	Tissue	Size	GenBank number	References
rNaβ1-1	Rat β1	*Rattus norvegicus*	Brain	218	M91808	Isom et al. (1992)
hNaβ1-1	Human β1, SCN1B	*Homo sapiens*	Brain	223	L10338	McClatchey et al. (1993)

Sumikawa et al., 1984). In addition, injection into oocytes of RNA transcribed *in vitro* from cDNA clones encoding the rat brain or muscle α-subunit results in functional channels (Noda et al., 1986b; Auld et al., 1988; Joho et al., 1990; Suzuki et al., 1988; Trimmer et al., 1989). However, the β_1-subunit may enhance sodium channel stability, as selective removal of the β_1-subunit from purified rat brain sodium channels in detergent solution is accompanied by loss of the ability to reconstitute neurotoxin-activated sodium influx (Messner and Catterall, 1986; Messner et al., 1986). The β_2-subunit may be involved in insertion of sodium channels into the cellular membrane, as suggested by the fact that disulfide linkage of the α- and β_2-subunits, insertion into the cell-surface membrane, and attainment of a functional conformation are closely related late events in the biogenesis of functional rat brain sodium channels (Schmidt et al., 1985; Schmidt and Catterall, 1986).

2.2.2.1 Rat β-Subunits

A cDNA clone has been obtained for the β_1-subunit from adult rat brain by Isom et al. (1992) (Table 2). This clone will be referred to as rNaβ1-1 in this review. Based on the cDNA sequence, rNaβ1-1 is a 199 amino acid protein with an additional 19 amino acid leader sequence at the amino-terminus and a single transmembrane segment 37 residues from the carboxy-terminus. It contains four potential N-linked glycosylation sites in regions predicted to be extracellular, which is consistent with biochemical data indicating three or four N-linked carbohydrate side chains (Messner and Catterall, 1985). RNA hybridizing to rNaβ1-1 cDNA was detected at high levels in rat brain and spinal cord, at moderate levels in rat heart, and at low but detectable levels in rat skeletal muscle, consistent with the results of antibody-binding experiments (Sutkowski and Catterall, 1990). The gene encoding the homologous mouse brain β_1-subunit (SCN1B) has been mapped to chromosome 7 (Tong et al., 1993). Coexpression of rNaβ1-1 with the rNaB2A α-subunit in *Xenopus* oocytes accelerated the macroscopic kinetics of inactivation of the sodium channels, shifted the $V_{1/2}$ of steady-state inactivation in the negative direction, and increased the level of current observed from the same quantity of α-subunit RNA by about 2.5-fold (Isom et al., 1992). All of these findings are similar to the effects observed following coinjection of low-mol wt rat brain RNA (Krafte et al., 1990), which suggests that the β_1-subunit does modify the functional properties of the sodium channel α-subunit expressed in *Xenopus* oocytes.

The interaction of the adult rat brain β_1-subunit is not specific to a single α-subunit, however. The rat brain β_1-subunit also modifies the functional properties of rNaB3 (Patton et al., 1994), rNaSk1 (Tong et al., 1993; Yang et al., 1993; Patton et al., 1994), rNaSk2/rNaH1 (Kyle et al., 1993), and hNaSk1 channels (Bennett et al., 1993). At least part of the explanation for this generality may be that the same β_1-subunit is expressed in multiple tissues. Complementary DNA clones encoding rNaβ1-1 have been isolated from rat skeletal muscle and heart (Tong et al., 1993; Bennett et al., 1993). In addition, rNaβ1-1 mRNA is expressed in skeletal muscle, and the level of expression during development correlates with the level of rNaSk1 mRNA (Yang et al., 1993). Therefore the adult skeletal muscle sodium channel may be associated with the same β_1-subunit as the adult rat brain rNaB2 channel. However, at least some of the sodium channels that are modified by the adult rat brain β_1-subunit are probably not associated with that subunit *in vivo*. The rNaB3 channel most likely represents a fetal subtype, and the fetal channel is not associated with a mature β_1-subunit (Beckh et al., 1989; Scheinman et al., 1989; Sutkowski and Catterall, 1990). Also, expression of rNaβ1-1 mRNA did not correlate with expression of rNaSk2 in rat skeletal muscle (Yang et al., 1993). Therefore the β_1-subunit appears to be capable of associating with α-subunits with which it is not normally colocalized.

2.2.2.2 Human β-Subunits

Using the rNaβ1-1 sequence as a probe, McClatchey et al. (1993) isolated a cDNA clone encoding a β_1-subunit from human brain (Table 2). The gene encoding this subunit was termed SCN1B, and the clone will be referred to as hNaβ1-1 in this review (Table 2). The predicted hNaβ1-1 protein is 223 amino acids in length, including any leader sequence, and it is 98% identical to rNaβ1-1 (Figure 3). High levels of RNA hybridizing to hNaβ1-1 were observed in human brainstem and cerebellum, with lower levels in skeletal muscle and heart. The SCN1B gene was localized to human chromosome 19. Coexpression of hNaβ1-1 with rNaSk1 in *Xenopus* oocytes accelerated the macroscopic kinetics of inactivation and shifted the $V_{1/2}$ of steady-state inactivation in the negative direction (Cannon et al., 1993), similar to the results observed when rNaβ1-1 was coexpressed with rNaB2 (Isom et al., 1992). These results indicate that the β_1-subunit is highly conserved between rats and humans, and suggest that the functional features important for interactions between α- and β_1-subunits are conserved among different tissues and species.

2.2.3 Evolutionary Relationships

Voltage-gated sodium, potassium, and calcium channels are clearly members of a single gene superfamily. Strong et al. (1993) extensively analyzed sequences from all three voltage-gated channels and the cyclic nucleotide-gated channels. They concluded that potassium channels arose as one lineage

```
hNaβ1-1   MGRLLALVVGAALVSSACGGCVEVDSETEAVYGMTFKILCISCKRRSETNAETFTEWTFRQKGTEEFVKILRYENEVLQLEEDERFEGRVVWNGSRGTKDLQDLSIFITN
rNaβ1-1   --T--------V-----W-------------------------------T------------------------------------------------------------
                   |         |         |         |         |         |         |         |         |         | 110

hNaβ1-1   VTYNHSGDYECHVYRLLFFENYEHNTSVVKKIHIEVVDKANRDMASIVSEIMMYVLIVVLTIWLVAEMIYCYKKIAAATETAAQENASEYLAITSESKENCTGVQVAE
rNaβ1-1   -------------------D-------------L----------------------------------V-----------A---------------------------
                   |         |         |         |         |         |         |         |         |         |   218
```

FIGURE 3. Alignment of the sodium channel β_1-subunit amino acid sequences. The alignment was generated using the program CLUSTALV (Higgins et al., 1992). Dashes indicate that the rNAβ1-1 sequence is identical to the hNaβ1-1 sequence

from the cyclic nucleotide-gated channels, while the sodium and calcium channels arose as a separate lineage. Hille (1989) has proposed that the ancestral sodium/calcium channel consisted of a single domain, which then duplicated twice to result in the current four domain structure. Domains I and III are most similar to each other, as are domains II and IV. Based on these similarities, the ancestral channel probably duplicated into a two subunit channel consisting of domains I–III and II–IV, and this channel then duplicated to result in the structure I, II, III, and IV. The evolution of the different types of sodium channels must have occurred after these initial duplications. According to the most parsimonious alignment of Strong et al. (1993), the *Drosophila* sodium channels diverged before the multiple different mammalian sodium channels, so that the distinct *Drosophila* genes would represent separate gene duplications from the mammalian genes. Based on this analysis the *Drosophila* sodium channel genes would not necessarily correspond to any of the mammalian sodium channel types. Since no functional studies have yet been carried out using the *Drosophila* cDNA clones, the only basis for comparison is sequence alignment. Of course, this conclusion may change if additional *Drosophila* sodium channel genes that are more similar to the different mammalian genes are isolated. The most parsimonious alignment utilizing conserved regions of eight sodium channel α-subunit sequences is shown in Figure 4, which was generously provided by Dr. George Gutman from the University of California, Irvine.

2.2.4 Structure–Function Correlations

An extensive amount of information has been obtained correlating the electrophysiological functions of the sodium channel with specific structural regions of the molecule (Stühmer, 1991; Catterall, 1991; Stühmer and Parekh, 1992). The basic approach has been to construct mutations of defined residues, and then to examine the electrophysiological effects of those mutations by expression in *Xenopus* oocytes or mammalian cells. All of these studies have been carried out using the various rat sodium channel cDNA clones. Some of the properties that have been examined include voltage-dependent activation, fast inactivation, and ion permeation.

2.2.4.1 Voltage-Dependent Activation

Sodium channel activation is thought to be initiated by the movement of a voltage sensor in the α-subunit, resulting in a conformational change that opens the pore (Armstrong, 1975). The most likely candidates for the voltage sensor are one or more of the S4 segments, with the positively charged amino acids at every third position moving in response to changes in membrane potential (Catterall, 1986). Consistent with this idea is the fact that the voltage-sensitive calcium and potassium channels, which also open in response to membrane depolarization, have S4 segments with positively charged amino acids at every third position (Jan and Jan, 1989; Catterall,

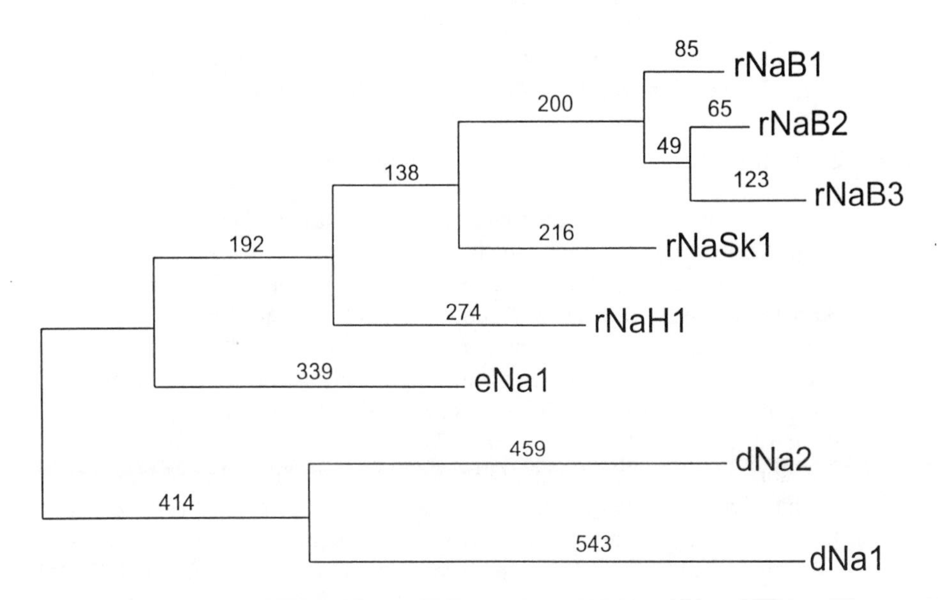

FIGURE 4. Evolutionary relationships among the sodium channel α-subunit genes. The most parsimonious evolutionary tree based on amino acid sequence alignments is shown. The tree was generated as described by Strong et al. (1993). The amino acid sequences for the different α-subunits are from the sources indicated in Table 1. The numbers represent branch lengths, with a total tree length of 3097 substitutions. This figure was generously provided by Dr. George Gutman from the University of California, Irvine.

1988). More direct evidence for the importance of the positive charges has come from the studies of Stühmer et al. (1989), who found that neutralizing positive charges in rNaB2 IS4 or IIS4 decreased the valence of the effective single-gate charge and shifted voltage-dependent activation in either the hyperpolarizing or depolarizing direction, depending on whether the neutralized amino acid was near the amino or carboxy end of S4. Similar mutagenesis studies using the *Shaker* B voltage-sensitive potassium channel have confirmed the importance of the positive charges in S4 for activation (Papazian et al., 1991). However, more detailed analyses of potassium channel charge neutralization mutants using limiting slope analysis have indicated that individual charges respond differently in the membrane electric field, so that purely electrostatic effects cannot explain all of the data (Logothetis et al., 1992; Liman et al., 1991). In addition, the nonpolar amino acids in the S4 regions of both the rNaB2A sodium channel (Auld et al., 1990) and potassium channels (Zagotta and Aldrich, 1990; Gautam and Tanouye, 1990; Lopez et al., 1991; McCormack et al., 1991; Schoppa et al., 1992) are critical for normal voltage-dependent activation. It is therefore apparent that the charged residues in the S4 regions are involved as part of the sodium channel voltage sensor, but the mechanism by which these regions are functioning is unknown. It is also unknown if the S4 regions in each of the four domains have different roles in the activation process, as suggested for domains I and II by the data of Stühmer et al. (1989).

2.2.4.2 Inactivation

Fast inactivation in the sodium channel is known to involve some intracellular protein component, since treatment of the inside of a squid axon with pronase markedly slows inactivation (Armstrong et al., 1973; Rudy, 1978). Armstrong and Bezanilla (1977) proposed that the intracellular protein functions like a ball on a string, swinging into place to occlude the pore following activation. Evidence consistent with this model has been obtained for the potassium channel by Hoshi et al. (1990) and Zagotta et al. (1990), who demonstrated that the amino terminus of the *Shaker* potassium channel functions as the inactivation particle. The region implicated in sodium channel inactivation is the cytoplasmic linker between domains III and IV. Antibodies against a peptide within this linker prevent inactivation of rNaB2A (Vassilev et al., 1988, 1989), and coinjection of rNaB2 I–III and IV as separate constructs results in a channel that inactivates very slowly (Stühmer et al., 1989). In addition, small insertions at the ends of the rNaB2A III–IV linker slow inactivation (Patton and Goldin, 1991), and phosphorylation by protein kinase C of a single serine residue in the rNaB2A linker slows inactivation of the channel (West et al., 1991).

The III–IV linker is 53 amino acids long, and its sequence is highly conserved among the different rat sodium channels. Although this linker contains numerous charged amino acids (12 positive charges and 3 negative charges), these charged residues are not essential for fast inactivation in either rNaB3 (Moorman et al., 1990) or rNaB2A (Patton et al., 1992). In fact, neutralization of the majority of the charges had very minimal effects on the kinetics of fast inactivation. On the other hand, deletion of the first 10 amino acids in the linker completely eliminated fast inactivation in the rNaB2A channel (Patton et al., 1992). Furthermore, substitution of glutamines for a conserved cluster of hydrophobic amino acids (isoleucine, phenylalanine, methionine) also eliminated fast inactivation in the same channel (West et al., 1992). The residue at the phenylalanine position was shown to be the most important of the three amino acids (West et al., 1992). Inactivation correlated with hydrophobicity of the amino acid at that position, suggesting that the phenylalanine residue might form the nucleus of an inactivation particle in the sodium channel (West et al., 1992; Scheuer et al., 1993). Consistent with this hypothesis, the III–IV linker was capable of inactivating a potassium channel when attached to the amino-terminus as part of a chimeric channel, and the same mutation that eliminated fast inactivation in the sodium channel also eliminated it in the chimeric channel (Patton et al., 1993). All of these results suggest that the linker between domains III and IV in the sodium channel functions as the inactivation particle originally suggested by Armstrong and Bezanilla (1977).

2.2.4.3 Permeation

The region of the sodium channel molecule that comprises the actual pore most likely resides between the S5 and S6 segments. The most convincing evidence for this conclusion has been obtained for voltage-gated potassium channels, for which sensitivity to the pore-blocking toxins charybdotoxin and tetraethylammonium ion, and conductance and selectivity have all been altered either by mutations within this region of by transfer of part of this region from one channel to another (MacKinnon and Miller, 1989; MacKinnon et al., 1990; MacKinnon and Yellen, 1990; Yool and Schwarz, 1991; Hartmann et al., 1991; Yellen et al., 1991; Kavanaugh et al., 1991; Heginbotham and MacKinnon, 1992; Kirsch et al., 1992; Choi et al., 1993).

The most informative probes for the sodium channel have been tetrodotoxin and saxitoxin, which are known to block the pore of the channel (Worley et al., 1986; Sigworth and Spalding, 1980; Heggeness and Starkus, 1986). Mutations that make the channel less sensitive to block by both toxins have been identified in the S5–S6 region of all four domains of rNaB2 (Noda et al., 1989; Terlau et al., 1991) and rNaB2A (Kontis and Goldin, 1993). Consistent with the hypothesis that this region lines the channel pore, two of the tetrodotoxin-resistant mutations in the S5–S6 region of rNaB2 domain I (D384N and E387Q) dramatically decrease ionic permeability without affecting gating of the channel, as demonstrated by the measurement of normal gating currents with greatly reduced ionic currents (Pusch et al., 1991).

Mutations in the S5–S6 region also affect sensitivity of the sodium channel to cadmium. Mutation of a cysteine to tyrosine in the S5–S6 region of rNaH1 domain I (C374Y) increases sensitivity to tetrodotoxin and decreases sensitivity to cadmium, making the channel appear more like the skeletal muscle rNaSk1 channel that contains a tyrosine at the comparable position (Satin et al., 1992). The converse mutation (tyrosine to cysteine in rNaSk1, Y401C) makes that channel more like the cardiac channel, with decreased sensitivity to tetrodotoxin and increased sensitivity to cadmium (Backx et al., 1992). A mutation of phenylalanine to cysteine (F385C) in the comparable region of rNaB2 has a similar effect, resulting in decreased sensitivity to tetrodotoxin and increased sensitivity to cadmium (Heinemann et al., 1992a). This particular cysteine residue is therefore a critical component of sodium channel sensitivity to tetrodotoxin and cadmium, both of which block the channel mouth, so it must be located very close to the actual pore of the channel.

Finally, one mutation in the S5–S6 region of rNaB2 domain III (K1422E) and one mutation in the comparable region of rNaB2 domain IV (A1714E) alter the selectivity of the sodium channel (Heinemann et al., 1992b). The channels containing either of these mutations have some of the selectivity properties of calcium channels, with loss of sodium–potassium selectivity and block by external calcium ions (Heinemann et al., 1992b). All of these results indicate that sequences in the S5–S6 regions of the four sodium channel domains comprise part of the permeation pathway. This region, termed SS1 and SS2 (short segments 1 and 2), has been predicted to make up part of the permeation pathway by Guy and co-workers (Guy and Conti, 1990; Durell and Guy, 1992; Guy, 1988). The results suggest that the comparable regions from the four domains probably do not have equivalent roles, however, with selectivity determined primarily by the residues in domains III and IV.

2.2.5 Clinical Correlations

Disorders of sodium channel inactivation have been implicated in the etiology of two human neuromuscular diseases, hyperkalemic periodic paralysis (HYPP) and paramyotonia congenita (PMC) (Rudel and Lehmann-Horn, 1985; Barchi, 1992; Ptacek et al., 1993b). HYPP is a disease in which increased levels of serum potassium lead to muscle hypoexcitability and paralysis. PMC patients experience cold-induced weakness and paralysis that is aggravated by increased muscle activity. These diseases are inherited in an autosomal dominant manner. Electrophysiological studies of cells from patients with these disorders have shown that sodium channel inactivation is slower than normal. In particular, sodium channels in muscle cells from HYPP patients demonstrated an aberrant gating mode in which the channels continued to reopen (Cannon et al., 1991; Lehmann-Horn et al., 1991).

It has since been shown that HYPP (Fontaine et al., 1990; Ebers et al., 1991; Ptacek et al., 1991b; Ptacek et al., 1991c) and PMC (Ptacek et al., 1991b; Ebers et al., 1991) are tightly linked to the SCN4A gene encoding the hNaSk1 skeletal muscle sodium channel. The actual mutations that cause these diseases have been identified in some families. As shown in Figure 1, two different mutations that result in HYPP have been identified, alteration of threonine to methionine (T704M) in domain IIS5 (Ptacek et al., 1991a) and methionine to valine (M1592V) in domain IVS6 (Rojas et al., 1991). Five mutations in three distinct regions have been identified as causing PMC (Figure 1). These mutations include alteration of arginine to either cysteine or histidine (R1448C/H) in domain IVS4 (Ptacek et al., 1992), alteration of leucine to arginine (L1433R) in domain IVS3 (Ptacek et al., 1993a), and two mutations in the III–IV linker, alteration of glycine to valine (G1306V) and threonine to methionine (T1313M) (McClatchey et al., 1992).

The functional effects of the two different mutations causing HYPP have been examined in rNaSk1 by expression in human embryonic kidney cells (Cannon and Strittmatter, 1993; Cummins et al., 1993). Both sets of investigators noted an increase in the sustained level of current, indicating that these mutations do appear to make sodium channel inactivation less complete. Cannon and Strittmatter (1993) also observed that the two mutations delay the kinetics of fast inactivation, but that activation was unaffected. In contrast, Cummins et al. (1993) observed a hyperpolarizing shift in the voltage dependence of activation with no effect on the kinetics of fast inactivation for the T704M mutation. This shift may account for the noninactivating sodium currents at negative potentials observed in muscles from HYPP patients

(Lehmann-Horn et al., 1987). Surprisingly, no effect of increasing extracellular potassium was observed. The functional effects of the mutations causing PMC are yet to be examined in an *in vitro* expression system. However, the results with the two HYPP mutations examined thus far indicate that very subtle changes in the electrophysiological properties of the sodium channel can have significant clinical implications.

2.2.6 Conclusion

The voltage-gated sodium channels from different tissues have previously been shown to be distinct based on pharmacological and physiological criteria. With the isolation of cDNA clones for the channels from a variety of tissues, it has become clear that this diversity reflects the expression of distinct sodium channel genes in each tissue. In fact, based on sequence comparisons between the channels from rat and human, there is greater similarity among the channels from the same tissue than from the same species. These results indicate that the tissue specialization of sodium channels preceded the separation of at least rats and humans. The tissue-specific diversity also suggests that each sodium channel type may be optimized either for specialized function or for subcellular localization in the tissue of origin. The divergence of the human heart 2 and rat glial 1 channels may reflect this specialization. There most likely exist genes encoding other sodium channels that are even more divergent than these two types, and once clones have been isolated it will be interesting to compare their sequences and properties with those of the sodium channels that have thus far been characterized.

Acknowledgments

I thank Dr. Kris Kontis, Dr. Dave Patton, Ray Smith, and Ted Shih for critical reading of the manuscript, and Dr. George Gutman for assistance in preparing the sequence alignments. Work in the author's laboratory is supported by grants from NIH, NSF, the Muscular Dystrophy Association, and the National Multiple Sclerosis Society.

REFERENCES

Agnew, W. S., Levinson, S. R., Brabson, J. S., and Raftery, M. A. (1978). Purification of the tetrodotoxin-binding component associated with the voltage-sensitive sodium channel from *Electrophorus electricus* electroplax membranes. *Proc. Natl. Acad. Sci. U.S.A.* 75, 2606–2610.

Agnew, W. S., Moore, A. C., Levinson, S. R., and Raftery, M. A. (1980). Identification of a large molecular weight peptide associated with a tetrodotoxin binding protein from the electroplax of *Electrophorus electricus*. *Biochem. Biophys. Res. Commun.* 92, 860–866.

Ahmed, C. M. I., Auld, V. J., Lester, H. A., Dunn, R., and Davidson, N. (1990). Both sodium channel II and IIA a subunits are expressed in rat brain. *Nucl. Acids Res.* 18, 5907.

Ahmed, C. M. I., Ware, D. H., Lee, S. C., Patten, C. D., Ferrer-Montiel, A. V., Schinder, A. F., McPherson, J. D., Wagner-McPherson, C. B., Wasmuth, J. J., Evans, G. A., and Montal, M. (1992). Primary structure, chromosomal localization, and functional expression of a voltage-gated sodium channel from human brain. *Proc. Natl. Acad. Sci. U.S.A.* 89, 8220–8224.

Anderson, P. A. V., Holman, M. A., and Greenberg, R. M. (1993). Deduced amino acid sequence of a putative sodium channel from the scyphozoan jellyfish *Cyanea capillata*. *Proc. Natl. Acad. Sci. U.S.A.* 90, 7419–7423.

Armstrong, C. M. (1975). Ionic pores, gates, and gating currents. *Q. Rev. Biophys.* 2, 179–210.

Armstrong, C. M. and Bezanilla, F. (1977). Inactivation of the sodium channel. II. Gating current experiments. *J. Gen. Physiol.* 70, 567–590.

Armstrong, C. M., Bezanilla, F., and Rojas, E. (1973). Destruction of sodium conductance inactivation in squid axons perfused with pronase. *J. Gen. Physiol.* 62, 375–391.

Auld, V. J., Goldin, A. L., Krafte, D. S., Marshall, J., Dunn, J. M., Catterall, W. A., Lester, H. A., Davidson, N., and Dunn, R. J. (1988). A rat brain Na^+ channel a subunit with novel gating properties. *Neuron* 1, 449–461.

Auld, V. J., Goldin, A. L., Krafte, D. S., Catterall, W. A., Lester, H. A., Davidson, N., and Dunn, R. J. (1990). A neutral amino acid change in segment IIS4 dramatically alters the gating properties of the voltage-dependent sodium channel. *Proc. Natl. Acad. Sci. U.S.A.* 87, 323–327.

Backx, P. H., Yue, D. T., Lawrence, J. H., Marban, E., and Tomaselli, G. F. (1992). Molecular localization of an ion-binding site within the pore of mammalian sodium channels. *Science* 257, 248–251.

Barchi, R. L. (1992). Sodium channel gene defects in the periodic paralyses. *Curr. Opinion Neurobiol.* 2, 631–637.

Barchi, R. L., Cohen, S. A., and Murphy, L. E. (1980). Purification from rat sarcolemma of the saxitoxin-binding component of the excitable membrane sodium channel. *Proc. Natl. Acad. Sci. U.S.A.* 77, 1306–1310.

Barres, B. A., Chun, L. L. Y., and Corey, D. P. (1989). Glial and neuronal forms of the voltage-dependent sodium channel: Characteristics and cell-type distribution. *Neuron* 2, 1375–1388.

Bean, B. P., Cohen, C. J., and Tsien, R. W. (1983). Lidocaine block of cardiac sodium channels. *J. Gen. Physiol.* 81, 613–642.

Beckh, S. (1990). Differential expression of sodium channel mRNAs in rat peripheral nervous system and innervated tissues. *FEBS Lett.* 262, 317–322.

Beckh, S., Noda, M., Lübbert, H., and Numa, S. (1989). Differential regulation of three sodium channel messenger RNAs in the rat central nervous system during development. *EMBO J.* 8, 3611–3636.

Bennett, P. B., Jr., Makita, N., and George, A. L., Jr. (1993). A molecular basis for gating mode transitions in human skeletal muscle Na^+ channels. *FEBS Lett.* 326, 21–24.

Bezanilla, F. (1985). Gating of sodium and potassium channel. *J. Membr. Biol.* 88, 97–111.

Brown, A. M., Lee, K. S., and Powell, T. (1981). Voltage clamp and internal perfusion of single rat heart muscle cells. *J. Physiol.* 318, 455–477.

Brysch, W., Creutzfeldt, O. D., Luno, K., Schlingensiepen, R., and Schlingensiepen, K.-H. (1991). Regional and temporal expression of sodium channel messenger RNAs in the rat brain during development. *Exp. Brain Res.* 86, 562–567.

Caffrey, J. M., Eng, D. L., Black, J. A., Waxman, S. G., and Kocsis, J. D. (1992). Three types of sodium channels in adult rat dorsal root ganglion neurons. *Brain Res.* 592, 283–297.

Canessa, C. M., Horisberger, J.-D., and Rossier, B. C. (1993). Epithelial sodium channel related to proteins involved in neurodegeneration. *Nature* 361, 467–470.

Cannon, S. C. and Strittmatter, S. M. (1993). Functional expression of sodium channel mutations identified in families with periodic paralysis. *Neuron* 10, 317–326.

Cannon, S. C., Brown, R. H., Jr., and Corey, D. P. (1991). A sodium channel defect in hyperkalemic periodic paralysis: Potassium-induced failure of inactivation. *Neuron* 6, 619–626.

Cannon, S. C., McClatchey, A. I., and Gusella, J. F. (1993). Modification of the Na^+ current conducted by the rat skeletal muscle α subunit by co-expression with a human brain β subunit. *Pflügers Arch.* 423, 155–157.

Casadei, J. M., Gordon, R. D., Lampson, L. A., Schotland, D. L., and Barchi, R. L. (1984). Monoclonal antibodies against the voltage-sensitive Na^+ channel from mammalian skeletal muscle. *Proc. Natl. Acad. Sci. U.S.A.* 81, 6227–6231.

Casadei, R. M., Gordon, R. D., and Barchi, R. L. (1986). Immunoaffinity isolation of Na^+ channels from rat skeletal muscle — analysis of subunits. *J. Biol. Chem.* 261, 4318–4323.

Catterall, W. A. (1980). Neurotoxins that act on voltage-sensitive sodium channels in excitable membranes. *Annu. Rev. Pharmacol. Toxicol.* 20, 15–43.

Catterall, W. A. (1986). Voltage-dependent gating of sodium channels: Correlating structure and function. *Trends Neurosci.* 9, 7–10.

Catterall, W. A. (1988). Structure and function of voltage-sensitive ion channels. *Science* 242, 50–61.

Catterall, W. A. (1991). Structure and function of voltage-gated sodium and calcium channels. *Curr. Opinion Neurobiol.* 1, 5–13.

Catterall, W. A. (1992). Cellular and molecular biology of voltage-gated sodium channels. *Physiol. Rev.* 72, S15–S48.

Chiu, S. Y., Schrager, P., and Ritchie, J. M. (1984). Neuronal-type Na^+ and K^+ channels in rabbit cultured Schwann cells. *Nature (London)* 311, 156–157.

Choi, K. L., Mossman, C., Aube, J., and Yellen, G. (1993). The internal quaternary ammonium receptor site of *Shaker* potassium channels. *Neuron* 10, 533–541.

Cooperman, S. S., Grubman, S. A., Barchi, R. L., Goodman, R. H., and Mandel, G. (1987). Modulation of sodium-channel mRNA levels in rat skeletal muscle. *Proc. Natl. Acad. Sci. U.S.A.* 84, 8721–8725.

Correa, A. M., Bezanilla, F., and Agnew, W. S. (1990). Voltage activation of purified eel sodium channels reconstituted into artificial liposomes. *Biochemistry* 29, 6230–6240.

Costa, M. R. and Catterall, W. A. (1984). Cyclic AMP-dependent phosphorylation of the α subunit of the sodium channel in synaptic nerve ending particles. *J. Biol. Chem.* 259, 8210–8218.

Costa, M. R., Casnellie, J. E., and Catterall, W. A. (1982). Selective phosphorylation of the α subunit of the sodium channel by cAMP-dependent protein kinase. *J. Biol. Chem.* 257, 7918–7921.

Cribbs, L. L., Satin, J., Fozzard, H. A., and Rogart, R. B. (1990). Functional expression of the rat heart I Na^+ channel isoform. Demonstration of properties characteristic of native cardiac Na^+ channels. *FEBS Lett.* 275, 195–200.

Cruz, L. J., Gray, W. R., Olivera, B. M., Zeikus, R. D., Kerr, L., Yoshikami, D., and Moczydlowski, E. (1985). *Conus geographus* toxins that discriminate between neuronal and muscle sodium channels. *J. Biol. Chem.* 260, 9280–9288.

Cummins, T. R., Zhou, J., Sigworth, F. J., Ukomadu, C., Stephan, M., Ptacek, L. J., and Agnew, W. S. (1993). Functional consequences of a Na^+ channel mutation causing hyperkalemic periodic paralysis. *Neuron* 10, 667–678.

Durell, S. R. and Guy, H. R. (1992). Atomic scale structure and functional models of voltage-gated potassium channels. *Biophys. J.* 62, 238–250.

Ebers, G. C., George, A. L., Jr., Barchi, R. L., Ting-Passador, S. S., Kallen, R. G., Lathrop, G. M., Beckmann, J. S., Hahn, A. F., Brown, W. F., Campbell, R. D., and Hudson, A. J. (1991). Paramyotonia congenita and hyperkalemic periodic paralysis are linked to the adult muscle sodium channel gene. *Ann. Neurol.* 30, 810–816.

Elmer, L. W., O'Brien, B. J., Nutter, T. J., and Angelides, K. J. (1985). Physicochemical characterization of the α-peptide of the sodium channel from rat brain. *Biochemistry* 24, 8128–8137.

Emerick, M. C. and Agnew, W. S. (1989). Identification of phosphorylation sites for adenosine 3′,5′-cyclic phosphate dependent protein kinase on the voltage-sensitive sodium channel from *Electrophorus electricus*. *Biochemistry* 28, 8367–8380.

Fakler, B., Ruppersberg, J. P., Spittelmeister, W., and Rüdel, R. (1990). Inactivation of human sodium channels and the effect of tocainide. *Pflügers Arch.* 415, 693–700.

Fontaine, B., Khurana, T. S., Hoffman, E. P., Bruns, G. A. P., Haines, J. L., Trofatter, J. A., Hanson, M. P., Rich, J., McFarlane, H., Yasek, D. M., Romano, D., Gusella, J. F., and Brown, R. H., Jr. (1990). Hyperkalemic periodic paralysis and the adult muscle sodium channel α-subunit gene. *Science* 250, 1000–1002.

Furman, R. E., Tanaka, J. C., Mueller, P., and Barchi, R. L. (1986). Voltage-dependent activation in purified reconstituted sodium channels from rabbit T-tubular membranes. *Proc. Natl. Acad. Sci. U.S.A.* 83, 488–492.

Furuyama, T., Morita, Y., Inagaki, S., and Takagi, H. (1993). Distribution of I, II and III subtypes of voltage-sensitive Na^+ channel mRNA in the rat brain. *Mol. Brain Res.* 17, 169–173.

Ganetzky, B. (1984). Genetic studies of membrane excitability in *Drosophila:* Lethal interaction between two temperature-sensitive paralytic mutations. *Genetics* 108, 897–911.

Ganetzky, B. (1986). Mutations affecting sodium channels in *Drosophila*. *BioEssays* 5, 11–14.

Gautam, M. and Tanouye, M. A. (1990). Alteration of potassium channel gating: Molecular analysis of the Drosophila Sh^5 mutation. *Neuron* 5, 67–73.

Gautron, S., Dos Santos, G., Pinto-Henrique, D., Koulakoff, A., Gros, F., and Berwald-Netter, Y. (1992). The glial voltage-gated sodium channel: Cell- and tissue-specific mRNA expression. *Proc. Natl. Acad. Sci. U.S.A.* 89, 7272–7276.

Gellens, M. E., George, A. L., Jr., Chen, L., Chahine, M., Horn, R., Barchi, R. L., and Kallen, R. G. (1992). Primary structure and functional expression of the human cardiac tetrodotoxin-insensitive voltage-dependent sodium channel. *Proc. Natl. Acad. Sci. U.S.A.* 89, 554–558.

George, A. L., Jr., Ledbetter, D. H., Kallen, R. G., and Barchi, R. L. (1991). Assignment of a human skeletal muscle sodium channel α-subunit gene *(SCN4A)* to 17q23.1-25.3. *Genomics* 9, 555–556.

George, A. L., Jr., Knittle, T. J., and Tamkun, M. M. (1992a). Molecular cloning of an atypical voltage-gated sodium channel expressed in human heart and uterus: Evidence for a distinct gene family. *Proc. Natl. Acad. Sci. U.S.A.* 89, 4893–4897.

George, A. L., Jr., Komisarof, J., Kallen, R. G., and Barchi, R. L. (1992b). Primary structure of the adult human skeletal muscle voltage-dependent sodium channel. *Ann. Neurol.* 31, 131–137.

George, A. L., Jr., Iyer, G. S., Kleinfield, R., Kallen, R. G., and Barchi, R. L. (1993). Genomic organization of the human skeletal muscle sodium channel gene. *Genomics* 15, 598–606.

Goldin, A. L., Snutch, T., Lübbert, H., Dowsett, A., Marshall, J., Auld, V., Downey, W., Fritz, L. C., Lester, H. A., Dunn, R., Catterall, W. A., and Davidson, N. (1986).

Messenger RNA coding for only the α subunit of the rat brain Na channel is sufficient for expression of functional channels in *Xenopus* oocytes. *Proc. Natl. Acad. Sci. U.S.A.* 83, 7503–7507.

Gonoi, T., Sherman, S. J., and Catterall, W. A. (1985). Voltage clamp analysis of tetrodotoxin-sensitive and -insensitive sodium channels in rat muscle cells developing in vitro. *J. Neurosci.* 5, 2559–2564.

Gordon, D., Merrick, D., Auld, V., Dunn, R., Goldin, A. L., Davidson, N., and Catterall, W. A. (1987). Tissue-specific expression of the R_I and R_{II} sodium channel subtypes. *Proc. Natl. Acad. Sci. U.S.A.* 84, 8682–8686.

Gordon, R. D., Fieles, W. E., Schotland, D. L., Hogue-Angeletti, R., and Barchi, R. L. (1987). Topographical localization of the C-terminal region of the voltage-dependent sodium channel from *Electrophorus electricus* using antibodies raised against a synthetic peptide. *Proc. Natl. Acad. Sci. U.S.A.* 84, 308–312.

Gordon, R. D., Li, Y., Fieles, W. E., Schotland, D. L., and Barchi, R. L. (1988). Topological localization of a segment of the eel voltage-dependent sodium channel primary sequence (AA 927–938) that discriminates between models of tertiary structure. *J. Neurosci.* 8, 3742–3749.

Greenblatt, R. E., Blatt, Y., and Montal, M. (1985). The structure of the voltage-sensitive sodium channel — inferences derived from computer-aided analysis of the *Electrophorus electricus* channel primary structure. *FEBS Lett.* 193, 125–134.

Guy, H. R. (1988). A model relating the structure of the sodium channel to its function. *Curr. Topics Membr. Transp.* 33, 289–308.

Guy, H. R. and Conti, F. (1990). Pursuing the structure and function of voltage-gated channels. *Trends Neurosci.* 13, 201–206.

Guy, H. R. and Seetharamulu, P. (1986). Molecular model of the action potential sodium channel. *Proc. Natl. Acad. Sci. U.S.A.* 83, 508–512.

Han, J., Lu, C.-M., Brown, G. B., and Rado, T. A. (1991). Direct amplification of a single dissected chromosomal segment by polymerase chain reaction: A human brain sodium channel gene is on chromosome 2q22-q23. *Proc. Natl. Acad. Sci. U.S.A.* 88, 335–339.

Hanke, W., Boheim, G., Barhanin, J., Pauron, D., and Lazdunski, M. (1984). Reconstitution of highly purified saxitoxin-sensitive Na^+-channels into planar lipid bilayers. *EMBO J.* 3, 509–515.

Hartmann, H. A., Kirsch, G. E., Drewe, J. A., Taglialatela, M., Joho, R. H., and Brown, A. M. (1991). Exchange of conduction pathways between two related K^+ channels. *Science* 251, 942–944.

Hartshorne, R. P. and Catterall, W. A. (1981). Purification of the saxitoxin receptor of the sodium channel from rat brain. *Proc. Natl. Acad. Sci. U.S.A.* 78, 4620–2464.

Hartshorne, R. P. and Catterall, W. A. (1984). The sodium channel from rat brain — purification and subunit composition. *J. Biol. Chem.* 259, 1667–1675.

Hartshorne, R. P., Messner, D. J., Coppersmith, J. C., and Catterall, W. A. (1982). The saxitoxin receptor of the sodium channel from rat brain — evidence for two nonidentical subunits. *J. Biol. Chem.* 257, 13888–13891.

Hartshorne, R. P., Keller, B. U., Talvenheimo, J. A., Catterall, W. A., and Montal, M. (1985). Functional reconstitution of the purified brain sodium channel in planar lipid bilayers. *Proc. Natl. Acad. Sci. U.S.A.* 82, 240–244.

Heggeness, S. T. and Starkus, J. G. (1986). Saxitoxin and tetrodotoxin. Electrostatic effects on sodium channel gating current in crayfish axons. *Biophys. J.* 49, 629–643.

Heginbotham, L. and MacKinnon, R. (1992). The aromatic binding site for tetraethylammonium ion on potassium channels. *Neuron* 8, 483–491.

Heinemann, S. H., Terlau, H., and Imoto, K. (1992a). Molecular basis for pharmacological differences between brain and cardiac sodium channels. *Pflügers Arch.* 422, 90–92.

Heinemann, S. H., Terlau, H., Stühmer, W., Imoto, K., and Numa, S. (1992b). Calcium channel characteristics conferred on the sodium channel by single mutations. *Nature (London)* 356, 441–443.

Higgins, D. G., Bleasby, A. J., and Fuchs, R. (1992). CLUSTALV: Improved software for multiple sequence alignment. *CABIOS* 8, 189–191.

Hille, B. (1989). Ionic channels: Evolutionary origins and modern roles. *Q. J. Exp. Physiol.* 74, 785–804.

Hirono, C., Yamagishi, S., Ohara, R., Hisanaga, Y., Nakayama, T., and Sugiyama, H. (1985). Characterization of mRNA responsible for induction of functional sodium channels in *Xenopus* oocytes. *Brain Res.* 359, 57–64.

Hodgkin, A. L. and Huxley, A. F. (1952). A quantitative description of membrane current and its application to conduction and excitation in nerve. *J. Physiol. (London)* 117, 500–544.

Hoshi, T., Zagotta, W. N., and Aldrich, R. W. (1990). Biophysical and molecular mechanisms of *Shaker* potassium channel inactivation. *Science* 250, 533–538.

Howe, J. R. and Ritchie, J. M. (1990). Sodium currents in Schwann cells from myelinated and non-myelinated nerves of neonatal and adult rabbits. *J. Physiol.* 425, 169–210.

Isom, L. L., DeJongh, K. S., Patton, D. E., Reber, B. F. X., Offord, J., Charbonneau, H., Walsh, K., Goldin, A. L., and Catterall, W. A. (1992). Primary structure and functional expression of the β_1 subunit of the rat brain sodium channel. *Science* 256, 839–842.

Jackson, F. R., Wilson, S. D., Strichartz, G. R., and Hall, L. M. (1984). Two types of mutants affecting voltage-sensitive sodium channels in *Drosophila melanogaster*. *Nature (London)* 308, 189–191.

Jackson, F. R., Gitschier, J., Strichartz, G. R., and Hall, L. M. (1985). Genetic modifications of voltage-sensitive sodium channels in *Drosophila:* Gene dosage studies of the seizure locus. *J. Neurosci.* 5, 1144–1151.

Jackson, F. R., Wilson, S. D., and Hall, L. M. (1986). The *tip-E* mutation of *Drosophila* decreases saxitoxin binding and interacts with other mutations affecting nerve membrane excitability. *J. Neurogenet.* 3, 1–17.

James, W. M. and Agnew, W. S. (1987). Multiple oligosaccharide chains in the voltage-sensitive Na channel from *Electrophorus electricus:* Evidence for α-2,8-linked polysialic acid. *Biochem. Biophys. Res. Commun.* 148, 817–826.

Jan, L. Y. and Jan, Y.-N. (1989). Voltage-sensitive ion channels. *Cell* 56, 13–25.

Joho, R. H., Moorman, J. R., VanDongen, A. M. J., Kirsch, G. E., Silberberg, H., Schuster, G., and Brown, A. M. (1990). Toxin and kinetic profile of rat brain type III sodium channel expressed in *Xenopus* oocytes. *Mol. Brain Res.* 7, 105–113.

Kallen, R. G., Sheng, Z.-H., Yang, J., Chen, L., Rogart, R. B., and Barchi, R. L. (1990). Primary structure and expression of a sodium channel characteristic of denervated and immature rat skeletal muscle. *Neuron* 4, 233–242.

Kauvar, L. M. (1982). Reduced [^{3}H]-tetrodotoxin binding in the *nap^{ts}* paralytic mutant of *Drosophila*. *Mol. Gen. Genet.* 187, 172–173.

Kavanaugh, M. P., Varnum, M. D., Osborne, P. B., Christie, M. J., Busch, A. E., Adelman, J. P., and North, R. A. (1991). Interaction between tetraethylammonium and amino acid residues in the pore of cloned voltage-dependent potassium channels. *J. Biol. Chem.* 266, 7583–7587.

Kayano, T., Noda, M., Flockerzi, V., Takahashi, H., and Numa, S. (1988). Primary structure of rat brain sodium channel III deduced from the cDNA sequence. *FEBS Lett.* 228, 187–194.

Keller, B. U., Hartshorne, R. P., Talvenheimo, J. A., Catterall, W. A., and Montal, M. (1986). Sodium channels in planar lipid bilayers — channel gating kinetics of purified sodium channels modified by batrachotoxin. *J. Gen. Physiol.* 88, 1–23.

Kernan, M. J., Kuroda, M. I., Kreber, R., Baker, B. S., and Ganetzky, B. (1991). *nap^{ts}*, a mutation affecting sodium channel activity in Drosophila, is an allele of *mle,* a regulator of X chromosome transcription. *Cell* 66, 949–959.

Kirsch, G. E., Drewe, J. A., Hartmann, H. A., Taglialatela, M., de Biasi, M., Brown, A. M., and Joho, R. H. (1992). Differences between the deep pores of K^+ channels determined by an interacting pair of nonpolar amino acids. *Neuron* 8, 499–505.

Koefoed-Johnson, V. and Ussing, H. H. (1958). The nature of the frog skin potential. *Acta Physiol. Scand.* 42, 298–308.

Kontis, K. J. and Goldin, A. L. (1993). Site-directed mutagenesis of the putative pore region of the rat IIA sodium channel. *Mol. Pharmacol.* 43, 635–644.

Kosower, E. M. (1985). A structural and dynamic molecular model for the sodium channel of *Electrophorus electricus*. *FEBS Lett.* 182, 234–242.

Kostyuk, P. G., Veselovsky, N. S., and Tsyandryenko, A. Y. (1981). Ionic currents in the somatic membrane of rat dorsal root ganglion neurons: I. Sodium currents. *J. Neurosci.* 6, 2423–2430.

Krafte, D. S., Snutch, T. P., Leonard, J. P., Davidson, N., and Lester, H. A. (1988). Evidence for the involvement of more than one mRNA species in controlling the inactivation process of rat brain Na channels expressed in *Xenopus* oocytes. *J. Neurosci.* 8, 2859–2868.

Krafte, D. S., Goldin, A. L., Auld, V. J., Dunn, R. J., Davidson, N., and Lester, H. A. (1990). Inactivation of cloned Na channels expressed in *Xenopus* oocytes. *J. Gen. Physiol.* 96, 689–706.

Kraner, S. D., Chong, J. A., Tsay, H.-J., and Mandel, G. (1992). Silencing the type II sodium channel gene: A model for neural-specific gene regulation. *Neuron* 9, 37–44.

Kulkarni, S. J. and Padhye, A. (1982). Temperature-sensitive paralytic mutations on the second and third chromosomes of *Drosophila melanogaster*. *Genet. Res.* 40, 191–199.

Kyle, J. W., Chang, S. Y., Satin, J., Fang, J. M., Fozzard, H. A., and Rogart, R. B. (1993). Rat brain β_1 Na channel subunit interacts with rat brain IIa, skeletal muscle (mu1), and rat heart (RH1) α-subunits. *Biophys. J.* 64, A88 (Abstr.).

Lehmann-Horn, F., Kuther, G., Ricker, K., Grafe, P., Ballanyi, K., and Rudel, R. (1987). Adynamia episodica hereditaria with myotonia: A non-inactivating sodium current and the effect of extracellular pH. *Muscle Nerve* 10, 363–374.

Lehmann-Horn, F., Iaizzo, P. A., Hatt, H., and Franke, C. (1991). Altered gating and conductance of Na^+ channels in hyperkalemic periodic paralysis. *Pflügers Arch.* 418, 297–299.

Levinson, S. R. and Ellory, J. C. (1973). Molecular size of the tetrodotoxin binding site estimated by irradiation inactivation. *Nature New Biol.* 245, 122–123.

Liman, E. R., Hess, P., Weaver, F., and Koren, G. (1991). Voltage-sensing residues in the S4 region of a mammalian K^+ channel. *Nature (London)* 353, 752–756.

Lingueglia, E., Voilley, N., Waldmann, R., Lazdunski, M., and Barbry, P. (1993). Expression cloning of an epithelial amiloride-sensitive Na^+ channel. A new channel type with homologies to *Caenorhabditis elegans* degenerins. *FEBS Lett.* 318, 95–99.

Logothetis, D. E., Movahedi, S., Satler, C., Lindpaintner, K., and Nadal-Ginard, B. (1992). Incremental reductions of positive charge within the S4 region of a voltage-gated K^+ channel result in corresponding decreases in gating charge. *Neuron* 8, 531–540.

Lopez, G. A., Jan, Y. N., and Jan, L. Y. (1991). Hydrophobic substitution mutations in the S4 sequence alter voltage-dependent gating in *Shaker* K^+ channels. *Neuron* 7, 327–336.

Loughney, K., Kreber, R., and Ganetzky, B. (1989). Molecular analysis of the *para* locus, a sodium channel gene in Drosophila. *Cell* 58, 1143–1154.

Lu, C.-M., Han, J., Rado, T. A., and Brown, G. B. (1992). Differential expression of two sodium channel subtypes in human brain. *FEBS Lett.* 303, 53–58.

MacKinnon, R. (1991). Determination of the subunit stoichiometry of a voltage-activated potassium channel. *Nature* 350, 232–235.

MacKinnon, R. and Miller, C. (1989). Mutant potassium channels with altered binding of charybdotoxin, a pore-blocking inhibitor. *Science* 245, 1382–1385.

MacKinnon, R. and Yellen, G. (1990). Mutations affecting TEA blockade and ion permeation in voltage-activated K^+ channels. *Science* 250, 276–279.

MacKinnon, R., Heginbotham, L., and Abramson, T. (1990). Mapping the receptor site for a pore-blocking potassium channel inhibitor. *Neuron* 5, 767–771.

Malo, D., Schurr, E., Dorfman, J., Canfield, V., Levenson, R., and Gros, P. (1991). Three brain sodium channel -subunit genes are clustered on the proximal segment of mouse chromosome 2. *Genomics* 10, 666–672.

Mandel, G. (1992). Tissue-specific expression of the voltage-sensitive sodium channel. *J. Membr. Biol.* 125, 193–205.

Maue, R. A., Goodman, R. H., and Mandel, G. (1990). Neuron-specific expression of the rat brain type II sodium channel gene is directed by upstream regulatory elements. *Neuron* 4, 223–231.

McClatchey, A. I., Van den Bergh, P., Pericak-Vance, M. A., Raskind, W., Verellen, C., McKenna-Yasek, D., Rao, K., Haines, J. L., Bird, T., Brown, R. H., Jr., and Gusella, J. F. (1992). Temperature-sensitive mutations in the III–IV cytoplasmic loop region of the skeletal muscle sodium channel gene in paramyotonia congenita. *Cell* 68, 769–774.

McClatchey, A. I., Cannon, S. C., Slaugenhaupt, S. A., and Gusella, J. F. (1993). The cloning and expression of a sodium channel β_1-subunit cDNA from human brain. *Hum. Mol. Genet.* 2, 745–749.

McCormack, K., Tanouye, M. A., Iverson, L. E., Lin, J.-W., Ramaswami, M., McCormack, T., Campanelli, J. T., Mathew, M. K., and Rudy, B. (1991). A role for hydrophobic residues in the voltage-dependent gating of Shaker K^+ channels. *Proc. Natl. Acad. Sci. U.S.A.* 88, 2931–2935.

Messner, D. J. and Catterall, W. A. (1985). The sodium channel from rat brain — separation and characterization of subunits. *J. Biol. Chem.* 260, 10597–10604.

Messner, D. J. and Catterall, W. A. (1986). The sodium channel from rat brain — role of the β_1 and β_2 subunits in saxitoxin binding. *J. Biol. Chem.* 261, 211–215.

Messner, D. J., Feller, D. J., Scheuer, T., and Catterall, W. A. (1986). Functional properties of rat brain sodium channels lacking the β_1 or β_2 subunit. *J. Biol. Chem.* 261, 14882–14890.

Miller, J. A., Agnew, W. S., and Levinson, S. R. (1983). Principal glycopeptide of the tetrodotoxin/saxitoxin binding protein from *Electrophorus electricus:* Isolation and partial chemical and physical characterization. *Biochemistry* 22, 470–476.

Moczydlowski, E., Olivera, B. M., Gray, W. R., and Strichartz, G. R. (1986). Discrimination of muscle and neuronal Na-channel subtypes by binding competition between [^{3}H]saxitoxin and mu-conotoxins. *Proc. Natl. Acad. Sci. U.S.A.* 83, 5321–5325.

Moorman, J. R., Kirsch, G. E., Brown, A. M., and Joho, R. H. (1990). Changes in sodium channel gating produced by point mutations in a cytoplasmic linker. *Science* 250, 688–691.

Mori, N., Schoenherr, C., Vandenbergh, D. J., and Anderson, D. J. (1992). A common silencer element in the SCG10 and type II Na^+ channel genes binds a factor present in nonneuronal cells but not in neuronal cells. *Neuron* 9, 45–54.

Munson, R., Jr., Westermark, B., and Glaser, L. (1979). Tetrodotoxin-sensitive sodium channels in normal human fibroblasts and normal human glia-like cells. *Proc. Natl. Acad. Sci. U.S.A.* 76, 6425–6429.

Nakayama, H., Withy, R. M., and Raftery, M. A. (1982). Use of a monoclonal antibody to purify the tetrodotoxin binding component from the electroplax of *Electrophorus electricus*. *Proc. Natl. Acad. Sci. U.S.A.* 79, 7575–7579.

Noda, M. and Numa, S. (1987). Structure and function of sodium channel. *J. Receptor Res.* 7, 467–497.

Noda, M., Shimizu, S., Tanabe, T., Takai, T., Kayano, T., Ikeda, T., Takahashi, H., Nakayama, H., Kanaoka, Y., Minamino, N., Kangawa, K., Matsuo, H., Raftery, M. A., Hirose, T., Inayama, S., Hayashida, H., Miyata, T., and Numa, S. (1984). Primary structure of *Electrophorus electricus* sodium channel deduced from cDNA sequence. *Nature (London)* 312, 121–127.

Noda, M., Ikeda, T., Kayano, T., Suzuki, H., Takeshima, H., Kurasaki, M., Takahashi, H., and Numa, S. (1986a). Existence of distinct sodium channel messenger RNAs in rat brain. *Nature (London)* 320, 188–192.

Noda, M., Ikeda, T., Suzuki, H., Takeshima, H., Takahashi, T., Kuno, M., and Numa, S. (1986b). Expression of functional sodium channels from cloned cDNA. *Nature (London)* 322, 826–828.

Noda, M., Suzuki, H., Numa, S., and Stühmer, W. (1989). A single point mutation confers tetrodotoxin and saxitoxin insensitivity on the sodium channel II. *FEBS Lett.* 259, 213–216.

Norman, R. I., Schmid, A., Lombet, A., Barhanin, J., and Lazdunski, M. (1983). Purification of binding protein for *tityus* γ toxin identified with the gating component of the voltage-sensitive Na^+ channel. *Proc. Natl. Acad. Sci. U.S.A.* 80, 4164–4168.

Numann, R., Catterall, W. A., and Scheuer, T. (1991). Functional modulation of brain sodium channels by protein kinase C phosphorylation. *Science* 254, 115–118.

O'Dowd, D. K., Germeraad, S. E., and Aldrich, R. W. (1989). Alterations in the expression and gating of Drosophila sodium channels by mutations in the *para* gene. *Neuron* 2, 1301–1311.

O'Dowd, D. K. and Aldrich, R. W. (1988). Voltage-clamp analysis of sodium channels in wild-type and mutant *Drosophila* neurons. *J. Neurosci.* 8, 3633–3643.

Papazian, D. M., Timpe, L. C., Jan, Y. N., and Jan, L. Y. (1991). Alteration of voltage-dependence of *Shaker* potassium channel by mutations in the S4 sequence. *Nature (London)* 349, 305–310.

Pappone, P. A. (1980). Voltage-clamp experiments in normal and denervated mammalian skeletal muscle fibres. *J. Physiol.* 306, 377–410.

Patlak, J. (1991). Molecular kinetics of voltage-dependent Na^+ channels. *Physiol. Rev.* 71, 1047–1080.

Patton, D. E. and Goldin, A. L. (1991). A voltage-dependent gating transition induces use-dependent block by tetrodotoxin of rat IIA sodium channels expressed in *Xenopus* oocytes. *Neuron* 7, 637–647.

Patton, D. E., West, J. W., Catterall, W. A., and Goldin, A. L. (1992). Amino acid residues required for fast sodium channel inactivation. Charge neutralizations and deletions in the III–IV linker. *Proc. Natl. Acad. Sci. U.S.A.* 89, 10905–10909.

Patton, D. E., West, J. W., Catterall, W. A., and Goldin, A. L. (1993b). A peptide segment critical for sodium channel inactivation functions as an inactivation gate in a potassium channel. *Neuron* 11, 967–974.

Patton, D. E., Isom, L. L., Catterall, W. A., and Goldin, A. L. (1994). The adult rat brain β_1 subunit modifies activation and inactivation gating of multiple sodium channel α subunits. *J. Biol. Chem.* 269, 17649–17655.

Pouyssegur, J., Jacques, Y., and Lazdunski, M. (1980). Identification of a tetrodotoxin-sensitive Na^+ channel in a variety of fibroblast lines. *Nature (London)* 286, 162–164.

Ptacek, L. J., George, A. L., Jr., Griggs, R. C., Tawil, R., Kallen, R. G., Barchi, R. L., Robertson, M., and Leppert, M. F. (1991a). Identification of a mutations in the gene causing hyperkalemic periodic paralysis. *Cell* 67, 1021–1027.

Ptacek, L. J., Trimmer, J. S., Agnew, W. S., Roberts, J. W., Petajan, J. H., and Leppert, M. (1991b). Paramyotonia congenita and hyperkalemic periodic paralysis map to the same sodium-channel gene locus. *Am. J. Hum. Genet.* 49, 851–854.

Ptacek, L. J., Tyler, F., Trimmer, J. S., Agnew, W. S., and Leppert, M. (1991c). Analysis in a large hyperkalemic periodic paralysis pedigree supports tight linkage to a sodium channel locus. *Am. J. Hum. Genet.* 49, 378–382.

Ptacek, L. J., George, A. L., Jr., Barchi, R. L., Griggs, R. C., Riggs, J. E., Robertson, M., and Leppert, M. F. (1992). Mutations in an S4 segment of the adult skeletal muscle sodium channel cause paramyotonia congenita. *Neuron* 8, 891–897.

Ptacek, L. J., Gouw, L., Kwiecinski, H., McManis, P., Mendell, J. R., Barohn, R. J., George, A. L., Jr., Barchi, R. L., Robertson, M., and Leppert, M. F. (1993a). Sodium channel mutations in paramyotonia congenita and hyperkalemic periodic paralysis. *Ann. Neurol.* 33, 300–307.

Ptacek, L. J., Johnson, K. J., and Griggs, R. C. (1993b). Genetics and physiology of the myotonic muscle disorders. *N. Engl. J. Med.* 328, 482–489.

Pusch, M., Noda, M., Stühmer, W., Numa, S., and Conti, F. (1991). Single point mutations of the sodium channel drastically reduce the pore permeability without preventing its gating. *Eur. Biophys. J.* 20, 127–133.

Ramaswami, M. and Tanouye, M. A. (1989). Two sodium-channel genes in *Drosophila:* Implications for channel diversity. *Proc. Natl. Acad. Sci. U.S.A.* 86, 2079–2082.

Recio-Pinto, E., Duch, D. S., Levinson, S. R., and Urban, B. W. (1987). Purified and unpurified sodium channels from eel electroplax in planar lipid bilayers. *J. Gen. Physiol.* 90, 375–395.

Recio-Pinto, E., Thornhill, W. B., Duch, D. S., Levinson, S. R., and Urban, B. W. (1990). Neuraminidase treatment modifies the function of electroplax sodium channels in planar lipid bilayers. *Neuron* 5, 675–684.

Rogart, R. B., Cribbs, L. L., Muglia, L. K., Kephart, D. D., and Kaiser, M. W. (1989). Molecular cloning of a putative tetrodotoxin-resistant rat heart Na^+ channel isoform. *Proc. Natl. Acad. Sci. U.S.A.* 86, 8170–8074.

Rojas, C. V., Wang, J., Schwartz, L. S., Hoffman, E. P., Powell, B. R., and Brown, R. H., Jr. (1991). A met-to-val mutation in the skeletal muscle Na^+ channel α-subunit in hyperkalaemic periodic paralysis. *Nature (London)* 354, 387–389.

Rosenberg, R. L., Tomiko, S. A., and Agnew, W. S. (1984a). Reconstitution of neurotoxin-modulated ion transport by the voltage-regulated sodium channel isolated from the electroplax of *Electrophorus electricus*. *Proc. Natl. Acad. Sci. U.S.A.* 81, 1239–1243.

Rosenberg, R. L., Tomiko, S. A., and Agnew, W. S. (1984b). Single-channel properties of the reconstituted voltage-regulated Na channel isolated from the electroplax of *Electrophorus electricus*. *Proc. Natl. Acad. Sci. U.S.A.* 81, 5594–5598.

Rosenthal, J. J. C. and Gilly, W. F. (1993). Amino acid sequence of a putative sodium channel expressed in the giant axon of the squid *Loligo opalescens*. *Proc. Natl. Acad. Sci. U.S.A.* 90, 10026–10030.

Rossie, S. and Catterall, W. A. (1987). Cyclic AMP-dependent phosphorylation of voltage-sensitive sodium channels in primary cultures of rat brain neurons. *J. Biol. Chem.* 262, 12735–12744.

Rossie, S. and Catterall, W. A. (1989). Phosphorylation of the α subunit of rat brain sodium channels by cAMP-dependent protein kinase at a new site containing ser^{686} and ser^{687}. *J. Biol. Chem.* 264, 14220–14224.

Rossie, S., Gordon, D., and Catterall, W. A. (1987). Identification of an intracellular domain of the sodium channel having multiple cAMP-dependent phosphorylation sites. *J. Biol. Chem.* 262, 17530–17535.

Rudel, R. and Lehmann-Horn, F. (1985). Membrane changes in cells from myotonia patients. *Physiol. Rev.* 65, 310–356.

Rudy, B. (1978). Slow inactivation of the sodium conductance in squid giant axons. Pronase resistance. *J. Physiol.* 283, 1–21.

Salkoff, L., Butler, A., Wei, A., Scavarda, N., Giffen, K., Ifune, C., Goodman, R., and Mandel, G. (1987). Genomic organization and deduced amino acid sequence of a putative sodium channel gene in *Drosophila*. *Science* 237, 744–748.

Sarao, R., Gupta, S. K., Auld, V. J., and Dunn, R. J. (1991). Developmentally regulated alternative RNA splicing of rat brain sodium channel mRNAs. *Nucl. Acids Res.* 19, 5673–5679.

Sariban-Sohraby, S. and Benos, D. J. (1986). The amiloride-sensitive sodium channel. *Am. J. Physiol.* 250, C175–C190.

Satin, J., Kyle, J. W., Chen, M., Bell, P., Cribbs, L. L., Fozzard, H. A., and Rogart, R. B. (1992). A mutant of TTX-resistant cardiac sodium channels with TTX-sensitive properties. *Science* 256, 1202–1205.

Sato, C. and Matsumoto, G. (1992). Primary structure of squid sodium channel deduced from the complementary DNA sequence. *Biochem. Biophys. Res. Commun.* 186, 61–68.

Sawaryn, A. and Drouin, H. (1991). Reevaluation of hydropathy profiles of voltage-gated ionic channels. *Experientia* 47, 962–964.

Schaller, K. L., Krzemien, D. M., McKenna, N. M., and Caldwell, J. H. (1992). Alternatively spliced sodium channel transcripts in brain and muscle. *J. Neurosci.* 12, 1370–1381.

Scheinman, R. I., Auld, V. J., Goldin, A. L., Davidson, N., Dunn, R. J., and Catterall, W. A. (1989). Developmental regulation of sodium channel expression in the rat forebrain. *J. Biol. Chem.* 264, 10660–10666.

Scheuer, T., West, J. W., Wang, Y. L., and Catterall, W. A. (1993). Effects of amino acid hydrophobicity at position 1489 on sodium channel inactivation. *Biophys. J.* 64, A88 (Abstr.).

Schmidt, J. W. and Catterall, W. A. (1986). Biosynthesis and processing of the α subunit of the voltage-sensitive sodium channel in rat brain. *Cell* 46, 437–445.

Schmidt, J. W. and Catterall, W. A. (1987). Palmitylation, sulfation, and glycosylation of the α subunit of the sodium channel: Role of post-translational modifications in channel assembly. *J. Biol. Chem.* 262, 13713–13723.

Schmidt, J. W., Rossie, S., and Catterall, W. A. (1985). A large intracellular pool of inactive Na channel α subunits in developing rat brain. *Proc. Natl. Acad. Sci. U.S.A.* 82, 4847–4851.

Schoppa, N. E., McCormack, K., Tanouye, M. A., and Sigworth, F. J. (1992). The size of gating charge in wild-type and mutant *Shaker* potassium channels. *Science* 255, 1712–1715.

Schwartz, A., Palti, Y., and Meiri, H. (1990). Structural and developmental differences between three types of Na channels in dorsal root ganglion cells of newborn rats. *J. Membr. Biol.* 116, 117–128.

Shenkel, S., Cooper, E. C., James, W., Agnew, W. S., and Sigworth, F. J. (1989). Purified, modified eel sodium channels are active in planar bilayers in the absence of activating neurotoxins. *Proc. Natl. Acad. Sci. U.S.A.* 86, 9592–9596.

Shrager, P., Chiu, S. Y., and Ritchie, J. M. (1985). Voltage-dependent sodium and potassium channels in mammalian cultured Schwann cells. *Proc. Natl. Acad. Sci. U.S.A.* 82, 948–952.

Siddiqi, O. and Benzer, S. (1976). Neurophysiological defects in temperature-sensitive mutants of *Drosophila melanogaster*. *Proc. Natl. Acad. Sci. U.S.A.* 73, 3253–3257.

Sigworth, F. J. and Spalding, B. C. (1980). Chemical modification reduces the conductance of sodium channels in nerve. *Nature (London)* 283, 293–295.

Sills, M. N., Xu, Y. C., Baracchini, E., Goodman, R. H., Cooperman, S. S., Mandel, G., and Chien, K. R. (1989). Expression of diverse Na^+ channel messenger RNAs in rat myocardium. Evidence for a cardiac-specific Na^+ channel. *J. Clin. Invest.* 84, 331–336.

Strong, M., Chandy, K. G., and Gutman, G. A. (1993). Molecular evolution of voltage-sensitive ion channel genes: On the origins of electrical excitability. *Mol. Biol. Evol.* 10, 221–242.

Stühmer, W. (1991). Structure-function studies of voltage-gated ion channels. *Annu. Rev. Biophys. Biophys. Chem.* 20, 65–78.

Stühmer, W. and Parekh, A. B. (1992). The structure and function of Na^+ channels. *Curr. Opinion Neurobiol.* 2, 243–246.

Stühmer, W., Conti, F., Suzuki, H., Wang, X., Noda, M., Yahagi, N., Kubo, H., and Numa, S. (1989). Structural parts involved in activation and inactivation of the sodium channel. *Nature (London)* 339, 597–603.

Sumikawa, K., Parker, I., and Miledi, R. (1984). Partial purification and functional expression of brain mRNAs coding for neurotransmitter receptors and voltage-operated channels. *Proc. Natl. Acad. Sci. U.S.A.* 81, 7994–7998.

Sutkowski, E. M. and Catterall, W. A. (1990). β1 subunits of sodium channels. Studies with subunit-specific antibodies. *J. Biol. Chem.* 265, 12393–12399.

Suzuki, D. T., Grigliatti, T. A., and Williamson, R. (1971). Temperature-sensitive mutations in *Drosophila melanogaster.* VII. A mutation *(para^{ts})* causing reversible adult paralysis. *Proc. Natl. Acad. Sci. U.S.A.* 68, 890–893.

Suzuki, H., Beckh, S., Kubo, H., Yahagi, N., Ishida, H., Kayano, T., Noda, M., and Numa, S. (1988). Functional expression of cloned cDNA encoding sodium channel III. *FEBS Lett.* 228, 195–200.

Tamkun, M. M. and Catterall, W. A. (1981). Reconstitution of the voltage-sensitive sodium channel of rat brain from solubilized components. *J. Biol. Chem.* 256, 11457–11463.

Tamkun, M. M., Talvenheimo, J. A., and Catterall, W. A. (1984). The sodium channel from rat brain — reconstitution of neurotoxin-activated ion flux and scorpion toxin binding from purified components. *J. Biol. Chem.* 259, 1676–1688.

Terlau, H., Heinemann, S. H., Stühmer, W., Pusch, M., Conti, F., Imoto, K., and Numa, S. (1991). Mapping the site of block by tetrodotoxin and saxitoxin of sodium channel II. *FEBS Lett.* 293, 93–96.

Thornhill, W. B. and Levinson, S. R. (1986). Biosynthesis of electroplax sodium channels. *Ann. N.Y. Acad. Sci.* 479, 356–363.

Thornhill, W. B. and Levinson, S. R. (1987). Biosynthesis of electroplax sodium channels in *Electrophorus* electrocytes and *Xenopus* oocytes. *Biochemistry* 26, 4381–4388.

Tomiko, S. A., Rosenberg, R. L., Emerick, M. C., and Agnew, W. S. (1986). Fluorescence assay for neurotoxin-modulated ion transport by the reconstituted voltage-activated sodium channel isolated from eel electric organ. *Biochemistry* 25, 2162–2174.

Tong, J., Potts, J. F., Rochelle, J. M., Seldin, M. F., and Agnew, W. S. (1993). A single β_1 subunit mapped to mouse chromosome 7 may be a common component of Na channel isoforms from brain, skeletal muscle and heart. *Biochem. Biophys. Res. Commun.* 195, 679–685.

Trimmer, J. S., Cooperman, S. S., Tomiko, S. A., Zhou, J., Crean, S. M., Boyle, M. B., Kallen, R. G., Sheng, Z., Barchi, R. L., Sigworth, F. J., Goodman, R. H., Agnew, W. S., and Mandel, G. (1989). Primary structure and functional expression of a mammalian skeletal muscle sodium channel. *Neuron* 3, 33–49.

Trimmer, J. S., Cooperman, S. S., Agnew, W. S., and Mandel, G. (1990). Regulation of muscle sodium channel transcripts during development and in response to denervation. *Dev. Biol.* 142, 360–367.

Ukomadu, C., Zhou, J., Sigworth, F. J., and Agnew, W. S. (1992). muI Na^+ channels expressed transiently in human embryonic kidney cells: Biochemical and biophysical properties. *Neuron* 8, 663–676.

Vassilev, P. M., Scheuer, T., and Catterall, W. A. (1988). Identification of an intracellular peptide segment involved in sodium channel inactivation. *Science* 241, 1658–1661.

Vassilev, P., Scheuer, T., and Catterall, W. A. (1989). Inhibition of inactivation of single sodium channels by a site-directed antibody. *Proc. Natl. Acad. Sci. U.S.A.* 86, 8147–8151.

Villegas, J., Sevcik, C., Barnola, F. V., and Villegas, R. (1976). Grayanotoxin, veratrine, and tetrodotoxin-sensitive sodium pathways in the Schwann cell membrane of squid nerve fibres. *J. Gen. Physiol.* 67, 369–380.

Waechter, C. J., Schmidt, J. W., and Catterall, W. A. (1983). Glycosylation is required for maintenance of functional sodium channels in neuroblastoma cells. *J. Biol. Chem.* 258, 5117–5123.

Wang, J., Rojas, C. V., Zhou, J., Schwartz, L. S., Nicholas, H., and Hoffman, E. P. (1992). Sequence and genomic structure of the human adult skeletal muscle sodium channel α subunit gene on 17q. *Biochem. Biophys. Res. Commun.* 182, 794–801.

Weigele, J. B. and Barchi, R. L. (1982). Functional reconstitution of the purified sodium channel protein from rat sarcolemma. *Proc. Natl. Acad. Sci. U.S.A.* 79, 3651–3655.

West, J. W., Numann, R., Murphy, B. J., Scheuer, T., and Catterall, W. A. (1991). A phosphorylation site in the Na^+ channel required for modulation by protein kinase C. *Science* 254, 866–868.

West, J. W., Patton, D. E., Scheuer, T., Wang, Y., Goldin, A. L., and Catterall, W. A. (1992). A cluster of hydrophobic amino acid residues required for fast Na^+ channel inactivation. *Proc. Natl. Acad. Sci. U.S.A.* 89, 10910–10914.

Westenbroek, R. E., Merrick, D. K, and Catterall, W. A. (1989). Differential subcellular localization of the R_I and R_{II} Na^+ channel subtypes in central neurons. *Neuron* 3, 695–704.

White, M. M., Chen, L., Kleinfield, R., Kallen, R. G., and Barchi, R. L. (1991). SkM2, a Na^+ channel cDNA clone from denervated skeletal muscle, encodes a tetrodotoxin-insensitive Na^+ channel. *Mol. Pharmacol.* 39, 604–608.

Wollner, D. A. and Catterall, W. A. (1985). Antigenic differences among the voltage-sensitive sodium channels in the peripheral and central nervous systems and skeletal muscle. *Brain Res.* 331, 145–149.

Worley, J. F., III, French, R. J., and Krueger, B. K. (1986). Trimethyloxonium modification of single batrachotoxin-activated sodium channels in planar bilayers. Changes in unit conductance and its block by saxitoxin and calcium. *J. Gen. Physiol.* 87, 327–349.

Wu, C.-F., Ganetzky, B., Jan, L. Y., Jan, Y.-N., and Benzer, S. (1978). A *Drosophila* mutant with a temperature-sensitive block in nerve conduction. *Proc. Natl. Acad. Sci. U.S.A.* 75, 4047–4051.

Yang, J. S.-J., Sladky, J. T., Kallen, R. G., and Barchi, R. L. (1991). TTX-sensitive and TTX-insensitive sodium channel mRNA transcripts are independently regulated in adult skeletal muscle after denervation. *Neuron* 7, 421–427.

Yang, J. S., Bennett, P. B., Makita, N., George, A. L., Jr., and Barchi, R. L. (1993). Expression of the sodium channel β1 subunit in rat skeletal muscle is selectively associated with the tetrodotoxin-sensitive α subunit isoform. *Neuron* 11, 915–922.

Yarowsky, P. J., Krueger, B. K., Olson, C. E., Clevinger, E. C., and Koos, R. D. (1991). Brain and heart sodium channel subtype mRNA expression in rat cerebral cortex. *Proc. Natl. Acad. Sci. U.S.A.* 88, 6453–6457.

Yellen, G., Jurman, M. E., Abramson, T., and MacKinnon, R. (1991). Mutations affecting internal TEA blockade identify the probable pore-forming region of a K^+ channel. *Science* 251, 939–942.

Yool, A. J., and Schwarz, T. L. (1991). Alteration of ionic selectivity of a K^+ channel by mutation of the H5 region. *Nature (London)* 349, 700–704.

Zagotta, W. N. and Aldrich, R. W. (1990). Alterations in activation gating of single *Shaker* A-type potassium channels by the SH^5 mutation. *J. Neurosci.* 10, 1799–1810.

Zagotta, W. N., Hoshi, T., and Aldrich, R. W. (1990). Restoration of inactivation in mutants of *Shaker* potassium channels by a peptide derived from ShB. *Science* 250, 568–571.

Zhou, X.-M. and Fishman, P. H. (1991). Desensitization of the human β_1-adrenergic receptor. Involvement of the cyclic AMP-dependent but not a receptor-specific protein kinase. *J. Biol. Chem.* 266, 7462–7468.

3

Voltage-Gated Calcium Channels

Anthony Stea, Tuck Wah Soong, and Terry P. Snutch

2.3.0 Introduction

Calcium entry into neurons affects many cellular activities including neuronal firing patterns, neurotransmitter release, gene expression, as well as a myriad of Ca-dependent enzyme-mediated effects. The rapid entry of Ca is mediated through voltage-dependent Ca channels and the elucidation of the structure and function of these channels has been a subject of great interest. Based on their electrophysiological and pharmacological characteristics, voltage-gated Ca channels have been classified into T, L, N, P, and Q types (see reviews Bean 1989a; Hess 1990; Zhang et al., 1993). T-type (low-voltage-activated) channels describe a broad class of molecules that transiently activate at negative potentials, generally in the range of –70 to –50 mV, and are quite sensitive to changes in resting potential. The L-, N-, Q-, and P-type (high voltage-activated) channels activate at more positive potentials (generally > –30 mV) and display diverse kinetics and voltage-dependent properties. There is some overlap in the biophysical properties of high-voltage-activated channels and it is useful to refer to their pharmacological profiles. L-type channels are sensitive to dihydropyridines, N-type channels are potently blocked by a *Conus geographus* peptide toxin (ω-conotoxin GVIA), and P-type channels are blocked by two fractions from the venom of the funnel web spider, *Agelenopsis aperta:* (1) nanomolar concentrations of the peptide ω-agatoxin IVA, and (2) micromolar concentrations of the polyamine FTX (Olivera et al., 1985; Bean, 1989a; Hess; 1990; McCleskey and Schroeder, 1991; Hillyard et al., 1992; Llinas et al., 1992; Mintz et al., 1992; Adams et al., 1993). Although generally used to diagnose Ca channel subtypes, the specificity of these compounds (and others such as ω-conotoxin MVIIC) is not absolute and a combination of electrophysiological and pharmacological characteristics should be considered. A complicating factor to

0-8493-8322-6/95/$0.00+$.50

the present classification scheme is that several types of Ca channels do not fall neatly into any of the categories. Furthermore, there is variability of properties even within a category (e.g., Artalejo et al., 1992; Huguenard and Prince, 1992; Forti and Pietrobon, 1993).

Biochemical purification and immunoprecipitation studies have shown that voltage-gated Ca channels are heteroligomeric complexes. The skeletal muscle L-type Ca channel/dihydropyridine receptor consists of five distinct protein subunits (α_1, α_2, β, δ, and γ; Campbell et al., 1988; Catterall et al., 1988). The α_1-subunit is the main pore-forming subunit and is the site of action of L-type Ca channel agonists and antagonists. In the past few years molecular cloning strategies have utilized the skeletal muscle sequence to identify a heterogeneous gene family of Ca channel α_1-subunits expressed in many tissues and cell types. Multiple β isoforms also exist and have been shown to have significant effects on the biophysical properties of cloned α_1-subunits. In the following sections we describe the primary structures and localization of the different classes of cloned Ca channel α_1-subunits. We also describe structure–function studies that have given some preliminary insight into the important motifs of the α_1-subunit. Finally, a brief overview of the primary structure and modulatory properties of Ca channel ancillary subunits is presented.

2.3.1 Primary Structure of Calcium Channel α_1-Subunits

To date, the primary structure of the main pore-forming (α_1)-subunit of six different types of Ca channels has been determined (Table 1). The structure of the first calcium channel α_1-subunit (α_{1S}) was determined from the skeletal muscle dihydropyridine receptor complex by Tanabe and co-workers (Tanabe et al., 1987). The α_{1S} cDNA coded for a protein of 1873 amino acids with a calculated molecular weight (M_r) of 212 kDa. A single nucleotide deletion in the α_{1S} coding region was shown to underlie the defect in excitation-contraction (E-C) coupling associated with muscular dysgenesis in mice (Chaudhari, 1992). The predicted structure of the α_{1S}-subunit contained four internal homologous repeats (I–IV) each having six putative α-helical membrane spanning segments (S1–S6) with one segment (S4) having positively charged residues every third or fourth amino acid (Tanabe et al., 1987). This structure is predicted for each of the classes of α_1-subunit except for an alternatively spliced α_1 subunit version that contains only two structural domains (Malouf et al., 1992). By analogy with the predicted structure of voltage-gated Na (Catterall, 1991) and K (MacKinnon, 1991) channels, most Ca channel models now also include a β-hairpin structure between each of the S5 and S6 segments and that likely forms part of the channel pore (Guy and Seetharamulu, 1986). The other five types of Ca channel α_1-subunits (classes A to E; using nomenclature of Snutch et al., 1990; Soong et al., 1993) were originally described from the central nervous system, although several of these classes are also expressed in other tissues.

Table 1. Molecular Properties of Cloned Calcium Channel α_1-Subunits

Class	cDNA	Source	Number of a.a.	Predicted molecular mass (kDA)	Distribution	References
α_{1S}		Rabbit skeletal muscle	1873	212	Skeletal Muscle	Tanabe et al. (1987)
	CSkm	Carp skeletal muscle	1852	210		Grabner et al. (1991)
α_{1A}	BI-1	Rabbit brain	2273	257	Brain, heart,	Mori et al. (1991)
	BI-2		2424	273	pituitary, GH4C1,	
	rbA-I	Rat brain	2212	252	PC12, C-cells	Starr et al. (1991)
α_{1B}	α_{1B-1}	Human neuroblastoma	2239	262	Brain, PC12,	Williams et al. (1992b)
	α_{1B-2}	cell line	2237	252	C-cells, IMR32	
	rbB-I	Rat brain	2336	262		Dubel et al. (1992)
	BIII	Rabbit brain	2339	261		Fujita et al. (1993)
	doe-4	*Discopyge ommata*	2326	265		Horne et al. (1993)
α_{1C}	pCARD3	Rabbit heart	2171	243	Brain, heart,	Mikami et al. (1988)
	pSCαL	Rabbit lung	2166	242	lung, pituitary,	Biel et al. (1990)
	rbC-I	Rat brain	2140	240	kidney, aorta,	Snutch et al. (1991)
	rbC-II		2143	240	GH4C1, PC12,	
	VSMα1	Rat aorta	2169	244	C-cells	Koch et al. (1990)
	mbC	Mouse brain	2139	240		Ma et al. (1992)
α_{1D}	α_{1D}	Human neuroblastoma cell line	2161	245	Brain, heart, pituitary,	Williams et al. (1992a)
	CACN 4	Human pancreatic islet	2181	248	pancreas, GH4C1,	Sieno et al. (1992)
	RBα1	Rat hippocampus	1634	187	PC12, C-cells,	Hui et al. (1991)
	HCa3a	Hamster insulin-secreting cell line	1610	182	IMR32, RIN5mF	Yaney et al. (1992)
α_{1E}	BII-1	Rabbit brain	2259	254	Brain, heart,	Niidome et al. (1992)
	BII-2		2178	245	testis, pituitary,	
	rbE-II	Rat brain	2222	252	C-cells,	Soong et al. (1993)
	doe-1	*Discopyge ommata*	2223	252	electric lobe	Horne et al. (1993)

2.3.1.1 Class A (α_{1A})

Class A (α_{1A}) subunit cDNAs have been described from both rabbit brain (BI-1, BI-2; Mori et al., 1991) and rat brain (rbA-I; Starr et al., 1991) and share ~86% amino acid identity. The α_{1A} cDNAs code for proteins of ~2200 to 2400 a.a. residues with M_r values ranging from 252 to 273 kDa (Mori et al., 1991; Starr et al., 1991). In addition, there is evidence for alternative splicing of the α_{1A} gene (Mori et al., 1991). The α_{1A} gene has been suggested to encode a P-type Ca channel (Mori et al., 1991; Starr et al., 1991), however, recent studies have challenged this idea (Tsien et al., 1991; Sather et al., 1993; Stea et al., 1994). Rabbit BI currents were sensitive to funnel-web spider toxin (Mori et al., 1991), but were only slightly sensitive to ω-agatoxin IVA (Sather et al., 1993), a potent blocker of the native P-type channels in Purkinje cells (Mintz et al., 1992). Expression of the cloned rat brain α_{1A} in *Xenopus* oocytes (rbA-I) produced channels that showed a partial blockade (~20%) by 200 n*M* ω-agatoxin IVA and almost complete blockade by 5μ*M* ω-conotoxin MVIIC (Stea et al., 1994). The kinetic and voltage-dependent properties of rbA-I were strongly affected by coexpression with β subunits and reflected similarities to both P- and Q-type currents. These results suggest that α_{1A} isoforms may encode P- and/or Q-type channels.

2.3.1.2 Class B (α_{1B})

Class B (α_{1B}) Ca channel clones have been isolated from rat brain (rbB-I; Dubel et al., 1992), human neuroblastoma (α_{1B-1}, α_{1B-2}; Williams et al., 1992b), and rabbit brain (BIII; Fujita et al., 1993) cDNA libraries. The α_{1B} cDNAs code for proteins of 2336 to 2339 a.a. with calculated M_r values of 261 to 262 kDa and share >90% a.a. identity. Molecular cloning (Williams et al., 1992b) and biochemical studies (Westenbroek et al., 1992) indicated that α_{1B} isoforms differing in their carboxyl-terminal regions are also expressed. Antibodies directed against rbB-I subunit immunoprecipitated radiolabeled high-affinity rat cortical ω-conotoxin GVIA (ω-CgTx) binding sites (Dubel et al., 1992), and transient expression of the human (Williams et al., 1992b), rabbit (Fujita et al., 1993) and rat (Stea et al., 1993) α_{1B} proteins resulted in high threshold currents that were potently and irreversibly blocked by ω-CgTx. Taken together, these studies indicate that α_{1B} code for N-type Ca channels. In HEK cells, the efficient expression of α_{1B} N-type channels required the coexpression of α_2- and β-subunits (Williams et al., 1992b). In *Xenopus* oocytes, functional N-type channels were induced by the expression of the rbB-I α_1-subunit alone, although channel properties, including the rates of activation and inactivation and the voltage-dependence of inactivation, were affected by coexpression with a β-subunit (Stea et al., 1993).

2.3.1.3 Class C (α_{1C})

The class C (α_{1C}) cDNA was first described from mammalian cardiac tissue (pCARD3; Mikami et al., 1989; Tanabe et al., 1990a) and subsequently from rabbit smooth muscle (pSCal; Biel et al., 1990), rat aorta (Vsmα1; Koch et al., 1990), rat brain (rbC-I, rbC-II; Snutch et al., 1991), and mouse brain (mbC; Ma et al., 1992). The α_{1C} cDNAs code for proteins of 2140 to 2170 a.a. with M_r values of ~240 kDa and share a high degree of amino acid identity (~90%). Functional studies have indicated that all expressed α_{1C} subunits encoded high-voltage-activated Ca currents that showed minimal inactivation during a depolarizing pulse and that were sensitive to dihydropyridine antagonists and agonists (Mikami et al., 1989; Biel et al., 1990; Tanabe et al., 1990a; Wei et al., 1991; Tomlinson et al., 1993). In rat, alternative splicing from a single α_{1C} gene generated multiple isoforms that were expressed in a variety of tissues, including brain, heart, adrenal, and pituitary glands (Perez-Reyes et al., 1990; Snutch et al., 1991). Using site-directed antibodies the rat neuronal α_{1C} protein was shown to exist in two size forms with apparent molecular masses of ~190 and ~210 kDa (Hell et al., 1993a,b). The larger form was specifically phosphorylated *in vitro* by protein kinase A, while both the long and short forms were substrates for protein kinase C, Ca- and calmodulin-dependent protein kinase II, and cGMP-dependent protein kinase (Hell et al., 1993a).

2.3.1.4 Class D (α_{1D})

The class D (α_{1D}) is more similar to the α_{1C} (~60–70%) and α_{1S} proteins (~65%) than to the α_{1A}, α_{1B} or α_{1E} (between 30 and 40% identity). Full length α_{1D} genes have been cloned from human neuroblastoma (α_{1D}; Williams et al., 1992a), human pancreatic β cells (CACN4; Seino et al., 1992), rat brain (rbα-1; Hui et al., 1991), and a hamster insulin-secreting cell line (HCa3a; Yaney et al., 1992). The predicted size of the α_{1D} proteins was quite variable, ranging from 1610 to 2180 a.a. with M_r values of between 182 and 247 kDa. The variability of the predicted size of the α_{1D} proteins is likely the result of alternative splicing (Hui et al., 1991). Similar to the α_{1S} and α_{1C} subunits, expression of the human α_{1D} resulted in functional dihydropyridine-sensitive L-type Ca channels (Williams et al., 1992a).

2.3.1.5 Class E (α_{1E})

A sixth type of mammalian Ca channel α_1-subunit, the α_{1E}, has been cloned from rabbit brain (BII-1, BII-2; Niidome et al., 1992) and rat brain (rbE-II; Soong et al., 1993). The α_{1E} proteins are predicted to be 2178 to 2259 a.a. with calculated M_r values of ~250 kDa. Expression of the rbE-II α_1-subunit in *Xenopus* oocytes resulted in currents that first activated at relatively negative potentials (> –50 mV), required a strong hyperpolarization to remove inactivation, and were not significantly affected by dihydropyridines, or ω-CgTx, and only slightly blocked by ω-agatoxin IVA (Soong et al., 1993). The rbE-II properties were not consistent with those of previously described high voltage-activated or prototypical T-type channels, leading to the suggestion that rbE-II represents a unique member of the low voltage-activated Ca channel family (Soong et al., 1993). A recent paper (Williams et al., 1994) described the cloning of α_{1E} subunits from human brain ($\alpha_{1E\text{-}1}$, $\alpha_{1E\text{-}3}$, 2251 and

2270 amino acids, respectively) and mouse brain (2272 amino acids). Expression of the human α_{1E} in HEK cells and *Xenopus* oocytes resulted in rapidly inactivating whole currents that were pharmacologically similar to rbE-II. The voltage-dependence of activation of the human α_{1E} was more positive compared to rbE-II, although different divalent ion concentrations were used in the two studies (Soong et al., 1993; Williams et al., 1994). Similar to that for rbe-II, the steady-state inactivation properties of the human α_{1E} were more negative compared to that for typical high threshold Ca chennels.

2.3.1.6 Comparison of Primary Structures

The six classes of Ca channel α_1-subunits are most similar in the putative membrane spanning regions (S1-S6) of domains I to IV but differ significantly in the putative cytoplasmic segment separating domains II and III, and in the carboxyl-terminal region (Figure 1; Snutch and Reiner, 1992). Members of the α_{1C}, α_{1D}, and α_{1S} classes have relatively short II–III segments (~100 to 150 a.a.), while this region is greater than 400 residues in members of α_{1A}, α_{1B}, and α_{1E} classes. The domain II–III segment has been identified as an important region in mediating excitation–contraction coupling in the α_{1S} Ca channel (Tanabe et al., 1990b). Utilizing a combination of photoaffinity labeling and site-directed antipeptide antibodies, the site of phenylalkylamine binding to the α_{1S} protein has been localized to a 42 amino acid segment extending from Glu-1349 to Trp-1391 (Striessnig et al., 1990). This region contains the domain IV S6 transmembrane region and adjacent carboxyl sequences modeled to be on the intracellular side of the Ca channel. A similar antipeptide antibody approach was used to identify regions involved in DHP binding to α_{1S} (Striessnig et al., 1991; Nakayama et al., 1991). The results showed that two regions participated in DHP binding, the putative extracellular domain III S5 to S6 loop and adjacent S6 transmembrane sequences together with the domain IV S6 region involved in phenylalkylamine binding. A consensus motif (QQ-E–L-GY–WI—E) found in all cloned α_1-subunits 24 amino acids downstream from the domain I S6 transmembrane segment has been identified as a binding site for the β-subunit (Pragnell et al., 1994).

The different classes of α_1-subunits all possess consensus sites for N-glycosylation in proposed extracellular regions. Glycosylation of the α_1-subunits may account for the differences in the molecular weights between that predicted from the primary amino acid sequence and that of the purified proteins. Other possibilities include proteolytic modification *in vivo* (Lai et al., 1990; Hell et al., 1993a,b), and alternative splicing of the RNA transcripts (Perez-Reyes et al., 1990; Mori et al., 1991; Snutch et al., 1991; Williams et al., 1992b). Consensus amino acid sequences for phosphorylation by cAMP-dependent protein kinase, cGMP-dependent protein kinase, protein kinase C, and Ca-calmodulin-dependent protein kinase have been identified in all of the cloned α_1-subunits and these sites may play a role in modulation of the calcium channels *in vivo* (Hell et al., 1993a).

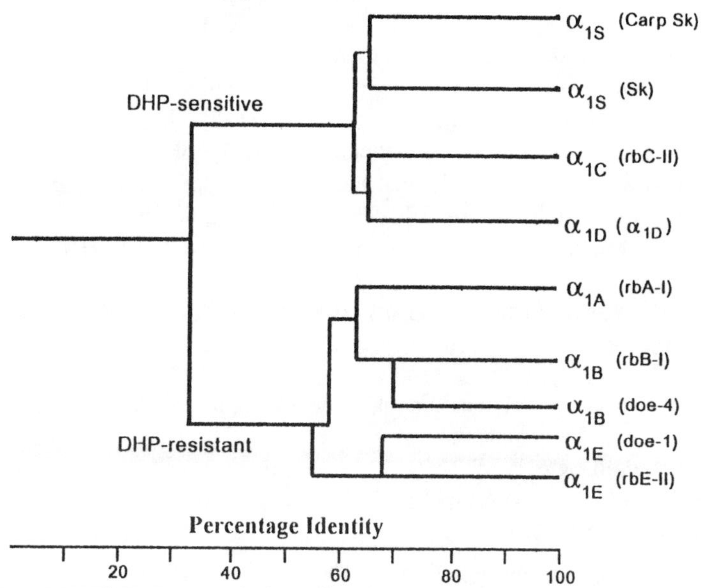

FIGURE 1. Similarity tree of cloned calcium channel α_1-subunits. The predicted amino acid sequences of the major classes of calcium channel α_1-subunits (Carp Sk, Grabner et al., 1991; Sk, Tanabe et al., 1987; rbC-II, Snutch et al., 1991; α_{1D}, Williams et al., 1992a; rbA-I, Starr et al., 1991; rbB-I, Dubel et al., 1992; doe-1, doe-4, Horne et al., 1993; rbE-II, Soong et al., 1993) were compared (Align version 1.0 software) and the percentage identity was plotted (see also Fujita et al., 1993; Horne et al., 1993).

2.3.1.7 Species Differences

In mammals, within a given class of α_1-subunit there is a greater degree of amino acid conservation between species (>85%) than between different classes of α_1-subunit within the same species (30–70%). While the majority of Ca channel α_1-subunits have been cloned from mammalian cDNA libraries, a few full-length α_1-subunits have been isolated from nonmammalian species. An α_1-subunit from carp *(Cyprinus carpio)* skeletal muscle (Grabner et al., 1991) coded for a protein of 1852 a.a. (M_r ~210 kDa) and shared ~65% identity with the rabbit skeletal muscle α_1-subunit (Tanabe et al., 1987). Two different full-length α_1-subunit cDNAs have been described from the marine ray, *Discopyge ommata* (Horne et al., 1993). The doe-1 α_1-subunit coded for a 2223 a.a. protein (252 kDa) that was ~68% identical to the α_{1E} proteins, while the doe-4 gene coded for a 2326 a.a. protein (265 kDa) and shared ~72% identity to the α_{1B} proteins (Horne et al., 1993). Expression of doe-1 with mammalian α_2- and β-subunits produced Ba currents that were sensitive to Ni^{2+} and insensitive to dihydropyridines, ω-CgTx, and ω-agatoxin IVA (Ellinor et al., 1993), similar to that observed for the α_{1E} member rbE-II (Soong et al., 1993). However, compared to rbE-II, the doe-1 current inactivated more rapidly and had a different sensitivity to holding potential. Although the evolutionary distance between the lower vertebrates and mammals is ~350 to 400 million years, there remains a high degree of identity (65–72%) between the carp and ray α_1-subunits and members of their respective mammalian classes (Figure 2). This conservation between evolutionary distant species is similar to that found for other highly conserved proteins such as the AChR α-subunit (human and *Torpedo*, ~80% identity) and cytochrome *c* (human and dogfish, ~77% identity) and is higher than for proteins such as hemoglobin, trypsin, albumin, and the immunoglobulins (Hille, 1992).

2.3.1.8 Chromosomal Location of Human Calcium Channel Subunit Genes

The chromosomal location of some Ca channel subunit genes has been reported. The gene coding for the human α_{1B} Ca channel has been localized to chromosome 9 (9q32-q34; Church et al., 1994) while the human α_{1E} gene is found on chromosome 1 (T. W. Soong and T. P. Snutch, unpublished observations). The chromosomal location for the genes encoding the DHP-sensitive L-type Ca channels have also been described. The human α_{1C} gene is located on chromosome 12 (12p13.3; Schultz et al., 1993), the human α_{1D} on cheomosome 3 (Chin et al., 1991), and the human α_{1S} on chromosome 1 (1q31-q32; Gregg et al., 1993a). In addition, the chromosomal location of several ancillary subunit genes has been found. The human α_2/δ gene is found on the proximal arm of chromosome 7 (7q; Iles et al., 1994). The human β_1 gene is on chromosome 17 (17q21-q22; Iles et al., 1993a; Gregg et al., 1993b) while the human β_3 gene is on chromosome 12 (12q13; Collin et al., 1994). The human gene coding for the skeletal muscle γ subunit is found on chromosome 17 (17q24; Iles et al., 1993a,b).

2.3.2 Spatial Distribution and Localization of Calcium Channel α_1-Subunit Expression

Many cell types express multiple types of Ca channels and a complete description of the spatial distribution of cloned Ca channels will aid in correlating their electrophysiological and pharmacological characteristics with those of Ca channels *in vivo*. The spatial distribution of Ca channel α_1-subunit expression has been determined by Northern blot, *in situ* hybridization, and RNA-PCR analyses. In addition, the cellular and subcellular distribution of α_1-subunit proteins has been examined by immunohistochemical methods using polyclonal antibodies directed against synthetic peptides.

2.3.2.1 Class A (α_{1A})

Northern blot analysis of the rat brain α_{1A} (rbA-I; Starr et al., 1991) mRNA revealed two transcripts, ~8.3 and 8.8 kb in size, which were highly expressed in the cerebellum and at lower levels in the spinal cord, pons/medulla, hypothalamus/thalamus, and olfactory bulb. In the striatum, hippocampus, and cortex, the 8.8 kb message appeared to predominate. PCR analysis of rbA

```
α1-C (rabbit)   Mlralvqpatpayqplpshlsaetestckgtvvheaqlnhfyispggsny   50
α1-C (rat)      M----vnentrmyvpeenh--------------------------qgsny   20
α1-D (human)    M---mmmmmmkkmqhqrqq--------------------------qadha   21


α1-A (rat)       MaRFgdempgrygaggggsgpaa--GvvvgaaggrgAggsrqggqpgaq   47
α1-A (rabbit)    MaRFgdemparyg--gggagaaa--GvvvgaaggrgAggsrqggqpgaq   45
α1-B (rat)       MvRFgdelggryg-gtgggErar--Gggaggaggpgqggl----ppgqr   42
α1-B (human)     MvRFgdelggryg-gpgggErar--Gggaggaggpgpggl----qpgqr   42
α1-E (ray)       MaRFgeavgslsadasseqgrsrhqvpvtgetavaaAaaavvagaaqgs   49
α1-C (rabbit)   gspRpahanmnanaaaglapEhiptpGaalswqaaidAarqaklmgsagn  100
α1-C (rat)      gspRpahanmnanaaaglapEhiptpGaalswqaaigAarqaklmgsagn   70
α1-D (human)    neanYargtrlplsgegptsQpnsskqtvlswqaaidAarqakaaqtmst   71
α1-S (rabbit)    M------------------Epsspqdegl--------------------   11


α1-A (rat)      rmykqsmaqraRtMALY--NPIPVrQNClTvNRSLFlFsEDNVVRKyAkk   95
α1-A (rabbit)   rmykqsmaqraRtMALY--NPIPVrQNClTvNRSLFlFsEDNVVRKyAkk   93
α1-B (rat)      vlykqsiaqraRtMALY--NPIPVkQNCfTvNRSLFvFsEDNVVRKyAkr   90
α1-B (human)    vlykqsiaqraRtMALY--NPIPVkQNCfTvNRSLFvFsEDNVVRKyAkr   90
α1-E (rat)                   MALY--NPIPVrQNCfTvNRSLFiFgEDNIVRKyAkk   35
α1-E (ray)      agfkqtraqraRtMALY--NPIPVrhNClTaNRSLFlFgEDNIVRKsArr   97
α1-C (rabbit)   atistvss-tqRkrqqYgkpkkqgsttatrppRaLlcltlkNpIRracis  149
α1-C (rat)      atistvss-tqRkrqqYgkpkkqggttatrppRaLlcltlkNpIRracis  119
α1-D (human)    sapppvgslsqRkrqqYakskkqgnssnsrpaRaLFclslNNpIRracis  121
α1-S (rabbit)   -----------Rkkq----pkkPlpEvlprppRaLFcltlQNplRKacis   46


                        ________I S1________             __________
α1-A (rat)      ItEWPPFEYMILATIIANCIVLALEQHLPDDDKTPMserLDdTEPYFIGI  145
α1-A (rabbit)   ItEWPPFEYMILATIIANCIVLALEQHLPDDDKTPMserLDdTEPYFIGI  143
α1-B (rat)      ItEWPPFEYMILATIIANCIVLALEQHLPDgDKTPMserLDdTEPYFIGI  140
α1-B (human)    ItEWPPFEYMILATIIANCIVLALEQHLPDgDKTPMserLDdTEPYFIGI  140
α1-E (rat)      lIDWPPFEYMILATIIANCIVLALEQHLPEDDKTPMsrrLEkTEPYFIGI   85
α1-E (ray)      VIEWPPFEYMILATIIANCVVLALEQHLPNgDKTPMaksLEqTEPYFIGI  147
α1-C (rabbit)   IVEWkPFEiiILlTIfANCVaLAiyipfPEDDsnatnsnLErvEylFliI  199
α1-C (rat)      IVEWkPFEiiILlTIfANCVaLAiyipfPEDDsnatnsnLErvEylFliI  169
α1-D (human)    IVEWkPFDifILlaIfANCVaLAiyipfPEDDsnstnhnLEkvEyaFliI  171
α1-S (rabbit)   IVEWkPFEtiILlTIfANCVaLAvylpMPEDDnnsLnlgLEklEyFFltV   96
```

FIGURE 2. Alignment of cloned calcium channel α_1-subunits. Representative members from each class of calcium channel α_1-subunits aligned using Multalin version 4.0 software (Corpet, 1988). Characters in bold and capitals indicate identical or conservative substitutions in >60% of the sequences. Dashes indicate gaps required to align the sequences and the amino acid number is given at the end of each line. Sequences were acquired from the following sources: α_{1A} (rat), Starr et al. (1991), GenBank Accession # M64373; α_{1A} (rabbit), Mori et al. (1991), #X57477; α_{1B} (rat), Dubel et al. (1992), #M92905; α_{1B} (human), Williams et al. (1992b), #M94172; α_{1E} (rat), Soong et al. (1993), #L15453; α_{1E} (ray), Horne et al. (1993), #L12531; α_{1C} (rabbit), Mikami et al. (1989), #X15539; α_{1C} (rat), Snutch et al. (1991), #M67515; α_{1D} (human), Williams et al. (1992a), #M76558; α_{1S} (rabbit), Tanabe et al. (1987), #L04684.

```
                   __I S2______                  ______I S3________
α1-A (rat)         FCFEAGIKIVALGFaFHKGSYLRNGWNVMDFVVVLtGILATV---------  187
α1-A (rabbit)      FCFEAGIKIIALGFaFHKGSYLRNGWNVMDFVVVLtGILATV---------  185
α1-B (rat)         FCFEAGIKIIALGFvFHKGSYLRNGWNVMDFVVVLteILATa---------  182
α1-B (human)       FCFEAGIKIIALGFvFHKGSYLRNGWNVMDFVVVLtGILATa---------  182
α1-E (rat)         FCFEAGIKIVALGFiFHKGSYLRNGWNVMDFIVVLsGILATa---------  127
α1-E (ray)         FCFEAGIKIVALGFvFHKGSYLRNGWNVMDFIVVLsGlLATa---------  189
α1-C (rabbit)      FtvEAflKVIAyGllFHpnaYLRNGWNlLDFIIVvvGlfsaIleqatk-a  248
α1-C (rat)         FtvEAflKVIAyGllFHpnaYLRNGWNlLDFIIVvvGlfsaIleqatk-a  218
α1-D (human)       FtvEtflKIIAyGlllHpnaYvRNGWNlLDFVIVivGlfsvIleqltket  221
α1-S (rabbit)      FsiEAamKIIAyGFlFHqdaYLRsGWNVLDFIIVflGVftaIleqvnviq  146

                                          ______I S4___________
α1-A (rat)         ---------GTeFD----LRTLRAVRVLRPLKLVSGIPSLQVVLKSIMKA  224
α1-A (rabbit)      ---------GTeFD----LRTLRAVRVLRPLKLVSGIPSLQVVLKSIMKA  222
α1-B (rat)         ---------GTdFD----LRTLRAVRVLRPLKLVSGIPSLQVVLKSIMKA  219
α1-B (human)       ---------GTdFD----LRTLRAVRVLRPLKLVSGIPSLQVVLKSIMKA  219
α1-E (rat)         ---------GThFNthvdLRTLRAVRVLRPLKLVSGIPSLQIVLKSIMKA  168
α1-E (ray)         ---------aThFN----LRTLRAVRVLRPLKLVSGIPSLQIVLKSIMKA  226
α1-C (rabbit)      dganalggkGagFD----vkaLRAfRVLRPLrLVSGVPSLQVVLnSIiKA  294
α1-C (rat)         dganalggkGagFD----vkaLRAfRVLRPLrLVSGVPSLQVVLnSIiKA  264
α1-D (human)       eggnhssgksggFD----vkaLRAfRVLRPLrLVSGVPSLQVVLnSIiKA  267
α1-S (rabbit)      sntapmsskGaglD----vkaLRAfRVLRPLrLVSGVPSLQVVLnSIfKA  192

                         ______I S5__________
α1-A (rat)         MIPLLQIGLLLFFAILIFAIIGLEFYMGKfHtTCF--eEgtddiQgEs-P  271
α1-A (rabbit)      MIPLLQIGLLLFFAILIFAIIGLEFYMGKfHtTCF--eEgtddiQgEs-P  269
α1-B (rat)         MVPLLQIGLLLFFAILmFAIIGLEFYMGKfHKaCF--pNstda-Epvg-d  265
α1-B (human)       MVPLLQIGLLLFFAILmFAIIGLEFYMGKfHKaCF--pNstda-Epvg-d  265
α1-E (rat)         MVPLLQIGLLLFFAILmFAIIGLEFYsGKLHraCF--mNnsgilEgfDpP  216
α1-E (ray)         MVPLLQIGLLLFFAILmFAIIGLEFYyGKLHrTCY--tDdaaa-EelDlq  273
α1-C (rabbit)      MVPLLhIaLLvlFvIiIYAIIGLElFMGKMHKTCYn-qEgvadvpaEDdP  343
α1-C (rat)         MVPLLhIaLLvlFvIiIYAIIGLElFMGKMHKTCYn-qEgiidvpaEEdP  313
α1-D (human)       MVPLLhIaLLvlFvIiIYAIIGLElFiGKMHKTCFf-aDs--divaEEdP  314
α1-S (rabbit)      MlPLfhIaLLvlFmViIYAIIGLElFkGKMHKTCYyigtdivatveNEkP  242

α1-A (rat)         aPCGt-EepaRtC-pNGTkCqpyWeGPNnGITQFDNILFAvLTVFQCITM  319
α1-A (rabbit)      aPCGt-EepaRtC-pNGTrCqpyWeGPNnGITQFDNILFAvLTVFQCITM  317
α1-B (rat)         fPCGk-EapaRlC-dsdTeCreyWpGPNfGITNFDNILFAiLTVFQCITM  313
α1-B (human)       fPCGk-EapaRlC-egdTeCreyWpGPNfGITNFDNILFAiLTVFQCITM  313
α1-E (rat)         hPCGv-Qg----C-paGyeCk-dWiGPNdGITQFDNILFAvLTVFQCITM  259
α1-E (ray)         fPCGt-QeptRlC-pNGTvCs-yWiGPNdGITQFDNILFAlLTVFQCITM  320
α1-C (rabbit)      sPCaletghgRqC-qNGTvCkpgWdGPkhGITNFDNfaFAmLTVFQCITM  392
α1-C (rat)         sPCaletghgRqC-qNGTvCkpgWdGPkhGITNFDNfaFAmLTVFQCITM  362
α1-D (human)       aPCaf-sgngRqCtaNGTeCrsgWvGPNgGITNFDNfaFAmLTVFQCITM  363
α1-S (rabbit)      sPCar-tgsgRpCtiNGseCrggWpGPNhGIThFDNfgFsmLTVYQCITM  291
```

FIGURE 2(2).

expression detected rbA-I transcripts in the rat brain, heart, and pituitary. There was no detectable rbA-I expression in the liver, spleen, and kidney (Starr et al., 1991), although another study found α_{1A} transcripts in the kidney distal convoluted tubule (Yu et al., 1992). In rabbit brain, two α_{1A} cDNA isoforms were reported (BI-1 and BI-2; Mori et al., 1991) but only a single

```
                                          __________I S6____________
α1-A (rat)       EGWTDlLYNsNDAsGntWNWLYFIPLIIIGSFFmLNLVLGVLSGEFAKER  369
α1-A (rabbit)    EGWTDlLYNsNDAsGntWNWLYFIPLIIIGSFFmLNLVLGVLSGEFAKER  367
α1-B (rat)       EGWTDILYNtNDAaGntWNWLYFIPLIIIGSFFmLNLVLGVLSGEFAKER  363
α1-B (human)     EGWTDILYNtNDAaGntWNWLYFIPLIIIGSFFmLNLVLGVLSGEFAKER  363
α1-E (rat)       EGWTtVLYNtNDAlGatWNWLYFIPLIIIGSFFVLNLVLGVLSGEFAKER  309
α1-E (ray)       EGWTtILYNtDDAlGamWNWLYFIPLIIIGSFFVLNLVLGVLSGEFAKER  370
α1-C (rabbit)    EGWTDVLYwmQDAmGyelpWvYFVsLVIfGSFFVLNLVLGVLSGEFsKER  442
α1-C (rat)       EGWTDVLYwmQDAmGyelpWvYFVsLVIfGSFFVLNLVLGVLSGEFsKER  412
α1-D (human)     EGWTDVLYwmNDAmGfelpWvYFVsLVIfGSFFVLNLVLGVLSGEFsKER  413
α1-S (rabbit)    EGWTDVLYwvNDAiGneWpWiYFVtLIllGSFFILNLVLGVLSGEFtKER  341

α1-A (rat)       ERVENRRaFLKLRRQQQIERELNGYMEWIsKAEEViLaEDEtDvEQrhPf  419
α1-A (rabbit)    ERVENRRaFLKLRRQQQIERELNGYMEWIsKAEEViLaEDEtDvEQrhPf  417
α1-B (rat)       ERVENRRaFLKLRRQQQIERELNGYLEWIfKAEEVmLaEEDkNaEEksPl  413
α1-B (human)     ERVENRRaFLKLRRQQQIERELNGYLEWIfKAEEVmLaEEDrNaEEksPl  413
α1-E (rat)       ERVENRRaFMKLRRQQQIERELNGYraWIdKAEEVmLaEENkNsgt-sal  358
α1-E (ray)       ERVENRRsFLKLRRQQQIERELNGYraWIdKAEEVmLlEENkNagEksal  420
α1-C (rabbit)    EkakaRgdFqKLRekQQlEeDLkGYLDWItqAEDIdpeNEDegmDEekP-  491
α1-C (rat)       EkakaRgdFqKLRekQQlEeDLkGYLDWItqAEDIdpeNEDegmDEdkP-  461
α1-D (human)     EkakaRgdFqKLRekQQlEeDLkGYLDWItqAEDIdpeNEEeggEEgk--  461
α1-S (rabbit)    EkaksRgtFqKLRekQQlEeDLrGYMsWItqgEvmdveDlreg-------  384

α1-A (rat)       dgaLRRAtlKkSkTDllnpEEaEDqlaDiasvGsPfARASIKSaKl-EnS  468
α1-A (rabbit)    dgaLRRAtiKkSkTDllhpEEaEDqlaDiasvGsPfARASIKSaKl-EnS  466
α1-B (rat)       davLkRAatKkSrnDlIhaEEgEDrfvDlcaaGsPfARASlKSgKt-EsS  462
α1-B (human)     d-vLkRAatKkSrnDlIhaEEgEDrfaDlcavGsPfARASlKSgKt-EsS  461
α1-E (rat)       e-vLRRAtiKrSrTEamtrDssDEhcvDissvGtPLARASIKStKv-Dga  406
α1-E (ray)       h-vLRRAtiKkgrmEmIqtEssEDqytEissvGsPLARASIKStKllEgS  469
α1-C (rabbit)    ----RnmsmptSeTEsVntENvaggdiEgencGarLAh-rIsksKf---S  533
α1-C (rat)       ----RnmsmptSeTEsVntENvaggdiEgencrarLAh-rIsksKf---S  503
α1-D (human)     ----RntsmptSeTEsVntENvsgeg-EnrgccgsLcq-aIsksKl---S  502
α1-S (rabbit)    -----klsleeggsDteslyEiEg-----------Lnk-iI---qf---i  411

                                                 ______II S1________
α1-A (rat)       tFFhkKERrmRfyIRrMVKtQaFYWtVLsLVALNTLwlAiVHYNQPeWLs  518
α1-A (rabbit)    sFFhkKERrmRfyIRrMVKtQaFYWtVLsLVALNTLcVAiVHYNQPeWLs  516
α1-B (rat)       sYFRRKEkmfRflIRrMVKaQsFYWvVLcvVALNTLcVAmVHYNQPqrLT  512
α1-B (human)     sYFRRKEkmfRffIRrMVKaQsFYWvVLcvVALNTLcVAmVHYNQPrrLT  511
α1-E (rat)       sYFRhKERllRisIRhMV,KSQvFYWiVLsvVALNTacVAiVHhNQPqWLT  456
α1-E (ray)       sYFRRKERmlRisIRhMVKShaFYWiVLgLVALNTvcVAvVHYDQPlWLs  519
α1-C (rabbit)    rYwRRwNRfcRrkcRaaVKSNvFYWlVifLVfLNTLtIAseHYNQPhWLT  583
α1-C (rat)       rYwRRwNRfcRrkcRaaVKSNvFYWlVifLVfLNTLtIAseHYNQPhWLT  553
α1-D (human)     rrwRRwNRfnRrrcRaaVKSvtFYWlVivLVfLNTLtIsseHYNQPdWLT  552
α1-S (rabbit)    rhwRqwNRvfRwkchdLVKSrvFYWlViliVALNTLsIAseHhNQPlWLT  461
```

FIGURE 2(3).

transcript of ~9.4 kb was detected by Northern blot analysis. *In situ* hybridization using cRNA or deoxyoligonucleotide probes demonstrated widespread signals throughout the rabbit (Fujita et al., 1993) or rat (Stea et al., 1994) brain. Regional localization of α_{1A} in rat and rabbit brain showed prominent expression in the dentate gyrus and CA fields of the hippocampus,

```
                         II S2                         II S3
α1-A (rat)      DfLYYAEFIFLGLFmsEMfiKMYGLGtRpYFhSSFNCFDCGVIIGSIFEV  568
α1-A (rabbit)   DfLYYAEFIFLGLFmsEMfiKMYGLGtRpYFhSSFNCFDCGVIIGSIFEV  566
α1-B (rat)      taLYFAEFVFLGLFltEMsLKMYGLGpRsYFrSSFNCFDfGVIVGSIFEV  562
α1-B (human)    ttLYFAEFVFLGLFltEMsLKMYGLGpRsYFrSSFNCFDfGVIVGSVFEV  561
α1-E (rat)      hlLYYAEFlFLGLFllEMsLKMYGMGpRlYFhSSFNCFDfGVtVGSIFEV  506
α1-E (ray)      NfLYYAEFtFLGLFssEMfLKMYGcGpRlYFhSSFNCFDCGVIIGSIFDV  569
α1-C (rabbit)   EvqdtANkalLaLFtaEMlLKMYsLGlqaYFvSlFNrFDCfIVcGgIlEt  633
α1-C (rat)      EvqdtANkalLaLFtaEMlLKMYsLGlqaYFvSlFNrFDCfIVcGgIlEt  603
α1-D (human)    QiqdiANkVlLaLFtcEMlvKMYsLGlqaYFvSlFNrFDCfVVcGgItEt  602
α1-S (rabbit)   hlqdiANrVlLsLFtiEMlLKMYGLGlRqYFmSiFNrFDCfVVcsgIlEl  511

                          II S4
α1-A (rat)      IWaviKPgTSFGISVLRALRLLRIFKVTKYWaSLRNLVVSLLNSMKSIIS  618
α1-A (rabbit)   IWaviKPgTSFGISVLRALRLLRIFKVTKYWaSLRNLVVSLLNSMKSIIS  616
α1-B (rat)      VWaaiKPgTSFGISVLRALRLLRIFKVTKYWnSLRNLVVSLLNSMKSIIS  612
α1-B (human)    VWaaiKPgsSFGISVLRALRLLRIFKVTKYWsSLRNLVVSLLNSMKSIIS  611
α1-E (rat)      VWaifrPgTSFGISVLRALRLLRIFKITKYWaSLRNLVVSLMsSMKSIIS  556
α1-E (ray)      VWtiirPeTSFGISVLRALRLLRIFKITKYWaSLRNLVVSLMsSMKSIIS  619
α1-C (rabbit)   IlvetKvmsplGISVLRcvRLLRIFKITrYWnSLsNLVaSLLNSvrSIaS  683
α1-C (rat)      IlvetKimsplGISVLRcvRLLRIFKITrYWnSLsNLVaSLLNSvrSIaS  653
α1-D (human)    IlveleimsplGISVfRcvRLLRIFKVTrhWtSLcNLVaSLLNSMKSsaS  652
α1-S (rabbit)   llvesgamTplGISVLRciRLLRlFKITKYWtSLsNLVaSLLNSirSIaS  561

                II S5
α1-A (rat)      LLFLLFLFIVVFALLGMQLFGGqFNFDE-gTPptNFDTFPAAIMTVFQIL  667
α1-A (rabbit)   LLFLLFLFIVVFALLGMQLFGGqFNFDE-gTPptNFDTFPAAIMTVFQIL  665
α1-B (rat)      LLFLLFLFIVVFALLGMQLFGGqFNFQD-ETPttNFDTFPAAILTVFQIL  661
α1-B (human)    LLFLLFLFIVVFALLGMQLFGGqFNFQD-ETPttNFDTFPAAILTVFQIL  660
α1-E (rat)      LLFLLFLFIVVFALLGMQLFGGrFNFND-gTPsaNFDTFPAAIMTVFQIL  605
α1-E (ray)      LLFLLFLFIVVFALLGMQLFGGqFNFEE-gTPptNFDTFPAAIiTVFQIL  668
α1-C (rabbit)   LLlLLFLFIIIFsLLGMQLFGGkFNFDEmQTrrstFDnFPqslLTVFQIL  733
α1-C (rat)      LLlLLFLFIIIFsLLGMQLFGGkFNFDEmQTrrstFDnFPqslLTVFQIL  703
α1-D (human)    LLlLLFLFIIIFsLLGMQLFGGkFNFDEtQTkrstFDnFPqAlLTVFQIL  702
α1-S (rabbit)   LLlLLFLFIIIFALLGMQLFGGrYDFEDtEvrrsNFDnFPqAlisVFQVL  611

                                                II S6
α1-A (rat)      TGEDWNeVMYDeIkSQGGVq-gGMvfsIYFIVLTLFGNYTLLNVFLAIAV  716
α1-A (rabbit)   TGEDWNeVMYDGIkSQGGVq-gGMvfsIYFIVLTLFGNYTLLNVFLAIAV  714
α1-B (rat)      TGEDWNaVMYhGIeSQGGVS-kGMfssfYFIVLTLFGNYTLLNVFLAIAV  710
α1-B (human)    TGEDWNaVMYhGIeSQGGVS-kGMfssfYFIVLTLFGNYTLLNVFLAIAV  709
α1-E (rat)      TGEDWNeVMYNGIrSQGGVS-sGMwsaIYFIVLTLFGNYTLLNVFLAIAV  654
α1-E (ray)      TGEDWNeVMYNGIkSQGGVn-sGMwssVYFIVLTLFGNYTLLNVFLAIAV  717
α1-C (rabbit)   TGEDWNsVMYDGImayGGpSfpGMlvcIYFIILficGNYiLLNVFLAIAV  783
α1-C (rat)      TGEDWNsVMYDGImayGGpSfpGMlvcIYFIILficGNYiLLNVFLAIAV  753
α1-D (human)    TGEDWNaVMYDGImayGGpSssGMivcIYFIILficGNYiLLNVFLAIAV  752
α1-S (rabbit)   TGEDWNsVMYNGImayGGpSypGvlvcIYFIILfvcGNYiLLNVFLAIAV  661
```

FIGURE 2(4).

the cerebral cortex, the pontine nucleus, the olfactory bulb, and the cerebrellar cortex (Stea et al., 1994; Fujita et al., 1993). Analysis of the cellular localization of α_{1A} in the rat brain showed concentrated signals from cerebrellar Purkinje cell bodies with lower levels in cerebellar granule cells (Stea et al., 1994). Other cells showing prominent expression of α_{1A} included CA1, CA2,

$\alpha_{1\text{-A}}$ *(rat)*	DNLANAQELTKDEQEEEEAaNQKlALQKAKEVaevSPLSAanmsiavkeq	766
$\alpha_{1\text{-A}}$ *(rabbit)*	DNLANAQELTKDEQEEEEAvNQKlALQKAKEVaevSPLSAanmsiamkeq	764
$\alpha_{1\text{-B}}$ *(rat)*	DNLANAQELTKDEEEmEEAaNQKlALQKAKEVaevSPMSAanisiaarqq	760
$\alpha_{1\text{-B}}$ *(human)*	DNLANAQELTKDEEEmEEAaNQKlALQKAKEVaevSPMSAanisiaarqq	759
$\alpha_{1\text{-E}}$ *(rat)*	DNLANAQELTKDEQEEEEAfNQKhALQKAKEV---SPMSApnmpsierdr	701
$\alpha_{1\text{-E}}$ *(ray)*	DNLANAQELTKEEQEEEEAiNQKhALQKAKEV---SPMSApgfpster--	762
$\alpha_{1\text{-C}}$ *(rabbit)*	DNLADAEsLTsaQkEEEEekErK---------------------------	806
$\alpha_{1\text{-C}}$ *(rat)*	DNLADAEsLTsaQkEEEEekErK---------------------------	776
$\alpha_{1\text{-D}}$ *(human)*	DNLADAEsLntaQkEEaEekErK---------------------------	775
$\alpha_{1\text{-S}}$ *(rabbit)*	DNLAEAEsLTsaQkakaEerkrr---------------------------	684

$\alpha_{1\text{-A}}$ *(rat)*	-----------------QknqkpaKSVWEqRTSEMRkqNllaSrEALyg-	798
$\alpha_{1\text{-A}}$ *(rabbit)*	-----------------QknqkpaKSVWEqRTSEMRkqNllaSrEALyse	797
$\alpha_{1\text{-B}}$ *(rat)*	-----------------Nsak--arSVWEqRaSQLRlqNlraScEALyse	791
$\alpha_{1\text{-B}}$ *(human)*	-----------------Nsak--arSVWEqRaSQLRlqNlraScEALyse	790
$\alpha_{1\text{-E}}$ *(rat)*	rrrhhmsmweprsshlrErrrrhhmSVWEqRTSQLRrhmqmsSqEALnke	751
$\alpha_{1\text{-E}}$ *(ray)*	-----------------EfrrhkhmSIWEaRTSQLRrrmqmsSrEALftd	795
$\alpha_{1\text{-C}}$ *(rabbit)*	------------------------Klar---TaspekkQevvgkpAL---	826
$\alpha_{1\text{-C}}$ *(rat)*	------------------------KlarpaRTaspekkQevmekpAv---	799
$\alpha_{1\text{-D}}$ *(human)*	------------------------Kiar---------------kEsL---	783
$\alpha_{1\text{-S}}$ *(rabbit)*	------------------------Kmsr----glpdktEe--eksvM---	701

$\alpha_{1\text{-A}}$ *(rat)*	--------daaErwpttyarplrpdvkthldrplvvdp--qenrnnntnk	838
$\alpha_{1\text{-A}}$ *(rabbit)*	-------mdpeErwkasyarhlrpdmkthldrplvvdp--qenrnnntnk	836
$\alpha_{1\text{-B}}$ *(rat)*	-------mdpeErlryastrhvrpdmkthmdrplvvepgrdglrgpagnk	834
$\alpha_{1\text{-B}}$ *(human)*	-------mdpeErlrfattrhlrpdmkthldrplvvelgrdgargpvggk	833
$\alpha_{1\text{-E}}$ *(rat)*	eappmnplnplNplsplnplnahpslyrrprpieglalglglekceeeri	801
$\alpha_{1\text{-E}}$ *(ray)*	---------alQglegsryrrhrsrife-aeslrrlaeqqaaeshqlgev	835
$\alpha_{1\text{-C}}$ *(rabbit)*	--	
$\alpha_{1\text{-C}}$ *(rat)*	--	
$\alpha_{1\text{-D}}$ *(human)*	--	
$\alpha_{1\text{-S}}$ *(rabbit)*	--	

$\alpha_{1\text{-A}}$ *(rat)*	srapea--------------lrqtarpresardp--------darrawps	866
$\alpha_{1\text{-A}}$ *(rabbit)*	srvaeptvdqrlgqqraedflrkqarhhdrardpsahaaagldarrpwag	888
$\alpha_{1\text{-B}}$ *(rat)*	skpegteate----------gadpprrhhrhrdr--------dktsastp	866
$\alpha_{1\text{-B}}$ *(human)*	arpeaaeape----------gvdpprrhhrhrdk--------dkt----p	861
$\alpha_{1\text{-E}}$ *(rat)*	srggslkgd-----------iggltsvldnqrsp----------------	824
$\alpha_{1\text{-E}}$ *(ray)*	grreafks-----------------rslrnswqp----------------	852
$\alpha_{1\text{-C}}$ *(rabbit)*	--	
$\alpha_{1\text{-C}}$ *(rat)*	--	
$\alpha_{1\text{-D}}$ *(human)*	--	
$\alpha_{1\text{-S}}$ *(rabbit)*	--	

FIGURE 2(5).

and CA3 hippocampal pyramidal cells, dentate granule cells, hilar interneurons, mitral cells of the olfactory bulb, cells of the red nucleus, and cells of the substantia nigra pars compacta (Stea et al., 1994).

While the exact relationship between α_{1A} cDNAs and P-type Ca channels remains to be precisely defined, the distribution of P-type channels in the rat central nervous system (CNS) has been examined immunohistochemically

```
α1-A (rat)       sperapgregpygresepqqrEhapprehvpwdadperakagdaprrhth  916
α1-A (rabbit)    sqeaelsregpygresdhqarEggleppgf-wegeaergkagdphrrhah  937
α1-B (rat)       aggeqdrtdcpkaestetgarEerarprr-----shskeapgadtqvrce  911
α1-B (human)     aagdqdraeapkaesgepgarEerprphr-----shskeaagp-pearse  905
α1-E (rat)       --lslgkreppwlprschgncDptqqetg-----ggetvvtfedrarhrq  867
α1-E (ray)       --agpdkrss-----sikvngEqgralgr-----sveagasf--rmaepi  888
α1-C (rabbit)    --------------------------------------------------
α1-C (rat)       --------------------------------------------------
α1-D (human)     --------------------------------------------------
α1-S (rabbit)    --------------------------------------------------

α1-A (rat)       rp----------------vaegeprrhRarrrpgDE-pDDrpErrprprd  949
α1-A (rabbit)    rqgvggsggsrsgsprtgtadgeprrhRvhrrpgEDgpDDkaErrgrhre  987
α1-B (rat)       rs-------------r--------rhhR--rgspEEatEreprrhrahrh  938
α1-B (human)     rg-------------rgpgpeggrrhhR--rgspEEaaEreprrhrahrh  940
α1-E (rat)       sq--------------------rrsrhRrvrtegkEsasasrsrsasqer  897
α1-E (ray)       ra--------------------rr-ryRslykeakmglEEsaEtslsrrp  917
α1-C (rabbit)    ----------------------------------EEakEEkiElksitad  842
α1-C (rat)       ----------------------------------EEskEEkiElksitad  815
α1-D (human)     ----------------------------------ENkkNNkpEvnqi-an  798
α1-S (rabbit)    ----------------------------------akklEQkpkgegi---  714

α1-A (rat)       atrParaadge--------gddgErkrrhrhgppa-------hddrerrh  984
α1-A (rabbit)    gsrParsgegeaegpdgggggggErrrrhrhgpppaydpdarrddrerrh  1037
α1-B (rat)       aqdsskegkegtapvl---vpkgErrarhrg-prtgpretenseeptrrh  984
α1-B (human)     -qdPskecaga----------kgErrarhrggpragpreaesgeeparrh  979
α1-E (rat)       sldegvsidge---------kehEpqsshrskepti--heeertqdlrrt  936
α1-E (ray)       gknkegrllqq---------lceEqesgqltqtpev--mdaq--gqmkaf  954
α1-C (rabbit)    gesPptt-------------------------------------------  849
α1-C (rat)       gesPptt-------------------------------------------  822
α1-D (human)     sdnkvti-------------------------------------------  805
α1-S (rabbit)    ---Ptta-------------------------------------------  718

α1-A (rat)       rrrkesqgsgvpmsgpnlsttrpiqqdlgrqDlplaedlDnmkNnkLatg  1034
α1-A (rabbit)    rrrkdtqgsgvpvsgpnlsttrpiqqdlsrqEpplaedmDnlkNsrLata  1087
α1-B (rat)       rakhkvpptlep-----------perevaekE---snvvEgdkEt--rNh  1018
α1-B (human)     rarhkaqpaheave------kettekeatekE---aeivEadkEkeLrNh  1020
α1-E (rat)       nslmvprgsglvga------ldeaetplvqpQ----pelEvgkDaaLtEq  976
α1-E (ray)       swqgephsssmtrt------pdvdtdp-sggN----lekEsgrtpengke  993
α1-C (rabbit)    -------------------------------------------kinMdDl  856
α1-C (rat)       -------------------------------------------kinMdDl  829
α1-D (human)     -------------------------------------------DdyreEd  812
α1-S (rabbit)    -------------------------------------------klkvdEf  725
```

FIGURE 2(6).

(Hillman et al., 1991; Llinas et al., 1992). Immunolabeling using a polyclonal antibody generated against an ~90 kDa bovine cerebellar protein capable of forming P-type channels in lipid bilayers showed strong reactivity in the molecular layer of the cerebellar cortex, especially at bifurcations of Purkinje cell dendrites. Strong labeling was also observed on cell bodies throughout the CNS, including the periglomerular cells of olfactory bulb, neurons in the

$\alpha_{1\text{-}A}$ *(rat)*	EPasphDslghsglppspakignstnpg---palatnPQNa-asrrtpNN	1080
$\alpha_{1\text{-}A}$ *(rabbit)*	EPvsphEnlshaglpqspakmgsstdpagptpataanPQNstasrrtpNN	1137
$\alpha_{1\text{-}B}$ *(rat)*	QPkEprcdleaiavtgvgslhmlpstcl---qkvdeqPEDadnqrNvtrm	1065
$\alpha_{1\text{-}B}$ *(human)*	QPrEphcdletsgtvtvgpmhtlpstcl---qkveeqPEDadnqrNvtrm	1067
$\alpha_{1\text{-}E}$ *(rat)*	EaegssEqallad--vqldvgrgisqse---pdlscmttNmdkatt-Est	1020
$\alpha_{1\text{-}E}$ *(ray)*	EsaNtsEqvneqsnwlnlqlnqqatpgd---relttgtrDtkqdktqEQt	1040
$\alpha_{1\text{-}C}$ *(rabbit)*	QPnEseDk--------------------------spyPNpettgeEdEEE	880
$\alpha_{1\text{-}C}$ *(rat)*	QPsEneDk--------------------------sphsNpntageEdEEE	853
$\alpha_{1\text{-}D}$ *(human)*	EdkDpypp--------------------------cdvPvgeeeeeEeEDE	836
$\alpha_{1\text{-}S}$ *(rabbit)*	Esnvnevk--------------------------dpyPsadfpgdDeEDE	749

$\alpha_{1\text{-}A}$ *(rat)*	PgnPsnPgPpktpEnsLivtnpsstqpnsaktarkpehmaveippacpp-	1129
$\alpha_{1\text{-}A}$ *(rabbit)*	PgnPsnPgPpktpEnsLivtnpstaqtnsaktarkpdhttveippacppp	1187
$\alpha_{1\text{-}B}$ *(rat)*	gsqP--sdPsttvhvpvtltgp-------------pgeatv-vpsa----	1095
$\alpha_{1\text{-}B}$ *(human)*	gsqP--PdPntivhipvmltgp-------------lgeatv-vpsg----	1097
$\alpha_{1\text{-}E}$ *(rat)*	svtvaiPdvdplvDstvvni---------------snktdgeaspl----	1051
$\alpha_{1\text{-}E}$ *(ray)*	eidvdcentetpmDslvt-----------------pgnaysssssv----	1069
$\alpha_{1\text{-}C}$ *(rabbit)*	PemPvgPrPrplsElhL---------------------------------	897
$\alpha_{1\text{-}C}$ *(rat)*	PemPvgPrPrplsElhL---------------------------------	870
$\alpha_{1\text{-}D}$ *(human)*	PevPagPrPrrisElnM---------------------------------	853
$\alpha_{1\text{-}S}$ *(rabbit)*	PeiPvsPrPrplaElqL---------------------------------	766

$\alpha_{1\text{-}A}$ *(rat)*	lnhtvvqvnknanpdplpkkeeekkeeeeadpgeDgpkPmPPySSMFIlS	1179
$\alpha_{1\text{-}A}$ *(rabbit)*	lnhtvvqvnknanpdplpkkedekkeevdegpgeDgpkPmPPySSMFIlS	1237
$\alpha_{1\text{-}B}$ *(rat)*	------------ntd-legqaegkkeaeaddvlrrgprPivPySSMFclS	1132
$\alpha_{1\text{-}B}$ *(human)*	------------nvd-lesqaegkkeveaddvmrsgprPivPySSMFclS	1134
$\alpha_{1\text{-}E}$ *(rat)*	------------keaetkeeeeevekkkqkkekrEtgkamvPhSSMFIfS	1089
$\alpha_{1\text{-}E}$ *(ray)*	------------kedekk------------------skaiiPytSMFlfr	1089
$\alpha_{1\text{-}C}$ *(rabbit)*	---------------------------------kEkavPmPeaSafFIfS	914
$\alpha_{1\text{-}C}$ *(rat)*	---------------------------------kEkavPmPeaSafFIfS	887
$\alpha_{1\text{-}D}$ *(human)*	---------------------------------kEkiaPiPegSafFIlS	870
$\alpha_{1\text{-}S}$ *(rabbit)*	---------------------------------kEkavPiPeaSSfFIfS	783

	III S1	
$\alpha_{1\text{-}A}$ *(rat)*	tTNplRrlCHYIlNLRYFEmcILmVIAMSSIALAAEDPVqpNapRNNVLr	1229
$\alpha_{1\text{-}A}$ *(rabbit)*	tTNplRrlCHYIlNLRYFEmcILmVIAMSSIALAAEDPVqpNapRNNVLr	1287
$\alpha_{1\text{-}B}$ *(rat)*	pTNllRrfCHYIVtMRYFEmvILvVIALSSIALAAEDPVrtDSfRNNaLk	1182
$\alpha_{1\text{-}B}$ *(human)*	pTNllRrfCHYIVtMRYFEvvILvVIALSSIALAAEDPVrtDSpRNNaLk	1184
$\alpha_{1\text{-}E}$ *(rat)*	tTNpiRkaCHYIVNLRYFEmcILlVIAaSSIALAAEDPVltNSeRNkVLr	1139
$\alpha_{1\text{-}E}$ *(ray)*	kTNpiRrvCHFIVNLRYFEmcILlVIAaSSVALAAEDPIhkDSaRNQVLr	1139
$\alpha_{1\text{-}C}$ *(rabbit)*	pnNrfRlqCHrIVNdtiFtnlILffIlLSSIsLAAEDPVqhtSfRNhILf	964
$\alpha_{1\text{-}C}$ *(rat)*	pnNrfRlqCHrIVNdtiFtnlILffIlLSSIsLAAEDPVqhtSfRNhILg	937
$\alpha_{1\text{-}D}$ *(human)*	kTNpiRvgCHklINhhiFtnlILvfImLSSaALAAEDPIrshSfRNtILg	920
$\alpha_{1\text{-}S}$ *(rabbit)*	pTNkvRvlCHrIVNatwFtnfILlfIlLSSaALAAEDPIraESvRNQILg	833

FIGURE 2(7).

deep layer of entorhinal and pyriform cortices, the brainstem, lateral habenula, and nuclei of the trapezoid body and inferior olive. Lower levels of labeling were found in layer II of the frontal cortex, hippocampal CA1 cells, substantia nigra, lateral reticular nucleus, and the horizontal cells of the retina (Hillman et al., 1991; Llinas et al., 1992). The relationship between the 90-kDa purified cerebellar protein and the larger sizes predicted for Ca channel α_1-subunits (~180–260 kDa) remains to be determined.

	III S2 ... III S3	
$\alpha_{1\text{-}A}$ *(rat)*	YFDYVFTGVFTFEMVIKMIDlGLvLHqGaYFRDlWNILDFIVVSGALVAF	1279
$\alpha_{1\text{-}A}$ *(rabbit)*	YFDYVFTGVFTFEMVIKMIDlGLvLHqGaYFaDlWNILDFIVVSGALVAF	1337
$\alpha_{1\text{-}B}$ *(rat)*	YmDYIFTGVFTFEMVIKMIDlGLlLHpGaYFRDlWNILDFIVVSGALVAF	1232
$\alpha_{1\text{-}B}$ *(human)*	YlDYIFTGVFTFEMVIKMIDlGLlLHpGaYFRDlWNILDFIVVSGALVAF	1234
$\alpha_{1\text{-}E}$ *(rat)*	YFDYVFTGVFTFEMVIKMIDqGLiLqdGsYFRDlWNILDFVVVvGALVAF	1189
$\alpha_{1\text{-}E}$ *(ray)*	YFDYVFTGVFTFEMVIKMIDiGLvfHeGsYFRDvWNILDFIVVSGALVAF	1189
$\alpha_{1\text{-}C}$ *(rabbit)*	YFDiVFTtIFTiEialKMtayGafLHkGsFcRNyfNILDllVVSvsLIsF	1014
$\alpha_{1\text{-}C}$ *(rat)*	naDYVFTsIFTlEiIlKMtayGafLHkGsFcRNyfNILDllVVSvsLIsF	987
$\alpha_{1\text{-}D}$ *(human)*	YFDYaFTaIFTvEillKMttfGafLHkGaFcRNyfNlLDmlVVgvsLVsF	970
$\alpha_{1\text{-}S}$ *(rabbit)*	YFDiaFTsVFTvEiVlKMttyGafLHkGsFcRNyfNILDllVVavsLIsm	883

	III S4	
$\alpha_{1\text{-}A}$ *(rat)*	Aft---gnskGkDINTIKSLRVLRVLRPLKTIKRLPKLKAVFDCVVnSLK	1326
$\alpha_{1\text{-}A}$ *(rabbit)*	Aft---gnskGkDINTIKSLRVLRVLRPLKTIKRLPKLKAVFDCVVnSLK	1384
$\alpha_{1\text{-}B}$ *(rat)*	AfssfmggskGkDINTIKSLRVLRVLRPLKTIKRLPKLKAVFDCVVnSLK	1282
$\alpha_{1\text{-}B}$ *(human)*	Afs----gskGkDINTIKSLRVLRVLRPLKTIKRLPKLKAVFDCVVnSLK	1280
$\alpha_{1\text{-}E}$ *(rat)*	AlanalgtnkGrDIkTIKSLRVLRVLRPLKTIKRLPKLKAVFDCVVtSLK	1239
$\alpha_{1\text{-}E}$ *(ray)*	AftnliggssGkDINTIKSLRVLRVLRPLKTIKRLPKLKAVFDCVVtSLK	1239
$\alpha_{1\text{-}C}$ *(rabbit)*	gi-------qssaINvVKiLRVLRVLRPLraInRakgLKhVvQCVfvair	1057
$\alpha_{1\text{-}C}$ *(rat)*	gi-------qssaINvVKiLRVLRVLRPLraInRakgLKhVvQCVfvair	1030
$\alpha_{1\text{-}D}$ *(human)*	gi-------qssaIsvVKiLRVLRVLRPLraInRakgLKhVvQCVfvair	1013
$\alpha_{1\text{-}S}$ *(rabbit)*	gl-------esstIsvVKiLRVLRVLRPLraInRakgLKhVvQCVfvair	926

	III S5	
$\alpha_{1\text{-}A}$ *(rat)*	NVfNILIVYmLFMFIFAVVAVQLFKGKFFhCTDeSKEfErDCRGkYllY-	1375
$\alpha_{1\text{-}A}$ *(rabbit)*	NVfNILIVYmLFMFIFAVVAVQLFKGKFFhCTDeSKEfEkDCRGkYllY-	1433
$\alpha_{1\text{-}B}$ *(rat)*	NVlNILIVYmLFMFIFAVIAVQLFKGKFFyCTDeSKElErDCRGQYldY-	1331
$\alpha_{1\text{-}B}$ *(human)*	NVlNILIVYmLFMFIFAVIAVQLFKGKFFyCTDeSKElErDCRGQYldY-	1329
$\alpha_{1\text{-}E}$ *(rat)*	NVfNILIVYkLFMFIFAVIAVQLFKGKFFyCTDsSKDtEkECiGNYvd-h	1288
$\alpha_{1\text{-}E}$ *(ray)*	NVfNILIVYkLFMFIFAVIAVQLFKGKFFyCTDsSKmtkqDCRGQFvlYr	1289
$\alpha_{1\text{-}C}$ *(rabbit)*	tIgNIvIVttLlqFmFAcIgVQLFKGKlYtCsDsSKQtEaECkGNYitYk	1107
$\alpha_{1\text{-}C}$ *(rat)*	tIgNIvIVttLlqFmFAcIgVQLFKGKlYtCsDsSKQtEaECkGNYitYk	1080
$\alpha_{1\text{-}D}$ *(human)*	tIgNIMIVttLlqFmFAcIgVQLFKGKFYrCTDeaKsnpeECRGlFilYk	1063
$\alpha_{1\text{-}S}$ *(rabbit)*	tIgNIvlVttLlqFmFAcIgVQLFKGKFFsCnDlSKmtEeECRGyYyvYk	976

$\alpha_{1\text{-}A}$ *(rat)*	--EkNevkardREWkkyDFHYDNVLWALLTLFTVSTGEGWPQVLkHSVDa	1423
$\alpha_{1\text{-}A}$ *(rabbit)*	--EkNevkardREWkkyEFHYDNVLWALLTLFTVSTGEGWPQVLkHSVDa	1481
$\alpha_{1\text{-}B}$ *(rat)*	--EkEeveaqpRQWkkyDFHYDNVLWALLTLFTVSTGEGWPmVLkHSVDa	1379
$\alpha_{1\text{-}B}$ *(human)*	--EkEeveaqpRQWkkyDFHYDNVLWALLTLFTVSTGEGWPmVLkHSVDa	1377
$\alpha_{1\text{-}E}$ *(rat)*	--EkNkmevkgREWkrhEFHYDNIiWALLTLFTVSTGEGWPQVLqHSVDv	1336
$\alpha_{1\text{-}E}$ *(ray)*	--QrtklsiengN--vttFHYDNVvWALLTLFTVSTGEGWPQVLqHSVDv	1335
$\alpha_{1\text{-}C}$ *(rabbit)*	dgEvDhpiiqpRsWenskFdFDNVLaAMMaLFTVSTfEGWPElLyrSIDs	1157
$\alpha_{1\text{-}C}$ *(rat)*	dgEvDhpiiqpRsWenskFdFDNVLaAMMaLFTVSTfEGWPElLyrSIDs	1130
$\alpha_{1\text{-}D}$ *(human)*	dgDvDspvvreRiWqnsDFnFDNVLsAMMaLFTVSTfEGWPalLykaIDs	1113
$\alpha_{1\text{-}S}$ *(rabbit)*	dgDptqmelrpRQWihnDFHFDNVLsAMMsLFTVSTfEGWPQlLyraIDs	1026

FIGURE 2(8).

2.3.2.2 Class B (α_{1B})

To date, α_{1B} expression has been found to be limited to the nervous system and to neuronally derived cell lines. At the level of Northern blot analysis, the relative distribution of BIII expression within the rabbit brain was found to

```
                           ______________III S6_____________
α1-A (rat)        TfENqGPspgYRMEMSIFYVVYFVVFPFFFVNIFVALIIITFQEQGDKmM  1473
α1-A (rabbit)     TfENqGPspgYRMEMSIFYVVYFVVFPFFFVNIFVALIIITFQEQGDKmM  1531
α1-B (rat)        TyEEqGPspgFRMELSIFYVVYFVVFPFFFVNIFVALIIITFQEQGDKvM  1429
α1-B (human)      TyEEqGPspgYRMELSIFYVVYFVVFPFFFVNIFVALIIITFQEQGDKvM  1427
α1-E (rat)        TeEDrGPsrsnRMEMSIFYVVYFVVFPFFFVNIFVALIIITFQEQGDKmM  1386
α1-E (ray)        TeaDqGPipgnRMEMSIFYIVYFVVFPFFFVNIFVALIIITFQEQGDKmL  1385
α1-C (rabbit)     htEDkGPiynYRvEiSIFFIIYiIIiaFFmmNIFVgfVIVTFQEQGEqey  1207
α1-C (rat)        htEDkGPiynYRvEiSIFFIIYiIIiaFFmmNIFVgfVIVTFQEQGEqey  1180
α1-D (human)      ngENiGPiynhRvEiSIFFIIYiIIvaFFmmNIFVgfVIVTFQEQGEKey  1163
α1-S (rabbit)     neEDmGPvynnRvEMaIFFIIYiIliaFFmmNIFVgfVIVTFQEQGEtey  1076

                                                             ______
α1-A (rat)        eEySLEKNERACIDFAISAKPLTRhMPQNKQsFQYrmWqFVVSPpFEYtI  1523
α1-A (rabbit)     eEySLEKNERACIDFAISAKPLTRhMPQNKQsFQYrmWqFVVSPpFEYtI  1581
α1-B (rat)        sECSLEKNERACIDFAISAKPLTRYMPQNKQsFQYktWtFVVSPpFEYfI  1479
α1-B (human)      sECSLEKNERACIDFAISAKPLTRYMPQNrQsFQYktWtFVVSPpFEYfI  1477
α1-E (rat)        eECSLEKNERACIDFAISAKPLTRYMPQNrhtFQYrvWhFVVSPsFEYtI  1436
α1-E (ray)        eEsSLEKNERACIDFAISAKPLTRYMPQNrQtFQYrvWqFVVSPsFEYtI  1435
α1-C (rabbit)     kNCeLDKNQRqCVEYAlkArPLrRYiP--KNqhQYkvWyvVnStyFEYlm  1255
α1-C (rat)        kNCeLDKNQRqCVEYAlkArPLrRYiP--KNqhQYkvWyvVnStyFEYlm  1228
α1-D (human)      kNCeLDKNQRqCVEYAlkArPLrRYiP--KNpYQYkfWyvVnSspFEYmm  1211
α1-S (rabbit)     kNCeLDKNQRqCVQYAlkArPLrcYiP--KNpYQYqvWyvVtSsyFEYlm  1124

                  _____IV S1____              _________IV S2________
α1-A (rat)        MAMIALNTIVLMMKfYgaSvaYEnALrVfNIvFTsLFsLECVLKVmAFGi  1573
α1-A (rabbit)     MAMIALNTIVLMMKfYgaSvaYDnALkVfNIvFTsLFsLEClLKVlAFGi  1631
α1-B (rat)        MAMIALNTVVLMMKfYdapyeYElmLkcLNIvFTsMFsLECILKIIAFGv  1529
α1-B (human)      MAMIALNTVVLMMKfYdapyeYElmLkcLNIvFTsMFsMECVLKIIAFGv  1527
α1-E (rat)        MAMIALNTVVLMMKyYsapwtYElALkyLNIaFTmvFsLECVLKVIAFGf  1486
α1-E (ray)        LtMIALNTVVLMMKhhspppgFasvLklMNIaFTitFtLECILKIIAFGf  1485
α1-C (rabbit)     fvLIlLNTIcLaMqhYgqSclFkiAMnILNmlFTgLFtvEmILKlIAFkp  1305
α1-C (rat)        fvLIlLNTIcLaMqhYgqSclFkiAMnILNmlFTgLFtvEmILKlIAFkp  1278
α1-D (human)      fvLImLNTlcLaMqhYeqSkmFNdAMdILNmvFTgvFtvEmVLKVIAFkp  1261
α1-S (rabbit)     fALImLNTIcLgMqhYhqSeemNhisdILNVaFTiiFtLEmILKllAFka  1174

                  _________IV S3______
α1-A (rat)        LNYFRDAWNIFDFVTVlGSItDIlVTEfg--------------------n  1603
α1-A (rabbit)     LNYFRDAWNIFDFVTVlGSItDIlVTEfg--------------------n  1661
α1-B (rat)        LNYFRDAWNVFDFVTVlGSItDIlVTEia--------------------n  1559
α1-B (human)      LNYFRDAWNVFDFVTVlGSItDIlVTEia------------------etn  1559
α1-E (rat)        LNYFRDtWNIFDFITVIGSItEIilTDsk--------------------l  1516
α1-E (ray)        LNYFRDsWNVFDFVTVVGSIsEIiVTEcn--------------------l  1515
α1-C (rabbit)     kgYFsDpWNVFDFliVIGSIiDVilsEtn----paehtqcspsm----na  1347
α1-C (rat)        khYFcDAWNtFDaliVVGSIvDIaITEvh----paehtqcspsm----sa  1320
α1-D (human)      kgYFsDAWNtFDsliVIGSIiDValsEad----ptesenvpvptatpgns  1307
α1-S (rabbit)     rgYFgDpWNVFDFliVIGSIiDVilsEidtflassgglyclgggcgnvdp  1224
```

FIGURE 2(9).

differ from that of rbB-I in rat brain. While BIII mRNA was expressed most abundantly in striatum and midbrain in rabbit brain, rbB-I transcripts were detected at relatively higher levels in the cerebellum, hippocampus, and thalamus–hypothalamus in the rat (Dubel et al., 1992; Fujita et al., 1993). Compared to the expression of BI (α_{1A}) in rabbit brain, BIII transcripts

```
                 IV S4
α1-A (rat)       Nfi---NLSFLRLFRAARLIKLLRQGyTIRILLWTFVQSFKALPYVCLLI  1650
α1-A (rabbit)    Nfi---NLSFLRLFRAARLIKLLRQGyTIRILLWTFVQSFKALPYVCLLI  1708
α1-B (rat)       Nfi---NLSFLRLFRAARLIKLcRQGyTIRILLWTFVQSFKALPYVCLLI  1606
α1-B (human)     Nfi---NLSFLRLFRAARLIKLLRQGyTIRILLWTFVQSFKALPYVCLLI  1606
α1-E (rat)       vntsgfNMSFLkLFRAARLIKLLRQGyTIRILLWTFVQSFKALPYVCLLI  1566
α1-E (ray)       kfv---NLSFLkLFRAARLIKLLRQGiTIRILLWTFVQSFKALPYVCLLI  1562
α1-C (rabbit)    EensrisitFfRLFRvmRLVKLLsrGegIRtLLWTFIkSFqALPYVaLLI  1397
α1-C (rat)       EensrisitFfRLFRvmRLVKLLsrGegIRtLLWTFIkSFqALPYVaLLI  1370
α1-D (human)     EesnrisitFfRLFRvmRLVKLLsrGegIRtLLWTFIkfFqALPYVaLLI  1357
α1-S (rabbit)    DesarissaFfRLFRvmRLIKLLsraegVRtLLWTFIkSFqALPYVaLLI  1274

                 IV S5
α1-A (rat)       AMLFFIYAIIGMQVFGNIgiDgededsdeDefQIteHNNFrTFfQALMLL  1700
α1-A (rabbit)    AMLFFIYAIIGMQVFGNIgiDmededsdeDefQIteHNNFrTFfQALMLL  1758
α1-B (rat)       AMLFFIYAIIGMQVFGNIALD--------DgTsINRHNNFrTFlQALMLL  1648
α1-B (human)     AMLFFIYAIIGMQVFGNIALD--------DdTsINRHNNFrTFlQALMLL  1648
α1-E (rat)       AMLFFIYAIIGMQVFGNIkLD--------EeshINRHNNFrsFfgsLMLL  1608
α1-E (ray)       AMLFFIYAIIGMQlFGNIgLD--------DhTpINRHNNFhTFfNALMLL  1604
α1-C (rabbit)    vMLFFIYAVIGMQVFGkIALN--------DtTEINRnNNFqTFpQAvLLL  1439
α1-C (rat)       vMLFFIYAVIGMQVFGkIALN--------DtTEINRnNNFqTFpQAvLLL  1412
α1-D (human)     AMLFFIYAVIGMQmFGkVAMr--------DnnQINRnNNFqTFpQAvLLL  1399
α1-S (rabbit)    vMLFFIYAVIGMQmFGkIALv--------DgTQINRnNNFqTFpQAvLLL  1316

α1-A (rat)       FRSATGEAWhNIMLsCLsGKpCDkNSgiqkpe-----CGnEFAYFYFVSF  1745
α1-A (rabbit)    FRSATGEAWhNIMLsCLsGKpCDkNSgiltpe-----CGnEFAYFYFVSF  1803
α1-B (rat)       FRSATGEAWhEIMLsCLgnraCDPhaN--ase-----CGsDFAYFYFVSF  1691
α1-B (human)     FRSATGEAWhEIMLsCLsnqaCDeQaN--ate-----CGsDFAYFYFVSF  1691
α1-E (rat)       FRSATGEAWQEIMLsCLgeKgCEPDttapsgqneserCGtDlAYvYFVSF  1658
α1-E (ray)       FRSATGEsWQEIMLaCLsGKeCE-gtreps-------CGtDvAYFYFVSF  1646
α1-C (rabbit)    FRcATGEAWQDIMLaCMpGKkCaPESEphnstegetpCGssFAvFYFISF  1489
α1-C (rat)       FRcATGEAWQDIMLaCMpGKkCaPESEpsnstkgetpCGssFAvFYFISF  1462
α1-D (human)     FRcATGEAWQEIMLaCLpGKlCDPESD--ynpgeehtCGsNFAivYFISF  1447
α1-S (rabbit)    FRcATGEAWQEILLaCsyGKlCDPESD--yapgeeytCGtNFAYYYFISF  1364

                 IV S6
α1-A (rat)       IFLCSFLMLNLFVAVIMDNFEYLTRDSSILGPHHLDEYVRVWAEYDPAAc  1795
α1-A (rabbit)    IFLCSFLMLNLFVAVIMDNFEYLTRDSSILGPHHLDEYVRVWAEYDPAAw  1853
α1-B (rat)       IFLCSFLMLNLFVAVIMDNFEYLTRDSSILGPHHLDEFIRVWAEYDPAAc  1741
α1-B (human)     IFLCSFLMLNLFVAVIMDNFEYLTRDSSILGPHHLDEFIRVWAEYDPAAc  1741
α1-E (rat)       IFfCSFLMLNLFVAVIMDNFEYLTRDSSILGPHHLDEFVRVWAEYDrAAc  1708
α1-E (ray)       IFLCSFLMLNLFVAVIMDNFEYLTRDSSILGPHHLDEFVRVWAEYDrAAc  1696
α1-C (rabbit)    ymLCaFLiiNLFVAVIMDNFDYLTRDwSILGPHHLDEFkRIWAEYDPeAk  1539
α1-C (rat)       ymLCaFLiiNLFVAVIMDNFDYLTRDwSILGPHHLDEFkRIWAEYDPeAk  1512
α1-D (human)     ymLCaFLiiNLFVAVIMDNFDYLTRDwSILGPHHLDEFkRIWsEYDPeAk  1497
α1-S (rabbit)    ymLCaFLiiNLFVAVIMDNFDYLTRDwSILGPHHLDEFkaIWAEYDPeAk  1414
```

FIGURE 2(10).

showed a similar distribution pattern but the relative levels of the two types of Ca channels differed (Fujita et al., 1993). PCR analysis utilizing the α_{1B-1} and α_{1B-2} sequences detected transcripts from human hippocampus, habenula, and thalamus, but not from skeletal muscle or aortic tissues (Williams et al., 1992b). In the marine ray, the level of doe-4 expression was higher in the electric lobe than in the brain (Horne et al., 1993).

$\alpha_{1\text{-}A}$ *(rat)*	GRIhYkDMYsLLRvisPPLGLGKkCPhRVAcKRLlrMDLPVaDDn-TVHF	1844
$\alpha_{1\text{-}A}$ *(rabbit)*	GRmlYrDMYaMLRhmpPPLGLGKnCPaRVAyKRLlrMDLPVaDDn-TVHF	1902
$\alpha_{1\text{-}B}$ *(rat)*	GRIsYnDMFeMLkhmsPPLGLGKkCPaRVAyKRLVrMNMPIsNEdmTVHF	1791
$\alpha_{1\text{-}B}$ *(human)*	GRIsYnDMFeMLkhmsPPLGLGKkCPaRVAyKRLVrMNMPIsNEdmTVHF	1791
$\alpha_{1\text{-}E}$ *(rat)*	GRIhYtEMYeMLtlmsPPLGLGKrCPskVAyKRLVlMNMPVaEDm-TVHF	1757
$\alpha_{1\text{-}E}$ *(ray)*	GRIhYtDMYqMLtlmsPPLGLGKkCPskVAyKRLVlMNMPVtEDk-TVHF	1745
$\alpha_{1\text{-}C}$ *(rabbit)*	GRIkhlDvvtLLRriqPPLGfGKlCPhRVAcKRLVsMNMPlnsDg-TVmF	1588
$\alpha_{1\text{-}C}$ *(rat)*	GRIkhlDvvtLLRriqPPLGfGKlCPhRVAcKRLVsMNMPlnsDg-TVmF	1561
$\alpha_{1\text{-}D}$ *(human)*	GRIkhlDvvtLLRriqPPLGfGKlCPhRVAcKRLVaMNMPlnsDg-TVmF	1546
$\alpha_{1\text{-}S}$ *(rabbit)*	GRIkhlDvvtLLRriqPPLGfGKfCPhRVAcKRLVgMNMPlnsDg-TVtF	1463

$\alpha_{1\text{-}A}$ *(rat)*	NSTLMALIRTALDIKiAkgGaDkQQmDaELRKEmmaIWpnLSQKtLDLLV	1894
$\alpha_{1\text{-}A}$ *(rabbit)*	NSTLMALIRTALDIKiAkgGaDkQQmDaELRKEmmaIWpnLSQKtLDLLV	1952
$\alpha_{1\text{-}B}$ *(rat)*	tSTLMALIRTALEIKlApaGtkqhQcDaELRKEIssVWanLpQKtLDLLV	1841
$\alpha_{1\text{-}B}$ *(human)*	tSTLMALIRTALEIKlApaGtkqhQcDaELRKEIsvVWanLpQKtLDLLV	1841
$\alpha_{1\text{-}E}$ *(rat)*	tSTLMALIRTALDIKiAkgGaDrQQlDsELqKEtlaIWphLSQKmLDLLV	1807
$\alpha_{1\text{-}E}$ *(ray)*	tSTLMgLIRTALQIKlArgGaDkQQlDaELRKEImtIWphLSQKtLDLLV	1795
$\alpha_{1\text{-}C}$ *(rabbit)*	NaTLfALVRTALrIK---tegNlEQaNeELRaiIkkIWkrtSmKlLDqvV	1635
$\alpha_{1\text{-}C}$ *(rat)*	NaTLfALVRTALrIK---tegNlEQaNeELRaiIkkIWkrtSmKlLDqvV	1608
$\alpha_{1\text{-}D}$ *(human)*	NaTLfALVRTALkIK---tegNlEQaNeELRavIkkIWkktSmKlLDqvV	1593
$\alpha_{1\text{-}S}$ *(rabbit)*	NaTLfALVRTALkIK---tegNfEQaNeELRaiIkkIWkrtSmKlLDqvI	1510

$\alpha_{1\text{-}A}$ *(rat)*	tPhkstDLTVGKIYAAMMImEYYRQsKakKlQaMre--eqnrtpL----m	1938
$\alpha_{1\text{-}A}$ *(rabbit)*	tPhkstDLTVGKIYAAMMImEYYRQsKakKlQaMre--eqnrtpL----m	1996
$\alpha_{1\text{-}B}$ *(rat)*	PPhkpDEMTVGKVYAALMIfDFYkQnKttrDQthqa--pgglsqMgpvsl	1889
$\alpha_{1\text{-}B}$ *(human)*	PPhkpDEMTVGKVYAALMIfDFYkQnKttrDQmqqa--pgglsqMgpvsl	1889
$\alpha_{1\text{-}E}$ *(rat)*	PmpkasDLTVGKIYAAMMImDYYkQsKvkKQrqqle--eqknapM-----	1850
$\alpha_{1\text{-}E}$ *(ray)*	PmhtysDLTVGKIYAAMMImDYYkQsKnkKyQkLqe--eqsrtpM-----	1838
$\alpha_{1\text{-}C}$ *(rabbit)*	PPagdDEvTVGKfYAtfLIqEYFRkfKkrKEQgLvgk-psqrnaLslqag	1684
$\alpha_{1\text{-}C}$ *(rat)*	PPagdDEvTVGKfYAtfLIqEYFRkfKkrKEQgLvgk-psqrnaLslqag	1657
$\alpha_{1\text{-}D}$ *(human)*	PPagdDEvTVGKfYAtfLIqDYFRkfKkrKEQgLvgkypakntTialqag	1643
$\alpha_{1\text{-}S}$ *(rabbit)*	PPigdDEvTVGKfYAtfLIqEhFRkfmkrqEE-yyg-yrpkkdtvqiqag	1558

$\alpha_{1\text{-}A}$ *(rat)*	FqrMeppsptQEggpsqna---L---pstqlDpggglmaQEssmkespsw	1982
$\alpha_{1\text{-}A}$ *(rabbit)*	FqrMeppp--DEggagqna---L---pstqlDpagglmahEdglksspsw	2038
$\alpha_{1\text{-}B}$ *(rat)*	FhpLkatleqtQpavlrgarvfLrQksatslsnggaiqtQEsgikeslsw	1939
$\alpha_{1\text{-}B}$ *(human)*	FhpLkatleqtQpavlrgarvfLrQksstslsnggaiqnQEsgikesvsw	1939
$\alpha_{1\text{-}E}$ *(rat)*	FqrMepsslPQEiisnaka---L---pylqqDpvsgl----sgrsgypsm	1890
$\alpha_{1\text{-}E}$ *(ray)*	FqrMeasslPpQiisstkg---L---pylqtgtgpdv----dsrseftpl	1878
$\alpha_{1\text{-}C}$ *(rabbit)*	lrtLhd-igPEirraisgdltaeeEldkamkEavsaaseDDifrragglf	1733
$\alpha_{1\text{-}C}$ *(rat)*	lrtLhd-igPEirraisgdltaeeEldkamkEavsaaseDDifrragglf	1706
$\alpha_{1\text{-}D}$ *(human)*	lrtLhd-igPEirraiscdl-qddEpeetkrEe-----eDDvfkrngall	1686
$\alpha_{1\text{-}S}$ *(rabbit)*	lrtieeeaaPEirrtisgdltaeeEleramvEa---ameErifrrtgglf	1605

FIGURE 2(11).

CNB-1, an antibody that recognizes the rbB-I α_1 subunit of the N-type calcium channels complex, immunoprecipitated ~43% of rat brain N-type calcium channels by ^{125}I-ω-CgTx and recognized two α_{1B} size forms of ~240 and ~210 kDa (Westenbroek et al., 1992). Throughout the rostral-caudal extent of the rat brain, rbB-I channels were localized predominantly along the entire length of dendrites of many types of neurons. In

```
α1-A (rat)      vtqraqemfQktgtwspergppiDMpnsqp----nsQsve--mremgtdg  2026
α1-A (rabbit)   vtqraqemfQktgtwsperappaDMadsqp----kpQsve--mremsqdg  2082
α1-B (rat)      Gtqrtqdvlyeara-plerghsaEipvgqp----galavdvqmqnmtlrg  1984
α1-B (human)    Gtqrtqdaphearp-plerghstEipvgrs----galavdvqmqsitrrg  1984
α1-E (rat)      splspqeifQla---cmdpaddgQfeqqs----lv------vtdpssm-  1926
α1-E (ray)      vplppv-mfQqg----rtssqgeEihkqrp----ke------lkkikle-  1912
α1-C (rabbit)   GnhvsyyqsDsrsafpqtfttqrpLhiskag-nnqgDtespsheklvdst  1782
α1-C (rat)      GnhvsyyqsDsrsnfpqtfatqrpLhinktg-nnqaDtespsheklvdst  1755
α1-D (human)    GnhvnhvnsDrrdslqqtntthrpLhvqrpsippasDtekplfppagnsv  1736
α1-S (rabbit)   Gqvdtfl--ErtnslppvmanqrpLqfaei---emeElespvf-------  1643

α1-A (rat)      ysdsehylpmEgQtraasMprLpaEnqrrrgrprgnnlsti---sDtSPM  2073
α1-A (rabbit)   ysdsehclpmEgQaraasMprLpaEnqrrrgrprgsdlsti---cDtSPM  2129
α1-B (rat)      -pdgepqpglEsQgraasMprLaaEtqp------------a---pNaSPM  2018
α1-B (human)    -pdgepqpglEsQgraasMprLaaEtqp------------v---tDaSPM  2018
α1-E (rat)      rrsfsti--rDkrsnsswLeefsmEr--------------s---sEnt-y  1956
α1-E (ray)      yphyghylpiEnQgravsMprLeiEs--------------a---eDtSPL  1945
α1-C (rabbit)   -ftpssysstgsNaninnanntalgrlprpa-gypstvstve--ghgSPL  1828
α1-C (rat)      -ftpssysstgsNaninnanntalgrfphpa-gysstvstve--ghgpPL  1801
α1-D (human)    chnhhnhnsigkQvptstnanLnnanmskaahgkrpsignlehvsEnghh  1786
α1-S (rabbit)   -----------------------lEdfp-----------------Qdart  1653

α1-A (rat)      Krsasvlgp------------------ka--rrLddyslErvppEE---n  2100
α1-A (rabbit)   Krsasvlgp------------------ka-srrLddyslErvppEE---n  2157
α1-B (rat)      Krsistlap------------------rphgtqLcntvlDrpppsQvs-h  2049
α1-B (human)    Krsistlaq------------------rprgthLcsttpDrpppsQassh  2050
α1-E (rat)      Ksrrrsyhs---------------------slrLsahrlNsdsghk----  1981
α1-E (ray)      Krslstfaan-----------------hsnstwLneyslEragpEDly-k  1977
α1-C (rabbit)   spavraqeaawklsskr---------chsqesqiamacqEgasqDDnydv  1869
α1-C (rat)      spavrvqeaawklsskr---------chsresqgatvsqDmfpdEtrssv  1842
α1-D (human)    sshkhdrepqrrssvkrtryyetyirsdsgdeqLpticrEdpeihgyfrd  1836
α1-S (rabbit)   nplarant-------------------nnananvaygnsNhsnnQmfssv  1684

α1-A (rat)      q-------------------ryh--qrrrdrghRtsERslg---------  2120
α1-A (rabbit)   q-------------------rhh--prrrerahRtsERslg---------  2177
α1-B (rat)      h-------------------hhhrchrrrdkkqRslEkgps---------  2071
α1-B (human)    h-------------------hhhrchrrrdrkqRslEkgps---------  2072
α1-E (rat)      -----------------------------sdthRsggRErg---------  1993
α1-E (ray)      r-------------------wsrrplrppsrssnagsRErg---------  1999
α1-C (rabbit)   r--iged----aecc----sepsllstemlsyqddeNRQlap--------  1901
α1-C (rat)      r--lsee----veyc----sepsllstdilsyqddeNRQltc--------  1874
α1-D (human)    phclgeqeyfsseecyeddssptwsrqnygyysRypgRNidserprgyhh  1886
α1-S (rabbit)   h-------------c---------------------EREfpg--------  1692
```

FIGURE 2(12).

addition, the large terminals of the mossy fibers of dentate gyrus granule neurons were heavily immunoreactive and sharply defined punctate localizations were detected in many regions. Low levels of immunolabeling were detected in the cell bodies of some pyramidal cells of the dorsal cortex, Purkinje cells, and other cell bodies elsewhere in the brain. The

```
α1-A (rat)       ----rytDvdtg-----lgtdlsmttqsgdl-ps---kdr--dqDRgrpk  2155
α1-A (rabbit)    ----rytDvdtg-----lgtdlsmttqsgdl-ps---rer--eqERgrpk  2212
α1-B (rat)       ----lsvDtega-----pStaagsglphgeg-stgcrrer--kqERgrsq  2109
α1-B (human)     ----lsaDmdga-----pSsavgpglppgeg-ptgcrrererrqERgrsq  2112
α1-E (rat)       ----rskErkhl-----lSpdvsrcnseerg-tqadwesperrqsRspse  2033
α1-E (ray)       ----rsrErkhl-----lSpersvcstgqca-hpsqhrgldqrlsRspsp  2039
α1-C (rabbit)    ----peeEkrdi----rlSpkkgflrsaslg-rrasfhleclkrQknqgg  1942
α1-C (rat)       ----leeDkrei----qpSpkrsflrsaslg-rrasfhleclkrQkdqgg  1915
α1-D (human)     pqgfledDdspvcydsrrSprrrllpptpashrrssfnfeclrrQssqee  1936
α1-S (rabbit)    -----eaEt--------paagrgalshs----hra------lgphskpca  1719


α1-A (rat)       D-----rkhrphhhhhhhhh------hppaPdr------------eryaq  2182
α1-A (rabbit)    D-----rkhrphhhhhhhhh------pgrgPgrvspgvsarrrrhgpvar  2251
α1-B (rat)       E-----rrqpsssssekqrfyscdrfgsrePpqpkpslsShpisptaale  2154
α1-B (human)     E-----rrqpsssssekqrfyscdrfggrePpkpkpslsShptsptagqe  2157
α1-E (rat)       g-----rsqtpnrqgtg---------slsessipsisdtStprrsrrqlp  2069
α1-E (ray)       g-----yshrpreqvns---------svsespvpsssgtSppkqgqrqlp  2075
α1-C (rabbit)    D-----isqktvlplhlvhh-qalavaglsPllqrshsptslprpcatpp  1986
α1-C (rat)       D-----isqktalplhlvhh-qalavaglsPllqrshspStfprprptpp  1959
α1-D (human)     vpsspifphrtalplhlmqq-qimavagldsskaqkyspShstrswatpp  1985
α1-S (rabbit)    g----------klngqlvq-----------Pgmpinqap---papcqqps  1745


α1-A (rat)       erPdtg--------------------------rarareqrwsr-------  2199
α1-A (rabbit)    vrParapalaharararaparllpelrlrrarrprprqrrrprrrrgggg  2201
α1-B (rat)       pgPhpqgsgsvngsplmstsgaStpgrggrrqlpqtpltprpsityktan  2204
α1-B (human)     pgPhpqgsgsvngspllstsgaStpgrggrrqlpqtpltprpsityktan  2207
α1-E (rat)       pvPpkprpllsysslmrhtggiSpppdgseggsplasqalesnsacltes  2119
α1-E (ray)       qtPskprplvsyspvaqr-gdvSghcspmcketryqslrvqpskalwses  2124
α1-C (rabbit)    atPgsrgwppqpiptlrlegadSseklnssfpsihcgswsgenspcrgds  2036
α1-C (rat)       vtPgsrgrplqpiptlrlegaeSseklnssfpsihcsswseettacsggs  2009
α1-D (human)     atPpyrdwtpcytpliqveqsealdqvngslpslhrsswytdepdis---  2032
α1-S (rabbit)    tdPperg--------------qrrtsltgslqdeapqrrssegstpr---  1778


α1-A (rat)       ----------------sPse-gre--------------------------  2206
α1-A (rabbit)    ralrrapgpreplaqdsPgr-grsvclara--arpagrQrLlpGprtgqa  2348
α1-B (rat)       sspvhfaegqsglpafsPgrlsrglsehnaLlQkeplsQpLasGsrigsd  2254
α1-B (human)     sspihfagaqtslpafsPgrlsrglsehnaLlQrdplsQpLapGsrigsd  2257
α1-E (rat)       snslhpqqgqhp----sPqhyise--------------------------  2139
α1-E (ray)       pgrsresesqhs----tPlryise--------------------------  2144
α1-C (rabbit)    saarrarpvsltvpsqagaqgrqfhgsassLvEavlisEgL--Gqfaqdp  2084
α1-C (rat)       smarrarpvsltvpsqagapgrqfhgsassLvEavlisEgL--Gqfaqdp  2057
α1-D (human)     --yrtftpasltvpssfrnknsdkqrsadsLvEavlisEgL--Gryardp  2078
α1-S (rabbit)    ---rpapatal-------------------LiQealvrggL--dtlaada  1804
```

FIGURE 2(13).

results suggest that rbB-I N-type channels play a significant role in mediating calcium influx into dendrites in many types of neurons and also in excitation–secretion of neurotransmitter at a subset of synapses (Westenbroek et al., 1992).

```
α1-A (rat)       ----------hathr-q                              2212
α1-A (rabbit)    prarlpqkparsvqr-errgLvlspppppgelapRahpartprPgpgqs  2397
α1-B (rat)       pYlgqrlDseAsahnLpeDtLtfEeAvatnsgrssRtsyvssltsqshpl  2304
α1-B (human)     pYlgqrlDseAsvhaLpeDtLtfEeAvatnsgrssRtsyvssltsqshpl  2307
α1-E (rat)       pYlalheDshAsdcg-eeEtLtfEaAvatsLg---RsntigsapPlrhsw  2185
α1-E (ray)       psltlhdgpgsldqaLgeEtLtfEaAvatsLg---RshtissapPlrqgw  2191
α1-C (rabbit)    kFievttQelAdacdLtiEeM--EnAaddiLsggaRqspngtllPfvnrr  2132
α1-C (rat)       nFievttQelAdacdMtiEeM--EnAadniLsggaqqspngtllPfvncr  2105
α1-D (human)     kFvsatkheiAdacdLtiDeM--EsAastlLngnvRprangdvgPlshrq  2126
α1-S (rabbit)    gFvtatsQalAdacqMepEev--EvAatelLkaresvqgmasv-Pgslsr  1851

α1-A (rabbit)    rsrrggrrwrasagk-g-----Gggprasapsp          2424
α1-B (rat)       rrvpngyhctlglst-g-----vrarhsyhhpdqdhwc     2336
α1-B (human)     rrvpngyhctlglss-g-----Grarhsyhhpdqdhwc     2339
α1-E (rat)       qmpnghyrrrrlggl-glammcGavsDllsdteeddkc     2222
α1-E (ray)       hlpngsyrtrmmqsg-a-----pstpDpythaeeddrc     2223
α1-C (rabbit)    dpgrdraggneqdasgacapgcGqseEaladrragvssl    2171
α1-C (rat)       dpgqdravvpedesc-vyalgrGrseEalpdsrsyvsnl    2143
α1-D (human)     dyelqdfgpgysde----epdpGrdeEdlademicittl    2161
α1-S (rabbit)    rsslgsldqvqg------------sqEtlipprp         1873
```

FIGURE 2(14).

2.3.2.3 Class C (α_{1C})

The α_{1C} subunits have been identified in rabbit cardiac muscle (pCARD3; Mikami et al., 1989), rabbit lung (pSCaL; Biel et al., 1990), rat brain (rbC-I and rbC-II; Snutch et al., 1991), rat aorta (VSmα1; Koch et al., 1990), and mouse brain (mbC; Ma et al., 1992). Northern blot analysis demonstrated two hybridizable RNA species of pCARD3 in rabbit heart, brain, and stomach with no RNA signal detected in kidney (Mikami et al., 1989). The two VSmα1 transcripts were detected in heart, brain, and in several types of smooth muscle including, uterus, lung, stomach, and small and large intestines (Koch et al., 1990). The rbC probe hybridized to ~8 and 10 kb transcripts expressed throughout the rat CNS including the olfactory bulb, cerebellum, striatum, hypothalamus–thalamus, hippocampus and cortex, pons–medulla, and spinal cord. The rbC-I and rbC-II isoforms were differentially expressed, with rbC-II detected at relatively higher levels in the olfactory bulb, cortex, hippocampus, hypothalamus, and striatum, while rbC-I predominated in spinal cord and trigeminal nerve. The rbC transcripts were also detected at moderate levels in rat heart, adrenal gland, pituitary, liver, and kidney, and at low levels in testes and spleen (Snutch et al., 1991). The mbC transcripts were highly expressed in mouse heart and at lower levels in the brain and spinal cord. No detectable mbC transcripts were found in liver or skeletal muscle (Ma et al., 1992).

Monoclonal antibodies (MANC-1 and MANC-3) against the α_2 δ-subunit of rabbit skeletal muscle Ca channel preferentially immunoprecipitated DHP-sensitive L-type Ca channels labeled with [^{3}H]PN 200-110 from rabbit skeletal muscle or rabbit brain (Ahlijanian et al., 1990). Immunostaining localized L-type Ca channels to the cell bodies and proximal dendrites of hippocampal pyramidal cells with high density clustering at the base of major dendrites (Westenbroek et al., 1990). A similar staining pattern was detected in many central neurons, including dentate granule cells, pyramidal neurons of the cortex, and cerebellar Purkinje and granule cells. The large spinal motor neurons of the spinal cord and the cell bodies located in ganglion cell, inner nuclear, and outer nuclear layers of the retina were also labeled by MANC-1 (Ahlijanian et al., 1990). An antibody (CNC1) that recognizes both rbC-I and rbC-II α_{1C} subunits immunoprecipitated ~75% of rat cerebral cortical [^{3}H] PN200-110 binding sites (Hell et al., 1993b). Immunocytochemical staining revealed widespread neuronal labeling, including strong staining on cell bodies and proximal dendrites throughout layers I to IV of the cerebral cortex and in the cerebellar granule and molecular layers. Cerebellar Purkinje cells showed strong CNC1 immunoreactivity on cell bodies and the major denritic shafts and branch points. In the hippocampus CNC1 immunoreactivity was mostly on the cell bodies of CA1 pyramidal cells and also dense on both the cell bodies and dendrites of the CA2 and CA3 pyramidal cells. While the dendrites of dentate gyrus granule cells were strongly labeled there was little immunoreactivity on the cell bodies (Hell et al., 1993b). On many central neurons CNC1 staining was distributed both generally on cell bodies and dendrites and also in highly punctate ~1.5 to 2 μm clusters on cell bodies and proximal dendrites, suggesting a synaptic localization for a subpopulation of α_{1C} L-type channels.

2.3.2.4 Class D (α_{1D})

The human β cell type α_1 subunit, cDNA (CACN4) probe hybridized to a single ~11-kb transcript present in rat brain and pancreatic islets (Sieno et al., 1992). Northern blot analysis did not detect a signal from human and monkey skeletal muscles, heart, kidney, spleen, liver, jejunum, and colon. RINm5F, an insulin-producing cell line, produced high levels of CACN4 mRNA while low levels of hybridization were detected in βTC-3 and HIT cells (Sieno et al., 1992). *In situ* hybridization using a rat β cell subunit antisense RNA probe showed specific hybridization to islets in β cells (Seino et al., 1992). Another α_{1D} subunit (HCa3a; Yaney et al., 1992) was isolated from an insulin-secreting cell line (HIT-T15) from the hamster. The HCa3a transcripts were detected in hamster pancreas, brain, heart, and skeletal muscle. PCR analysis revealed the message was also present in the rat CNS including the cortex, cerebellum, hypothalamus, and brain stem (Yaney et al., 1992). Northern blot analysis showed that RBα1 was expressed as an 8.6-kb transcript in rat brain, heart, aorta, uterus, lung, and as a 6.5-kb transcript in aorta and skeletal muscle (Hui et al., 1991). *In situ* hybridization to rat brain sections revealed prominent α_{1D} expression in most regions, including the olfactory bulb, cerebral cortex, hippocampal CA fields and dentate gyrus, cerebellar Purkinje cells, suprachiasmatic nucleus, supraoptic nucleus, pituitary gland, and pineal gland (Chin et al., 1992). PCR identified α_{1D} transcripts in IMR32 cells and human brain regions including the hippocampus, basal ganglia, habenula, and thalamus (Williams et al., 1992a). No α_{1D} transcripts were detected in human skeletal muscle RNA.

A polyclonal antisera generated against the rat brain a_{1D} sequence immunoprecipitated ~20% of rat cerebral cortical [^{3}H]PN200-110 binding sites (Hell et al., 1993b). The CND1 antibody recognized two a_{1D} size forms with apparent molecular masses of ~180 and 200 kDa. Immunocytochemical staining revealed strong labeling on most central neurons, including those in layers I through VI of the cerebral cortex, the hippocampal CA1–CA3 pyramidal and dentate gyrus areas, and the granule layer of the cerebellum. Strong labeling was also detected on cerebellar Purkinje cell bodies. Throughout the rat CNS, the densest CND1 labeling was found on cell bodies and proximal dendrites, although confocal microscopy also revealed weak CND1 immunoreactivity on more distal portions of some dendrites (Hell et al., 1993b). In contrast to the punctate pattern observed for a_{1C} L-type channels, the a_{1D} L-type channels were not clustered but were distributed evenly over the cell body with accumulations at the base of major dendrites. In general, the high concentration of a_{1C} and a_{1D} L-type channels on cell bodies and proximal dendrites was complementary to that of a_{1B} N-type channels, which were highly expressed along the entire length of dendrites and at much lower levels on cell bodies (Westenbroek et al., 1992).

2.3.2.5 Class E (α_{1E})

Northern blot analysis demonstrated much higher levels of BII mRNA transcripts (~10.5 and 11 kb) in the rabbit cerebral cortex, hippocampus, and corpus striatum compared to the olfactory bulb, midbrain, cerebellum, and pons–medulla (Niidome et al., 1992). No detectable hybridization was found in skeletal muscle, heart, stomach, or kidney. Interestingly, while both BII transcripts were detected in hippocampus and striatum, only the 11 kb transcript was found in the cerebral cortex and only the 10.5 kb transcript found in the cerebellum. *In situ* hybridization showed that rbE-II transcripts were detected in neurons throughout the rat brain and spinal cord, and were most abundant in the hippocampus and olfactory bulb (Soong et al., 1993). Strong rbE-II signals were also detected in specific nuclei in the brain including intralaminar, parafasicular, and reticular thalamic nuclei, pontine nuclei, the inferior olive, and nucleus of solitary tract. Labeling was also detected in pineal gland, the ganglion cell layer and inner nuclear layer of the retina, sensory ganglia, and the anterior and intermediate lobes of the pituitary (Soong et al., 1993). The *in situ* localization of α_{1E} transcripts in mouse brain revealed a similar distribution to that of rbE-II (Williams et al., 1994). PCR analyses revealed that the human α_{1E} was expressed in human kidney, mouse retina, spleen and pancreatic islets cells, and also in several neuronal and endocrine cell lines (Williams et al., 1994).

It is apparent that within brain regions of a given species, each of the classes of Ca channel α_1-subunits shows a distinct expression pattern, presumably reflecting the unique physiological roles of the different Ca channels. In addition, comparison of the expression patterns between species shows that individual classes of Ca channels exhibit somewhat different patterns. While there are several possible explanations for these results, including species-specific developmental differences, they raise an interesting possibility that a single type of Ca channel may mediate different Ca-dependent functions.

2.3.3 Structure and Function Studies

Expression studies in both *Xenopus* oocytes and mammalian cells demonstrate that the α_1-subunit is the major voltage-sensitive, pore-forming molecule of calcium channel complexes (Mikami et al., 1989; Perez-Reyes et al., 1989; Biel et al., 1990; Mori et al., 1991; Soong et al., 1993). While determination of the specific regions of the α_1-subunits that are responsible for Ca channel functional properties has lagged behind studies of Na and K channels, some recent progress has been made in this area.

In cardiac and skeletal muscle, L-type calcium channels serve both to mediate voltage-dependent Ca influx and excitation–contraction (E–C) coupling. However, the properties of both the L-type currents and the mechanisms of E–C coupling differ between cardiac and skeletal muscle (Adams et al., 1990). Tanabe and co-workers have elegantly utilized transient expression in myotubes from mice with muscular dysgenesis to examine functional components of L-type channel α_1-subunits. Nuclear injection of the skeletal muscle α_{1S} cDNA into myotubes resulted in L-type currents and E–C coupling that closely resembled that of native skeletal muscle, while injection of the cardiac α_{1C} cDNA resulted in L-type currents and E–C coupling that exhibited properties similar to that of cardiac muscle (Tanabe et al., 1990a). To dissect regions of the L-type channel α_1-subunits responsible for the respective cardiac and skeletal muscle characteristics, Tanabe and co-workers constructed a series of chimeric cardiac–skeletal muscle cDNAs. In the first set of experiments, chimeras were constructed such that segments encoding putative cytoplasmic segments of the cardiac α_{1C} were replaced by the corresponding region of the skeletal muscle α_{1S} (Tanabe et al., 1990b). Examination of electrically evoked contractions in microinjected myotubes showed that chimeras containing the domain II to III segment of the α_{1C} had cardiac-type E–C coupling while a chimera containing the α_{1S} II to III segment and all other cytoplasmic and transmembrane regions from the α_{1C} exhibited skeletal muscle-type E–C coupling. In addition to illustrating that the domain II to III segment is the major determinant of skeletal muscle E–C coupling, these studies also suggested that several putative cytoplasmic regions (the amino- and carboxyl-termini, and the domain I to II and domain III to IV segments) do not play a major role in determining the biophysical properties of L-type calcium currents expressed in dysgenic cells (Tanabe et al., 1990b). In another study, the functional role of the carboxyl-terminal region of the α_{1S} was examined by removing sequences encoding 211 residues and injecting the truncated cDNA into myotubes. The results showed that the resulting truncated α_1-subunit could still function both as an L-type channel and in mediating E–C coupling (Beam et al., 1992). Chimeric cDNAs were also used to examine regions involved in channel activation (Tanabe et al., 1991). These experiments showed that the nature of domain I in the chimeric α_1-subunits determines whether the calcium current exhibits rapid (cardiac-like) or slow (skeletal muscle-like) activation. Switching of domains II, III, and IV did not significantly affect calcium channel activation kinetics (Tanabe et al., 1991). In more recent experiments, differences in amino acid sequence of the S3 segment and the S3–S4 linker were found to be critical in determining skeletal muscle activation kinetics (Nakai et al., 1994).

By analogy with Na and K channels, the SS1–SS2 region within the S5 to S6 segment of each domain has been hypothesized to form the ion conduction pore. Mutagenesis of this region in Na and K channels has shown that substitution of single amino acids can dramatically alter both the affinity of pharmacological agents and ion conduction across the membrane (Catterall, 1991; MacKinnon, 1991). In the SS1–SS2 region of domain III, calcium channels have a glutamic acid residue at the corresponding position of a lysine at residue 1422 in sodium channels. Similarly, in SS1–SS2 of domain IV calcium channels have a glutamic acid in the corresponding position (1714) of an alanine residue in sodium channels. Changing either lysine 1422 or alanine 1714 of the rat sodium channel II to glutamic acid altered the permeation properties of the sodium channel to resemble those more of a calcium channel (Heinemann et al., 1992). The analogous experiment of changing glutamic acids 1469 of the brain BI (α_{1A}) calcium channel to glutamine altered the selectivity for divalent cations over monovalent cations and also decreased the sensitivity to block by cadmium (Kim et al., 1993). Substitution of glutamic acid 1469 for glutamine did not seem to be as critical in affecting ion selectivity of the BI calcium channel. The sequential mutation of glutamate residues 393, 736, 1145, and 1446 in the cardiac α_{1C} showed that conserved glutamates in all four domains each made unique contributions to high-affinity ion binding, selectivity, and permeation (Yang et al., 1993).

2.3.4 Primary Structure and Expression of Calcium Channel Ancillary Subunits

The skeletal muscle dihydropyridine receptor is known to be a heteroligomeric structure consisting of the pore-forming α_1-subunit (α_{1S}), disulfide-linked α_2- and δ-subunits, a β-subunit, and a γ-subunit (Campbell et al., 1988; Catterall et al., 1988). Similarly, the purified ω-conotoxin GVIA receptor (N-type channel) consists of four (Witcher et al., 1993) or five (McEnery et al., 1991) different subunits, and it is also likely that neuronal L-type channels are multimeric complexes (Ahlijanian et al., 1990; Westenbroek et al., 1990). While some types of α_1-subunits can form functional Ca channels when expressed alone (Mikami et al., 1989; Perez-Reyes et al., 1989; Biel et al., 1990; Soong et al., 1993; Stea et al., 1993; Tomlinson et al., 1993), other α_1-subunits have been found to require additional subunits to generate functional calcium channels (Williams et al., 1992a, 1992b; Ellinor et al., 1993). Of particular significance, the biophysical properties of α_1-subunits can be modulated by coexpression with Ca channel ancillary subunits. The major modulatory subunit appears to be the β-subunit, affecting both kinetic and voltage-dependent properties. In addition to the ancillary subunits discussed below, a 95-kDa glycoprotein has been shown to be a structural component of the mammalian N-type channel (Witcher et al., 1993) and several other smaller proteins (25–40 kDa) have also been found to associate with N-type channels (Saisu et al., 1991; Gundersen and Umbach, 1992).

2.3.4.1 α_2- and δ-Subunits

Biochemical purification of the skeletal muscle DHP receptor complex showed the α_2- (~140 kDa) and δ-(~30 kDa) subunits were distinct polypeptides that were covalently attached via a disulfide linkage (Campbell et al., 1988; Catterall et al., 1988). Molecular cloning revealed that both the α_2- and δ-subunits were encoded in the same primary transcript and were generated posttranslationally by cleavage between alanine 934 and alanine 935 (DeJongh et al., 1990; Jay et al., 1991). To date, purification of neuronal Ca channel complexes has not revealed separate α_2 and δ polypeptides (McEnery et al., 1991; Witcher et al., 1993). The α_2 δ-subunit was first isolated from rabbit skeletal muscle and encoded a protein consisting of 1079 amino acids with a calculated M_r of 125 kDa (Ellis et al., 1988). The α_2 δ cDNAs have also been cloned from human brain (α_2b, Williams et al., 1992a) and rat brain (rBα_2, Kim et al., 1992). The cDNAs are nearly identical (95–-97%) and have a predicted protein structure containing three transmembrane regions and several consensus phosphorylation and glycosylation sites. Coexpression of the α_2 δ-subunit with α_{1S} and α_{1A}, α_{1B}, α_{1C}, and α_{1D}-subunits indicated a functional interaction between these subunits (Mikami et al., 1989; Mori et al., 1991; Singer et al., 1991; Williams et al., 1992a, 1992b; Stea et al., 1993; Tomlinson et al., 1993). The modulatory effects of coexpression of α_2/δ varied from modest potentiation of the peak whole cell currents to highly synergistic effects when coexpressed with β-subunits.

2.3.4.2 β-Subunit

Similar to that for the α_1-subunits, a heterogeneous family of β-subunit cDNAs have been recently described (Table 2; Figure 3). The nomenclature of Perez-Reyes and co-workers named the individual β-subunit genes numerically while giving lower case subscripts to splice variants (Perez-Reyes et al., 1992). The first β-subunit was described from rabbit skeletal muscle (β_{1a}, Ruth et al., 1989) but has also been isolated from human heart (Collin et al., 1993) and human skeletal muscle (Powers et al., 1993). The β_{1a}-subunit consisted of a 524 a.a. protein (57.9 kDa) that was highly hydrophilic and possessed no typical membrane spanning regions. Coexpression of β_{1a} with the α_{1S}-subunit increased the rate of channel activation to values more consistent with that of L-type channels in muscle (Lacerda et al., 1991; Varadi et al., 1991). An alternatively spliced form of the skeletal muscle β_1 cDNA (β_{1b} in the Perez-Reyes et al., nomenclature) expressed in brain, heart, and spleen has also been described (Pragnell et al., 1991; Powers et al., 1992;

Table 2. Molecular Properties of Cloned Calcium Channel β-Subunits

Class	cDNA	Source	Number of a.a.	Predicted molecular mass (kDA)	Distribution	References
β_{1a}	Ca_{B-1}	Rabbit skeletal muscle	524	58	Skeletal muscle, brain, heart	Ruth et al. (1989)
	β_c	Human heart	522	58		Collin et al. (1993)
	β_{1M}	Human skeletal muscle	523	58		Powers et al. (1992)
β_{1b}	β_{1b}	Rat brain	597	65	Brain, heart, pancreatic islet cells, RIN5mF cells, βTC3 cells	Pragnell et al. (1991)
	β_a	Human heart	597	65		Collin et al. (1993)
	β_{1B-2}	Human hippocampus	596	65		Powers et al. (1992)
β_{1c}	β_2	Human hippocampus	478	53	Brain, heart, spleen	Williams et al. (1992a)
	β_b	Human heart	477	53		Collin et al. (1993)
	β_{1B-1}	Human hippocampus	478	53		Powers et al. (1992)
β_2	β_2	Rat brain	604	68	Brain, aorta, heart, lung, trachea, HIT cells, RIN5mF cells	Perez-Reyes et al. (1992)
	Ca_{B-2a}	Rabbit heart	606	68		Hullin et al. (1992)
	Ca_{B-2b}		632	71		
β_3	β_3	Rat brain	484	55	Brain, aorta, trachea, lung, kidney, spleen, skeletal muscle, HIT cells, RIN5mF cells	Castellano et al. (1993a)
	Ca_{B-3}	Rabbit brain	477	54		Hullin et al. (1992)
β_4	β_4	Rat brain	519	58	Brain, kidney	Castellano et al. (1993b)

```
β1a (rabbit)   MvqktsmsrgpyppsqeipmevfDpSpqgkyskRkgRFKrSDGSTssdTt   50
β1c (human)    MvqktsmsrgpyppsqeipmgvfDpSpqgkyskRkgRFKrSDGSTssdTt   50
β1b (rat)      MvqksgmsrgpyppsqeipmevfDpSpqgkyskRkgRFKrSDGSTssdTt   50
β2a (rat)      M----------------------Qcc----------glvhrr--------   10
β2a (rabbit)   M---------------------lD-----------rhlaaph--------   10
β2b (rabbit)   M-------------nqasgldllkiS-ygkgarRknRFKgSDGSTssdTt   36
β3 (rat)       M---------------------yDdS-------yvpgFedSE--------   14
β4 (rat)       M-sssyakngaadgphspssqvargt-----ttRrsRlKrSDGST---Ts   41

β1a (rabbit)   sNsfVRQGSAESYTSRPS-DSDVSLEEDREAlRkEAERQAlAQLEKAKTK   99
β1c (human)    sNsfVRQGSAESYTSRPS-DSDVSLEEDREAlRkEAERQAlAQLEKAKTK   99
β1b (rat)      sNsfVRQGSAESYTSRPS-DSDVSLEEDREAlRkEAERQAlAQLEKAKTK   99
β2a (rat)      -rvrVsyGSADSYTSRPS-DSDVSLEEDREAvRrEAERQAqAQLEKAKTK   58
β2a (rabbit)   tQglVlEGSADSYTSRPS-DSDVSLEEDREAvRrEAERQAqAQLEKAKTK   59
β2b (rabbit)   sNsfVRQGSADSYTSRPS-DSDVSLEEDREAvRrEAERQAqAQLEKAKTK   85
β3 (rat)       ------aGSADSYTSRPSlDSDVSLEEDREsaRrEvEsQAqqQLErAKhK   58
β4 (rat)       tsfilRQGSADSYTSRPS-DSDVSLEEDREAiRqErEqQAaiQLErAKsK   90

β1a (rabbit)   PVAFAVRTNVgYnpspgDEVPVeGvAItFEpKDFLHIKEKYNNDWWIGRL  149
β1c (human)    PVAFAVRTNVgYnpspgDEVPVqGvAItFEpKDFLHIKEKYNNDWWIGRL  149
β1b (rat)      PVAFAVRTNVgYnpspgDEVPVqGvAItFEpKDFLHIKEKYNNDWWIGRL  149
β2a (rat)      PVAFAVRTNVrYsaaqEDDVPVpGmAIsFEAKDFLHVKEKFNNDWWIGRL  108
β2a (rabbit)   PVAFAVRTNVsYsaahEDDVPVpGmAIsFEAKDFLHVKEKFNNDWWIGRL  109
β2b (rabbit)   PVAFAVRTNVsYsaahEDDVPVpGmAIsFEAKDFLHVKEKFNNDWWIGRL  135
β3 (rat)       PVAFAVRTNVsYcgvlDEEcPVqGsgVnFEAKDFLHIKEKYsNDWWIGRL  108
β4 (rat)       PVAFAVkTNVsYcgalDEDVPVpstAIsFDAKDFLHIKEKYNNDWWIGRL  140

β1a (rabbit)   VKEGCEVGFIPSPVKLDsLRLlQEQklrQsrlsSSKSGdNSSSSLGDVVt  199
β1c (human)    VKEGCEVGFIPSPVKLDsLRLlQEQklrQnrlgSSKSGdNSSSSLGDVVt  199
β1b (rat)      VKEGCEVGFIPSPVKLDsLRLlQEQtlrQnrlsSSKSGdNSSSSLGDVVt  199
β2a (rat)      VKEGCEIGFIPSPVKLEnMRLqhEQrakQgkfySSKSGgNSSSSLGDIVp  158
β2a (rabbit)   VKEGCEIGFIPSPVKLEnMRLqhEQrakQgkfySSKSGgNSSSSLGDIVp  159
β2b (rabbit)   VKEGCEIGFIPSPVKLEnMRLqhEQrakQgkfySSKSGgNSSSSLGDIVp  185
β3 (rat)       VKEGgDIaFIPSPqrLEsiRLkQEQkarrs---------gnpSSLsDI--  147
β4 (rat)       VKEGCEIGFIPSPlrLEniRiqQEQ--krgrfhggKSsgNSSSSLGEmVs  188

β1a (rabbit)   GtRRpTPPaSgnEmtnlafelepldleedeaelgeqsgsaktsvssvttp  249
β1c (human)    GtRRpTPPaSAkQkQ-----------------------------------  214
β1b (rat)      GtRRpTPPaSAkQkQ-----------------------------------  214
β2a (rat)      ssRksTPPsSAiDiD-------atgldaeendipanhrspkpsansvtsp  201
β2a (rabbit)   ssRksTPPsSAiDiD-------atgldaeendipanhrspkpsansvtsp  202
β2b (rabbit)   ssRksTPPsSAiDiD-------atgldaeendipanhrspkpsansvtsp  228
β3 (rat)       GnRRspPPslAkQkQ-----------------------------------  162
β4 (rat)       GtfRaTPtttAkQkQ-----------------------------------  203
```

FIGURE 3. Alignment of cloned calcium channel β-subunits. Representative members from each class of calcium channel β-subunits aligned using Multalin version 4.0 software (Corpet, 1988). Characters in bold and capitals indicate identical or conservative substitutions in >60% of the sequences. Dashes indicate gaps required to align the sequences and the amino acid number is given at the end of each line. Sequences were derived from the following sources: β_{1a} (rabbit), Ruth et al. (1989), GenBank Accession #M25817; β_{1c} (human), Williams et al. (1992a), #M76560; β_{1b} (rat), Pragnell et al. (1991), #X61394; β_{2a} (rat), Perez-Reyes et al. (1992), #M80545; β_{2a}, β_{2b} (rabbit), Hullin et al. (1992), #X64298; β_3, β_4 (rat), Castellano et al. (1993a,b), #M88751, #L02315.

```
β1a (rabbit)    pphgtripffKkTEHVPPYDVVPSMRPIILVGPSLKGYEVTDMMQKALFD  299
β1c (human)     ----------KsTEHVPPYDVVPSMRPIILVGPSLKGYEVTDMMQKALFD  254
β1b (rat)       ----------KsTEHVPPYDVVPSMRPIILVGPSLKGYEVTDMMQKALFD  254
β2a (rat)       hskekrmpffKkTEHtPPYDVVPSMRPVVLVGPSLKGYEVTDMMQKALFD  251
β2a (rabbit)    hskekrmpffKkTEHtPPYDVVPSMRPVVLVGPSLKGYEVTDMMQKALFD  252
β2b (rabbit)    hskekrmpffKkTEHtPPYDVVPSMRPVVLVGPSLKGYEVTDMMQKALFD  278
β3 (rat)        ----------KqaEHVPPYDVVPSMRPVVLVGPSLKGYEVTDMMQKALFD  202
β4 (rat)        ----------KvTEHIPPYDVVPSMRPVVLVGPSLKGYEVTDMMQKALFD  243

β1a (rabbit)    FLKHlFDGRISITRVTADISLAKRSVLNNPSKHiIIERSNTRSSLAEVQS  349
β1c (human)     FLKHRFDGRISITRVTADISLAKRSVLNNPSKHiIIERSNTRSSLAEVQS  304
β1b (rat)       FLKHRFDGRISITRVTADISLAKRSVLNNPSKHiIIERSNTRSSLAEVQS  304
β2a (rat)       FLKHRFEGRISITRVTADISLAKRSVLNNPSKHaIIERSNTRSSLAEVQS  301
β2a (rabbit)    FLKHRFEGRISITRVTADISLAKRSVLNNPSKHaIIERSNTRSSLAEVQS  302
β2b (rabbit)    FLKHRFEGRISITRVTADISLAKRSVLNNPSKHaIIERSNTRSSLAEVQS  328
β3 (rat)        FLKHRFDGRISITRVTADlSLAKRSVLNNPgKrtIIERSsaRSSiAEVQS  252
β4 (rat)        FLKHRFDGRISITRVTADISLAKRSVLNNPSKraIIERSNTRSSLAEVQS  293

β1a (rabbit)    EIERIFELARTLQLVaLDADTINHPAQLSKTSLAPIIVYIKItSPKVLQR  399
β1c (human)     EIERIFELARTLQLVaLDADTINHPAQLSKTSLAPIIVYIKItSPKVLQR  354
β1b (rat)       EIERIFELARTLQLVaLDADTINHPAQLSKTSLAPIIVYIKItSPKVLQR  354
β2a (rat)       EIERIFELARTLQLVVLDADTINHPAQLSKTSLAPIIVYVKISSPKVLQR  351
β2a (rabbit)    EIERIFELARTLQLVVLDADTINHPAQLSKTSLAPIVVYVKISSPKVLQR  352
β2b (rabbit)    EIERIFELARTLQLVVLDADTINHPAQLSKTSLAPIVVYVKISSPKVLQR  378
β3 (rat)        EIERIFELAksLQLVVLDADTINHPAQLaKTSLAPIIVFVKVSSPKVLQR  302
β4 (rat)        EIERIFELARsLQLVVLDADTINHPAQLiKTSLAPIIVhVKVSSPKVLQR  343

β1a (rabbit)    LIKSRGKSQsKHLNVQiaAsEKLAQCPP-EMFDIILDENQLEDACEHLAE  448
β1c (human)     LIKSRGKSQsKHLNVQiaAsEKLAQCPP-EMFDIILDENQLEDACEHLAE  403
β1b (rat)       LIKSRGKSQsKHLNVQiaAsEKLAQCPP-EMFDIILDENQLEDACEHLAE  403
β2a (rat)       LIKSRGKSQaKHLNVQMvAaDKLAQCPPqEsFDVILDENQLEDACEHLAD  401
β2a (rabbit)    LIKSRGKSQaKHLNVQMvAaDKLAQCPP-ELFDVILDENQLEDACEHLAD  401
β2b (rabbit)    LIKSRGKSQaKHLNVQMvAaDKLAQCPP-ELFDVILDENQLEDACEHLAD  427
β3 (rat)        LIrSRGKSQmKHLtVQMmAyDKLvQCPP-EsFDVILDENQLDDACEHLAE  351
β4 (rat)        LIKSRGKSQsKHLNVQLvAaDKLAQCPP-EMFDVILDENQLEDACEHLgE  392

β1a (rabbit)    YLEAYWKATHPPSStPPNPLLnRTMATaALaaSPapvSNlQvq-------  491
β1c (human)     YLEAYWKATHPPSStPPNPLLnRTMATaALaaSPapvSNlQvq-------  446
β1b (rat)       YLEAYWKATHPPSrtPPNPLLnRTMATaALavSPapvSNlQgpylvsgDq  453
β2a (rat)       YLEAYWKATHPPSSnlPNPLLsRTLATstLplSPtlaSNsQgs---qgDq  448
β2a (rabbit)    YLEAYWKATHPPSSnlPNPLLsRTLATsALpvSPtlaSNsQgs---qgDq  448
β2b (rabbit)    YLEAYWKATHPPSSnlPNPLLsRTLATsALpvSPtlaSNsQgs---qgDq  474
β3 (rat)        YLEvYWrATHhPapgPg--MLgppsAipgLqnqqllgergEeh-------  392
β4 (rat)        YLEAYWrATHtsSStPmtPLLgRnvgstALspyPtaiSglQsqrmrhsNh  438
```

FIGURE 3(2).

Collin et al., 1993). The human and rat β_{1b} isoforms differ from β_{1a} in a 45–50 a.a. deletion at residue 209 and in a longer carboxyl-terminus (~120 a.a.). Coexpression of either the β_{1a}- or β_{1b}-subunits with different L-type Ca channel α_1-subunits resulted in significant changes to the biophysical properties (Singer et al., 1991; Perez Reyes et al., 1992; Mori et al., 1991; Tomlinson et al., 1993). A third spliced form of the β_1 gene, β_{1c}, was isolated from human heart and hippocampal neurons (Williams et al., 1992a; Powers et al., 1992; Collin et al., 1993) and had the same deletion (at a.a. 209) as β_{1D} but had a carboxyl-terminus similar to β_{1a}. Coexpression of β_{1c} with α_{1D} (Williams et al., 1992a) or α_{1B} (Williams et al., 1992b) resulted in significant changes in the currents expressed.

```
β1a (rabbit)   ------------------------------------------vltsLrRN  499
β1c (human)    ------------------------------------------vltsLrRN  455
β1b (rat)      plDratgehasvhey------pgelgqppglypsnhppgRagtlwaLsRQ  497
β2a (rat)      rtDrsap-rsasqaeeepclepvkksqhrsssat-hqnhRsgtgrgLsRQ  496
β2a (rabbit)   rtDrsaparsasqaeeepclepakksqhrssssaphhnhRsgtsrgLsRQ  498
β2b (rabbit)   rtDrsaparsasqaeeepclepakksqhrssssaphhnhRsgtsrgLsRQ  524
β3 (rat)       -----------------------------------spleRdslmpsdeas  407
β4 (rat)       stEnspierrslm----------------------tsdenyhnerarksr  470

β1a (rabbit)   lsFwgglEtSqrgggavPqQQ-----------------------------  520
β1c (human)    lgFwgglEsSqrgs-vvPqEQ-----------------------------  474
β1b (rat)      DtFdadtpgSRNsvYtePgDscvdmetD---Psegpgpgdpagggtppar  544
β2a (rat)      EtFdsetQeSRDsaYvePkEDYshehvDrYvPHREhnhreeshssnghrh  546
β2a (rabbit)   EtFdsetQeSRDsaYvePkEDYshehvDhYaPHRDhnhrdethrssdhrh  548
β2b (rabbit)   EtFdsetQeSRDsaYvePkEDYshehvDhYaPHRDhnhrdethrssdhrh  574
β3 (rat)       EssrqawtgSsQrssrhleEDYadayqDlYqPHRQhtsglpsanghdp--  455
β4 (rat)       NrlssssQhSRDh-YplveEDYpdsyqDtYkPHRNrgspggcshds----  515

β1a (rabbit)   ------ehaM                                          524
β1c (human)    ------ehaM                                          478
β1b (rat)      qgsweeeedyeeEmtdN-----rnRgrnkarycaegggpvlgrnkneleg  589
β2a (rat)      reprhrtrdMgrDqdhNecskqrsRhkskdrycdkege-viskrrseage  595
β2a (rabbit)   retrhrsrdMdrEqdhNecnkqrsRhkskdrycdkdge-viskkrneage  597
β2b (rabbit)   retrhrsrdMdrEqdhNecnkqrsRhkskdrycdkdge-viskkrneage  623
β3 (rat)       -----qdrlLaqDsehDhndrnwqRnrp----------------------  478
β4 (rat)       ------rhrL                                          519

β1b (rat)      WgqgvYir      597
β2a (rat)      WnrdvYirq     604
β2a (rabbit)   WnrdvYirq     606
β2b (rabbit)   WnrdvYirq     632
β3 (rat)       WpkdsY        484
```

FIGURE 3(3).

A second β-subunit gene, (β_2), coded for a protein of 604 a.a. (M_r ~68 kDa) and was expressed in rat heart, brain, and lung (Perez-Reyes et al., 1992). Alternatively spliced forms of the β_2 subunit were also detected in rabbit cardiac tissue and brain (Hullin et al., 1992) and coexpression of the β_2-subunit altered the functional properties of the cardiac a_{1C}-subunit (Perez-Reyes et al., 1992; Hullin et al., 1992) and the α_{1A} subunit (Stea et al., 1994). The β_2 subunit caused a significant slowing of the inactivation of both α_{1A} (Stea et al., 1994) and the endogenous oocyte calcium current (Lacerda et al., 1994). Two additional β-subunit genes, β_3 (M_r ~55 kDa) and β_4 (M_r ~58), have also been described (Hullin et al., 1992; Castellano et al., 1993a,b). The β_4-subunit was predominantly expressed in brain, while transcripts for the β_3-subunit were detected in brain, lung, spleen, heart, and skeletal muscle (Hullin et al., 1992; Castellano et al., 1993a). Coexpression studies showed that both the β_3 and β_4 subunits modulated the properties of α_1 subunits by increasing the whole cell current, increasing the rate of activation, and shifting the voltage dependence of activation and inactivation (Castellano et al., 1993a,b; Sather et al., 1993; Stea et al., 1994).

In general, coexpression of β with α_1-subunits resulted in increased whole-cell currents, increased rates of activation, and hyperpolarizing shifts in the voltage-dependence of activation and inactivation (for examples, Hullin et al., 1992; Perez-Reyes et al., 1992; Castellano et al., 1993a,b; Ellinor et al., 1993; Stea et al., 1993; Soong et al., 1993; Sather et al., 1993). However, different β-subunits have distinct effects on the rate of inactivation of α_1-subunits. For example, while β_1- and β_3-subunits accelerated the rate of inactivation (Sather et al., 1993; Stea et al., 1994) the β_2-subunit causes a significant slowing of the inactivation rate (Lacerda et al., 1994; Stea et al., 1994). Coexpression of a β-subunit increased the number of binding sites for ω-CgTx (Williams et al., 1992b) and for dihydropyridines (Nishimura et al., 1993), suggesting that the β-subunit increased the number of functional Ca channels in the plasma membrane. However, while coexpression of α_{1C} with β_{1a} showed an increase in both dihydropyridine binding and whole-cell current, there was no corresponding increase in the amount of α_1-subunit protein expressed at the cell surface (Nishimura et al., 1993). In addition, examination of the effects of β-subunit coexpression on gating charge showed there was no significant increase in the total gating charge in oocytes expressing the α_{1C}- and β-subunits compared to that of α_{1C} alone (Neely et al., 1993). One possibility is that the β-subunit alters the conformation of the α_1-subunit affecting both the probability of opening and the binding of ligands to the Ca channel complex (Nishimura et al., 1993, Neely et al., 1993). In support of this, coexpression of a β-subunit was found to increase the single channel open probability of a human α_{1C} channel (Wakamori et al., 1993). The site of β-subunit interaction with the α_1-subunit has been localized to a conserved motif (QQ-E–LGY–WI—E) in the putative cytoplasmic linker between domains I and II of all cloned α_1-subunits (Pragnell et al., 1994). Single amino acid substitutions within the motif affected the affinity of β-subunit binding, altered the level of peak whole cell currents, and affected several voltage-dependent and kinetic properties (Pragnell et al., 1994). These changes were consistent with previous studies examining α_1- and β-subunit coexpression (see Mori et al., 1991; Sather et al., 1993; Stea et al., 1993; Tomlinson et al., 1993; Castellano et al., 1993a,b).

2.3.4.3 γ-Subunit

The γ-subunit has been cloned from rabbit and human skeletal muscle and codes for a 222 a.a. protein (M_r ~25 kDa) with four proposed transmembrane regions (Jay et al., 1990; Powers et al., 1993). To date, γ-subunit transcripts have been detected only in skeletal muscle and lung (Jay et al., 1990; Powers et al., 1993). Coexpression of the skeletal muscle γ-subunit with the cardiac α_{1C}-subunit did not have significant effects on channel properties (Wei et al., 1991). In contrast, coexpression of the γ-subunit with the cardiac α_1-and the skeletal muscle β_{1a}-subunits increased both whole cell currents and the rate of activation at negative potentials, suggesting that the γ-subunit has synergistic effects when coexpressed with other subunits (Wei et al., 1991).

2.3.5 Summary

The past few years has seen the molecular nature of Ca channel diversity beginning to be revealed. A major goal of the present research is to correlate the properties of cloned Ca channels with those of the diverse types of Ca currents found in native cells. Based on amino acid similarities, two broad groups of Ca channel α_1-subunits can be identified. One group of related α_1-subunits defines the dihydropyridine-sensitive L-type Ca channels (α_{1S}, α_{1C} and α_{1D}), while a second group of more distantly related proteins encodes Ca channels with diverse electrophysiological and pharmacological characteristics (α_{1A}, α_{1B} and α_{1E}). The α_{1B}-subunits clearly define the major structural component of N-type channels, while the α_{1A} and α_{1E} proteins appear to define novel types of Ca channels. It is also likely that additional Ca channel types remain to be described both at the physiological and molecular genetic levels. The availability of cloned Ca channel subunits will allow the mechanistic dissection of a number of important Ca channel properties including voltage- and Ca-dependent inactivation (Brehm and Eckert, 1978), Ca permeation and selectivity (Hess and Tsien, 1984; Almers and McCleskey, 1984), facilitation (Fenwick et al., 1982; Artalejo et al., 1991; Pietrobon and Hess, 1990), gating mode shifting (Hess et al., 1984), and modulation by second messengers and G-proteins (Marchetti et al., 1986; Bean, 1989b; Bernheim et al., 1991). Expression of different combinations of cloned Ca channel subunits results in currents with distinct properties, and a significant issue will be to define the specific coexpression patterns of α_1-subunits with that of ancillary subunits, and then to correlate subunit expression patterns with those of native Ca currents in identified cell types. In addition, and perhaps the most important issue to be addressed with respect to the heterogeneous family of neuronal Ca channels, concerns the elucidation of their individual contributions to neuronal physiology. Since neuronal Ca channels do not act in isolation, insight into this problem will require a combined knowledge of the physiological properties of the Ca channel subtypes and their subcellular locations, together with an understanding of their modulation by neurotransmitters and second messengers and the nature of the intracellular targets of calcium-dependent processes.

REFERENCES

Adams, B. A., Tanabe, T., Mikami, A., Numa, S., and Beam, K. G. (1990). Intramembrane charge movement restored in dysgenic skeletal muscle by injection of dihydropyridine receptor cDNAs. *Nature* 346, 569–572.

Adams, M. E., Myers, R. A., Imperial, J.S., and Olivera, B. M. (1993). Toxityping rat brain calcium channels with ω-toxins from spider and cone snail venoms. *Biochemistry* 32, 12566–12570.

Ahlijanian, M. K., Westenbroek, R. E., and Catterall, W. A. (1990). Subunit structure and localization of dihydropyridine-sensitive calcium channels in mammalian brain, spinal cord, and retina. *Neuron* 4, 819–832.

Almers, W. and McCleskey, E. W. (1984). Non-selective conductance in calcium channels of frog muscle: Calcium selectivity in a single-file pore. *J. Physiol.* 353; 585–608.

Artalejo, C. R., Mogul, D. J., Perlman, R. L., and Fox, A. P. (1991). Three types of bovine chromaffin cell Ca^{2+} channels: Facilitation increases the opening probability of a 27 pS channel. *J. Physiol.* 444, 213–240.

Artalejo, C. R., Perlman, R. L., and Fox, A. P. (1992). ω-Conotoxin GVIA blocks a Ca^{2+} current in bovine chromaffin cells that is not the classic N type. *Neuron* 8, 85–95.

Beam, K. G., Adams, B. A., Niidome, T., Numa, S., and Tanabe, T. (1992). Function of a truncated dihydropyridine receptor as both voltage sensor and calcium channel. *Nature* 360, 169–171.

Bean, B. P. (1989a). Classes of calcium channels in vertebrate cells. *Annu. Rev. Physiol.* 51, 367–384.

Bean, B. P. (1989b). Neurotransmitter inhibition of neuronal calcium currents by changes in channel voltage dependence. *Nature* 340, 153–156.

Bernheim, L., Beech, D. J., and Hille, B. (1991). A diffusible second messenger mediates one of the pathways coupling receptors to calcium channels in rat sympathetic neurons. *Neuron* 6, 859–867.

Biel, M., Ruth, P., Bosse, E., Hullin, R., Stuhmer, W., Flockerzi, V., and Hofmann, F. (1990). Primary structure and functional expression of a high voltage activated calcium channel from rabbit lung. *FEBS Lett.* 269, 409–412.

Brehm, P. and Eckert, R. (1978). Calcium entry leads to inactivation of calcium channels in Paramecium. *Science* 202, 1203–1206.

Campbell, K. P., Leung, A. T., and Sharp, A. H. (1988). The biochemistry and molecular biology of the dihydropyridine-sensitive calcium channel. *Trends Neurosci.* 11, 425–430.

Castellano, A., Wei, X., Birmbaumer, L., and Perez-Reyes, E. (1993a). Cloning and expression of a third calcium channel β subunit. *J. Biol. Chem.* 268, 3450–3455.

Castellano, A., Wei, X., Birnbaumer, L, and Perez-Reyes, E. (1993b). Cloning and expression of a neuronal calcium channel β subunit. *J. Biol. Chem.* 268, 12359–12366.

Catterall, W. A. (1991). Structure and function of voltage-gated sodium and calcium channels. *Current Opin. Neurobiol.* 1, 5–13.

Catterall, W. A., Seagar, M. J., and Takahashi, M. (1988). Molecular properties of the dihydropyridine-sensitive calcium channel. *J. Biol. Chem.* 263, 3535–3538.

Chaudhari, N. (1992). A single nucleotide deletion in the skeletal muscle-specific calcium channel transcript of muscular dysgenesis (*mdg*) mice. J. Biol. Chem. 267, 25636–25639.

Chin, H., Kozak, C. A., Kim, H.-L., Mock, B., and McBride, O. W. (1991). A brain L-type calcium channel α_1 subunit gene (CCHL1A2) maps to mouse chromosome 14 and human chromosome 3. *Genomics* 11, 914–919.

Chin, H., Smith, M. A., Kim, H.-L., and Kim, H. (1992). Expression of dihydropyridine-sensitive brain calcium channels in the rat central nervous system. *FEBS* 299, 69–74.

Church, D. M., Stotler, C. J., Rutter, J. L., Murrell, J. R., Trofatter, J. A., and Buckler, A. J., (1994). Isolation of genes from complex sources of mammalian genomic DNA using exon amplification. *Nature Genetics* 6, 98–105.

Collin, T., Wang, J.-J., Nargeot, J., and Schwartz, A. (1993). Molecular cloning of three isoforms of the L-type voltage-dependent calcium channel β subunit from normal human heart. *Circ. Res.* 72, 1337–1344.

Collin, T., Lory, P., Taviaux, S., Courtieu, C., Guilbault, P., Berta, P., and Nargeot, J. (1994). Cloning, chromosomal location, and functional expression of the human voltage-dependent calcium channel β_3 subunit. *Eur. J. Biochem.* 220, 257–262.

Corpet, F. (1988). Multiple sequence alignment with hierarchical clustering. *Nucl. Acids Res.* 16, 10881–10890.

DeJongh, K. S., Warner, C., and Catterall, W. A. (1990). Subunits of purified calcium channels α_2 and δ are encoded by the same gene. *J. Biol. Chem.* 265, 14738–14741.

Dubel, S. J., Starr, T. V. B., Hell, J., Ahlijanian, M. K., Enyeart, J. J., Catterall, W. A., and Snutch, T. P. (1992). Molecular cloning of the α_1 subunit of an ω-conotoxin-sensitive calcium channel. *Proc. Natl. Acad. Sci. U.S.A.* 89, 5058–5062.

Ellinor, P. T., Zhang, J.-F., Randall, A. D., Zhou, M., Schwarz, T. L., Tsien, R. W., and Horne, W. A. (1993). Functional expression of a novel neuronal voltage-dependent calcium channel. *Nature* 363, 455–458.

Ellis, S. B., Williams, M. E., Ways, N. R., Brenner, R., Sharp, A. A., Leung, A. T., Campbell, K. P., McKenna, E., Koch, N. J., Hui, A., Schwartz, A., and Harpold, M. M. (1988). Sequence and expression of mRNAs encoding the α_1 and α_2 subunits of a DHP-sensitive calcium channel. *Science* 241, 1661–1664.

Fenwick, E. M., Marty, A., and Neher, E. (1982). Sodium and calcium channels in bovine chromaffin cells. *J. Physiol.* 331, 599–635.

Forti, L. and Pietrobon, D. (1993). Functional diversity of L-type calcium channels in rat cerebellar neurons. *Neuron* 10, 437–450.

Fujita, Y., Mynlieff, M., Dirksen, R. T., Kim, M., Niidome, T., Nakai, J., Friedrich, T., Iwabe, N., Miyata, T., Furuichi, T., Furutama, D., Mikoshiba, K., Mori, Y., and Beam, K. G. (1993). Primary structure and functional expression of the ω-conotoxin-sensitive N-type calcium channel from rabbit brain. *Neuron* 10, 585–598.

Grabner, M., Friedrich, K., Knaus, H.-G., Striessnig, J., Scheffauer, F., Staudinger, R., Koch, W. J., Schwartz, A., and Glossmann, H. (1991). Calcium channel from Cyprinus carpio skeletal muscle. *Proc. Natl. Acad. Sci. U.S.A.* 88, 727–731.

Gregg, R. G., Couch, F., Hogan, K., and Powers, P. A. (1993a). Assignment of the human gene for the α_1 subunit of the skeletal muscle DHP-sensitive calcium channel (CACNL1A3) to chromosome 1q31-q32. *Genomics* 15, 107–112.

Gregg, R. G., Powers, P. A., and Hogan, K. (1993b). Assignment of the human gene for the β subunit of the skeletal muscle DHP-sensitive calcium channel (CACNLB1) to chromosome 17 using somatic cell hybrids and linkage mapping. *Genomics* 15, 185–187.

Gundersen, C. B. and Umbach, J. A. (1992). Suppression cloning of the cDNA candidate for a subunit of a presynaptic calcium channel. *Neuron* 9, 527–537.

Guy, H. R. and Seetharamulu, P. (1986). Molecular model of the action potential sodium channel. *Proc. Natl. Acad. Sci. U.S.A.* 84, 5469–5473.

Heinemann, S. H., Terlau, H., Stuhmer, W., Imoto, K., and Numa, S. (1992). Calcium channel characteristics conferred on the sodium channel by single mutations. *Nature* 356, 441–443.

Hell, J. W. Yokoyama, C. T., Wong, S. T., Warner, C., Snutch, T. P., and Catterall, W. A. (1993a). Differential phosphorylation of two size forms of the neuronal class C L-type calcium channel α_1 subunit. *J. Biol. Chem.*, 268, 19451–19457.

Hell, J. W., Westenbroek, R. E., Warner, C., Ahlijanian, M. A., Prystay, W., Gilbert, M. M., Snutch, T. P., and Catterall, W. A. (1993b). Identification and differential subcellular localization of the neuronal class C and class D L-type calcium channel α_1 subunits. *J. Cell Biol.* 123, 949–962.

Hess, P. (1990). Calcium channels in vertebrate cells. *Annu. Rev. Neurosci.* 13, 337–356.

Hess, P. and Tsien, R. W. (1984). Mechanism of ion permeation through calcium channels. *Nature* 309, 453–456.

Hess, P., Lansman, J. B., and Tsien, R. W. (1984). Different modes of Ca channel gating behavior favoured by dihydropyridine Ca agonists and antagonists. *Nature* 311, 538–544.

Hille, B. (1992). *Ionic Channels of Excitable Membranes,* Sinauer Associates, Sunderland, MA.

Hillman, D., Chen, S., Aung, T. T., Cherksey, B., Sugimori, M., and Llinas, R. R. (1991). Localization of P-type calcium channels in the central nervous system. *Proc. Natl. Acad. Sci. U.S.A.* 88, 7076–7080.

Hillyard, D. R., Monje, V. D., Mintz, I. M., Bean, B. P., Nadasdi, L., Ramachandran, J., Miljanich, G., Azimi-Zoonooz, A., McIntosh, J. M., Cruz, L. J., Imperial, J. S., and Olivera, B. M. (1992). A new conus peptide ligand for mammalian presynaptic Ca channels. *Neuron* 9, 69–77.

Horne, W. A., Ellinor, P. T., Inman, I., Zhou, M., Tsien, R. W., and Schwarz, T. L. (1993). Molecular diversity of Ca channel α_1 subunits from the marine ray Discopyge ommata. *Proc. Natl. Acad. Sci. U.S.A.* 90, 3787–3791.

Huguenard, J. R. and Prince, D. A. (1992). A novel T-type current underlies prolonged Ca^{2+}-dependent burst firing in GABAergic neurons of rat reticular nucleus. *J. Neurosci.* 12, 3804–3817.

Hui., A., Ellinor, P. T., Krizanova, O., Wang, J.-J., Diebold, R. J., and Schwartz, A. (1991). Molecular cloning of multiple subtypes of a novel rat brain isoform of the α_1 subunit of the voltage-dependent calcium channel. *Neuron* 7, 35–44.

Hullin, R., Singer-Lahat, D., Freichel, M., Biel, M., Dascal, N., Hofmann, F., and Flockerzi, V. (1992). Calcium channel β subunit heterogeneity: Functional expression of cloned cDNA from heart, aorta, and brain. *EMBO J.* 11, 885–890.

Iles, D. E., Segers, B., Sengers, R. C. A., Monsieurs, K., Heytens, L., Halsall, P. J., Hopkins, P. M., Ellis, F. R., Ahll-curran, J. L., Stewart, A. D., and Wieringa, B. (1993a). Genetic mapping of the β_1- and γ-subunits of the human skeletal muscle L-type voltage-dependent calcium channel on chromosome 17q and exclusion as candidate genes for malignant hyperthermia susceptibility. *Human Mol. Gen.* 2, 863–868.

Iles, D. E., Segers, B., Weghuis, D.O., Suijkerbuijk, R., and Wieringa, B. (1993b). Localization of the γ-subunit of the skeletal muscle L-type voltage-dependent calcium channel gene (CACNLG) to human chromosome band 17q24 by in situ hybridization and identification of a polymorphic repetitive DNA sequence at the gene locus. *Cytogenet. Cell Genet.* 64, 227–230.

Iles, D. E., Lehmann-Horn, F., Scherer, S. W., Tsui, L.-C., Weghuis, D. O., Suijkerbuijk, R. F., Heytens, L., Mikala, G., Schwartz, A., Ellis, F. R., Stewart, A. D., Deufel, A. D., and Wieringa, B. (1994). Localization of the gene encoding the α_2/δ-subunits of the L-type voltage-dependent calcium channel to chromosome 7q and analysis of the segregation of flanking markers in malignant hyperthermia susceptible families. *Human Mol. Gen.* 3, 969–975.

Jay, S. D., Ellis, S. B, McCue, A. F., Williams, M. E., Vedvick, T. S., Harpold, M. M., and Campbell, K. P. (1990). Primary structure of the γ subunit of the DHP-sensitive calcium channel from skeletal muscle. *Science* 248, 490–492.

Jay, S. D., Sharp, A. H., Kahl, S. D., Vedvick, T. S., Harpold, M. M., and Campbell, K. P. (1991). Structural characterization of the dihydropyridine-sensitive calcium channel α_2-subunit and the associated δ peptides. *J. Biol. Chem.* 266, 3287–3293.

Kim, H. L., Kim, H., Lee, P., King, R. G., and Chin, H. (1992). Rat brain expresses an alternatively spliced form of the dihydopyridine-sensitive L-type calcium channel α_2 subunit. *Proc. Natl. Acad. Sci. U.S.A.* 89, 3251–3255.

Kim, M.-S., Morii, T., Sun, L.-X., Imoto, K., and Mori, Y. (1993). Structural determinants of ion selectivity in brain calcium channel. *FEBS Lett.* 318, 145–148.

Koch, W. J., Ellinor, P. T., and Schwartz, A. (1990). cDNA cloning of a dihydropyridine-sensitive calcium channel from rat aorta. *J. Biol. Chem.* 265, 17786–17791.

Lacerda, A. E., Kim, H. S., Ruth, P., Perez-Reyes, E., Flockerzi, V., Hofman, F., Birnbaumer, L., and Brown, A. M. (1991). Normalization of current kinetics by interaction between the α_1 and β subunits of the skeletal muscle dihydropyridine-sensitive Ca^{2+} channel. *Nature* 352, 527–530.

Lacerda, A. E., Perez-Reyes, E., Wei, X., Castellano, A., and Brown, A. M. (1994). T-type and N-type calcium channels of Xenopus oocytes: Evidence for specific interactions with β subunits. *Biophys. J.* 66, 1833–1843.

Lai, Y., Seagar, M. J., Takahashi, M., and Catterall, W. A. (1990). Cyclic AMP-dependent phosphorylation of two size forms of α_1 subunits of L-type calcium channels in rat skeletal muscle. *J. Biol. Chem.* 265, 20839–20848.

Llinas, R., Sugimori, M., Hillman, D. E., and Cherksey, B. (1992). Distribution and functional significance of the P-type, voltage-dependent Ca^{2+} channels in the mammalian central nervous system. *TINS* 15, 351–355.

Ma, W.-J., Holz, W., and Uhler, M. D. (1992). Expression of a cDNA for a neuronal calcium channel α_1 subunit enhances secretion from adrenal chromaffin cells. *J. Biol. Chem.* 267, 22728–22732.

MacKinnon, R. (1991). New insights into the structure and function of potassium channels. *Current Opin. Neurobiol.* 1, 14–19.

Malouf, N. N., McMahon, D. K., Hainsworth, C. N., and Kay, B. K. (1992). A two-motif isoform of the major calcium channel subunit in skeletal muscle. *Neuron* 8, 899–906.

Marchetti, C., Carbone, E., and Lux, H. D. (1986). Effects of dopamine and noradrenaline on Ca channels of cultured sensory and sympathetic neurons of chick. *Pflügers Arch.* 406, 104–111.

McCleskey, E. W. and Schroeder, J. E. (1991). Functional properties of voltage-dependent calcium channels. *Curr. Topics Membr.* 39, 295–326.

McEnery, M. W., Snowman, A. M., Sharp, A. H., Adams, M. E., and Snyder, S. H. (1991). Purified ω-conotoxin GVIA receptor of rat brain resembles a dihydropyridine-sensitive L-type calcium channel. *Proc. Natl. Acad. Sci. U.S.A.* 88, 11095–11099.

Mikami, A., Imoto, K., Tanabe, T., Niidome, T., Mori, Y., Takeshima, H., Narumiya, S., and Numa, S. (1989). Primary structure and functional expression of the cardiac dihydropyridine-sensitive calcium channel. *Nature* 340, 230–233.

Mintz, I. M., Venema, V. J., Swiderek, K. M., Lee, T. D., Bean, B. P., and Adams, M. E. (1992). P-type calcium channels blocked by the spider toxin ω-Aga-IVA. *Nature* 355, 827–829.

Mori, Y., Friedrich, T., Kim, M.-S., Mikami, A., Nakai, J., Ruth, P., Bosse, E., Hofmann, F., Flockerzi, V., Furuichi, T., Mikoshiba, K., Imoto, K., Tanabe, T., and Numa, S. (1991). Primary structure and functional expression from complementary DNA of a brain calcium channel. *Nature* 350, 398–402.

Nakai, J., Adams, B. A., Imoto, K., and Beam, K. G. (1994). Critical roles of the S3 segment and S3–S4 linker of repeat I in activation of L-type channels. *Proc. Natl. Acad. Sci. U.S.A.* 91, 1014–1018.

Nakayama, H., Taki, M., Striessnig, J., Glossmann, H., Catterall, W. A., and Kanaoka, Y. (1991). Identification of 1,4-dihydropyridine binding regions within the α1 subunit of skeletal muscle Ca channels by photoaffinity labeling with diazipine. *Proc. Natl. Acad. Sci. U.S.A.* 88, 9203–9207.

Neely, A., Wei, X., Olcese, R., Birnbaumer, L., and Stefani, E. (1993). Potentiation by the β subunit of the ratio of the ionic current to the charge movement in the cardiac calcium channel. *Science* 262, 575–578.

Niidome, T., Kim, M.-S., Friedrich, T., and Mori, Y. (1992). Molecular cloning and characterization of a novel calcium channel from rabbit brain. *FEBS Lett.* 308, 7–13.

Nishimura, S., Takeshima, H., Hofmann, F., Flockerzi, V., and Imoto, K. (1993). Requirement of the calcium channel β subunit for functional conformation. *FEBS Lett.* 324, 283–286.

Olivera, B. M., Gray, W. R., Zeikus, R., McIntosh, J. M., Varga, J., Rivier, J., de Santos, V., and Cruz, L. J. (1985). Peptide neurotoxins from fish-hunting cone snails. *Science* 230, 1338–1343.

Perez-Reyes, E., Kim, H. S., Lacerda, A. E., Horne, W., Wei, X., Rampe, D., Campbell, K. P., Brown, A. M., and Birnbaumer, L. (1989). Induction of calcium currents by the expression of the α_1-subunit of the dihydropyridine receptor from skeletal muscle. *Nature* 340, 233–236.

Perez-Reyes, E., Wei, X., Castellano, A., and Birnbaumer, L. (1990). Molecular diversity of L-type calcium channels. *J. Biol. Chem.* 265, 20430–20436.

Perez-Reyes, E., Castellano, A., Kim, H. S., Bertrand, P., Baggstrom, E., Lacerda, A. E., Wei, X., and Birnbaumer, L. (1992). Cloning and expression of a cardiac/brain β subunit of the L-type calcium channel. *J. Biol. Chem.* 267, 1792–1797.

Pietrobon, D. and Hess, P. (1990). Novel mechanism of voltage-dependent gating in L-type calcium channels. *Nature* 346, 651–655.

Powers, P. A., Liu, S., Hogan, K., and Gregg, R. G. (1992). Skeletal muscle and brain isoforms of a β-subunit of human voltage-dependent calcium channels are encoded by a single gene. *J. Biol. Chem.* 267, 22967–22972.

Powers, P. A., Lui, S. L., Hogan, K., and Gregg, R. G. (1993). Molecular characterization of the gene encoding the γ subunit of the human skeletal muscle 1,4-dihydropyridine-sensitive Ca^{2+} channel (CACNLG), cDNA sequence, gene structure, and chromosomal location. *J. Biol. Chem.* 268, 9275–9279.

Pragnell, M., Sakamoto, J., Jay, S. D., and Campbell, K. P. (1991). Cloning and tissue-specific expression of the brain calcium channel β-subunit. *FEBS Lett.* 291, 253–258.

Pragnell, M., De Waard, M., Mori, Y., Tanabe, T., Snutch, T. P., and Campbell, K. P. (1994). Calcium channel β subunit binds to a conserved motif in the I–II cytoplasmic linker of the α_1 subunit. *Nature* 368, 67–70.

Ruth, P., Rohrkasten, A., Biel, M., Bosse, E., Regulla, S., Meyer, H. E., Flockerzi, V., Hofmann, F. (1989). Primary structure of the β subunit of the DHP-sensitive calcium channel from skeletal muscle. *Science* 245, 1115–1118.

Saisu, H., Ibaraki, K., Yamaguchi, T., Sekine, Y., and Abe, T. (1991). Monoclonal antibodies immunoprecipitating ω-conotoxin-sensitive calcium channel molecules recognize two novel proteins localized in the nervous system. *Biochem. Biophys. Res. Commun.* 181, 59–66.

Sather, W. A., Tanabe, T., Zhang, J.-F., Mori, Y., Adams, M. E., and Tsien, R. W. (1993). Distinctive biophysical and pharmacological properties of class A (BI) calcium channel α_1 subunits. *Neuron* 11, 291–303.

Schultz, D., Mikala, G., Yatani, A., Engle, D. B., Iles, D. E., Segers, B., Sinke, R. J., Weghuis, D. O., Klockner, U., Wakamori, M., Wang, J.-J., Melvin, D., Varadi, G., and Schwartz, A. (1993). Cloning, chromosomal location, and functional expression of the α_1 subunit of the L-type voltage-dependent calcium channel from normal human heart. *Proc. Natl. Acad. Sci. U.S.A.* 90, 6228–6232.

Seino, S., Chen, L., Seino, M., Blondel, O., Takeda, J., Johnson, J. H., and Bell, G. I. (1992). Cloning of the α_1 subunit of a voltage-dependent calcium channel expressed in pancreatic β cells. *Proc. Natl. Acad. Sci. U.S.A.* 89, 584–588.

Singer, D., Biel, M., Lotan, I., Flockerzi, V., Hofman, F., and Dascal, N. (1991). The roles of the subunits in the function of the calcium channel. *Science* 253, 1553–1557.

Snutch, T. P. and Reiner, P. B. (1992). Ca^{2+} channels: Diversity of form and function. *Current Opinion Neurobiol.* 2, 247–253.

Snutch, T. P., Leonard, J. P., Gilbert, M. M., Lester, H. A., and Davidson, N. (1990). Rat brain expresses a heterogeneous family of calcium channels. *Proc. Natl. Acad. Sci. U.S.A.* 87, 3391–3395.

Snutch, T. P., Tomlinson, W. J., Leonard, J. P., and Gilbert, M. M. (1991). Distinct calcium channels are generated by alternative splicing and are differentially expressed in the mammalian CNS. *Neuron* 7, 45–57.

Soong, T. W., Stea, A., Hodson, C. D., Dubel, S. J., Vincent, S. R., and Snutch, T. P. (1993). Structure and functional expression of a member of the low voltage-activated calcium channel family. *Science* 260, 1133–1136.

Starr, T. V. B., Prystay, W., and Snutch, T. P. (1991). Primary structure of a calcium channel that is highly expressed in the rat cerebellum. *Proc. Natl. Acad. Sci. U.S.A.* 88, 5621–5625.

Stea, A., Dubel, S. J., Pragnell, M., Leonard, J. P., Campbell, K. P., and Snutch, T. P. (1993). A β subunit normalizes the electrophysiological properties of a cloned N-type Ca channel α_1 subunit. *Neuropharmacology* 32, 1103–1116.

Stea, A., Tomlinson, W. J., Soong, T. W., Bourinet, E., Dubel, S. J., Vincent, S. R., and Snutch, T. P. (1994). The localization and functional properties of a rat brain α_{1A} calcium channel reflect similarities to neuronal Q- and P-type channels. *Proc. Natl. Acad. Sci. U.S.A.*, in press.

Striessnig, J., Glossmann, H., and Catterall, W. A. (1990). Identification of a phenylalkylamine binding region within the α_1 subunit of skeletal muscle Ca channels. *Proc. Natl. Acad. Sci. U.S.A.* 87, 9108–9112.

Striessnig, J., Murphy, B. J., and Catterall, W. A. (1991). Dihydropyridine receptor of L-type Ca channels: Identification of binding domains for [^{3}H](+)-PN200-110 and [^{3}H]azidopine within the α_1 subunit. *Proc. Natl. Acad. Sci. U.S.A.* 88, 10769–10773.

Tanabe, T., Takeshima, H., Mikami, A., Flockerzi, V., Takahashi, H., Kangawa, K., Kojima, M., Matsuo, H., Hirose, T., and Numa, S. (1987). Primary structure of the receptor for calcium channel blockers from skeletal muscle. *Nature* 328, 313–318.

Tanabe, T., Mikami, A., Numa, S., and Beam, K. G. (1990a). Cardiac-type excitation-contraction coupling in dysgenic skeletal muscle injected with cardiac dihydropyridine receptor cDNA. *Nature* 344, 451–453.

Tanabe, T., Beam, K. G., Adams, B. A., Niidome, T., and Numa, S. (1990b). Regions of the skeletal muscle dihydropyridine receptor critical for excitation-contraction coupling. *Nature* 346, 567–569.

Tanabe, T., Adams, B. A., Numa, S., and Beam, K. G. (1991). Repeat I of the dihydropyridine receptor is critical in determining calcium channel activation kinetics. *Nature* 352, 800–803.

Tomlinson, W. J., Stea, A., Bourinet, E., Charnet, P., Nargeot, J., and Snutch, T. P. (1993). Functional properties of a neuronal class C L-type channel. *Neuropharmacology* 32, 1117–1126.

Tsien, R. W., Ellinor, P. T., and Horne, W. A. (1991). Molecular diversity of voltage-dependent Ca^{2+} channels. *Trends Pharmacol.* 12, 349–354.

Varadi, G., Lory, P., Schultz, D., Varadi, M., and Schwartz, A. (1991). Acceleration of activation and inactivation by the β subunit of the skeletal muscle calcium channel. *Nature* 352, 159–162.

Wakamori, M., Mikala, G., Schwartz, A., and Yatani, A. (1993). Single-channel analysis of a cloned human heart L-type Ca channel α_1 subunit and the effects of a cardiac β subunit. *Biochem. Biophys. Res. Commun.* 196, 1170–1176.

Wei, X., Perez-Reyes, E., Lacerda, A. E., Schuster, G., Brown, A. M., and Birnbaumer, L. (1991). Heterologous regulation of the cardiac Ca^{2+} channel α_1 subunit by skeletal muscle β and γ subunits. *J. Biol. Chem.* 266, 21943–21947.

Westenbroek, R. E., Ahlijanian, M. K., and Catterall, W. A. (1990). Clustering of L-type Ca2+ channels at the base of major dendrites in hippocampal pyramidal neurons. *Nature* 347, 281–284.

Westenbroek, R. E., Hell, J. W., Warner, C., Dubel, S. J., Snutch, T. P., and Catterall, W. A. (1992). Biochemical properties and subcellular distribution of an N-type calcium channel α_1 subunit. *Neuron* 9, 1–20.

Williams, M. E., Feldman, D. H., McCue, A. F., Brenner, R., Velicelebi, G., Ellis, S. B., and Harpold, M. M. (1992a). Structure and functional expression of α_1, α_2, and β subunits of a novel human neuronal calcium channel subtype. *Neuron* 8, 71–84.

Williams, M. E., Brust, P. F., Feldman, D. H., Patthi, S., Simerson, S., Maroufi, A., McCue, A. F., Velicelebi, G., Ellis, S. B., and Harpold, M. (1992b). Structure and functional expression of an ω-conotoxin-sensitive human N-type calcium channel. *Science* 257, 389–395.

Williams, M. E., Marubio, L. M., Deal, C. R., Hans, M., Brust, P. F., Philipson, L. H., Miller, R. J., Johnson, E. C., Harpold, M. M., and Ellis, S. B. (1994). Structural and functional characterization of neuronal α_{1E} Ca channel subtypes. *J. Biol. Chem.*, in press.

Witcher, D. R., De Waard, M., Sakamoto, J., Franzini-Armstrong, C., Pragnell, M., Kahl, S. D., and Campbell, K. P. (1993). Subunit identification and reconstitution of the N-type Ca channel complex purified from brain. *Science* 261, 486–489.

Yaney, G. C., Wheeler, M. B., Wei, X., Perez-Reyes, E., Birnbaumer, L., Boyd III, A. E., and Moss, L. G. (1992). Cloning of a novel α_1-subunit of the voltage-dependent calcium channel from the β-cell. *Mol. Endocrinol.* 6, 2143–2152.

Yang, J., Ellinor, P. T., Sather, W. A., Zhang, J.-F., and Tsien, R. W. (1993). Molecular determinants of Ca selectivity and ion permeation in L-type Ca channels. *Nature* 366, 158–161.

Yu, A. S. L., Hebert, S. C., Brenner, B. M., and Lytton, J. (1992). Molecular characterization and nephron distribution of a family of transcripts encoding the pore-forming subunit of Ca^{2+} channels in the kidney. *Proc. Natl. Acad. Sci. U.S.A.* 89, 10494–10498.

Zhang, J.-F., Randall, A. D., Ellinor, P. T., Horne, W. A., Sather, W. A., Tanabe, T., Schwartz, T. L., and Tsien, R. W. (1993). Distinctive pharmacology and kinetics of cloned neuronal Ca channels and their possible counterparts in mammalian CNS neurons. *Neuropharmacology* 32, 1075–1088.

4

Nicotinic Acetylcholine Receptors

Jon M. Lindstrom

2.4.0 Introduction

Nicotinic acetylcholine receptors (AChRs) are members of a gene superfamily of ligand-gated ion channels that includes $GABA_A$ receptors, glycine receptors, and $5HT_3$ serotonin receptors (Betz, 1990; Barnard, 1992), but does not include glutamate receptors (Seeburg, 1993). AChRs are the primary excitatory ligand-gated cation channel in the skeletal muscle and peripheral nervous system of vertebrates, but in the central nervous system glutamate receptors are the primary excitatory receptors and nicotinic AChRs are present in much smaller amounts. The situation is reversed in insects, where nicotinic AChRs are the primary excitatory receptor and neuromuscular transmission employs amino acid receptors (Schoss et al., 1991; Schuster et al., 1991; Gundelfinger, 1992). The primary inhibitory ligand-gated anion channels in the central nervous system of vertebrates are members of the gene superfamily that includes AChRs. All of the receptors in this superfamily are thought to be formed from multiple homologous subunits oriented around a central cation channel to which each subunit contributes a component like a barrel stave. Confirmation of the generality of this assumption about the general homologies between the various excitatory and inhibitory receptors in the gene superfamily was recently provided by the observation that alteration of only three amino acids could convert an AChR cation channel into an anion channel (Galzi et al., 1992). The synthesis, structure, and function of AChRs are the best characterized of the receptors in the superfamily, and thus they serve as a model for the understanding of the other receptors in this superfamily, and also to some extent as a model for the excitatory amino acid receptor superfamily.

There are three branches of the AChR gene family: (1) AChRs of skeletal muscles and fish electric organs, (2) neuronal AChRs that, unlike muscle

0-8493-8322-6/95/$0.00+$.50

AChRs, do not bind the snake venom toxin α-bungarotoxin (αBgt), and (3) neuronal AChRs that do bind αBgt. There are subtypes of each of these AChRs that differ in the combinations of subunits that they employ.

2.4.0.1 Muscle-Type AChRs

By far, the most is known about the cell biology, structure, and function of muscle-type AChRs. This is because muscle has provided a simple homogeneous system for electrophysiological, morphological, and developmental studies, because αBgt has provided a highly specific ligand for locating, quantitating, and purifying these AChRs, and because fish electric organs have provided an abundant source of purified AChR protein for biochemical and structural studies (reviewed in Changeux, 1990, 1991; Karlin, 1991, 1993). AChRs of fetal muscle prior to innervation are thought to be composed of four kinds of subunits in the stoichiometry $(\alpha 1)_2\beta 1\gamma\delta$; at mature neuromuscular junctions a second subtype predominates that has shorter channel open time, a much longer half life, and differs by the substitution of one subunit resulting in the stoichiometry $(\alpha 1)_2\beta 1\varepsilon\delta$ (Witzmann et al., 1990). The structure of *Torpedo* electric organ AChR has recently been solved to a resolution of 9 Å by electron microscopic analysis of two-dimensional crystalline arrays of AChRs in their native membranes (Unwin, 1993a). Although not yet atomic resolution, this has provided details of structure unprecedented for any other neurotransmitter receptor or ion channel, and has challenged accepted structural interpretations of some aspects of subunit amino acid sequences.

Muscle AChRs are the target of an antibody-mediated autoimmune response in myasthenia gravis and in an animal model of this disease in which the characteristic muscle weakness is produced by immunization with purified AChR (Lindstrom et al., 1988).

2.4.0.2 Neuronal AChRs That Do Not Bind αBgt

The availability of monoclonal antibodies (mAbs) and cDNAs initially developed for studies of muscle-type AChRs subsequently led to them being applied to neuronal tissues, resulting in the identification of a group of AChRs that did not bind αBgt (reviewed in Lindstrom et al., 1990; Deneris et al., 1991; Heinemann et al., 1991; Role, 1992; Sargent, 1993). Subunit cDNAs homologous to muscle AChR subunits were identified and termed α2, α3, α4, α5, α6 and β2, β3, and β4. It was found that pairwise combinations of α2-, α3-, or α4-subunits with β2- or β4-subunits could produce ACh-gated cation channels when coexpressed in *Xenopus* oocytes. This suggests the potential existence of a large number of neuronal AChR subtypes of unknown functional significance, but there do seem to be some simplifying generalizations that apply. The predominant subtype of neuronal AChR in brain that does not bind αBgt and that is responsible for the high-affinity binding of nicotine is composed of AChRs with the subunit stoichiometry

$(\alpha4)_2(\beta2)_3$ (Whiting and Lindstrom, 1988; Anand et al., 1991; Cooper et al., 1991; Flores et al., 1992), while the primary subtype of ganglionic postsynaptic AChRs that do not bind αBgt are composed of a combination of α3-, β4-, and α5-subunits (Conroy et al., 1992; Vernalis et al., 1993; Corriveau and Berg, 1993).

Two confusions of nomenclature may complicate reading the "older" literature on neuronal AChRs. First, when these AChRs were first purified (Whiting and Lindstrom, 1986, 1987a; Whiting et al. 1987a,b) their subunits were named α and β in order of increasing molecular weight according to the convention for electric organ and muscle AChRs, but affinity labeling (Whiting and Lindstrom, 1987b) revealed that the ACh binding subunit had a higher molecular weight than the structural subunit (the opposite of muscle), and N-terminal protein sequencing revealed that the ACh binding subunit initially termed β corresponded to the cDNA termed α4 and the structural subunit corresponded to the cDNA termed β2 (Whiting et al., 1987b; Schoepfer et al., 1988). Now the cDNA nomenclature has been adopted (Whiting and Lindstrom, 1988). Initially the Patrick and Heinemann groups termed structural subunits β2, β3 ... (Heinemann et al., 1991) while the Ballivet group termed them non-α1, non-α2 ... (Nef et al., 1988). Now the β terminology is used for euphony.

The major known pathological significance of neuronal nicotinic AChRs is that their interaction with nicotine is responsible for the addiction to tobacco (Bock and Marsh, 1990). Upregulation of α4β2 AChRs is primarily responsible for the increase in nicotine binding, which is characteristic of nicotine addiction (Flores et al., 1992; Marks et al., 1992). Decrease in brain AChRs is associated with both Alzheimer's and Parkinson's diseases, but this loss may be secondary to cell death rather than primarily responsible for it (Whitehouse et al., 1988; Wells et al., 1993; Lange et al. 1993).

Ganglionic α3α5β4 AChRs function in a postsynaptic role similar to that of α1β1εδ AChRs of muscle, but the functional roles of brain AChRs are less clear, and many may be in presynaptic locations where their function is to modulate the release of various transmitters rather than to provide the critical postsynaptic link in signaling between one nerve and another (Clarke et al., 1986; Grady et al., 1992; Harsing et al., 1992). AChR subunit mRNAs have been localized by *in situ* hybridization, and AChR subunit proteins have been localized using mAbs, [^{3}H]nicotine, [^{3}H]ACh, and ^{125}I-labeled αBgt (e.g., Clarke et al., 1985; Jacob et al., 1986; Swanson et al., 1987; Wada et al., 1989; Keyser et al., 1988; Sargent et al., 1989; Morris et al., 1990; Hill et al., 1993). Higher resolution localization studies, probably in combination with electrophysiology, will be required to determine specific functional roles and whether any particular subtypes are always associated with particular functional roles. Electrophysiological properties of neuronal AChRs have been studied, and they often appear to be more permeable to Ca^{2+} than are muscle AChRs (e.g., Lipton et al., 1987; Mulle et al., 1991, 1992), but much more is known about the electrophysiological properties of combinations of cDNAs expressed in *Xenopus* oocytes or fibroblasts than is known about native neuronal AChRs (e.g., Papke et al., 1989; Charnat et al., 1992; Whiting et al., 1991; Vernino et al., 1992).

Clearly, much remains to be learned about the structure, function, normal functional roles, and pathological significance of neuronal AChRs. Developmental effects may turn out to be especially interesting in terms of atypical effects of AChR function. For examples, nicotinic AChRs have been shown to control neuronal process outgrowth through Ca^{2+} influx in a process that may be relevant to mechanisms for specific synapse formation (Lipton et al., 1988; Lipton and Kater, 1989), and nicotine has been shown to alter growth regulation on lung carcinoma cells in a process that may have direct pathological significance (Maneckjie and Minna, 1990).

2.4.0.3 Neuronal AChRs That Bind αBgt

The most recently cloned and least well known of the types of AChR are those which bind αBgt (reviewed in Clarke, 1992). cDNAs have been cloned for α7- and α8-subunits of these AChRs (Schoepfer et al., 1990; Couturier et al., 1990). Initially there was some confusion in the nomenclature of these subunits. When they were first cloned they were termed "αBgt binding protein subunits 1 and 2" (Schoepfer et al., 1990), but now the α7 and α8 convention has been adopted for euphony. In brain there are about as many α7 AChRs as α4β2 AChRs (≈1–2 pmol/g brain) (Whiting and Lindstrom, 1987a). Subtypes have been identified that contain α7-, α8-, or both α7- and α8-subunits (Schoepfer et al., 1990; Keyser et al., 1993). Some of these AChRs have been purified, but the full subunit composition or stoichiometry of none of these subtypes is known (Gotti et al., 1991; 1992). α7 AChRs have lower affinity for αBgt than do muscle type α1β1γδ AChRs, but still reasonably high affinity (K_D = 2 n*M*), whereas α8 AChRs have such low affinity for αBgt (K_D = 20 n*M*) that they would have escaped detection with the low concentrations of αBgt used in previous binding studies (Keyser et al., 1993). By contrast, α8 AChRs have much higher affinities for most small cholinergic ligands than do α7 AChRs (Anand et al., 1993b).

The functional roles of these AChRs have been mysterious because at least some are located extrasynaptically (Jacob and Berg, 1983), although a specific synaptic role has been found for one such AChR (Fuchs and Murrow, 1992). The function of these AChRs has been difficult to detect; they desensitize very quickly and may exhibit inward rectification, and they are sometimes found in cells containing α3α5β4 AChRs, which dominate the depolarizing response to ACh (Vijayaraghavan et al., 1992; Devillers-Thiéry et al., 1992; Alkondon and Albuquerque, 1993). It has recently been appreciated that they are especially permeable to calcium ions (Fuchs and Murrow, 1992; Vijayaraghavan et al., 1992) and may even exceed NMDA receptors in this regard (Séguéla et al., 1993). The high Ca^{2+} permeability of α7 and α8 AChRs can permit Ca^{2+} entering through the AChR channel to act as a second messenger to modulate Ca^{2+}-sensitive properties. When α7 and α8 homomers are expressed in *Xenopus* oocytes, Ca^{2+} entering through these channels activates a Ca^{2+}-sensitive Cl^- channel, which greatly amplifies the current (Devillers-Thiéry et al., 1992; Séguéla et al., 1993; Gerzanich et al., 1994). This amplification can be pragmatically useful, but also confusing, with the result that in early studies of α7 homomers the Ca^{2+}-sensitive Cl^- current was mistaken for current through the homomers. The Cl^- current can be eliminated by Ca^{2+} chelating agents such as BAPTA. A physiologically significant *in vivo* example of Ca^{2+} entering through an αBgt-sensitive AChR and acting as a second messenger is the especially interesting case where brief AChR activation of an excitatory AChR results in a prolonged inhibitory effect on chick cochlear hair cells through activation of a Ca^{2+}-sensitive K^+ channel (Fuchs and Murrow, 1992).

The special property of these AChRs that has proven most experimentally useful is that, unlike any other AChR subunit, α7 cDNA expressed in *Xenopus* oocytes results in the formation of α7 homomers that function as ACh-gated cation channels (Couturier et al., 1990). α7 homomers exhibit pharmacological properties not very different from native α7 AChRs (Bertrand et al., 1992b; Anand et al., 1993b,c). α8-subunits are much less efficient at forming homomers than are α7-subunits, although they will form functional homomers with electrophysiological properties like α7 homomers and pharmacological properties not very different from native α8 AChRs (Gerzanich et al., 1994; Anand et al., 1993b). The ability to efficiently express functional α7 homomers has been a terrific boon for studies testing the functional effects of mutagenesis of specific amino acids (Revah et al., 1991; Galzi et al., 1991, 1992; Devillers-Thiéry et al., 1992; Bertrand et al., 1992a,b, 1993). Such studies have lead to a number of conclusions relevant to the structure and function of all of the receptors in the gene family.

α7-subunit mRNA has been localized by *in situ* hybridization (Couturier et al., 1990; Séguéla et al., 1993) and α7- and α8-subunit proteins have been localized using mAbs (Keyser et al., 1993; Britto et al., 1992a,b; Hamassaki-Britto et al., 1994). It is known that there are large changes in the amount of brain αBgt-binding AChRs during development (Wang and Schmidt, 1976), and studies of α7 regulatory sequences during development have begun (Matter-Sadzinski et al., 1992) modeled on the studies of regulatory elements of neuronal AChRs that do not bind αBgt (Matter et al., 1990) and the extensive studies of this type of muscle AChR subunits (e.g., Martinou et al., 1991; Changeux, 1991; Neville et al., 1992).

Much remains to be learned about neuronal AChRs that bind αBgt, starting with their complete subunit composition and extending through the details of their functional roles, not to mention their pathological roles. It is already clear that α7 and α8 AChRs are a distinct branch of the nicotinic AChR gene family, that they exhibit some unique properties that may permit them to function in unusual roles, and that there should be many interesting developments in this area in the next few years as a result of the many studies now underway on this long neglected branch of the AChR gene family.

2.4.1 Structural and Functional Significance of AChR Subunit Amino Acid Sequences

Figure 1 compares the amino acid sequences of all known subunits of AChRs. All of the subunits exhibit basic homologies throughout their sequences, which indicates that they all evolved from a common ancestor and that they all exhibit some basic homologies in shape and function (Raftery et al., 1980; Noda et al., 1983). Sequences for β3 and α6 are included on the basis of the homologies of their cDNAs, although neither has yet been shown to be part of functional AChRs *in vitro* or *in vivo*. Sequences for rat subunits are shown, except in the case of α8, which has so far been identified only in chickens. Sequences of some subunit cDNAs are also known from many other species, including humans [for example, Luther et al., (1989) compares the sequences of human α1- and δ-subunits with those of five other species]. Amino acid sequence variation between species is usually very small, typically <10%. The most variable region between subunits and between species is the putative large cytoplasmic domain, with the result that subunit-specific mAbs and species-specific mAbs often map to this region (Das and Lindstrom, 1991).

In subsequent sections the subunit amino acid sequences will be briefly discussed, starting with the N-terminus and large extracellular domain, then moving progressively in the direction of the C-terminus considering next three closely spaced putative transmembrane domains in the middle of the subunit, then the large cytoplasmic domain, and finally the fourth putative transmembrane domain, which occurs shortly before the C-terminus on the extracellular surface.

2.4.1.1 Large N-Terminal Extracellular Domain

All subunits exhibit signal sequences; these are presumably cleaved cotranslationally and serve to target the N-terminal domain of the mature subunit to the extracellular surface as it synthesized in the endoplasmic reticulum. Synthesis and maturation of α1-subunits have been studied in greatest detail (Merlie et al., 1983; Blount and Merlie, 1988, 1989, 1990, 1991a,b; Blount et al., 1990). The N-terminal sequences of the mature subunits are known from the N-terminal sequences only of α-, β-, γ-, and δ-subunits purified from electric organ and fetal calf AChR (Raftery et al., 1980; Conti-Tronconi et al., 1982) and of α4- and β2-subunits purified from rat and chicken AChRs (Whiting et al., 1987; Schoepfer et al., 1988).

Putative sites of N-glycosylation are found throughout the N-terminal half of the subunits that precedes the first putative transmembrane domain M1. For example, neuronal α-subunits, but not muscle α-subunits, have a putative N-glycosylation site at position 24 of the consensus sequence shown in Figure 1. Virtually all AChR subunits, except α7 and α8, have a glycosylation site at position 141 of the consensus sequence. In α1-subunits, at least, this site is known to be glycosylated and is known to be located

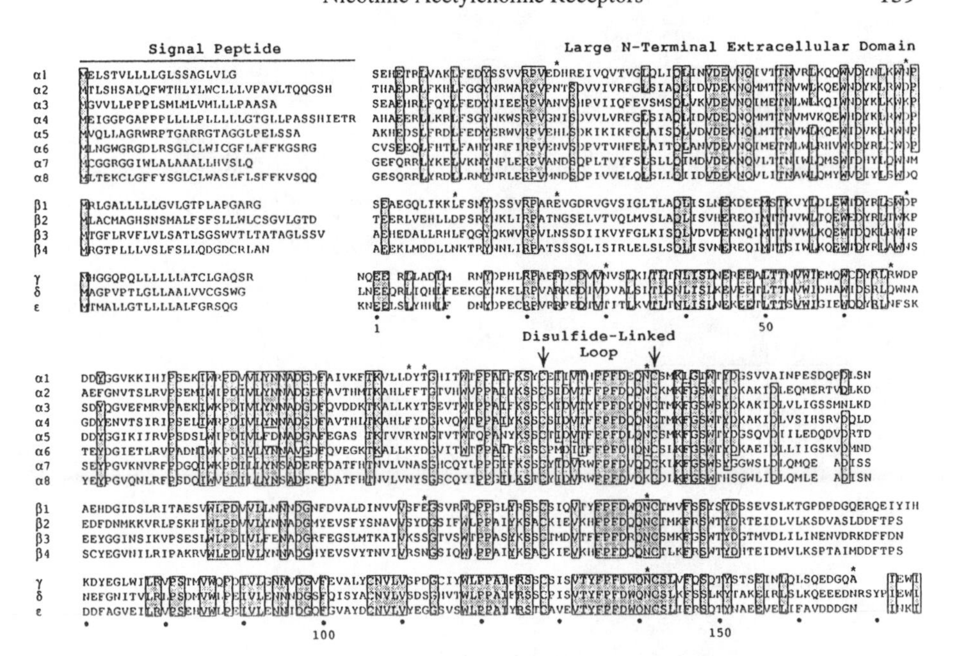

FIGURE 1. Homologies among nicotinic acetylcholine receptor subunits. Amino acid sequences predicted from cDNAs for all rat ACHRs are shown, except in the case of α8, which is known only from chickens. Highlighted sequences emphasize homologies between groups of these subunits. An asterisk indicates potential sites of N-glycosylation for some or all of the subunits in that column. Sequences for α1, α2, α3, α4, and α5 are from Boulter et al. (1990). Sequence for α6 is from J. Boulter, unpublished Genbank accession number L08227. Sequence for α7 is from Seguela et al. (1993). Sequence for α8 is from Schoepfer et al. (1990). Sequences for β1, β2, β3, and β4 are from Boulter et al. (1990). Sequence for γ is from Witzemann et al. (1987). Sequence for δ is from Witzemann et al. (1990). Sequence for ε is from Criado et al. (1988).

within a highly conserved disulfide-linked loop that appears to be present in all subunits (Kallaris et al., 1989; Gehle and Sumikawa, 1991).

The main immunogenic region of α1-subunits is a conformation-dependent epitope that is responsible for provoking more than half of the autoantibodies to muscle AChRs, which cause the weakness characteristic of myasthenia gravis and its animal model, experimental autoimmune myasthenia gravis (Lindstrom et al., 1988; Tzartos et al., 1991; Protti et al., 1993). Amino acids 68 and 71 in the α1 sequence are particularly important to the binding of antibodies to the main immunogenic region (Saedi et al., 1990). The endogenous function of this region, if any, is not known. mAbs to this region do not block AChR function, but are especially effective at crosslinking AChRs on muscle cell surfaces, which increases the rate of AChR turnover, and they are effective at fixing complement, which results in focal lysis of the postsynaptic membrane (Lindstrom et al., 1988; Tzartos et al., 1991). Sequences homologous to the main immunogenic region are also found in the α5-subunits of

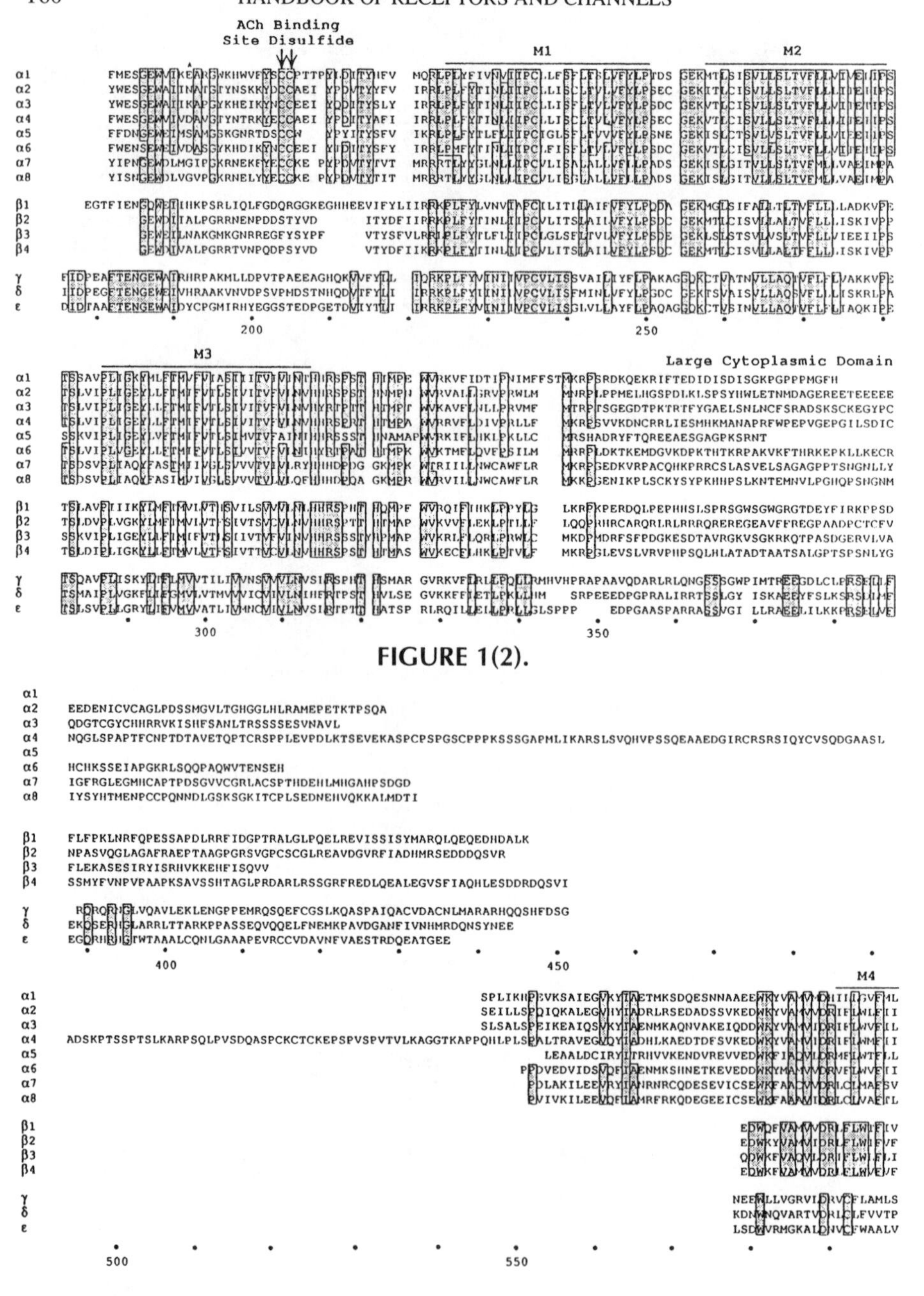

FIGURE 1(2).

FIGURE 1(3).

ganglionic AChRs and in β3-subunits, but it is unknown if this has a pathological significance. There are numerous other B lymphocyte epitopes on AChR subunits (Das and Lindstrom, 1991). There are also numerous T lymphocyte epitopes on AChR subunits, but none predominates as does the main immunogenic region for antibodies on α-subunits (Protti et al., 1993).

```
α1    VCLIGTLAVFAGRLIELHQQG
α2    VCFLGTIGLFLPPFLAGMI
α3    VCILGTAGLFLQPLMARDDT
α4    VCLLGTVGLFLPPWLAAC
α5    VSIIGTLGLFVPVIYKWANIIVPVHIGNTIK
α6    VCVFGTVGLFLQPLLGNTGAS
α7    FTIICTIGILMSAPNFVEAVSKDFA
α8    FAIICTFTILMSAPNFIEAVSKDFT

β1    FTSVGTLVIFLDATYHLPPPEPFP
β2    VCVFGTVGMFLQPLFQNYTATTFLHPDHSAPSSK
β3    ASVLGSILIFIPALKMWIHRFH
β4    VCILGTMGLFLPPLFQIHAPSKD

γ     LFICGTAGIFLMAHYNQMPDLPFFPGDPRPYLPLPD
δ     VMVVGTAWIFLQGVYNQPPPQPFPGDPFSYDEQDRRFI
ε     LFSVGSTLIFLGGYFNQMPDLPYPPCIQP
      •        •
      600
```

FIGURE 1(4).

Specific association of AChR subunits is thought to be mediated primarily by sequences in the N-terminal extracellular domain (Verrall and Hall, 1992). In particular, there is evidence that positions 106 and 115 of ε-subunits are involved in their association with α-subunits (Gu et al., 1990). Evidence from several systems suggests that α1-subunits assemble first with γ-, δ-, or ε-subunits followed by association with β-subunits (Blount and Merlie, 1989, 1991b; Blount et al., 1990; Saedi et al., 1991; Gu et al., 1991). α1-subunits go through a conformational maturation prior to assembly during which the main immunogenic region conformation matures and they acquire the ability to bind αBgt (Merlie et al., 1983; Blount and Merlie, 1988). The ability to bind small cholinergic ligands is acquired with association of α1- with γ-, δ-, or ε-subunits, and at this time the differences in ligand binding associated with the two ACh binding sites of the mature muscle AChR are also acquired (Blount and Merlie, 1989; Saedi et al., 1991). This is consistent with affinity labeling evidence for the idea that the ACh binding site occurs at the interfaces between α-subunits and γ-, δ-, or ε-subunits (Pederson and Cohen, 1990; Middleton and Cohen, 1991). Muscle AChRs are expressed on the cell surface efficiently only when all subunits are present, though γ-less or δ-less AChRs may be expressed on the cell surface less efficiently (Gu et al., 1990). Similarly, α4-subunits are expressed on the surface of *Xenopus* oocytes only when β2-subunit cDNAs are coexpressed (Wong and Lindstrom, unpublished). Pairwise combinations of α3- with β2- or β4-subunits are efficiently expressed in oocytes (Deneris et al., 1991; Heinemann et al., 1991), although native ganglionic AChRs appear to also contain α5-subunits (Conroy et al., 1992; Vernalis et al., 1993). α7-Subunits assemble into homomers and are as efficiently expressed on the surface of *Xenopus* oocytes (about 40% of the total) as are α1β1γδ or α4β2 AChRs (Anand et al., 1993c). Unassembled α7-subunits do not bind αBgt, much as unassembled α1-subunits do not bind small cholinergic ligands. α8-Subunits are much less efficient than α7-subunits in forming homomers, suggesting that, like α1β1γδ AChRs or α4β2 AChRs, a structural subunit is required for efficient surface expression (Gerzanich et al., 1994). Unlike α7-subunits, α8-subunits also bind αBgt when assembled into arrays differ in size from native AChRs.

All AChR subunits exhibit a disulfide-linked loop on their extracellular surface with a conserved sequence usually containing an N-glycosylation site and cysteines at positions corresponding to positions 128 and 142 of the α1 sequence. Formation of this disulfide bond may be associated with the conformational maturation of α1-subunits, which precedes assembly with other subunits (Blount and Merlie, 1990). Disruption of this loop by *in vitro* mutagenesis prevents binding of αBgt to α-subunits and inhibits expression of assembled AChRs on the cell surface (Sumikawa and Gehle, 1992).

The ACh binding site is formed in part by amino acids near a disulfide-linked cysteine pair at positions 192 and 193 of α-subunits, which was the first site at which AChRs were affinity labeled (Kao et al., 1984; Kao and Karlin, 1986). Subsequent affinity labeling experiments have identified other residues in this region including α Tyr-190, α Tyr-198 (Abramson et al., 1989; Middleton and Cohen, 1991), and, to a lesser degree, at more distant sites, α Tyr-93 and α Trp-151 as contributing to the ACh binding site (Galzi et al., 1990; Cohen et al., 1991). None of the amino acids labeled on α corresponds to an acidic amino acid that might form an anionic subsite that would bind the quaternary amine of ACh and its analogs. *In vitro* mutagenesis experiments with α-subunits also failed to reveal a candidate for the anionic subsite, but provided additional evidence for the importance of aromatic residues, e.g., Tyr α92, Trp α148, Tyr α187, Tyr α190, and Tyr α198 (Tomaselli et al., 1991; Galzi et al., 1991; Filatov et al., 1993; O'Leary and White, 1992). These results lead to the ideas that the ACh binding site might be formed from amino acids from several parts of the α sequence and that binding of the quaternary amine of ACh might occur through π-cation interactions with aromatic amino acids. This idea seemed especially appealing because the crystal structure of ACh esterase revealed that its catalytic site was composed of several amino acids that were distant in the sequence and that in the native esterase were located at the bottom of a long, narrow gorge lacking critical acidic amino acids, but lined with many aromatic amino acids that might interact with ACh by π-cation interactions (Sussman et al., 1991). However, clever affinity labeling experiments using a bivalent reagent anchored by one end at α192, 193 and able to reach out over a 9 Å distance known to encompass the ACh binding site and to react with carboxyl groups, revealed labeling of amino acids somewhere within the sequence 164 to 224 of δ-subunits (Czajkowski and Karlin, 1991). Subsequently, mutagenesis of these amino acids to remove their charge (by converting them to glutamine or asparagine) revealed that mouse Asp δ180 (δ178 in the rat consensus sequence shown in Figure 1) and Glu δ189 (δ187 in the consensus sequence) were especially important for binding of ligands. This suggests that this region of all subunits that interface with α-subunits to form ACh binding sites may provide critical residues for forming the anionic subsite and permitting binding of small cholinergic ligands. This concept and evidence suggesting that the subunits are arrayed around the central cation channel in the order α1γα1β1δ (Karlin et al., 1983; Blount and Merlie, 1989) were combined to provide a speculative model of the amino acids at the interfaces between subunits in the region of the ACh binding sites (Czajkowski et al., 1993). There is some evidence that, under some conditions, it may also be possible to gate the AChR channel from another site on the extracellular surface including lys α125 (Okonjo et al., 1991; Pereira et al., 1993) or even by phosphorylation on the cytoplasmic surface (Ferrier-Montiel et al., 1991).

2.4.1.2 Transmembrane Domains M1, M2, and M3

Immediately C-terminal of the sequences contributing to the ACh binding site is the first of four hydrophobic sequences termed M1–M4 that are thought to form transmembrane domains. These were initially identified by hydropathy plots and were generally assumed to be α-helical (Noda et al., 1983), but more recent structural evidence suggests that the transmembrane domains are primarily not α-helical (Unwin, 1993a; Akabas et al., 1992). M1–M3 are rather highly conserved sequences, probably reflecting their functional importance. Because of its location between the ACh binding site on the extracellular surface and M2, which is thought to line the cation channel (Hucho et al., 1986; Giraudat et al., 1986), M1 may be in a uniquely good position to act as a linkage between ligand binding and channel gating. The M1 sequence can be photolabeled by the noncompetitive inhibitor quinacrine azide during the active conformation of the AChR (DiPaola et al., 1990). *In vitro* mutagenesis evidence suggests that during activation a conformation change takes place that involves movement around the proline, which is conserved in the center of M1 of all subunits (Lo et al., 1991). The sorts of conformational changes involved in activation or desensitization of AChRs must be propagated over the long distances (~50 Å) between the ACh binding sites and the narrowest part of the channel. These conformation changes may involve subtle changes in the angle of tilt between each of the homologous rod-like subunits organized around the channel to produce an iris-like opening or closing of the narrowest part of the channel by a few angstrom, but details of these conformation changes are not known yet (Unwin et al., 1988; Unwin, 1993b).

The M2 sequence is thought to line a central cation channel. Three lines of evidence suggest that this is so: (1) homologous positions of M2 from all subunits can be affinity labeled by noncompetitive antagonists that are thought to act by plugging the channel (Hucho et al., 1986; Giraudat et al., 1986), (2) *in vitro* mutagenesis studies have revealed three critical sets of acidic residues bordering either end of M2 [e.g. in the consensus sequence of Figure 1 Asp α252 and Glu α256 on the cytoplasmic surface and Glu α277 on the extracellular surface (Imoto et al., 1988; Konno et al., 1991)], which are thought to form charged rings lining the channel lumen that serve as ion selectivity filters, and (3) certain M2 amino acids were shown to be accessible from within the channel (Akabas et al., 1993) and critical to its function (Lester, 1992; Cohen et al., 1992). In anion-selective receptors from this gene superfamily these acidic amino acids are replaced by neutral or positively charged amino acids (Betz, 1990; Barnard, 1992). M2 sequences contain several serine and threonine residues whose hydroxyl-containing side chains are thought to provide a hydrophilic environment for the passage of hydrated cations if they were located on one side of an amphipathic helix (Lester, 1992) or on the exposed part of a β barrel structure (Akabas et al., 1992). It has been speculated that a ring of leucines (i.e., Leu α266 in the consensus sequence of Figure 1) may be involved in forming the actual gate (Unwin, 1993a).

In vitro mutagenesis studies of the M2 region of α7 homomers have proven especially interesting. Introduction of three amino acids from the M2 segment of the anion-selective glycine or $GABA_A$ receptors into the α7 homomer (i.e., in the consensus sequence shown in Figure 1, after α7 glycine 255 insert a proline, replace α7 glutamate 256 with alanine, and replace α7 valine 268 with theonine) was sufficient to convert α7 channels from cation selectivity to anion selectivity (Galzi et al., 1992)! The single mutation of replacing (in the consensus sequence) α7 glutamate 256 with alanine on the putative cytoplasmic lumen of the channel reduces the characteristic high Ca^{2+} permeability of α7 homomers without affecting other properties (Bertrand et al., 1993). Changing two rings of leucines at the extracellular end of α7 M2 (in the consensus sequence of Figure 1 leucines α7 273, 274) to threonine or some other amino acids also abolishes Ca^{2+} permeability, but in addition also increases the apparent affinity for ACh (Bertrand et al., 1993). Single mutations at any of three sites (consensus sequence in Figure 1 changing α7 threonine 263 to glutamine, or α7 leucine 266 to threonine, or α7 valine 270 to threonine) apparently cause what would normally be desensitized conformations of α7 homomers typically characterized by high agonist affinity and a closed channel to behave as activatable channels with altered pharmacology (Revah et al., 1991; Bertrand et al., 1992a; 1993).

M3 is located immediately C-terminal of M2. Its sequence is rather highly conserved, although not so much as M2, suggesting that it may play an important role along with parts of M1 and M2 in forming the core of the central transmembrane domain of AChRs.

2.4.1.3 Large Cytoplasmic Domain

A large cytoplasmic domain is located C-terminal of M3. Its cytoplasmic orientation has been shown by locating binding sites for sequence-specific mAbs to parts of this sequence (Ratnam et al., 1986a,b), by the use of reporter epitopes (Anand et al., 1993a), by expression of fusion proteins (Chavez and Hall, 1992), or by using glycosylation consensus sequences as reporter domains (Chavez and Hall, 1991). This sequence also contains phosphorylation sites that may be involved in regulating AChR function, turnover, or location (Miles and Huganir, 1988; Huganir and Greengard, 1990; Wallace et al., 1991). The cytoplasmic surfaces of all muscle AChR subunits are thought to contain regions for interacting with a 43-kDa extrinsic membrane protein that is thought to be associated in a 1:1 ratio with muscle AChRs and responsible for mediating their association with cytoskeletal elements that position muscle AChRs at the tips of folds in the postsynaptic membrane (Froehner et al., 1990; Maimone and Merlie, 1993). It is currently unknown whether similar extrinsic membrane proteins are associated with neuronal AChR subtypes. Subtype-specific AChR-associated extrinsic membrane proteins might facilitate axonal transport or location of AChRs at subtype-specific presynaptic, extrasynaptic, or postsynaptic sites. Some of the variability in sequence between subunits in this large cytoplasmic domain may reflect specific variations for particular functional roles. Other of the sequence variability characteristic of this large cytoplasmic domain may reflect a relative looseness and lack of criticality of parts of this sequence, as suggested by the observation that mAbs that efficiently recognize both native and denatured subunits tend to be directed at this region (Das and Lindstrom, 1991; Ratnam et al., 1986a,b). Subunit-specific antibodies made to this region of bacterially expressed subunit peptides often react well with native AChRs (Schoepfer et al., 1989, 1990).

2.4.1.4 Transmembrane Domain M4 and the C-Terminus

M4 is located at the C-terminal end of the large cytoplasmic domain. It is the most hydrophobic and least conserved of the four putative transmembrane domains, and may be on the surface of the AChR where it could interact with the lipid annulus surrounding the AChR. M4 can even be replaced by an extraneous transmembrane domain and retain function of the AChR (Tobimatsu et al., 1987). However, Cys γ592 (in the consensus sequence in Figure 1) has recently been shown by labeling and mutagenesis to affect channel function, so the role of M4 may be more subtle than has previously been appreciated (Li et al., 1990).

All AChR subunits terminate shortly after M4. Several lines of evidence suggest that the C-terminal ends of the subunits are on the extracellular surface (McCrea et al., 1987; DiPaola et al., 1989; Dwyer, 1991; Chavez and Hall, 1992; Anand et al., 1993a).

2.4.2 Three-Dimensional Structure of the AChR

High resolution microscopy has provided images of the densely packed muscle-type AChRs of *Torpedo* electric organ membranes (Mitra et al., 1989; Unwin et al., 1993a,b). It is assumed that because of sequence homologies of

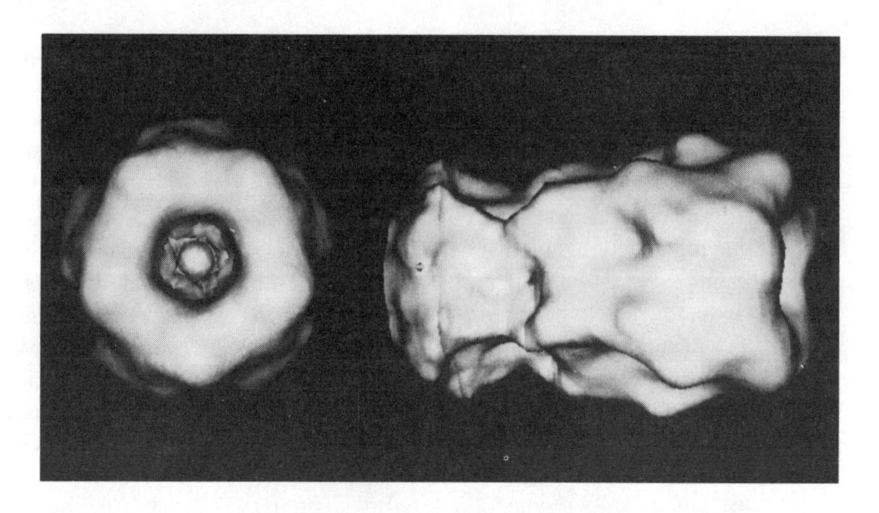

FIGURE 2. Three-dimensional structure of AChRs from *Torpedo* electric organ. These images are based on data obtained by diffraction analysis at 9 Å resolution of electron micrographs of frozen two-dimensional helical crystalline arrays (Unwin, 1993a,b). The side view of the 120 Å-long AChR molecule shows a band across the structure corresponding to the part of the AChR in contact with the lipid bilayer. The AChR is embedded in the postsynaptic membrane with the larger part of the AChR on the extracellular surface. The putative ACh binding sites are suspected to be within clefts in the two α-subunits about 30 Å above the lipid bilayer (about half-way up the extracellular domain). The top view of the AChR peers into the central cation channel. In this view, the pentagonal arrangement of the five subunits around the channel is obvious. The rod-like subunits are thought to be organized like barrel staves, perhaps in the order αγαδβ. The diameter of the vestibule (about 25 Å) of the channel narrows abruptly near the level of the lipid bilayer near the location of the rings of negatively charged amino acids at the extracellular ends of the M2 segments of each subunit, which are thought to line the conducting pore.

their subunits other AChR subtypes will have a basically similar appearance. Diffraction analysis of two-dimensional crystalline arrays of AChRs in fragments of their native membranes has provided resolution to 9 Å (Unwin, 1993a), which is sufficient to resolve feature like a-helices, but is not yet atomic resolution.

Figure 2 shows images based on the 9 Å resolution analysis of *Torpedo* AChR structure (Unwin, 1993a,b). The five subunits are clearly oriented around the central channel in a pentagonal array. The AChR is about 120 Å long with about 60 Å extending on the extracellular surface and about 20 Å on the cytoplasmic surface. In the tubular crystalline arrays, but not shown in Figure 2, a 43K protein is centered below each AChR and in contact with its large cytoplasmic domain.

The extracellular mouth of the channel has an internal diameter of about 25 Å, but the channel narrows abruptly near the level of the lipid bilayer in the region where the extracellular ring of negatively charged amino acids is expected at the outer border of M2. On the cytoplasmic surface the channel flares open again. Instead of observing four α-helices crossing the membrane of every subunit, as would be expected if M1–M4 were all α–helices, only a single α-helix crossing the membrane was observed per subunit, and it was observed to kink in the middle toward the narrowest part of the channel. The proline in the middle of M1 might be expected to produce such a kink, if M1 were an α-helix. Unwin (1993a) proposes that this α-helix is M2 and that the kink corresponds to leucine 266 on the consensus sequence shown in Figure 1, the leucine that mutagenesis has shown to be important to the α7 channel (Revah et al., 1991; Bertrand et al., 1992a), and that this ring of leucines forms the channel gate. Akabas et al. (1992), on the other hand, use a combination of *in vitro* mutagenesis and labeling techniques to argue that some or all of M2 is in a β conformation. The diffraction results indicate that no more than one of the putative transmembrane domains is α-helical and suggest that the rest may be in a β barrel arrangement.

The ACh binding site has not been precisely localized (Unwin, 1993a). However, half way up the extracellular domain each subunit contains a cavity surrounded by three α-helices. The two α-subunits have previously been identified by labeling with subunit-specific mAbs and by binding of αBgt (Kubalek et al., 1987). The α-subunit cavities are larger and located in the region where αBgt binding was observed, so it was proposed that these clefts lead to the ACh binding site in a similar way that the 20 Å gorge in the ACh esterase molecule leads to its binding site (Sussman et al., 1991).

Clearly more remains to be learned about the three-dimensional structure of the AChR. What has been learned has provided remarkable images and has challenged some convenient assumptions based on casual hydropathy analysis of subunit sequences. What will be learned in the future by this approach should further challenge old assumptions and provide a detailed model for understanding the structures of all of the receptors in this gene superfamily.

REFERENCES

Abramson, S., Culver, Y., and Taylor, P. (1989). An analog of lophotoxin reacts covalently with Tyr 190 in the α subunit of the nicotinic acetylcholine receptor. *J. Biol. Chem.* 264, 1266–1267.

Akabas, M., Stauffer, D., Xu, M., and Karlin, A. (1992). Acetylcholine receptor channel structure probed in cysteine-substitution mutants. *Science* 258, 307–310.

Alkondon, M. and Albuquerque, E. (1993). Diversity of nicotinic acetylcholine receptors in rat hippocampal neurons: I Pharmacological and functional evidence for distinct structural subtypes. *J. Pharmacol. Exp. Ther.* 265, 1455–1473.

Anand, R., Conroy, W. G., Schoepfer, R., Whiting, P., and Lindstrom, J. (1991). Chicken neuronal nicotinic acetylcholine receptors expressed in *Xenopus* oocytes have a pentameric quaternary structure. *J. Biol. Chem.* 266, 11192–11198.

Anand, R., Bason, L., Saedi, M., Gerzanich, V., Peng, X., and Lindstrom, J. (1993a). Reporter epitopes: A novel approach to examine transmembrane topology of integral membrane proteins applied to the α1 subunit of the nicotinic acetylcholine receptor. *Biochemistry* 32, 9975–9985.

Anand, R., Peng, X., Ballesta, J., and Lindstrom, J. (1993b). Pharmacological characterization of α bungarotoxin sensitive AChRs immunoisolated from chick retina: Contrasting properties of α7 and α8 subunit-containing subtypes. *Mol. Pharmacol.* 44, 1046–1050.

Anand, R., Peng, X., and Lindstrom, J. (1993c). Homomeric and native α7 acetylcholine receptors exhibit remarkably similar but nonidentical pharmacological properties suggesting that the native receptor is a heteromeric protein complex. *FEBS Lett.* 327, 241–246.

Barnard, E. (1992). Receptor classes and the transmitter-gated ion channels. *Trends Biochem. Sci.* 17, 368–374.

Bertrand, D., Devillers-Thiery, A., Revah, F., Galzi, J.-L., Hussy, N., Mulle, C., Bertrand, S., Ballivet, M., and Changeux, J.-P. (1992a). Unconventional pharmacology of a neuronal nicotinic receptor mutated in the channel domain. *Proc. Natl. Acad. Sci. U.S.A.* 89, 1291–1265.

Bertrand, D., Bertrand, S., and Ballivet, M (1992b). Pharmacological properties of the homomeric α7 receptor. *Neurosci. Lett.* 146, 87–90.

Bertrand, D., Galzi, J.-L., Devillers-Thiery, A., Bertrand, S., and Changeux, J.-P. (1993). Mutations at two distinct sites within the channel domain M2 alter calcium permeability of neuronal α7 nicotinic receptor. *Proc. Natl. Acad. Sci. U.S.A.* 90, 6971–6975.

Betz, H. (1990). Ligand-gated ion channels in the brain: The amino acid receptor superfamily. *Neuron* 5, 383–392.

Blount, P. and Merlie, J. (1988). Native folding of an acetylcholine receptor α subunit expressed in the absence of other receptor subunits. *J. Biol. Chem.* 262, 4367–4376.

Blount, P. and Merlie, J. P. (1989). Molecular basis of the two nonequivalent ligand binding sites of the muscle nicotinic acetylcholine receptor. *Neuron* 3, 349–357.

Blount, P. and Merlie, J. P. (1990). Mutational analysis of muscle nicotinic acetylcholine receptor subunit assembly. *J. Cell Biol.* 111, 2612–2622.

Blount, P. and Merlie, J. (1991a). BIP associates with newly synthesized subunits of the mouse muscle nicotinic receptor. *J. Cell Biol.* 113, 1125–1132.

Blount, P. and Merlie, J. (1991b). Characterization of an adult muscle acetylcholine receptor subunit by expression in fibroblasts. *J. Biol. Chem.* 266, 14692–14696.

Blount, P., Smith, M., and Merlie, J. (1990). Assembly intermediates of the mouse muscle nicotinic acetylcholine receptor in stably transfected fibroblasts. *J. Cell Biol.* 111, 2601–2611.

Bock, G. and Marsh, J., Eds. (1990). *The Biology of Nicotine Dependence, Ciba Foundation Symposium 152.* John Wiley & Sons, New York.

Boulter, J., O'Shea-Greenfield, A., Duvoisin, R., Connolly, J., Wada, E., Jensen, A., Gardner, P., Ballivet, M., Deneris, E., McKinnon, D., Heinemann, S., and Patrick, J. (1990). α3, α5, and β4: Three members of the rat neuronal nicotinic acetylcholine receptor-related gene family form a gene cluster. *J. Biol. Chem.* 265, 4472–4482.

Britto, L., Hamassaki-Britto, D., Ferro, E., Keyser, K., Karten, H., and Lindstrom, J. (1992a). Neurons of the chick brain and retina expressing both α-bungarotoxin-sensitive and α-bungarotoxin insensitive nicotinic acetylcholine receptors: An immunohistochemical analysis. *Brain Res.* 590, 193–200.

Britto, L., Keyser, K., Lindstrom, J., and Karten, H. (1992b). Immunohistochemical localization of nicotinic acetylcholine receptor subunits in the mesencephalon and diencephalon of the chick *(Gallus gallus). J. Comp. Neurol.* 317, 325–340.

Changeux, J.-P. (1990). Functional architecture and dynamics of the nicotinic acetylcholine receptor: An allosteric ligand-gated ion channel. *1988–1989 Fidia Res. Found. Neurosci. Award Lect.* Vol. 4, 21–168.

Changeux, J.-P. (1991). Compartmentalized transcription of acetylcholine receptor genes during motor endplate epigenesis. *New Biol.* 3, 413–429.

Charnat, P., Labarca, C., Cohen, B., Davidson, N., Lester, H., and Pilar, G. (1992). Pharmacological and kinetic properties of α4β2 neuronal nicotinic acetylcholine receptors expressed in *Xenopus* oocytes. *J. Physiol.* 450, 375–394.

Chavez, R. and Hall, Z. (1991). The transmembrane topology of the amino terminus of the α subunit of the nicotinic acetylcholine receptor. *J. Biol. Chem.* 266, 15532–15538.
Chavez, R. and Hall, Z. (1992). Expression of fusion proteins of the nicotinic acetylcholine receptor from mammalian muscle identifies the membrane-spanning regions in the α and δ subunits. *J. Cell Biol.* 116, 385–393.
Clarke, P. B. S. (1992). The fall and rise of neuronal α-bungarotoxin binding proteins. *Trends Pharmacol. Sci.* 13, 407–413.
Clarke, P., Schwartz, R., Paul, S., Pert, C., and Pert, A. (1985). Nicotinic binding in rat brain: Autoradiographic comparison of [^{3}H]acetylcholine, [^{3}H]nicotine, and [^{125}I]α-bungarotoxin. *J. Neurosci.* 5, 1307–1315.
Clarke, P., Hamill, G., Nadi, N., Jacobowitz, D., and Pert, A. (1986). ^{3}H-Nicotine and ^{125}Iα-bungarotoxin-labeled nicotinic receptors in the interpeduncular nucleus of rats. II. Effects of habenular deafferentation. *J. Comp. Neurol.* 251, 407–413.
Cohen, B., Labarca, C., Davidson, N., and Lester, H. (1992). Mutations in M2 alter the selectivity of the mouse nicotinic acetylcholine receptor for organic and alkali metal cations. *J. Gen. Physiol.* 100, 373–400.
Cohen, J., Sharp, S., and Liu, W. (1991). Structure of the agonist binding site of the nicotinic acetylcholine receptor. *J. Biol. Chem.* 266, 23354–23364.
Conroy, W., Vernallis, A., and Berg, D. (1992). The α5 gene product assembles with multiple acetylcholine receptor subunits to form distinctive receptor subtypes in brain. *Neuron* 9, 1–20.
Conti-Tronconi, B., Gotti, G., Hunkapiller, M., and Raftery, M. (1982). Mammalian muscle acetylcholine receptor: A supramolecular structure formed by four related proteins. *Science* 218, 1227–1229.
Cooper, E., Couturier, S., and Ballivet, M. (1991). Pentameric structure and subunit stoichiometry of a neuroanl nicotinic acetylcholine receptor. *Nature* 350, 235–238.
Corriveau, R. and Berg, D. (1993). Coexpression of multiple acetylcholine receptor genes in neurons: Quantification of transcripts during development. *J. Neurosci.* 13, 2662–2671.
Couturier, S., Bertrand, D., Matter, J., Hernandez, M., Bertrand, S., Millar, N., Valera, S., Barkas, T., and Ballivet, M. (1990). A neuronal nicotinic acetylcholine receptor subunit (α7) is developmentally regulated and forms a homoligomeric channel blocked by α-bungarotoxin. *Neuron* 5, 847–856.
Criado, M., Witzmann, V., Koenen, M., and Sakmann, B. (1988). Nucleotide sequence of rat muscle acetylcholine receptor ε subunit. *Nucl. Acids Res.* 16, 10920.
Czajkowski, C. and Karlin, A. (1991). Agonist binding site of *Torpedo* electric tissue nicotinic acetylcholine receptor. *J. Biol. Chem.* 266, 22603–22612.
Czajkowski, C., Kaufmann, C., and Karlin, A. (1993). Negatively charged amino acid residues in the nicotinic receptor δ subunit that contribute to the binding of acetycholine. *Proc. Natl. Acad. Sci. U.S.A.* 90, 6285–6289.
Das, M. and Lindstrom, J. (1991). Epitope mapping of antibodies to acetylcholine receptors. *Biochemistry* 30, 2470–2477.
Deneris, E., Connolly, J., Rogers, S., and Duvoisin, R. (1991). Pharmacological and functional diversity of neuronal nicotinic acetylcholine receptors. *Trends Phamacol. Sci.* 12, 34–40.
Devillers-Thiéry, A., Galzi, J.-L., Bertrand, S., Changeux, J.-P., and Bertrand, D. (1992). Stratified organization of the nicotinic acetylcholine receptor channel. *NeuroReport* 3, 1001–1004.
DiPaola, M., Czajkowski, C., and Karlin, A. (1989). The sideness of the COOH terminus of the acetylcholine δ subunit. *J. Biol. Chem.* 264, 15457–15463.
DiPaola, M., Kao, P., and Karlin, A. (1990). Mapping the subunit site photolabeled by the non-competitive inhibitor ^{3}H quinacrine azide in the active state of the nicotinic acetylcholine receptor. *J. Biol. Chem.* 265, 11017–11029.
Dwyer, B. (1991). Topological dispositions of lysine α380 and lysine γ486 in the acetylcholine receptor from *Torpedo californica. Biochemistry* 30, 4105–4112.

Ferrier-Montiel, A., Montal, M., Diaz-Muñoz, M., and Montal, M. (1991). Agonist-independent activation of acetylcholine receptor channels by protein kinase A phosphorylation. *Proc. Natl. Acad. Sci. U.S.A.* 88, 10213–10217.

Filatov, G., Aylwin, M., and White, M. (1993). Selective enhancement of the interaction of curare with the nicotinic receptor. *Mol. Pharmacol.* 44, 237–241.

Flores, C., Rogers, S., Pabreza, L., Wolfe, B., and Kellar, K. (1992). A subtype of nicotinic cholinergic receptor in rat brain is composed of α4 and β2 subunits and is upregulated by chronic nicotine treatment. *Mol. Pharmacol.* 41, 31–37.

Froehner, S., Luetje, C., Scotland, P., and Patrick, J. (1990). The postsynaptic 43K protein clusters muscle nicotinic receptors in *Xenopus* oocytes. *Neuron* 5, 403–410.

Fuchs, P. and Murrow, B. (1992). Cholinergic inhibition of short (outer) hair cells of the chick's cochlea. *J. Neurosci.* 12, 800–809.

Galzi, J.-L., Revah, F., Black, D., Goeldner, M., Hirth, C., and Changeux, J.-P. (1990). Identification of a novel amino acid α tyrosine 93 within the cholinergic ligand-binding sites of the acetylcholine receptor by photoaffinity labeling. *J. Biol. Chem.* 265, 10430–10437.

Galzi, J.-L., Bertrand, D., Devillers-Thiery, A., Revah, F., Bertrand, S., and Changeux, J.-P. (1991). Functional significance of aromatic amino acids from three peptide loops of the α7 neurnonal nicotinic receptor site investigated by site directed mutagenesis. *FEBS Lett.* 294, 198–202.

Galzi, J.-L., Devillers-Theiry, A., Hussy, N., Bertrand, S., Changeux, J.-P., and Bertrand, D. (1992). Mutations in the channel domain of a neuronal nicotinic receptor convert ion selectivity from cationic to anionic. *Nature* 359, 500–505.

Gehle, V. and Sumikawa, K. (1991). Site directed mutagenesis of the conserved N-glycosylation site on the nicotinic acetylcholine receptor subunits. *Mol. Brain Res.* 11, 17–25.

Gerzanich, V., Anand, R., and Lindstrom, J. (1994). Homomers of α8 and α7 subunits of nicotinic receptors exhibit similar channel but contrasting binding site properties. *Mol. Pharmacol.* 45, 212–220.

Giraudat, J., Dennis, M., Heidmann, T., Chang, J., and Changeux, J. P. (1986). Structure of the high-affinity binding site for noncompetitive blockers of the acetylcholine receptor: Serine 262 of the δ subunit is labeled by [^{3}H] chlorpromazine. *Proc. Natl. Acad. Sci. U.S.A.* 83, 2719–2723.

Gotti, C., Ogando, A., Hanke, W., Schlue, R., Moretti, M., and Clementi, F. (1991). Purification and characterization of an α-bungarotoxin receptor that forms a functional nicotinic channel. *Proc. Natl. Acad. Sci. U.S.A.* 88, 3258–3262.

Gotti, C., Hanke, W., Schlue, R., Moretti, M., and Clementi, F. (1992). A functional α-bungarotoxin receptor is present in chick cerebellum: Purification and characterization. *Neuroscience* 50, 117–127.

Grady, S., Marks, M., Wonnacott, S., and Collins, A. (1992). Characterization of nicotinic receptor-mediated ^{3}H dopamine release from synaptosomes prepared from mouse striatum. *J. Neurochem.* 59, 848–856.

Gu, Y., Camacho, P., Gardner, P., and Hall, Z. (1990). Identification of two amino acid residues in the ε subunit that promote mammalian muscle acetylcholine receptor assembly in COS cells. *Neuron* 6, 879–887.

Gu, Y., Forsayeth, J., Verrall, S., Yu, X., and Hall, Z. (1991). Assembly of the mammalian muscle acetylcholine receptor in transfected COS cells. *J. Cell Biol.* 114, 799–807.

Gundelfinger, E. (1992). How complex is the nicotinic receptor system of insects? *Trends Neurosci.* 15, 206–211.

Hamassaki-Britto, D., Brzozowska-Prechtl, A., Karten, H., and Lindstrom, J. (1994). Bipolar cells of the chick retina containing α-bungarotoxin-sensitive nicotinic acetylcholine receptors. *Visual Neurosci.* 11, 63–70.

Harsing, L., Sershen, H., and Lajtha, A. (1992). Dopamine efflux from striatum after chronic nicotine: Evidence for autoreceptor desensitization. *J. Neurochem.* 59, 48–54.

Heinemann, S., Boulter, J., Connolly, J., Deneris, E., Duvoisin, R., Hartley, M., Hermans-Borgmeyer, I., Hollmann, M., O'Shea-Greenfield, A., Papke, R., Rogers, S., and Patrick, J. (1991). The nicotinic receptor genes. *Clin. Neuropharmacol.* 14, S45–S61.

Hill, J., Zoli, M., Bourgeois, J.-P., and Changeux, J.-P. (1993). Immunocytochemical localization of a neuronal nicotinic receptor: The β2 subunit. *J. Neurosci.* 13, 1551–1568.

Hucho, F., Oberthür, and Lottspeich, F. (1986). The ion channel of the nicotinic acetylcholine receptor is formed by the homologous helices MII of the receptor subunits. *FEBS Lett.* 205, 137–142.

Huganir, R., and Grungard, P. (1990). Regulation of neurotransmitter receptor desensitization by protein phosphorylation. *Neuron* 5, 555–567.

Imoto, K., Busch, C., Sakmann, B., Mishina, M., Konno, T., Nakai, J., Bujo, H., Mori, Y., Fukuda, K., and Numa, S. (1988). Rings of negatively charged amino acids determine the acetylcholine receptor channel conductance. *Nature* 335, 645–648.

Jacob, M. and Berg, D. (1983). The ultrastructural localization of α-bungarotoxin binding sites in relation to synapses on chick ciliary ganglion neurons. *J. Neurosci.* 3, 260–271.

Jacob, M., Lindstrom, J., and Berg, D. (1986). Surface and intracellular distribution of a putative neuronal nicotinic acetylcholine receptor. *J. Cell Biol.* 103, 205–214.

Kao, P. and Karlin, A. (1986). Acetylcholine receptor binding site contains a disulfide crosslink between adjacent half-cystinyl residues. *J. Biol. Chem.* 261, 8085–8088.

Kao, P., Dwork, A., Kaldany, R., Silver, M., Wideman, J., Stein, S., and Karlin, A. (1984). Identification of the α subunit half cysteine specifically labeled by an affinity reagent for the acetylcholine receptor binding site. *J. Biol. Chem.* 259, 11662–11665.

Karlin, A. (1991). Exploration of the nicotinic acetylcholine receptor. *Harvey Lect. Ser.* 85, 71–107.

Karlin, A. (1993). Structure of nicotinic acetylcholine receptors. *Curr. Opin. Neurobiol.* 3, 299–309.

Karlin, A., Holtzman, E., Yodh, N., Label, P., Wall, J., and Hainfeld, J. (1983). The arrangement of the subunits of the acetylcholine receptor of *Torpedo californica. J. Biol. Chem.* 258, 6678–6681.

Kellaris, K., Ware, D., Smith, S., and Kyte, J. (1989). Assessment of the number of free cysteines and isolation and identification of cysteine-containing peptides from acetylcholine receptor. *Biochemistry* 28, 3469–3482.

Keyser, K. T., Hughes, T. E., Whiting, P. J., Lindstrom, J. M., and Karten, H. J. (1988). Cholinoceptive neurons in the retina on the chick: An immunohistochemical study of the nicotinic acetylcholine receptors. *Vis. Neurosci.* 1, 349–366.

Keyser, K., Britto, L., Schoepfer, R., Whiting, P., Cooper, J., Conroy, W., Brozozowska-Prechtl, A., Karten, H., and Lindstrom, J. (1993). Three subtypes of α-bungarotoxin sensitive nicotinic acetylcholine receptors are expressed in chick retina. *J. Neurosci.* 13, 442–454.

Konno, T., Busch, C., VonKitzing, E., Imoto, K., Wang, F., Nakai, J., Mishina, M., Numa, S., and Sakmann, B. (1991). Rings of anionic amino acids as structural determinants of ion selectivity in the acetylcholine receptor channel. *Proc. R. Soc., London Ser. B* 244, 69–79.

Kubalek, E., Ralston, S., Lindstrom, J., and Unwin, N. (1987). Location of subunits within the acetylcholine receptor: Analysis of tubular crystals from *Torpedo* marmorata. *J. Cell Biol.* 105, 9–18.

Lange, K., Wells, F., Jenner, P., and Marsden, P. (1993). Altered muscarinic and nicotinic receptor densities in cortical and subcortical regions in Parkinson's disease. *J. Neurochem.* 60, 197–203.

Lester, H. (1992). The permeation pathway of neurotransmitter-gated ion channels. *Annu. Rev. Biophys. Biomol. Struc.* 21, 267–292.

Li, L., Schuchard, M., Palma, A., Pradier, L., and McNamee, M. (1990). Functional role of the cystein 451 thiol group in the M4 helix of the γ subunit of *Torpedo californica* acetylcholine receptor. *Biochemistry* 29, 5428–5436.

Lindstrom, J., Shelton, G. D., and Fujii, Y. (1988). Myasthenia gravis. *Adv. Immunol.* 42, 233–284.

Lindstrom, J., Schoepfer, R., Conroy, W. G., and Whiting, P. (1990). Structural and functional heterogeneity of nicotinic receptors. In *The Biology of Nicotine Dependence. Ciba Foundation Symposium 152,* John Wiley & Sons, New York, 43.

Lipton, S. and Kater, S. (1989). Neurotransmitter regulation of neuronal outgrowth, plasticity, and survival. *Trends Neurosci.* 12, 265–270.

Lipton, S., Aizenman, E., and Loring, R. (1987). Neural nicotinic acetylcholine responses in solitary mammalian retinal ganglion cells. *Pflügers Arch.* 410, 37–43.

Lipton, S., Frosch, M., Phillips, M., Tauck, D., and Aizenman, E. (1988). Nicotinic antagonists enhance process outgrowth by rat retinal ganglion cells in culture. *Science* 239, 1293–1296.

Lo, D., Pinkham, J., and Stevens, C. (1991). Role of a key cysteine residue in the gating of the acetylcholine receptor. *Neuron* 6, 31–40.

Luther, M., Schoepfer, R., Whiting, P., Blatt, Y., Montal, M. S., Montal, M., and Lindstrom, J. (1989). A muscle acetylcholine receptor is expressed in the human cerebellar medulloblastoma cell line TE671. *J. Neurosci.* 9, 1082–1096.

Maimone, M. and Merlie, J. (1993). Interaction of the 43kd postsynaptic protein with all subunits of the muscle nicotinic acetylcholine receptor. *Neuron* 11, 53–66.

Maneckjie, R. and Minna, J. (1990). Opioid and nicotine receptors affect growth regulation of human lung cancer cell lines. *Proc. Natl. Acad. Sci. U.S.A.* 87, 3294–3298.

Marks, M., Pauly, J., Gross, D., Deneris, E., Hermans-Borgmeyer, I., Heinemann, S., and Collins, A. (1992). Nicotine binding and nicotinic receptor subunit RNA after chronic nicotine treatment. *J. Neurosci.* 12, 2765–2784.

Martinou, J.-C., Falls, D., Fischbach, G., and Merlie, J.-P. (1991). Acetylcholine receptor-inducting activity stimulates expression of the ε subunit gene of the muscle acetylcholine receptor. *Proc. Natl. Acad. Sci. U.S.A.* 88, 7669–7673.

Matter, J., Matter-Sadzinski, L., and Ballivet, M. (1990). Expression of neuronal nicotinic acetylcholine receptor genes in the developing chick visual system. *EMBO J.* 9, 1021–1026.

Matter-Sadzinski, L., Hernandez, M.-C., Roztocil, T., Ballivet, M., and Matter, J.-M. (1992). Neuronal specificity of the α7 nicotinic acetylcholine promoter develops during morphogenesis of the central nervous system. *EMBO J.* 11, 4529–4538.

McCrea, P. D., Popot, J.-L., and Engelman, D. M. (1987). Transmembrane topography of the nicotinic acetylcholine receptor δ subunit. *EMBO J.* 6, 3619–3626.

Merlie, J., Sebbane, R., Gardner, S., Olson, E., and Lindstrom, J. (1983). The regulation of acetylcholine receptor expression in mammalian muscle. *Cold Spring Harbor Symp. Quant. Biol.* 48, 135–146.

Middleton, R. and Cohen, J. (1991). Mapping of the acetylcholine binding site of the nicotinic acetylcholine receptor: ^{3}H nicotine as an agonist photoaffinity label. *Biochemistry* 30, 6987–6997.

Miles, K. and Huganir, R. (1988). Regulation of nicotinic acetylcholine receptors by protein phosphorylation. *Mol. Neurobiol.* 2, 91–124.

Mitra, A., McCarthy, M., and Stroud, R. (1989). Three-dimensional structure of the nicotinic acetylcholine receptor and location of the major associated 43kD cyto-skeletal protein, determined at 22 Å by low-dose electron microscopy and X-ray diffraction to 12.5Å. *J. Cell Biol.* 109, 755–774.

Morris, B., Hicks, A., Wisden, W., Darlison, M., Hunt, S., and Barnard, E. (1990). Distinct regional expression of nicotinic acetylcholine receptor genes in chick brain. *Mol. Brain Res.* 7, 305–315.

Mulle, C., Vidal, C., Benoit, P., and Changeux, J. P. (1991). Existence of different subtypes of nicotinic acetylcholine receptors in the rat habenulo-interpenduncular system. *J. Neurosci.* 11, 2588–2597.

Mulle, C., Choquet, D., Korn, H., and Changeux, J.-P. (1992). Calcium influx through nicotinic receptor in rat central neurons: Its relevance to cellular regulation. *Neuron* 8, 135–143.

Nef, P., Oneyser, C., Alliod, C., Couturier, S., and Ballivet, M. (1988). Genes expressed in the brain define three distinct neuronal nicotinic acetylcholine receptors. *EMBO J.* 7, 595–601.

Neville, C., Schmidt, M., and Schmidt, J. (1992). Response of myagenic determination factors to cessation and resumption of electrical activity in skeletal muscle: A possible role for myogenin in denervation supersensitivity. *Cell. Mol. Neurobiol.* 12, 511–527.

Noda, M., Takahashi, H., Tanabe, T., Toyosato, M., Kikyotani, S., Furutani, Y., Hirose, T., Takashima, H., Inayama, S., Miyata, T., and Numa, S. (1983). Structural homology of *Torpedo californica* acetylcholine receptor subunits. *Nature* 302, 528–532.

Okonjo, K., Kuhlmann, J., and Maelicke, A. (1991). A second pathway of activation of the *Torpedo* acetylcholine receptor channel. *Eur. J. Biochem.* 200, 671–677.

O'Leary, M. and White, M. (1992). Mutational analysis of ligand-induced activation of the *Torpedo* acetylcholine receptor. *J. Biol. Chem.* 267, 8360–8365.

Papke, R., Boulter, J., Patrick, J., and Heinemann (1989). Single-channel currents of rat neuronal nicotinic acetylcholine receptors expressed in *Xenopus laevis* oocytes. *Neuron* 3, 589–596.

Pedersen, S. and Cohen, J. (1990). *d*-Tubocurarine binding sites are located at αγ and αδ subunit interfaces of the nicotinic acetylcholine receptor. *Proc. Natl. Acad. Sci. U.S.A.* 87, 2785–2789.

Pereira, E., Reinhardt-Maelicke, S., Schrattenholz, A., Maelicke, A., and Albuquerque, E. (1993). Identification and functional characterization of a new agonist site on nicotinic acetylcholine receptors of cultured hippocampal neurons. *J. Pharmacol. Exp. Ther.* 265, 1474–1491.

Protti, M., Manfredi, A., Horton, R., Bellone, M., and Conti-Tronconi, B. (1993). Myasthenia gravis: Recognition of a human autoantigen at the molecular level. *Immunol. Today* 14, 363–368.

Raftery, M., Hunkapillar, M., Strader, C., and Hood, L. (1980). Acetylcholine receptor: Complex of homologous subunits. *Science* 208, 1454–1457.

Ratnam, M., Le Nguyen, D., Rivier, J., Sargent, P. B., and Lindstrom, J. (1986a). Transmembrane topography of nicotinic acetylcholine receptor: Immunochemical tests contradict theoretical predictions based on hydrophobicity profiles. *Biochemistry* 25, 2633–2643.

Ratnam, M., Sargent, P. B., Sarin, V., Fox, J. L., Le Nguyen, D., Rivier, J., Criado, M., and Lindstrom, J. (1986b). Location of antigenic determinants on primary sequences of subunits of nicotinic acetylcholine receptor by peptide mapping. *Biochemistry* 25, 2621–2632.

Revah, F., Bertrand, D., Galzi, J.-L., Devillers-Theiry, A., Mulle, C., Hussy, N., Bertrand, S., Ballivet, M., and Changeux, J.-P. (1991). Mutations in the channel domain alter desensitization of a neuronal nicotinic receptor. *Nature* 353, 846–849.

Role, L. (1992). Diversity in primary structure and function of neuronal nicotinic acetylcholine receptor channels. *Curr. Opin. Neurobiol.* 2, 254–262.

Saedi, M., Anand, R., Conroy, W. G., and Lindstrom, J. (1990). Determination of amino acids critical to the main immunogenic region of intact acetylcholine receptors by *in vitro* mutagenesis. *FEBS Lett.* 267, 55–59.

Saedi, M., Conroy, W. G., and Lindstrom, J. (1991). Assembly of *Torpedo* acetylcholine receptor in *Xenopus* oocytes. *J. Cell Biol.* 112, 1007–1015.

Sargent, P. (1993). The diversity of neuronal nicotinic acetylcholine receptors. *Annu. Rev. Neurosci.* 16, 403–443.

Sargent, P. B., Pike, S. H., Nadel, S. B., and Lindstrom, J. M. (1989). Nicotinic acetylcholine receptor-like molecules in the retina, retinotectal pathway, and optic tectum of the frog. *J. Neurosci.* 9, 565–573.

Schloss, P., Betz, H., Schröder, C., and Gundelfinger, E. (1991). Neuronal nicotinic acetylcholine receptors in Drosophila: Antibodies against an α-like and a non-α-subunit recognize the same high affinity α-bungarotoxin binding complex. *J. Neurochem.* 57, 1556–1562.

Schoepfer, R., Whiting, P., Esch, F., Blacher, R., Shimasaki, S., and Lindstrom, J. (1988). cDNA clones coding for the structural subunit of a chicken brain nicotinic acetylcholine receptor. *Neuron* 1, 241–248.

Schoepfer, R., Halvorsen, S., Conroy, W. G., Whiting, P., and Lindstrom, J. (1989). Antisera against an α3 fusion protein bind to ganglionic but not to brain nicotinic acetylcholine receptors. *FEBS Lett.* 257, 393–399.

Schoepfer, R., Conroy, W. G., Whiting, P., Gore, M., and Lindstrom, J. (1990). Brain α-bungarotoxin-binding protein cDNAs and mAbs reveal subtypes of this branch of the ligand-gated ion channel superfamily. *Neuron* 5, 35–48.

Schuster, C., Ultsch, A., Schloss, P., Cox, J., Schmidt, B., and Betz, H. (1991). Molecular cloning of an invertebrate glutamate receptor subunit expressed in *Drosophila* muscle. *Science* 254, 112–114.

Seeburg, P. (1993). The molecular biology of mammalian glutamate receptor channels. *Trends Neurosci.* 16, 359–364.

Séguéla, P., Wadiche, J., Dinelly-Miller, K., Dani, J., and Patrick, J. (1993). Molecular cloning, functional properties, and distribution of rat brain α7: A nicotinic cation channel highly permeable to calcium. *J. Neurosci.* 13, 596–604.

Sumikawa, K. and Gehle, V. (1991). Assembly of mutant subunits of the nicotinic acetylcholine receptor lacking the conserved disulfide loop structure. *J. Biol. Chem.* 267, 6286–6290.

Sussman, J., Harel, M., Frolow, F, Oefner, C., Goldman, A., Toker, L., and Silman, I. (1991). Atomic structure of acetylcholinesterase from *Torpedo californica:* A prototypic acetylcholine binding protein. *Science* 253, 872–879.

Swanson, L., Simmons, D., Whiting, P., and Lindstrom, J. (1987). Immunohistochemical localization of neuronal nicotinic receptors in the rodent central nervous system. *J. Neurosci.* 7, 3334–3342.

Tobimatsu, T., Fujita, Y., Fukuda, K., Tanaka, K., Mori, Y., Konno, T., Mishina, M., and Numa, S. (1987). Effects of substitution of putative transmembrane segments on nicotinic acetylcholine receptor function. *FEBS Lett.* 222, 56–62.

Tomaselli, G., McLaughlin, J., Jurman, M., Hawrot, E., and Yellen, G. (1991). Mutations affecting agonist sensitivity of the nicotinic acetylcholine receptor. *Biophys. J.* 60, 721–727.

Tzartos, S., Cung, M., Demange, P., Loutrari, H., Mamalaki, A., Marraud, M., Papadouli, I., Sakarellos, C., and Tsikaris, V. (1991). The main immunogenic region (MIR) of the nicotinic acetylcholine receptor and the anti-MIR antibodies. *Mol. Neurobiol.* 5, 1–29.

Unwin, N. (1993a). Nicotinic acetylcholine receptor at 9Å resolution. *J. Mol. Biol.* 229, 1101–1124.

Unwin, N., (1993b). Neurotransmitter action: Opening of ligand-gated ion channels. *Cell* 10, 31–41.

Unwin, N., Toyoshima, C., and Kubalek, E. (1988). Arrangement of the acetylcholine receptor subunits in the resting and desensitized states determined by cryoelectron microscopy of crystallized *Torpedo* postsynaptic membranes. *J. Cell Biol.* 107, 1123–1138.

Vernalis, A., Conroy, W., and Berg, D. (1993). Neurons assemble acetylcholine receptors with as many as three kinds of subunits while maintaining subunit segregation among receptor subtypes. *Neuron* 10, 451–464.

Vernino, S., Amador, M., Luetje, C., Patrick, J., and Dani, J. (1992). Calcium modulation and high calcium permeability of neuronal nicotinic acetylcholine receptors. Neuron 8, 127–134.

Verrall, S. and Hall, Z. (1992). The N-terminal domains of acetylcholine receptor subunits contain recognition signals for the initial steps of receptor assembly. *Cell* 68, 23–31.

Vijayaraghavan, S., Rathouz, M., Pugh, P., and Berg, D. (1992). Nicotinic receptors that bind α-bungarotoxin on neurons raise intracellular free Ca^{++}. *Neuron* 8, 353–362.

Wada, E., Wada, K., Boulter, J., Deneris, E., Heinemann, S., Patrick, J., and Swanson, L. (1989). The distribution of α2, α3, α4, and β2 neuronal nicotinic receptor subunit mRNAs in the central nervous system: A hybridization histochemical study in the rat. *J. Comp. Neurol.* 284, 314–335.

Wallace, B., Qu, Z., and Huganir, R. (1991). Agrin induces phosphorylation of the nicotinic acetylcholine receptor. *Neuron* 6, 869–878.

Wang, G. and Schmidt, J. (1976). Receptors for α-bungarotoxin in the developing visual system of the chick. *Brain Res.* 114, 524–529.

Whitehouse, P., Martino, A., Marcus, K., Zweig, R., Singer, H., Price, D., and Kellar, K. (1988). Reductions in acetylcholine and nicotine binding in several degenerative diseases. *Arch. Neurol.* 45, 722–724.

Whiting, P. J. and Lindstrom, J. M. (1986). Purification and characterization of a nicotinic acetylcholine receptor from chick brain. *Biochemistry* 25, 2082–2093.

Whiting, P. J. and Lindstrom, J. M. (1987a). Purification and characterization of a nicotinic acetylcholine receptor from rat brain. *Proc. Natl. Acad. Sci. U.S.A.* 84, 595–599.

Whiting, P. and Lindstrom, J. (1987b). Affinity labeling of neuronal acetylcholine receptors localizes the neurotransmitter binding site to the β subunit. *FEBS Lett.* 213, 55–60.

Whiting, P. J. and Lindstrom, J. M. (1988). Characterization of bovine and human neuronal nicotinic acetylcholine receptors using monoclonal antibodies. *J. Neurosci.* 8, 3395–3404.

Whiting, P., Esch, F., Shimasaki, S., and Lindstrom, J. (1987b). Neuronal nicotinic acetylcholine receptor β subunit is coded for by the cDNA clone α4. *FEBS Lett.* 219, 459–463.

Whiting, P., Liu, R., Morley, B. J., and Lindstrom, J. (1987a). Structurally different neuronal nicotinic acetylcholine receptor subtypes purified and characterized using monoclonal antibodies. *J. Neurosci.* 7, 4005–4016.

Whiting, P., Schoepfer, R., Lindstrom, J., and Priestly (1991). Structural and pharmacological characterization of the major brain nicotinic acetylcholine receptor subtype stably expressed in mouse fibroblasts. *Mol. Pharmacol.* 40, 463–472.

Witzemann, V., Barg, B., Nishikawa, Y., Sakmann, B., and Numa, S. (1987). Differential regulation of muscle acetylcholine receptor γ and ε subunit mRNAs. *FEBS Lett.* 223, 104–112.

Witzemann, V., Stein, E., Barg, B., Konno, T., Koenen, M., Kues, W., Criado, M., Hofmann, M., and Sakmann, B. (1990). Primary structure and functional expression of the α-, β-, γ-, δ-, and ε-subunits of the acetylcholine receptor from rat muscle. *Eur. J. Biochem.* 194, 437–448.

5

5-HT_3 Receptors

Jeremy J. Lambert, John A. Peters, and Anthony G. Hope

2.5.0 Introduction

Receptors for the neurotransmitter 5-hydroxytryptamine (5-HT) constitute a broad family of cell surface glycoproteins that regulate diverse physiological functions (Frazer et al., 1990). The classification of 5-HT receptors is complex, but one scheme that incorporates pharmacological (operational), functional (transductional), and structural data recognizes the existence of four major classes; 5-HT_1, 5-HT_2, 5-HT_3, and 5-HT_4 (Humphrey et al., 1993). The 5-HT_1 and 5-HT_2 categories subsume five and three receptor subtypes, respectively. In addition, pharmacological and structural data indicate the existence of several distinct 5-HT receptors that remain to be classified definitively. In mammals, 12 5-HT receptors have been cloned to date. With the exception of the 5-HT_3 receptor, all belong to the G-protein-linked superfamily. The 5-HT_3 class, however, occupies a unique position among vertebrate monoamine receptors in that it is a ligand-gated ion channel (Maricq et al., 1991).

5-HT_3 receptors are thought to be located exclusively upon neurons of the peripheral and central nervous systems. Activation of the receptor involves the transient opening of an integral, cation-selective, ion channel giving rise to neuronal depolarization and excitation. The current underlying the response, which is carried predominantly by the inward movement of Na ions under physiological conditions, demonstrates rapid activation and desensitization kinetics (Higashi and Nishi, 1982; Derkach et al., 1989; Lambert et al., 1989). The 5-HT_3 receptor appears to be distributed ubiquitously in the peripheral nervous system, occurring in autonomic, enteric, and primary visceral and somatic afferent neurons (Wallis, 1989). At such locations, 5-HT_3 receptor-mediated excitation may evoke neurotransmitter release (Gaddum and Picarelli, 1957; Fozard and Mwaluko, 1976), trigger reflex bradycardia and hypotension (Fozard and Host, 1982), or give rise to

0-8493-8322-6/95/$0.00+$.50

the sensation of pain (Armstrong et al., 1953; Richardson et al., 1985). Such responses have frequently been exploited in the development and evaluation of 5-HT_3 receptor ligands (Fozard, 1989). Centrally, the 5-HT_3 receptor occurs in association with cortical, limbic, and brainstem structures (Laporte et al., 1992). Within the lateral amygdala, postsynaptically located 5-HT_3 receptors mediate a rapid excitatory postsynaptic potential in response to synaptically released 5-HT, establishing a novel, fast, excitatory transmitter role for 5-HT in the central nervous system (Sugita et al., 1992). Centrally located 5-HT_3 receptors also modulate the release of neurotransmitters, a stimulatory or facilitatory influence being documented for dopamine in the striatum and nucleus accumbens (Blandina et al., 1989; Jiang et al., 1990), cholecystokinin in the cerebral cortex and nucleus accumbens (Paudice and Raiteri, 1991), noradrenaline and GABA in the hippocampus (Feuerstein and Hertting, 1988; Ropert and Guy, 1991), and 5-HT itself in the frontal cortex and hippocampus (Martin et al., 1992; Blier and Bouchard, 1993). By contrast, 5-HT_3 receptor activation is reported to depress the evoked release of noradrenaline in the hypothalamus (Blandina et al., 1991; Goldfarb et al., 1993) and to inhibit the release of acetylcholine from the cerebral cortex (Barnes et al., 1989; Bianchi et al., 1990) or synaptosomes isolated from the latter (Maura et al., 1992). However, others have not observed such inhibitory influences upon acetylcholine release (Johnson et al., 1993). The effects observed with 5-HT_3 receptor antagonists in behavioral studies performed upon laboratory animals, which in some cases are predictive of potential anxiolytic, antipsychotic, and cognitive-enhancing properties in man (Costall et al., 1990; Greenshaw, 1993) presumably derive, at least in part, from a negation of the influence of 5-HT_3 receptors upon neurotransmitter release. To date, however, the approved use of 5-HT_3 receptor antagonists in man is restricted to the prevention of emesis and nausea in patients undergoing anticancer chemotherapy or radiotherapy (Sanger, 1992; Costall and Naylor, 1992) and the treatment of postoperative nausea and vomiting (Russell and Kenny, 1992). Whether the antagonists act at central, or peripherally located 5-HT_3 receptors to exert their antiemetic effect is still under debate (Sanger, 1992). In addition to occurring upon central and peripheral neurons, 5-HT_3 receptors are also expressed at high density in several neuronal clonal cell lines (Peters and Lambert, 1989). As will be described in greater detail in subsequent sections of this chapter, the cell lines have proved to be an exceptionally useful model system for the electrophysiological, biochemical, and molecular characterization of the 5-HT_3 receptor.

In keeping with the theme of this volume, the scope of this chapter will be limited to what is known of the structure and functioning of the 5-HT_3 receptor and its relevance to other ligand-gated ion channels. Details of the receptors distribution and its potential involvement in a variety of physiological and pathological processes can be found in a recently published book devoted to the topic of central and peripheral 5-HT_3 receptors (Hamon, 1992).

2.5.1 Operational Definition of the 5-HT_3 Receptor

5-HT_3 receptors may be defined pharmacologically according to the scheme proposed by Humphrey et al. (1993). First, the receptor should be resistant to antagonism by compounds acting at the 5-HT_1, 5-HT_2 and 5-HT_4 classes of receptor. Second, the response elicted by 5-HT should be mimicked by agonists reportedly selective for the 5-HT_3 receptor, namely 2-methyl-5-HT, 1-phenylbiguanide, and *m*-chlorophenylbiguanide. As discussed elsewhere, these compounds are not ideal (Fozard, 1990; Peters et al., 1994). 2-Methyl-5-HT is not entirely selective, and all may act as partial agonists in some test systems. Additionally, 1-phenylbiguanide possesses neither affinity nor efficacy when tested upon guinea pig tissues (Butler et al., 1990). A recently described compound, SR 57227A (Bachy et al., 1993) appears to act as a full and selective agonist but has yet to be evaluated in a wide range of preparations. Third, the receptor should be antagonized by selective, high-affinity, competitively acting 5-HT_3 receptor antagonists. In contrast to the dearth of selective agonists, numerous antagonists have been described (King, 1994), but those most commonly used in classification are the first generation agents MDL-72222 (bemesetron; Fozard, 1984), ICS 205-930 (tropisetron; Richardson et al., 1985), ondansetron (Butler et al., 1988), and granisetron (Sanger and Nelson, 1989).

2.5.2 Electrophysiological Characterization

2.5.2.1 5-HT_3 Receptor Activation

Before the recent cloning of a 5-HT_3 receptor subunit established unequivocally that it is a member of the ligand-gated, ion channel family (Maricq et al., 1991), the rapidity of activation of the 5-HT-induced current (Higashi and Nishi, 1982; Yang, 1990; Yakel and Jackson, 1988) precluded the direct involvement of G proteins or second messengers in the response. The long-term stability of 5-HT-evoked macroscopic and single channel currents recorded with nucleotide-free internal solutions (Neijt et al., 1988, 1989; Yakel et al., 1988; Derkach et al., 1989; Lambert et al., 1989; Yang, 1990; Peters et al., 1994) and the lack of effect of G-protein activators and inhibitors supported this conclusion. However, a regulatory role for G proteins or second messengers is not ruled out by these observations.

The relationship between agonist concentration and effect has been investigated in voltage-clamp studies of neuronal clonal cell lines (Boess et al., 1992a; Neijt et al., 1988; Lovinger, 1991; Yakel et al., 1991), rabbit nodose ganglion neurons (Higashi and Nishi, 1982), and *Xenopus* oocytes injected with mRNA encoding the cloned murine 5-HT_3 R-A or 5-HT_3 R-A_S subunits (Maricq et al., 1991; Hope et al., 1993; Yakel et al., 1993; Downie et al., 1994). In these studies, the apparent dissociation constant for the receptor for 5-HT shows little variation (ca. 1–4 μM) although slightly higher values have been reported for voltage-clamped guinea pig coeliac ganglion and submucous neurons (Surprenant, 1990; Surprenant et al., 1991). These latter results may be due to species differences in 5-HT$_3$ receptor properties (see Peters et al., 1992, 1994). In all cases, the amplitude of the 5-HT$_3$-mediated current increases dramatically with 5-HT concentration, suggesting a cooperative receptor model in which the binding of agonist to a subunit facilitates the binding of additional agonist molecules to other subunits (Boess et al., 1992a; Furukawa et al., 1992; Higashi and Nishi, 1982; Lovinger, 1991; Neijt et al., 1988; Sepulveda et al., 1991; Surprenant, 1990; Yakel et al., 1991). Interestingly, cooperativity is also evident in experiments performed with homooligomeric receptors (Downie et al., 1994; Hope et al., 1993; Maricq et al., 1991; Yakel et al., 1993).

2.5.2.2 5-HT$_3$ Receptor Desensitization

Consistent with other ligand-gated ion channels 5-HT$_3$ receptors desensitize in the continued presence of the agonist (see Peters et al., 1993a; Yakel, 1992). Characteristically, in voltage-clamp experiments, this is manifest as a decrease of the 5-HT-induced current to continually perfused agonist, or a progressive decline of the response to the repetitive application of 5-HT. The rate of onset of desensitization increases with increasing agonist concentration until the kinetics of this process saturate (Neijt et al., 1989; Sepulveda et al., 1991; Yakel, 1992). Across neuronal cell lines (Neijt et al., 1989; Sepulveda et al., 1991; Shao et al., 1991; Yang, 1990), PC12 cells (Furukawa et al., 1992), and rat superior cervical ganglion neurons (Yang et al., 1992) for saturating concentrations of 5-HT, desensitization develops to a steady state within 1 to 10 s. However, differences in the actual rates of onset of desensitization, and whether or not the process is best described by one or two exponential functions, have been reported. Although methodological differences may contribute, these results suggest heterogeneity of desensitization kinetics across cells, and even within the same cell type. Similarly, no consensus of opinion has emerged on the influence of holding potential on receptor desensitization kinetics. Hence, in N1E-115 cells holding potential has little influence on desensitization (Neijt et al., 1989; Sepulveda et al., 1991). However, in rat superior cervical ganglion neurons (Yang et al., 1992), mouse hippocampal neurons (Yakel and Jackson, 1988), differentiated NG 108-15 cells (Shao et al., 1991), and oocytes expressing the 5-HT$_3$ R-A clone (Yakel et al., 1993), the overall rate of onset of desensitization is clearly enhanced by membrane hyperpolarization. Desensitization of the nicotinic receptor at the frog neuromuscular junction is enhanced by Ca^{2+} entry through the associated ion channel, to act at an as yet unidentified intracellular site of action (Cachelin and Colquhoun, 1989). Evidence is now accumulating that the 5-HT$_3$ ion channel may also conduct Ca^{2+} (see Section 2.5.2.4) and recent studies have demonstrated that Ca^{2+} greatly enhances the desensitization of 5-HT-mediated currents recorded from oocytes expressing the 5-HT$_3$ R-A clone (Yakel et al., 1993). These effects were attenuated by the intracellular injection of the Ca^{2+} chelator BAPTA, suggesting that as for nicotinic receptors, Ca^{2+} may act at an intracellular locus (Yakel et al., 1993). If so, then Ca^{2+} could bind to a cytoplasmic facing site on the ion channel itself, or, alternatively, it could act indirectly via second messenger systems. The influence of such systems on 5-HT_3 receptor kinetics is not at all clear (see Peters et al., 1994; Yakel, 1992). However, it should be borne in mind that the cloning of 5-HT_3 receptor subunits (Maricq et al., 1991; Hope et al., 1993) has established that, as for other members of the ligand-gated ion channel family, the potential for biochemical modulation of the receptor certainly exists (Section 2.5.4.3).

Recently, both tetraethylammonium (TEA) and the 5-HT moiety 5-hydroxyindole have been shown to modify receptor desensitization kinetics. In voltage-clamped N1E-115 cells the preapplication of 5-HT (1.5 μM) greatly reduced the current induced by a subsequent application of 10 μM 5-HT (Kooyman et al., 1993a). However, when TEA (1–10 m*M*) was coapplied with the low dose of 5-HT, then the subsequent desensitization was prevented (Kooyman et al., 1993a). TEA could not, however, reverse receptor desensitization once it was established (Kooyman et al., 1993a). Also, in voltage-clamped N1E-115 cells, 5-hydroxyindole (10 μM–10 m*M*) enhances the amplitude of 5-HT-evoked currents, and greatly retards the rate of current decay that occurs in the continued presence of 5-HT (Kooyman, 1993; Kooyman et al., 1993b). By contrast to TEA, 5-hydroxyindole can reverse 5-HT receptor desensitization even in the continued presence of the agonist (Kooyman, 1993). To date the influence of other positive allosteric modulators of the 5-HT_3 receptor such as ethanol (Lovinger and White, 1991), trichloroethanol (Downie et al., 1993; Lovinger and Zhou, 1993), and ketamine (Peters et al., 1991b) on desensitization kinetics has not been determined. Recently the technique of site-directed mutagenesis has identified amino acids in the M2 region of a nicotinic receptor subunit that influence desensitization kinetics (Bertrand et al., 1992; Galzi et al., 1992). The influence of these amino acids (Yakel et al., 1993) on 5-HT_3 receptor desensitization is considered elsewhere (Section 2.5.4.3).

2.5.2.3 The Voltage Dependence of 5-HT_3 Receptor-Mediated Currents

Some studies report a linear relationship between the 5-HT_3 receptor-mediated current and the holding potential for rabbit nodose ganglion neurons (Higashi and Nishi, 1982) and N1E-115 cells (Neijt et al., 1989). However, other studies on N1E-115 cells (Peters et al., 1988; Lambert et al., 1989, Sepulveda et al., 1991), NCB-20, N18, and NG108-15 cells (Yakel and Jackson, 1988, Lambert et al., 1989; Yang 1990; Yakel et al., 1990; Shao et al., 1991; Lovinger, 1991), rabbit (Peters et al., 1991c, 1993), rat (Lovinger and White, 1991), and guinea pig (Gill, Peters, and Lambert, unpublished observations) nodose ganglion neurons and rat dorsal root (Robertson and Bevan, 1991), and superior cervical ganglion neurons (Yang et al., 1992), consistently report inward rectification.

The molecular mechanisms underlying the rectification are not clear. The phenomenon is observed whether Na^+, K^+, or Cs^+ is the predominant internal cation (Lambert et al., 1989; Yang 1990; Yakel et al., 1990). Ion channel block by internal divalent cations, as proposed for some neuronal nicotinic receptors (Hille, 1992), is unlikely, as reducing the internal free Ca^{2+} and Mg^{2+} to nanomolar concentrations does not influence rectification (Yang, 1990). For rat superior cervical ganglion neurons, a small inward rectification of the single channel conductance may contribute to the rectification of the whole-cell current, but for rabbit nodose ganglion neurons (Peters et al., 1993), guinea pig submucous plexus neurons (Derkach et al., 1989), and NG 108-15 cells (Shao et al., 1991), the single channel current–voltage relationship is essentially linear or only rectifies modestly. Further experiments investigating particularly the voltage dependence of single channel kinetics may be instructive. Note that the current–voltage relationship for the cloned 5-HT_3 subunits will be discussed elsewhere (see Section 2.5.4.2).

2.5.2.4 The Ionic Selectivity of the 5-HT_3 Receptor–Channel Complex

In agreement with earlier investigations (see Peters et al., 1994 for references), the results of patch clamp studies on the neuronal cell lines NG 108-15 (Yakel et al., 1990), N18 (Yang, 1990), N1E-115 (Neijt et al., 1989; Lambert et al., 1989), and NCB-20 (Lambert et al., 1989) demonstrate that the 5-HT receptor–channel complex selectively conducts monovalent cations. The same conclusion has been reached for the 5-HT_3 receptors of the guinea pig submucous plexus (Derkach et al., 1989), the rat superior cervical (Yang et al., 1992) and dorsal root ganglia (Robertson and Bevan, 1991), and rabbit nodose ganglia (Peters et al., 1991c, 1993; Malone et al., 1994). By determining the 5-HT reversal potential in solutions with known intra- and extracellular ionic concentrations, the permeability of potassium ions relative to sodium ions (P_K/P_{Na}) can be calculated from the Goldman–Hodgkin–Katz voltage equation (Hille, 1992; see Table 1). The majority of these studies suggest these ions to be equipermeant (Table 1).

To further characterize the 5-HT_3 ion channel, a number of studies have compared the permeation properties of a range of metallic and organic monovalent cations. Similar to the nicotinic receptor (Adams et al., 1980; Dwyer et al., 1980), the 5-HT-activated ion channel of N18 neuroblastoma cells (Yang, 1990), differentiated PC12 cells (Furukawa et al., 1992), rat superior cervical ganglion neurons (Yang et al., 1992), and rabbit nodose ganglion neurons (Malone et al., 1994) discriminate poorly between monovalent cations, giving a weak selectivity sequence of $Cs^+ > K^+ > Li^+ > Na^+ > Rb^+$. In N18 neuroblastoma cells (Yang, 1990) and rabbit nodose ganglion neurons (Malone et al., 1994), the permeability of organic cations decreases as a function of their geometric mean diameter. Similar observations had previously been made for the nicotinic receptor of the frog neuromuscular junction (Adams et al; 1980; Dwyer et al., 1980). Hence, like the nicotinic receptor,

Table 1. The Electrophysiological Properties of 5-HT_3 Receptors[a]

Cell type	Recording technique	$E_{5\text{-}HT}$ (mV)	P_K/P_{Na}	Channel conductance (pS)	References
N1E-115 neuroblastoma	WCR/FA	−2.1	1.09	0.31	Lambert et al. (1989)
N1E-115 neuroblastoma	WCR	+20.0	0.42	—	Neijt et al. (1989)
N18 neuroblastoma	WCR/FA	−1.6	1.10	0.59	Yang (1990)
NG 108-15 hybrid cells	WCR/FA FA/SCR	+3.5	0.89	3.6–4.4[1] 7.2–12.0[2]	Yakel et al. (1990); Shao et al. (1991)
NCB-20 hybrid cells	WCR	−2.0	1.09	—	Lambert et al. (1989)
Rabbit nodose ganglion cells	VC	+7.0	0.43	—	Higashi and Nishi (1982)
Rabbit nodose ganglion cells	WCR/SCR	−1.6	1.06	16.5,[3] 19.3[4]	Peters et al. (1993b)
Rat superior cervical ganglion cells	WCR/FA SCR	−4.7	1.24	2.6 11.1	Yang et al. (1992)
Guinea pig submucous plexus neurons	WCR/SCR	+3.3	0.44	15.0/9.2	Derkach et al. (1989)
Guinea pig coeliac ganglion neurons	WCR/SCR	−1.0	—	10	Surprenant et al. (1991)

[a] A summary of some of the characteristics of 5-HT_3 receptors determined using electrophysiological techniques. WCR, whole cell recording; FA, fluctuation analysis; SCR, single channel recording; VC, two-electrode voltage clamp; $E_{5\text{-}HT}$, reversal potential of the 5-HT-induced current; P_K/P_{Na}, permeability of potassium relative to sodium. Notes: determined from differentiated (1) and undifferentiated (2) cells; chord conductance (3), slope conductance (4). The ratios presented for rabbit nodose ganglion neurones (Higashi and Nishi, 1982) and guinea pig submucous plexus neurons (Derkach et al., 1989; R. A. North, personal communication) are conductance, rather than permeability, ratios.

the ion channel associated with the 5-HT_3 receptor can be viewed as a large water-filled pore (Adams et al., 1980; Dwyer et al., 1980). A plot of the square root of the permeability ratio P_x/P_{Na} versus the geometric mean diameter of the permeant species X gives a linear relationship, the *x* axis intercept of which estimates the size of the channel at its narrowest region (Cohen et al., 1992). By treating the channel as a simple cylinder, these plots reveal the minimum pore size of the frog neuromuscular junction nicotinic receptor channel (Adams et al., 1980; Dwyer et al., 1980) and the 5-HT_3 receptor channel of N18 cells (Yang, 1990) and rabbit nodose ganglion neurons (Malone et al., 1994) to be very similar (0.76, 0.75, and 0.82 nm, respectively; see Figure 1).

Patch-clamp and voltage-clamp studies on NG 108-15 hybrid cells (Yakel et al., 1990), rabbit nodose ganglion cells (Higashi and Nishi, 1992), and N1E-115 neuroblastoma cells (Peters et al., 1988) suggested that the 5-HT_3 channel was relatively impermeable to Ca^{2+}. However, the latter two studies were performed over a limited Ca^{2+} concentration range, and in the presence

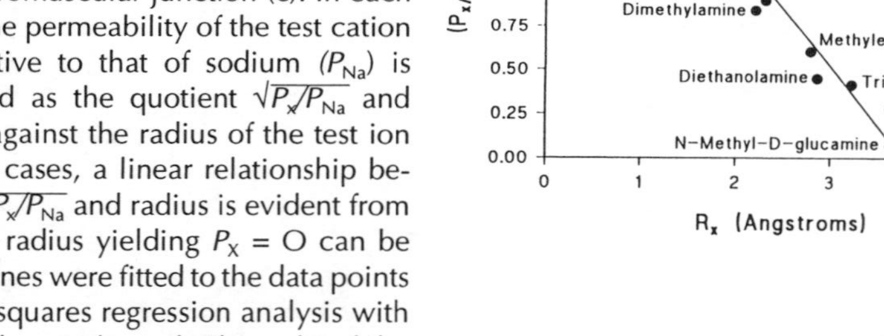

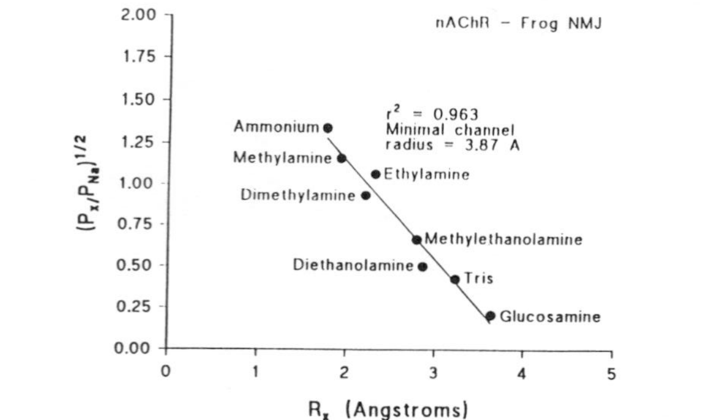

FIGURE 1. The minimum pore size of the 5-HT$_3$ receptor channel. Estimates of the minimum pore size of the 5-HT$_3$ receptor channel complex of (a) N18 neuroblastoma cells and (b) rabbit nodose ganglion neurons compared with the nicotinic acetylcholine receptor of the frog neuromuscular junction (c). In each panel, the permeability of the test cation (P_x) relative to that of sodium (P_{Na}) is expressed as the quotient $\sqrt{P_x/P_{Na}}$ and plotted against the radius of the test ion X. In all cases, a linear relationship between $\sqrt{P_x/P_{Na}}$ and radius is evident from which a radius yielding P_X = O can be found. Lines were fitted to the data points by least-squares regression analysis with the r^2 values indicated. The radii of the organic cations were kindly calculated by Dr. R. McGuire, Organon Laboratories, Lanarkshire, Scotland, using volume maps of energy minimized conformations of the compounds constructed in Chem-X. Data were obtained from (a) Yang (1990), (b) Malone et al. (1994) and Malone, Peters, and Lambert (unpublished observations), and (c) Dwyer et al. (1980). Data in c were extracted from a large base of data, some of which has been omitted in the interests of clarity.

of highly permeant monovalent cations, conditions that may mask a possible Ca^{2+} permeability. Recent work has revealed substantial P_{Ca}/P_{Na} ratios of 0.37, 0.55, and 1.12 for rabbit nodose ganglion neurons (Peters et al., 1994), rat superior cervical ganglion neurons (Yang et al., 1992), and N18 neuroblastoma cells (Yang, 1990), respectively. In addition to Ca^{2+}, the 5-HT_3 channel of N18 cells is also permeable to Mg^{2+} and Ba^{2+}. Consistent with the view that the channel can be considered as a large water-filled pore, the divalent cations are approximately equipermeant (Yang, 1990).

An alternative approach to study calcium permeability is provided by the use of the fluorescent calcium dye Fura 2. In NG 108-15 cells, 5-HT induces a concentration-dependent increase of Fura-2 fluorescence, an effect that is blocked by selective 5-HT_3 antagonists (Reiser, 1992). However, in that study, the fluorescent signal detected might also have reflected the entry of calcium through calcium channels activated by the 5-HT-induced membrane depolarization. In similar experiments performed on undifferentiated N1E-115 cells (Hargreaves et al., 1994), depolarization per se was insufficient to elicit an increase in fluorescence. By contrast, the 5-HT_3 receptor selective agonist *m*-chlorophenylbiguanide evoked a clear increase in intracellular Ca^{2+} concentration that could be blocked by pretreatment with ondansetron or the omission of Ca^{2+} from the extracellular solution. We have recently stably transfected a functional 5-HT_3 receptor subunit cloned from N1E-115 neuroblastoma cells (Hope et al., 1993) into the human embryonic kidney epithelial cell line HEK 239. Such cells have functional 5-HT_3 receptors, as assessed by patch-clamp techniques, but do not possess voltage-activated calcium channels (Hope, 1993). On such cells, in a physiological salt solution, 5-HT induces an increase in Fura-2 fluorescence, an effect that is blocked by the removal of extracellular calcium or by the appropriate concentrations of selective 5-HT receptor antagonists (Hope, 1993; Cholewinski, Hope and Lambert, unpublished observations). Collectively, the above studies suggest that a small influx of Ca^{2+} through 5-HT_3 receptor channels will occur under physiological conditions and may contribute to the influence of 5-HT_3 receptors on neurotransmitter release (e.g., Blandina et al., 1989), guanylate cyclase activation (Reiser, 1992), and phosphatidylinositol trisphosphate generation (Edwards et al., 1991; Reiser, 1992). Recently neuronal nicotinic (Ferrer-Montiel and Montal, 1993; Vernino et al., 1992; Mulle et al., 1992) and amino acid receptors (Keller et al., 1992) have been identified with relatively high permeabilities to calcium. Whether a similar diversity exists for 5-HT_3 receptors remains to be determined.

At physiological concentrations, both calcium and magnesium depress the 5-HT-evoked response recorded from N1E-115 cells (Peters and Usherwood, 1983; Peters et al., 1988), NG 108-15 cells (Christian et al., 1978; Yakel et al., 1990), N18 cells (Yang, 1990), and rabbit nodose ganglion neurons (Peters et al., 1993). In *Xenopus laevis* oocytes (Maricq et al., 1991; Eiselé et al., 1993) and HEK 293 cells (Gill et al., 1993) expressing the 5-HT_3 R-A subunit divalent cations similarly block current responses evoked by 5-HT. The mechanism of this effect remains to be determined, although some possibilities may be excluded. The effect occurs without a concomitant change in the reversal potential and, at least in cells other than the oocyte expression system, blockade is voltage independent. Therefore, it is unlikely that an alteration of ion channel selectivity, or a voltage-dependent ion channel blocking mechanism underlies this phenomenon. The reported voltage-dependent block of the 5-HT_3 R-A ion channel by Mg^{2+} and Ca^{2+} (Maricq et al., 1991; Eiselé et al., 1993) will be considered elsewhere (Section 2.5.4.2).

Divalent cations such as Mg^{2+} and Ca^{2+} would be expected to screen the negative charges located on the extracellular surface of the membrane and may also interact with negatively charged amino acids proposed to be located near the extracellular entrance of the 5-HT_3 ion channel (Hope et al., 1993; Imoto et al., 1988; Maricq et al., 1991). Such effects would reduce electrostatic interactions that enhance the accumulation of permeant cations at the mouth of the channel. As a consequence, a reduced single channel conductance and a decrease of the 5-HT-induced whole cell current would be anticipated. Clearly, single channel experiments are required to examine this possibility.

In addition to inhibiting the amplitude of the 5-HT-induced response, Mg^{2+} and Ca^{2+} also increase the current decay rate in N1E-115 cells (Peters et al., 1988) and N18 cells (Yang, 1990) although, in NG 108-15 cells, Ca^{2+} is reported to have the opposite effect (Yakel et al., 1990). The observations in N1E-115 cells and N18 cells suggest an effect of the divalent cations on channel kinetics or receptor desensitization. Consistent with the latter suggestion, Ca^{2+} greatly enhances the desensitization of 5-HT-mediated currents recorded from oocytes expressing the 5-HT_3 R-A clone (Yakel et al., 1993; see Section 2.5.2.2). Finally, the influence of Mg^{2+} and Ca^{2+} on 5-HT-mediated responses could result from an interaction of the divalent cations with the agonist binding site. In support of this suggestion, the specific binding of the 5-HT_3 selective agonist [^{3}H]*m*-chlorophenylbiguanide to a membrane preparation of N1E-115 cells is inhibited by physiological concentrations of Mg^{2+} and Ca^{2+} (Lummis et al., 1993). In summary, physiological concentrations of Mg^{2+} and Ca^{2+} inhibit 5-HT_3-mediated responses but the mechanism(s) of this effect requires further investigation.

In NCB-20 cells, the divalent cations Zn^{2+}, Cd^{2+}, and Cu^{2+} are potent antagonists of the 5-HT_3 receptor-mediated current (Lovinger, 1991). However, in contrast to Mg^{2+} and Ca^{2+}, the antagonism by the group IIb cations (Zn^{2+} and Cd^{2+}), but not the group Ib (Cu^{2+}) cation, was enhanced by membrane hyperpolarization (Lovinger, 1991). The voltage-dependent block by the group IIb cations may suggest that they are binding within the ion channel lumen. However, an open channel blocking mechanism is not easily reconciled with the observation that divalent inhibition is reduced by increasing the agonist concentration (Lovinger, 1991). Perhaps distinct binding sites exist for the different divalent cations (Lovinger, 1991). Support for this suggestion comes from experiments utilizing the whole cell clamp technique on HEK 293 cells stably transfected with the 5-HT_3 R-A clone (Gill et al., 1993).

In such cells, the antagonist actions of Ca^{2+} and Mg^{2+} are indistinguishable from those reported for N1E-115 cells (Peters et al., 1988) but the effect of Zn^{2+} is drastically altered. In comparison to the antagonism (Lovinger, 1991) produced by Zn^{2+} ($IC_{50} = 20\ \mu M$) in NCB-20 cells (from which the 5-HT_3 R-A subunit was cloned), this divalent cation exerts a biphasic effect upon the 5-HT-induced response mediated by the 5-HT_3 R-A, with low concentrations (1–10 μM) potentiating the 5-HT-induced current and only higher concentrations (>30 μM) producing inhibition (Gill et al., 1993). This result might also suggest a different subunit composition between the 5-HT_3 R-A and the 5-HT_3 receptor native to the NCB-20 cell (see below).

2.5.2.5 Single Channel Properties of 5-HT_3 Receptors

The techniques of single channel recording (from outside-out membrane patches) and fluctuation analysis of whole cell 5-HT-induced current noise have been utilized to determine the conductance of the channel integral to the 5-HT_3 receptor. Such studies have revealed that the conductance can vary considerably (ca. 60-fold) across cell types (Peters et al., 1991a, 1992, 1993) (see Table 1). Hence, on outside-out membrane patches made from rabbit nodose ganglion (Malone et al., 1991a; Peters et al., 1991c, 1993), guinea pig coeliac ganglion (Surprenant et al., 1991), and submucous plexus neurons (Derkach et al., 1989), clearly discernible 5-HT activated single channels of 17–19, 10, and 15 and 9 pS, respectively, have been observed. In rat superior cervical ganglion neurons, 5-HT gates an ion channel with a chord conductance of approximately 15 pS (determined on outside out patches held at –104 mV). However, the single channel properties of this 5-HT_3 receptor are complex. The channel rectifies inwardly such that the chord conductance determined at –54 mV is reduced to 10 pS (Yang et al., 1992), a value identical to the mainstate conductance reported for 5-HT_3 receptors in mouse superior cervical ganglion neurons (Hussy and Jones, 1993). Fluctuation analysis of 5-HT-induced whole cell current noise in rat superior cervical ganglion neurons gives a much lower conductance estimate of 2.6 pS (Yang et al., 1992). Although the technique of fluctuation analysis is inevitably less exact than the direct determination of the single channel conductance from isolated membrane patches, it seems unlikely that methodological differences alone are responsible for this discrepancy. Alternatively, as fluctuation analysis gives a weighted mean estimate of channel conductance, the possibility exists that the discrepancy arises from the activation of a heterogeneous population of 5-HT_3 receptors, including the resolvable conductance observed on outside-out membrane patches, and a small conductance, indistinguishable from the background noise (Yang et al., 1992). Indeed, on outside-out membrane patches of rat dorsal root ganglion neurons, N1E-115 and NCB-20 cells, 5-HT evokes a rapidly rising, smoothly decaying, "mini whole-cell" current. Such currents have been interpreted as resulting from the summed activity of many 5-HT_3-activated channels of low conductance (Lambert et al., 1989; Peters et al., 1991c; Robertson and Bevan, 1991).

Consistent with this interpretation, fluctuation analysis of 5-HT-induced whole cell current noise estimates the mean single channel conductance of NIE-115 cells (Lambert et al., 1989) and N18 cells (Yang, 1990) to be below 1 pS, although at least in some studies somewhat larger values have been reported for the related cell line NG108-15 (Yakel et al., 1990, Shao et al., 1991, but also see Hussy and Jones, 1993). Interestingly, by reducing the external concentration of Ca^{2+} and substituting external Na^+ and K^+ with the more permeant NH_4 ion (Malone et al., 1991a, 1994; Yang, 1990), the conductance of the 5-HT_3 receptor of N1E-115 cells was dramatically increased, such that clearly discernible single channel events could now be observed (Kooyman, 1993). A comparison of the amino acid sequences of 5-HT_3 clones exhibiting divergent single channel conductances, allied with site directed mutagenesis, might clarify the relationship between amino acid sequence of the M2 region and the single channel conductance for the 5-HT_3 receptor (Sections 2.5.4.2 and 2.5.4.3).

2.5.3 Biochemical Characterization

2.5.3.1 Solubilization and Purification of 5-HT_3 Receptors

5-HT_3 receptors have been solubilized from membrane homogenates derived from several tissues and cell types, namely, rabbit small bowel muscularis (Gordon et al., 1990), rat brain (McKernan et al., 1990a), and the clonal cell lines NCB-20 (McKernan et al., 1990b,c), NG 108-15 (Miquel et al., 1990; Boess et al., 1992b), and N1E-115 (Lummis and Martin, 1992). Detergents that have proved suitable for this purpose include sodium cholate, sodium deoxycholate, CHAPS (3-[(3-cholamidopropyl)dimethyl ammonio]-1-propane sulfonic acid), *n*-octylglucoside, and Triton X-100. Under optimal conditions, such detergents are reported to solubilize between 23 and 72% of 5-HT_3 receptors present in membrane homogenates. However, several detergents, particularly those with high critical micellar concentration (CMC) values such as *n*-octylglucoside and CHAPS, may exert an inhibitory influence in radioligand binding assays performed upon 5-HT_3 receptors (McKernan, 1992). Consequently, in several studies solubilized receptors have been exchanged into lubrol, a detergent that does not present this interference at concentrations several times CMC, prior to conducting binding assays (McKernan et al., 1990a; Boess et al., 1992b). Radioligand binding assays indicate solubilized 5-HT_3 receptors to retain the pharmacological profile of the membrane-bound receptor. Thus, with few exceptions, the binding affinities of agonists and antagonists are not greatly altered upon solubilization and their rank order of potency in displacing radiolabeled antagonists is generally unaffected. These observations indicate the receptor to be solubilized in an intact form, as recently confirmed by the incorporation of functional purified 5-HT_3 receptors into liposomes (Lummis and Martin, 1992).

Several strategies have been employed in the purification of 5-HT$_3$ receptors (see McKernan, 1992). Modest purification has been achieved using ion exchange chromatography with either QAE-sepharose or hydroxyapatite. Similarly, lectin affinity chromatography with a lentil lectin-agarose column produced a sevenfold purification of 5-HT$_3$ receptors solubilized from NCB-20 cells (McKernan, 1992), whereas a tenfold purification was reported for receptors solubilized from NG108-15 cells using a wheat germ agglutin-agarose column (Miquel et al., 1990). However, to date, the most extensive purification of 5-HT$_3$ receptors has been achieved by affinity chromatography employing specific, high-affinity, 5-HT$_3$ receptor ligands immobilized on a suitable matrix. McKernan et al. (1990b) obtained a 1,700-fold purification of receptors solubilized from NCB-20 cells using a column constructed from an analogue of tropisetron (L-680,652) linked to agarose. Employing GR 119566X (a derivative of GR 67330) linked to Affi-Gel 15, a similar degree of purification of 5-HT$_3$ receptors solubilized from NG 108-15 cells has been reported (Boess et al., 1992b), as has the purification of receptors from N1E-115 cells (Lummis and Martin, 1992). Radioligand binding studies performed on purified receptors show a rank order of antagonist and agonist potencies similar to those observed with intact cells, crude membrane homogenates, and solubilized preparations. However, at least in some studies, the absolute affinity of certain agents [e.g., ondansetron, *m*-chlorophenylbiguanide (Boess et al., 1992b)] for membrane-bound, solubilized and purified receptors may vary considerably.

2.5.3.2 Molecular Size of the 5-HT$_3$ Receptor

The molecular size of the intact oligomeric 5-HT$_3$ receptor has been estimated by the techniques of gel-exclusion chromatography and sucrose-density-gradient centrifugation either alone, or in combination. Miquel et al. (1990), using the gel-filtration technique, obtained an apparent molecular mass of 600 kDa for the oligomeric complex solubilized from NG 108-15 cells. While this value lies within the range 443–669 kDa reported by Gordon et al. (1990) for the receptor of rabbit small bowel, it is much larger than the estimate of 370 kDa provided by Boess et al. (1992b) for the 5-HT$_3$ receptor/detergent complex purified from NG 108-15 cells. The latter value is in reasonable agreement with estimates of molecular mass derived from sucrose-gradient-density centrifugation. Svedberg coefficients in the range 11 to 12.85 have been determined for the 5-HT$_3$ receptor, values consistent with a globular protein of approximately 300 kDa molecular mass (Miquel et al., 1990; McKernan et al., 1990b; Boess et al., 1992b). The most accurate determination of the molecular weight of the 5-HT$_3$ receptor, where the influence of detergent binding has been taken into account, suggests a value of 249 kDa (McKernan et al., 1990c). This mass is typical of members of the ligand-gated ion channel family that includes the nicotinic, GABA$_A$, and glycine receptors (Strange, 1988).

Sodium dodecyl sulfate-polyacrylamide gel electrophoresis (SDS-PAGE) of 5-HT$_3$ receptors purified from NCB-20 and NG 108-15 cells suggests the oligomeric complex to be comprised of subunits of differing apparent molecular masses. McKernan et al. (1990b) observed two major protein bands, corresponding to apparent molecular masses of 38 and 54 kDa, for the denatured 5-HT$_3$ receptor isolated from NCB-20 cells. Boess et al., (1992b) reported broad bands corresponding to polypeptides of 36, 40, 50, and 76 kDa following SDS-PAGE of the receptor purified from NG 108-15 cells. Only a single band at 54.7 kDa was reproducibly discerned in studies performed upon the receptor isolated from N1E-115 cells (Lummis and Martin, 1992). Collectively, these results might indicate the existence of at least two distinct 5-HT$_3$ receptor subunits with molecular masses in the order of 54 and 38 kDa. Several independent lines of evidence provide some support for this proposition. First, cDNAs encoding 5-HT$_3$ receptor subunits have recently been cloned from the murine clonal cell lines NCB-20 and N1E-115 (Maricq et al., 1991; Hope et al., 1993). The predicted size of these subunits, derived from the deduced amino acid sequence, has been reported as 55,966 and 53,189 Da, respectively. The former estimate includes the contribution of the signal peptide, whereas the latter refers to the mature protein, which largely accounts for the difference in the published molecular weights of these subunits, which appear to exist as splice variants (Hope et al., 1993; Kawashima et al., 1993, see below). The molecular size of the mature protein corresponds closely to the estimate (54.7 kDa) made for the single protein band observed by Lummis and Martin (1992) and the larger of the putative receptor components (54 kDa) found by McKernan et al. (1990b). Furthermore, antibodies raised to either the amino terminal domain or the putative large intracellular loop of the cloned 5-HT$_3$ receptor subunit (Maricq et al., 1991) cross-reacted with the 54 kDa subunit affinity purified from NCB-20 cells (Turton et al., 1993). Second, evaluation of the target size of 5-HT$_3$ receptor in rat frontal cortex by radiation inactivation yielded a molecular mass of 49.1 kDa for a site binding [^{3}H]GR 65630 (Lummis et al., 1990). Although this value compares favorably with the predicted molecular mass of cloned 5-HT$_3$ receptor subunits, a smaller target size of 35 kDa has been reported for a site labeled by [^{3}H]zacopride in both NG 108-15 cells and rat cerebral cortex (Gozlan et al., 1989; Miquel et al., 1990). The latter estimate correlates with the apparent molecular masses of 36–38 kDa detected in some studies where purified 5-HT$_3$ receptor have been subjected to SDS-PAGE (e.g., McKernan et al., 1990b; Boess et al., 1992b).

In summary, the purification of 5-HT$_3$ receptors from neuronal clonal cells lines and their subsequent analysis by SDS-PAGE, together with limited supporting evidence, suggests the existence of structurally distinct 5-HT$_3$ receptor subunits. As discussed by several authors of such studies, the variability seen in SDS-PAGE of purified receptor obtained from different cell lines may simply reflect a variable subunit composition of the receptor across the cell lines examined. More mundanely, it might also indicate the presence of aggregated or degraded material or a protein copurifying with the 5-HT_3 receptor. Direct evidence from cloning studies, as discussed later, is largely still lacking.

2.5.4 Molecular Characterization

2.5.4.1 Cloning of 5-HT_3 Receptor Subunits

A cDNA clone encoding a functional 5-HT_3 receptor (5-HT_3 R-A) was first obtained by screening, for functional expression, a cDNA library constructed from size fractionated mRNA isolated from the hybridoma cell line NCB-20 (Maricq et al., 1991). RNA transcripts derived from this clone, when injected into *Xenopus laevis* oocytes, drive the expression of a homooligomeric receptor with properties that closely resemble those of 5-HT_3 receptors native to the NCB-20 and other murine clonal cell lines. Sequence analysis of the 2131 base pair cDNA encoding the 5-HT_3 R-A predicts a protein of 487 amino acids length (Figure 2) with a deduced mol wt of 55,966 Da. Taking the signal peptide sequence into account, the mol wt of the mature protein is predicted to be 53,509 Da. The 5-HT_3 R-A exhibits sequence similarity to members of the ligand-gated ion channel family. For example, the α subunit of the nicotinic acetylcholine receptor (nAChR) of *Torpedo* and the α7 subunit of the nAChR of chick brain are, respectively, 27 and 30% identical in sequence to the 5-HT_3 R-A (Maricq et al., 1991; Couturier et al., 1990). A sequence identity of 22% exists between the 5-HT_3 R-A and either the β1 subunit of the bovine $GABA_A$ receptor or the 48K subunit of the rat glycine receptor. Hydrophobicity analysis, and the presence of an N-terminus signal peptide in the precursor polypeptide, suggests the 5-HT_3 R-A subunit to have a topological organization typical of the ligand-gated ion channel family, exemplified by the nAChR. Four hydrophobic transmembrane regions (M1–M4), with a long cytoplasmic loop between M3 and M4 that contains potential regulatory sites and a large extracellular N-terminal domain, have been postulated (Maricq et al., 1991). The latter, which presumably contains the binding site for agonists and some antagonists, presents three putative N-glycosylation sites, consistent with the glycoprotein nature of the purified 5-HT_3 receptor (Miquel et al., 1990; McKernan, 1992). By analogy to the nAChR (Galzi et al., 1991), it has been assumed that the M2 region of the 5-HT_3 R-A exists as an α-helix, which contributes to the lining of the cation selective ion channel.

A recent study (Eiselé et al., 1993) upon a chimaeric protein construct embodying the amino-terminal domain of the nAChR α7 subunit and the putative transmembrane and carboxy-terminal domains of the 5-HT_3 R-A, strongly supports the general features of the topological organization outlined above. In brief, the chimara functioned as a homooligomeric complex when expressed in *Xenopus* oocytes and convincingly combined the pharmacological features of the α7 nicotinic receptor with the distinctive channel proper-

```
MRLCIPQVLLALFLSMLTAPGEGSRRRATQEDTTQPALLRLSDHLLANYKKGVRPVRDW 59
        Signal Sequence
RKPTTVSIDVIMYAILNVDEKNQVLTTYIWYRQYWTDEFLQWTPEDFDNVTKLSIPTDS 118
                                                ▼
IWVPDILINEFVDVGKSPNIPYVYVHHRGEVQNYKPLQLVTACSLDIYNFPFDVQNCSL 177
                                          S              ▼S
TFTSWLHTIQDINITLWRSPEEVRSDKSIFINQGEWELLEVFPQFKEFSIDISNSYAEM 236
            ▼
KFYVIIRRRPLFYAVSLLLPSIFLMVVDIVGFCLPPDSGERVSFKITLLLGYSVFLIIV 295
                   M1               ▲            M2
SDTLPATIGTPLIGVYFVVCMALLVISLAETIFIVRLVHKQDLQRPVPDWLRHLVLDRI 354
               M3
AWILCLGEQPMAHRPPATFQANKTDDCSGSDLLPAMGNHCSHVGGPQDLEKTPRGRGSP 413
                                                         ◆
LPPPREASLAVRGLLQELSSIRHFLEKRDEMREVARDWLRVGYVLDRLLFRIYLLAVLA 472
                                          ◇
YSITLVTLWSIWHYS                                             487
 M4
```

FIGURE 2. The primary structure of a cloned murine 5-HT_3 receptor subunit. The amino acid sequence (single letter code) of the 5-HT_3 R-A subunit, cloned from the murine hybridoma cell line NCB-20 (Maricq et al., 1991) is illustrated with numbering from the first methionine residue. The proposed signal sequence as well as the putative transmembrane domains (M1–M4) are underlined with a solid line. In the large extracellular loop, the putative N-linked glycosylation sites (Asn-X-Ser/Thr, see Pless and Lennarz, 1977) are indicated (▼), as well as the position of the cysteine–cysteine loop (S–S). Also highlighted are three potential phosphorylation sites located within the proposed intracellular domains: (▲) a casein kinase phosphorylation site, (◆) a protein kinase A phosphorylation site, and (◇) a tyrosine kinase phosphorylation site. The six amino acids deleted from the intracellular loop of a possible splice variant of the cloned murine 5-HT_3 receptor subunit, 5-HT_3 R-A (Hope et al., 1993), are indicated in a box.

ties of the 5-HT_3 R-A. The significance of this study is considered further below (Section 2.5.4.2).

A functional 5-HT_3 receptor ion channel probably consists of a pentameric arrangement of receptor subunits. At least for the 5-HT_3 receptor purified from NG 108-15 cells, there is some evidence for such a multimeric arrangement, since electron microscopy of uranyl acetate-stained material reveals rosette structures of 8 to 9 nm diameter that contain a centrally located stain-filled pore of approximately 2 nm diameter (Boess et al., 1992b). Such a pattern is reminiscent of that seen for the nAChR of *Torpedo* (Cartaud et al., 1978).

Unlike most other ligand-gated ion channels, for which a large number of structurally distinct subunits are known, the number of 5-HT_3 receptor subunits cloned to date is very limited. Hope et al. (1993) employed the polymerase chain reaction (PCR) to clone a cDNA encoding a receptor subunit from the neuroblastoma cell line N1E-115. RNA transcripts derived from the cloned DNA were shown to direct the synthesis of functional homooligomeric 5-HT_3 receptors in the *Xenopus* oocyte expression system. Sequence analysis of the cloned DNA predicted a mature peptide of 460 amino acids, with a molecular mass of 53,178 Da, and a putative signal sequence of 23 amino acids. When aligned with 5-HT_3 R-A, the clone exhibited 98% sequence identity. The principal difference between the two cloned receptors was found to lie in the deletion of 6 consecutive amino acid residues from the second cytoplasmic loop of the subunit cloned from the N1E-115 cell line (Hope et al., 1993; see Figure 2). Subsequent analysis of RNA isolated from NCB-20 and N1E-115 cells revealed that both forms of the receptor subunit are expressed in both cell lines. Recently, the two forms of the clone were detected in the hybridoma cell line NG 108-15 (Werner et al., 1993). It is likely that both the full length subunit and the subunit deleted of the six amino acids (termed 5-HT_3 R-A_s; Hope et al., 1993) arise from the alternative splicing of RNA transcripts from a single gene. The functional significance of the deletion is discussed below. Sequence and PCR analysis of 5-HT_3 receptor cDNAs isolated from NG 108-15 cells suggests that approximately 80% encode the "short" (i.e., 5-HT_3 R-A_s) form of the receptor subunit. A similar conclusion was reached from an analysis (using reverse transcriptase and PCR) of mRNA extracted from cortex and brain stem (Werner et al., 1993).

In addition to the receptor subunit isolated from the neuronal cell lines, a cDNA encoding a 5-HT_3 receptor, again functional as a homooligomer when expressed in *Xenopus laevis* oocytes, has been cloned from a rat superior cervical ganglion cDNA library (Johnson and Heinemann, 1992; Isenberg et al., 1993). The rat subunit is very similar in overall structure to previously isolated clones because the mature peptide comprising 461 amino acids exhibits 95% sequence homology with the mouse 5-HT_3 R-A_s and in common with the latter lacks the six consecutive amino acids found in the mouse 5-HT_3 R-A (Isenberg et al., 1993). The putative N-glycosylation and potential phosphorylation sites reported for the 5-HT_3 R-A are conserved within the rat homologue.

2.5.4.2 Functional Studies Performed on Cloned 5-HT_3 Receptors

The 5-HT_3 R-A heterologously expressed in *Xenopus laevis* oocytes shares many of the properties of the 5-HT_3 receptor native to the NCB-20 cell line from which the cDNA was cloned. Thus, 5-HT elicited a rapidly activating and desensitizing inward current response that reversed in sign at a potential close to zero millivolts (Maricq et al., 1991). Additionally, the large negative shift in $E_{5\text{-HT}}$ obtained upon substitution of extracellular Na by Tris suggests the homooligomeric complex to retain selectivity toward cations. Similar conclusions have been drawn from ion substitution experiments performed on HEK 293 cells stably transfected with the 5-HT_3 R-A since substitution of extracellular Na^+ by *N*-methyl-D-glucamine, or internal replacement of internal K^+ by TEA, resulted in negative and positive shifts in $E_{5\text{-HT}}$, respectively (Gill, Peters, and Lambert, unpublished observations).

Pharmacologically, the cloned receptor displayed the profile of activation and blockade by selective agonists and antagonists, respectively, that typify the 5-HT_3 receptor. Interestingly, (+)-tubocurarine, which is an unusually potent antagonist of 5-HT_3 receptors expressed in murine clonal cell lines (Yakel and Jackson, 1988; Peters et al., 1990; Yang, 1990) and mouse hippocampal and nodose ganglion neurons (Yakel and Jackson, 1988; Malone et al., 1991b), also blocked, at low nanomolar concentrations, the inward current response mediated by the 5-HT_3 R-A (Maricq et al., 1991; Downie et al., 1994). Furthermore, trichloroethanol enhanced 5-HT_3 R-A-mediated currents in a manner similar to that reported for native 5-HT_3 receptors (Downie et al., 1993; Lovinger and Zhou, 1993).

The apparent splice variant of the 5-HT_3 R-A (i.e., the 5-HT_3 R-A_s) cloned from N1E-115 cells (Hope et al., 1993) shares the above properties (Downie et al., 1994). Table 2 compares quantitatively the influence of agonists and antagonists upon the 5-HT_3 R-A and 5-HT_3 R-A_s expressed in *Xenopus* oocytes and the receptor native to N1E-115 cells. Briefly, the potencies of all agonist and antagonist compounds tested upon the 5-HT_3 R-A were very similar to those determined under identical conditions for the 5-HT_3 R-A_s. However, unlike the other agonists examined, 2-methyl-5-HT did exert a differential effect at the two homooligomer complexes. Acting at the 5-HT_3 R-A, a maximally effective concentration of 2-methyl-5-HT evoked a current response that amounted to 63% of the maximum effect obtained with 5-HT. The corresponding value obtained with the 5-HT_3 R-A_s was only 9% of the 5-HT-induced maximum (Downie et al., 1994). Qualitatively similar results have been obtained with 2-methyl-5-HT in experiments performed on 5-HT_3 R-A and 5-HT_3 R-A_s subunits expressed in HEK 293 cells (Sepúlveda and Lummis, 1994).

In preliminary experiments where the conductance of the 5-HT_3 R-A expressed in HEK 293 cells has been assessed by fluctuation analysis (Peters, Gill, Lambert, and Julius, unpublished observations), a value (<1 pS) similar to that found for native 5-HT_3 receptors in murine cell lines was obtained (see Section 2.5.2.5). It would appear then that homooligomeric receptors formed from either the 5-HT_3 R-A or 5-HT_3 R-A_s subunits reproduce, in a large part, the properties of the relevant native 5-HT_3 receptors. However, some potentially important differences have also been noted. In the original report of Maricq et al. (1991), Ca^{2+} and Mg^{2+} were found to suppress the response to 5-HT in a voltage-dependent manner, introducing a region of negative slope conductance into the current–voltage relationship at hyperpolarized potentials. A similar effect of Ca^{2+} upon currents mediated by the 5-HT_3 R-A subunit has been reported by Eisele et al. (1993) and was in fact considered a reliable indicator 5-HT_3 receptor channel properties in the chimaeric $\alpha7$ nictotinic/5-HT_3 R-A receptor constructs discussed above. Blockade by divalent cations also occurs at native 5-HT_3 receptors (see Section 2.5.2.4), but here the effect is voltage independent (Peters et al., 1988; Yang, 1990). Unfortunately, the significance of this difference in voltage sensitivity is difficult to assess. Thus, although homooligomeric receptors formed from the

Table 2. A comparison of the Pharmacological Properties of the Homooligomeric 5-HT_3 R-A_S Expressed in *Xenopus* Oocytes with the 5-HT_3 Receptor Native to N1E-115 Neuroblastoma Cells[a]

	5-HT_3 R-A_S[b]			5-HT_3 N1E-115[c]		
Compound	pEC_{50} or pIC_{50}	nH	α	pEC_{50} or pIC_{50}	nH	α
5-HT	5.63*	2.2	1	5.38*	2.3	1
2-Methyl-5HT	4.83*	2.2	0.09	4.76*	1.2	0.2
m-Chlorophenylbiguanide	6.06*	1.7	0.84	5.85*	1.5	1
Phenylbiguanide	4.52*	1.7	0.76	ND	ND	ND
Granisetron	9.85	—	—	9.90		
Ondansetron	8.92	—	—	9.60		
(+)-Tubocurarine	8.85	—	—	9.09		
Metoclopramide	7.14	—	—	7.62		
Cocaine	5.63	—	—	5.14		

Note: * = pEC_{50} values

[a] $pEC_{50} = -\log_{10}EC_{50}(M)$; $pIC_{50} = -\log_{10}IC_{50}(M)$; nH, Hill coefficient; α, efficacy (5-HT = 1); ND = not determined.

[b] Data from Hope et al. (1993) and Downie, Peters, and Lambert (unpublished observations).

[c] Data from Sepulveda et al. (1991) and Peters et al. (1991b, 1993a).

subunit cloned from rat sympathetic neurons are also reported to be voltage dependently blocked by Ca^{2+} (Johnson and Heinemann, 1992), no such effect was observed by Yakel et al. (1993) for the 5-HT_3 R-A expressed in *Xenopus* oocytes. Indeed, in the latter study Ca^{2+} had no consistent influence upon the peak amplitude of the 5-HT-induced current but did accelerate its desensitization in a voltage-dependent manner (see Section 2.5.2.2). Adding further complexity, for the 5-HT_3 R-A expressed in HEK 293 cells, blockade by both Ca^{2+} and Mg^{2+} has been observed, but the effect is voltage independent (Gill et al., 1993). It might also be pertinent to note one additional discrepancy observed with Zn^{2+}, which at the native NCB-20 cell 5-HT_3 receptor produces voltage-dependent blockade (Lovinger, 1991), yet acting at low concentration at the 5-HT_3 R-A in HEK 293 cells enhances the response, blockade being observed only with higher concentrations of Zn^{2+} (Gill et al., 1993). Establishing whether or not cloned homooligomeric 5-HT_3 receptors truly differ from native receptors with regard to modulation by divalent cations warrants further investigation.

2.5.4.3 The Relationship of 5-HT_3 Receptor Amino Acid Sequence to Function

The techniques of site-directed mutagenesis and affinity labeling have greatly advanced our understanding of how the amino acid sequence of ligand-gated ion channels relates to their function. With the exception of one recent study (Yakel et al., 1993), these approaches have not been applied to the 5-HT_3

receptor. Hence, the relationship between sequence and function must largely be inferred from results obtained with related receptors, particularly nicotinic subunits.

Experiments conducted upon a chimaric subunit comprising the N-terminal region of the nicotinic α7-subunit and the C-terminal and transmembrane domains of the 5-HT_3R-A have demonstrated the expressed receptor to possess nicotinic pharmacology and to be insensitive to 5-HT (Eiselé et al., 1993). This result strongly suggests that at least a part of the agonist binding site of the 5-HT_3 receptor is located within the hydrophilic N-terminal domain. This region of all nicotinic, $GABA_A$, glycine, and 5-HT_3 receptor subunits presents two invariant cysteine residues separated by 13 amino acid residues (Figure 3). The two cysteine residues are thought to form a disulfide bond resulting in a loop structure that may be involved in the tertiary folding of the protein. It has been argued from modeling studies (Cockcroft et al., 1990) that the Cys-Cys loop may form the agonist binding site of ligand-gated ion channels, but direct evidence for this proposal is lacking.

Studies involving the chemical modification of 5-HT_3 receptors native to NG 108-15 cells point to the involvement of tryptophan residues in ligand binding (Miquel et al., 1991). The density, but not affinity, of sites labeled with the 5-HT_3 receptor antagonist [^{3}H]zacopride was greatly reduced in membrane homogenates that had been pretreated with *N*-bromosuccinimide, an agent that selectively oxidizes tryptophan residues. Binding was preserved in membranes pre-exposed to saturating concentrations of some 5-HT_3 receptor agonists and antagonists prior to and during incubation with the reducing agent. However, not all compounds tested afforded such protection, which may indicate subtle differences in ligand binding mechanisms at the 5-HT_3 receptor. In preliminary studies, agents modifying cystine, cysteine, aspartate, and glutamate residues had a minimal impact upon ligand binding whereas those targeted against histidine, tyrosine, and arginine residues exerted a modest depression of labeling by [^{3}H]zacopride (Miquel et al., 1991).

While chemical modification studies suggest the contribution of certain classes of amino acid to the formation of the ligand binding site, the precise amino acid residues involved are unknown. Broad clues as to their identity might, however, be obtained by comparing the primary amino acid sequences of the extracellular N-terminal of 5-HT_3, nicotinic, $GABA_A$, and glycine receptor subunits, since recent evidence suggests that agonist and antagonist recognition sites of these ligand gated ion channels may exist in homologous domains. Such a comparison (Changeux et al., 1992; Figure 4) indicates the presence of strictly conserved canonical amino acids that have been postulated to exist within, or in close proximity to, three loops of amino acids involved in ligand binding.

A frequently cited example is provided by the carbamylcholine-sensitive covalent labeling of *Torpedo* nicotinic α-subunits by the photolabile antagonist *p*-N,N-dimethylaminobenzene diazonium fluroborate (DDF), which occurs within three distinct regions (loops A, B, and C) of the N-terminal domain. Residues strongly labeled within these areas are Tyr-93 (loop A), Trp-149 (loop B), Tyr-190, Cys-192, and Cys-193 (loop C), and, more

A

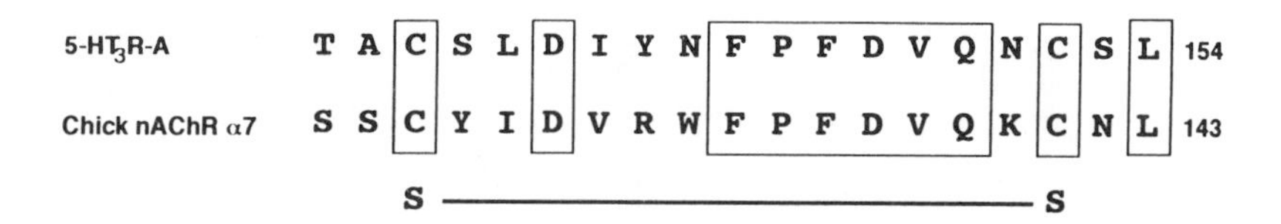

B

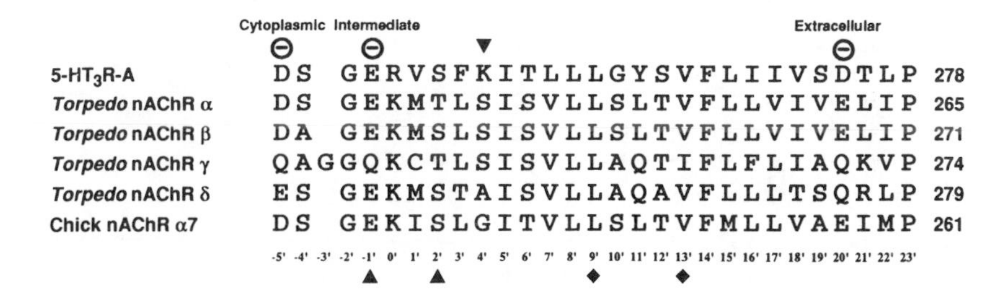

FIGURE 3. Sequence similarities within the "cysteine–cysteine loop" and the putative transmembrane domain of the 5-HT$_3$ R-A subunit. (A) The amino acid sequence of the cysteine–cysteine loop (S–S) region of the cloned α7-subunit of the chick nACh receptor (Couturier et al., 1990) is compared with that of the cloned 5-HT$_3$ R-A subunit (Maricq et al., 1991). This domain represents a signature feature of ligand-gated receptor ion-channel proteins and is highly conserved in both the subunits illustrated. Regions of sequence identity are highlighted in boxes. (B) The amino acid sequence (single letter code) from the M2 domain of the 5-HT$_3$ R-A subunit is shown aligned with the corresponding regions of cloned nAChR subunits. The relative positions of individual amino acids are numbered underneath as by Hille (1992). The conserved negatively charged amino acids (Imoto et al., 1988) are indicated by (⊖). Amino acid residues thought to constitute the selectivity filter of neurotransmitter-gated ion channels are indicated by (▲) (Konno et al., 1991, Villaroel et al., 1992). The conserved leucine residues, which in the chick α7 homooligomeric AChR are implicated in determining both desensitization kinetics and pharmacological properties (Revah et al., 1991, Bertrand et al., 1992), are represented by (♦). Also indicated (♦), is a conserved valine residue that in the nAChR α7 also contributes to agonist-induced desensitization (Galzi et al., 1992). A positively charged lysine residue that is unique to the 5-HT$_3$ R-A subunit is shown (▼).

weakly, Trp-86 (loop A), Tyr-151 (loop B), and Tyr-198 (loop C). The contribution of these residues, either individually or in groups, to the ligand binding site has been amply confirmed by their labeling with structurally distinct ligands or by functional studies performed upon receptors modified

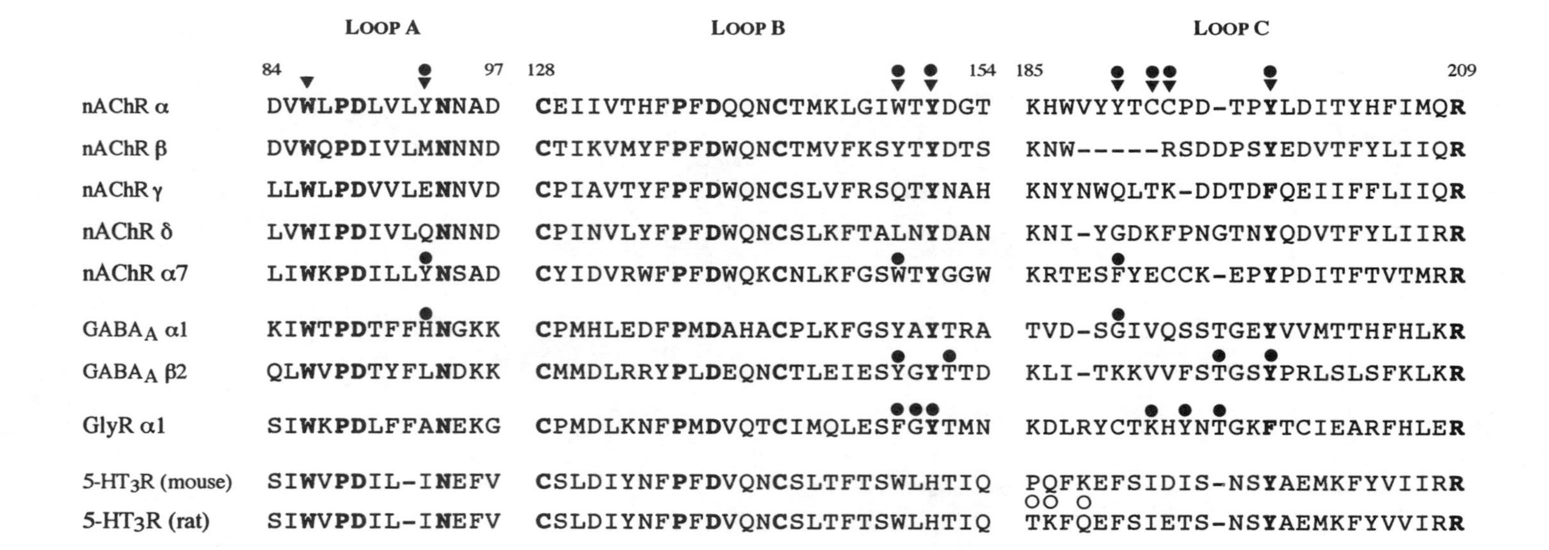

FIGURE 4. A comparison of the primary amino acid sequences of nicotinic, $GABA_A$, glycine, and 5-HT_3 receptor subunits over regions of the hydrophilic N-terminal domain (loops A, B, and C) implicated in ligand binding. The sequences illustrated (from top to bottom) are *Torpedo* nicotinic, α-, β-, γ-, and δ-subunits, chick nicotinic α7-subunit; rat $GABA_A$ α1- and β2-subunits, human glycine α1-subunit, and mouse and rat 5-HT_3 R-A subunits. Conserved canonic amino acids are shown in bold lettering. The positions of DDF-labeled residues in the *Torpedo* α-subunit are indicated by the inverted triangles. Filled circles highlight residues that have been demonstrated, from mutagenesis studies, to influence ligand binding (or receptor activation). Open circles indicate differences in the sequences of the mouse and rat 5-HT_3 R-A in loop C. Numbering applies to the *Torpedo* α-subunit. The sequence comparison is based on that performed by Changeux et al. (1992).

by site-directed mutagenesis (see Changeux et al., 1992 for a review).

A number of recent investigations conducted upon $GABA_A$ and glycine receptor subunits implicate residues located within domains corresponding to the A, B, and C loops of nicotinic α-subunits in ligand binding phenomena. Indeed, in some cases, the residues identified are homologous to those labeled by DDF (Figure 3). For example, a histidine residue (His-101) in α1 and at an equivalent position in α2, α3, and α5 $GABA_A$ receptor subunits is the homologue of Tyr-93, and is a major determinant of the high-affinity binding of benzodiazepines to the $GABA_A$ receptor (Wieland et al., 1992). The characteristics of benzodiazepine binding are also influenced by Gly-200 in the α1 $GABA_A$ subunit, which is homologous to Tyr-190 (Pritchett and Seeberg, 1991). Amino acids present in the β2-subunit of the $GABA_A$ receptor that influence the activation of a heterooligomeric complex of α1-, β2-, and γ2-subunits by GABA and muscimol have been identified as Tyr-157, Thr-160, Thr-202, and Tyr-205 (Amin and Weiss, 1993). These residues reside in binding domains that correspond to the B and C loops of nicotinic α-subunits, and two (Tyr-157 and Tyr-205) are the homologues of DDF-labeled Trp-149 and Tyr-198. Amino acids that act as important determinants of ligand binding in glycine α1 receptor subunits occur at locations that correspond to loops B and C and, once more, several of these (i.e., Phe-159, Tyr-161, and Lys-200) can be viewed as homologues of residues that are affinity labeled in nicotinic α-subunits (Vandenberg et al., 1992; Schmieden et al., 1993).

The knowledge gained from mutagenesis studies performed upon other ligand-gated ion channels may guide initial attempts to map the agonist and antagonist binding site(s) of the 5-HT_3 receptor, especially since three of the DDF-labeled residues (i.e., Trp-86, Trp-149, and Tyr-198) are conserved within the 5-HT_3 R-A (i.e., Trp-98, Trp-160, and Tyr-211; 5-HT_3 R-A_s numbering) and tryptophan residues are already implicated in ligand binding (Miquel et al., 1991). The pharmacological properties of the 5-HT_3 receptor are highly species dependent (Peters et al., 1992, 1994). The isolation and sequencing of pharmacologically diverse 5-HT_3 receptor subunits from different species should also assist in the identification of those amino acids that constitute the agonist and antagonist binding. For example, a comparison of the mouse and rat homologues of the 5-HT_3 R-A_s, which differ at least 100-fold in their sensitivity to blockade by (+)-tubocurarine (Johnson and Heinemann, 1992; Hope et al., 1993), reveals only 16 differences in primary amino acid sequence within the N-terminal domain (Hope et al., 1993; Isenberg et al., 1993). Notably, three of these cluster in a region corresponding to the loop C of nicotinic α-subunits (Figure 4). This domain, along with residues contained within the δ- and γ-subunits of the *Torpedo* receptor (Pedersen and Cohen, 1990), has been specifically implicated in the binding of (+)-tubocurarine (Cohen et al., 1991, Filatov et al., 1993). While none of the amino acid differences between the mouse and rat homologues of the 5-HT_3 receptor correspond exactly to residues currently known to influence the binding of (+)-tubocurarine to nicotinic receptors (Figure 4), this domain represents an obvious target for mutagenesis studies.

Mutagenesis, allied with the electrophysiological recording techniques, has been instrumental in furthering our understanding of the relationship between the amino acid sequence of the M2 region of nicotinic and glutamate receptor channels and their operation. For the 5-HT_3 receptor, these studies are at an early stage (Yakel et al., 1993). However, the alignment of the amino acid sequence of the putative M2 region of the 5-HT_3 R-A or 5-HT_3 R-A_s with nicotinic receptor subunits reveals considerable sequence homology (Figure 4). Hence, by analogy to the nicotinic receptor, predictions can be made as to the likely role of amino acids conserved across these receptor subunits. Whether the results obtained with nicotinic receptors will indeed generalize to the 5-HT_3 receptor clearly remains to be determined.

The chick α7 nicotinic subunit, like the 5-HT_3 receptor subunit, expresses well as a homooligomeric complex (Gasic and Heinemann, 1992). Features common to the M2 regions of the two receptors include the presence of acidic residues which, in the homooligomeric complex, would constitute cytoplasmic (labeled –5′), intermediate (–1′) and extracellularly (20′) located rings of negative charge bracketing the M2 region, and the occurrence of two rings (2′, 6′) of amino acids with side chains bearing hydroxyl groups. A third hydroxyl ring located at position 10′ of the nicotinic subunit is not evident in the 5-HT_3 R-A (Figure 4).

Mutations of the homologous acidic residues found in nicotinic AChR subunits of *Torpedo* (–5′, –1′, and 20′), that result in a decrease in the net negative charge associated with each ring, depress single channel conductance and alter the pattern of rectification observed in the current–voltage relationship (Imoto, 1993; Imoto et al., 1988). These effects, which are observed in the absence of divalent cations, are most pronounced for mutations directed toward the intermediate ring and preferentially affect the conduction of large alkali metal cations (i.e., Cs^+ and Rb^+; Konno et al., 1991). A recent study, performed on recombinant nicotinic receptors assembled from murine (BC3H-1) α-, β-, δ- and γ-subunits, has confirmed the initial results obtained with the *Torpedo* receptor but has also revealed that a simple, unmodified, electrostatic mechanism cannot adequately account for the effects of the mutations upon channel conductance (Kienker et al., 1994).

In rat muscle nicotinic receptor subunits, substitution by glycine, valine, or alanine of polar amino acids corresponding to the ring denoted 2′ (Figure 4) increases or decreases channel conductance to K^+ relative to wild-type receptors (Villaroel et al., 1992). The effect upon conductance correlates with the volume of the side chain of the substituted amino acid and, in the case of reduction of conductance, is emphasized when the permeant species is an alkali cation possessing a relatively large crystal radius (i.e., Cs^+ and Rb^+). Such observations are consistent with the notion that residues at the 2′ position contribute to a constriction of the channel that limits ion flow, and to some extent acts as a selectivity filter. This idea is supported by recent reports that have demonstrated the permeability of Tris and some large organic cations, relative to Na^+, to be dependent upon the residues located in the 2′ position. However, in these studies, emphasis was placed on the chemical nature of the side chain, rather than its volume, as the major influence upon ion flow (Cohen et al., 1992). As mentioned above the, $5\text{-}HT_3$ receptor–channel complex is permeable to divalent cations (Section 2.5.2.4). Recent work has demonstrated that for the α7 nicotinic subunit, the permeability of divalent cations relative to sodium is greatly increased by the mutation of a threonine residue (6′) in the M2 region, to a negatively charged aspartic acid residue (Ferrer-Montiel and Montal, 1993; Figure 4). It is, therefore, of interest that a threonine residue also occupies this position (6′) in the $5\text{-}HT_3$ receptor subunit (Figure 4).

The similarities in location of key residues in the $5\text{-}HT_3$ and nicotinic receptor subunits may well explain why these two receptor classes share many functional characteristics. However, one striking difference, the exceptionally low conductance of the $5\text{-}HT_3$ receptor native to the N1E-115 and N18 cell lines (Lambert et al., 1989; Yang 1990) and of HEK 293 cells stably transfected with the $5\text{-}HT_3$ R-A (Peters, Lambert, Gill, and Julius, unpublished observations), is not readily explained by such comparisons. It is intriguing that a basic lysine residue resides at the 4′ position of the putative M2 domain of the $5\text{-}HT_3$ R-A. Assuming that M2 exists as an α-helix, and further that the negatively charged and polar residues discussed above face the channel lumen, this would result in the energetically unfavorable placement of lysine away from the pore. It has been argued for the $5\text{-}HT_3$ R-A that an interaction between the lysine residue and an appropriately located acidic aspartate residue within the adjacent M1 domain would perhaps stabilize this orientation (Maricq et al., 1991). However, if the M2 domain were to exist (perhaps locally) as a β-strand, as has recently been suggested for a mouse nAChR subunit (Akabas et al., 1992), the likely location of lysine within the conduction pathway would be expected to influence channel conductance substantially. Clearly, replacement of lysine by amino acids occupying homologous positions in nicotinic subunits would be of interest. As discussed elsewhere (Section 2.5.2.5) heterogeneity in the biophysical properties of the $5\text{-}HT_3$ receptor, which may or may not be species dependent, is evident from the measurements of the conductance of the channel integral to the receptor (e.g., an approximately 60-fold difference between the $5\text{-}HT_3$ receptors of N1E-115 cells and rabbit nodose ganglion neurons). The future isolation of cDNAs from preparations expressing diverse single channel properties should clarify the molecular determinants of ion transport through this channel.

Recent studies upon the chick α7 nicotinic subunit have produced the unexpected finding that the replacement by threonine of either a leucine residue located at position 9′, or a valine residue at position 13′, decreases the rate of onset of desensitization, and increases the apparent affinity of acetylcholine (Bertrand et al., 1992; Galzi et al., 1992). Remarkably, antagonists of the wild-type receptor act as agonists at the mutated α7 homooligomer by preferentially activating a novel high conductance channel state that is also elicited by low concentrations of acetylcholine. Such results have been interpreted as revealing a desensitized state of the receptor with high affinity to antagonists that, as a result of the leucine–threonine or valine–threonine exchange, now produces a functional response (Bertrand et al., 1992; Galzi et al., 1992). Both the valine (13′) and the leucine (9′) residues are conserved in the 5-HT_3 receptor subunits (Figure 3). To date the possible role of valine (13′) has not been investigated, however, a recent study has determined the influence of leucine (9′) (Yakel et al., 1993). The substitution of leucine (9′) for threonine had little effect on 5-HT_3 affinity, nor did 5-HT_3 receptor antagonists activate the mutant receptor (Yakel et al., 1993). However, as for the α7 nicotinic subunit, the rate of desensitization was reduced by this mutation, and enhanced by the replacement of leucine 9′ by phenylalanine (Yakel et al., 1993). These results suggest that the conformational changes leading to receptor desensitization may be similar for these two ligand-gated ion channels.

It is now established that the functioning of many ligand-gated ion channels including nicotinic, glutamate, glycine, and $GABA_A$ can be modulated by phosphorylation (Raymond et al., 1993; Swope et al., 1992). Such receptors are proposed to contain a large cytoplasmic loop that links transmembrane regions three and four and contains a number of consensus sites for protein phosphorylation (Raymond et al., 1993; Swope et al., 1992). To date the functional influence on the 5-HT_3 receptor of second messengers such as cAMP is not clear (Peters et al., 1994; Yakel, 1992; Yang, 1990). However, the recent cloning of 5-HT_3 receptor subunits (Maricq et al., 1991; Hope et al., 1993) has identified on the large intracellular loop consensus sequences for phosphorylation by protein kinase A and tyrosine kinase. In addition, there is a putative phosphorylation site for caesin kinase on the short cytoplasmic loop between transmembrane regions M1 and M2. The operational consequences, if any, of such sites should be established by future mutagenesis studies.

2.5.5 Conclusion

A significant development in the field of 5-HT_3 receptor research was made by the cloning of a functional 5-HT_3 receptor subunit (Maricq et al., 1991). The combination of electrophysiology and molecular biology has greatly enhanced our understanding of the pharmacological and biophysical properties of nicotinic (Karlin, 1993), $GABA_A$ (Luddens and Wisden, 1991), and glutamate (Sommer and Seeburg, 1992) receptors and we can now look forward to their successful application to the 5-HT_3 receptor. In this respect, it is of interest that a number of amino acids known to play an essential role in the functioning of the closely related nicotinic receptor subunits (Bertrand et al., 1992, Galzi et al., 1992, Karlin, 1993) occupy equivalent positions in the 5-HT_3 R-A clone.

The pharmacological, and possibly the biophysical, properties of 5-HT_3 receptors are highly species dependent (Peters et al., 1992, 1994). From a therapeutic stand point, such studies emphasize the importance of cloning and functionally expressing human 5-HT_3 receptor subunits. However, by comparing the amino acid sequences of functionally diverse subunits, they also present us with a unique opportunity to identify amino acid domains involved in ligand binding and ion conductance.

Molecular biology has revealed a rich subunit diversity for nicotinic, $GABA_A$, and glutamate receptors that was not evident from studies utilizing more traditional pharmacological, physiological, and biochemical techniques. Similarly, using these conventional approaches, the evidence for $5\text{-}HT_3$ receptor subtypes within a species is fragmentary (Peters et al., 1992; Hussy and Jones, 1993), but would seem likely in view of the heterogeneity exhibited by other members of the ligand-gated ion channel family. The application of the experimental approaches utilized so successfully for other ligand-gated ion channels should ensure that the $5\text{-}HT_3$ receptor will not remain the poor relation of the family for much longer.

Acknowledgments

Some of the work reported here was supported by the Wellcome Trust (U.K.), SmithKline Beecham (Marlow, U.K.), and Glaxo (Ware, U.K.). A. G. Hope was supported by an S.E.R.C. studentship. We thank Gillian Thomson and Denise Lamond for typing the manuscript, Dr. Ross McGuire, Organon (Newhouse, U.K.) for calculating the radii of organic cations, and Drs. H. Vijverberg, A. Kooyman, and H. Malone for providing unpublished observations.

REFERENCES

Adams, D. J., Dwyer, T. M., and Hille, B. (1980). The permeability of endplate channels to monovalent and divalent metal cations. *J. Gen. Physiol.* 75, 493–510.

Akabas, M. H., Stauffer, D. A., Xu, M., and Karlin, A. (1992). Acetylcholine receptor channel structure probed in cysteine-substitution mutants. *Science* 258, 307–310.

Amin, J. and Weiss, D. S. (1993). $GABA_A$ receptor needs two homologous domains of the β-subunit for activation by GABA but not by pentobarbital. *Nature* 366, 565–569.

Armstrong, D., Dry, R. M. L., Keele, C. A., and Markham, J. W. (1953). Observations on chemical excitants of cutaneous pain in man. *J. Physiol.* 120, 326–351.

Bachy, A., Heaulme, M., Giudice, A., Michaud, J.-C., Lefevre, I. A., Souilhac, J., Manara, L., Emerit, M. B., Gozlan, H., Hamon, M., Keane, P. E., Soubrie, P., and LeFure, G. (1993) SR 57227A: A potent and selective agonist at central and peripheral $5\text{-}HT_3$ receptors *in vitro* and *in vivo*. *Eur. J. Pharmacol.* 237, 299–309.

Barnes, J. M., Barnes, N. M., Costall, B., Naylor, R. J., and Tyers, M. B. (1989). $5\text{-}HT_3$ receptors mediate inhibition of acetylcholine release in cortical tissue. *Nature* 338, 762–763.

Bertrand, D., Devillers-Thiéry, A., Revah, F., Galzi, J.-L., Hussy, N., Mulle, C., Bertrand, S., Ballivet, M., and Changeux, J.-P. (1992). Unconventional pharmacology of a neuronal nicotinic receptor mutated in the channel domain. *Proc. Natl. Acad. Sci. U.S.A.* 89, 1261–1265.

Bianchi, C., Siniscalchi, A., and Beani, L. (1990). $5\text{-}HT_{1A}$ agonists increase and $5\text{-}HT_3$ agonists decrease acetylcholine efflux from cerebral cortex of freely-moving guinea-pigs. *Br. J. Pharmacol.* 101, 448–452.

Blandina, P., Goldfarb, J., Craddock-Royal, B., and Green, J. P. (1989). Release of endogenous dopamine by stimulation of 5-hydroxytryptamine$_3$ receptors in rat striatum. *J. Pharmacol. Exp. Ther.* 251, 803–809.

Blandina, P., Goldfarb, J., Walcott, J., and Green, J. P. (1991). Serotonergic modulation of the release of endogenous norepinephrine from rat hypothalamic slices. *J. Pharmacol. Exp. Ther.* 256, 341–347.

Blier, P. and Bouchard, C. (1993). Functional characterization of a 5-HT$_3$ receptor which modulates the release of 5-HT in the guinea-pig brain. *Br. J. Pharmacol.* 108, 13–22.

Boess, F. G., Sepulveda, M.-I., Lummis, S. C. R., and Martin, I. L. (1992a). 5-HT$_3$ receptors in NG 108-15 neuroblastoma × glioma cells: Effect of the novel agonist 1-(m-chlorophenyl)-biguanide. *Neuropharmacology* 31, 561–564.

Boess, F. G., Lummis, S. C. R., and Martin, I. L. (1992b). Molecular properties of 5-hydroxytryptamine$_3$ receptor-type binding sites purified from NG108-15 cells. *J. Neurochem.* 59, 1692–1701.

Butler, A., Hill, J. M., Ireland, S. J., Jordan, C. C., and Tyers, M. B. (1988). Pharmacological properties of GR38032F, a novel antagonist at 5-HT$_3$ receptors. *Br. J. Pharmacol.* 94, 397–412.

Butler, A., Elswood, C. J., Burridge, J., Ireland, S. J., Bunce, K. T., Kilpatrick, G. J., and Tyers, M. B. (1990). The pharmacological characterization of 5-HT$_3$ receptors in three isolated preparations derived from guinea-pig tissues. *Br. J. Pharmacol.* 101, 591–598.

Cachelin, A. B. and Colquhoun, D. (1989). Desensitization of the acetylcholine receptor of frog end-plates measured in a Vaseline-gap voltage-clamp. *J. Physiol.* 415, 159–188.

Cartaud, J., Benedetti, E. L., Sobel, A., and Changeux, J.-P. (1978). A morphological study of the cholinergic receptor protein from *Torpedo mamorata* in its membrane environment and in its detergent-extracted form. *J. Cell. Sci.* 29, 313–337.

Changeux, J.-P., Galzi, J.-L., Devillers-Thiéry, A., and Bertrand, D. (1992). The functional architecture of the acetylcholine nicotinic receptor explored by affinity labelling and site-directed mutagenesis. *Q. Rev. Biophys.* 25, 395–432.

Christian, C. N., Nelson, P. G., Bullock, P., Mullinax, D., and Nirenberg, M. (1978). Pharmacological responses of cells of a neuroblastoma × glioma hybrid clone and modulation of synapses between hybrid cells and mouse myotubes. *Brain Res.* 147, 261–276.

Cockcroft, V. B., Osguthorpe, D. J., Barnard, E. A., and Lunt, G. G. (1990). Modeling of agonist binding to the ligand gated ion channel superfamily of receptors. *Proteins: Struct. Funct. Genet.* 8, 386–397.

Cohen, B. N., Labarca, C., Davidson, N., and Lester, H. A. (1992). Mutations in M2 alter the selectivity of the mouse nicotinic acetylcholine receptor for organic and alkali metal cations. *J. Gen. Physiol.* 100, 373–400.

Costall, B. and Naylor, R. J. (1992). Neuropharmacology of emesis in relation to clinical response. *Br. J. Cancer* 66, (Suppl. 19), S2–S8.

Costall, B., Naylor, R. J., and Tyers, M. B, (1990). The psychopharmacology of 5-HT$_3$ receptors. *Pharmacol. Ther.* 47, 181–202.

Couturier, S., Bertrand, D., Matter, J.-M., Hernandez, M.-C., Bertrand, S., Millar, N., Valera, S., Barkas, T., and Ballivet, M. (1990). A neuronal nicotinic acetylcholine receptor subunit (α_7) is developmentally regulated and forms a homo-oligomeric channel blocked by α-BTX. *Neuron* 5, 847–856.

Derkach, V., Surprenant, A., and North, R. A. (1989). 5-HT$_3$ receptors are membrane ion channels. *Nature* 339, 706–709.

Downie, D. L., Hope, A. G., Lambert, J. J., Peters, J. A., Burchell, B., and Julius, D. (1993). Enhancement by trichloroethanol of agonist-induced currents mediated by a cloned murine 5-HT$_3$ receptor subunit (5-HT$_3$ R-A) expressed in *Xenopus* oocytes. *Br. J. Pharmacol.* 109, 53p.

Downie, D. L., Hope, A. G., Lambert, J. J., Peters, J. A., Blackburn, T. P., and Jones, B. J. (1994). Pharmacological characterization of the apparent splice variants of the murine 5-HT_3 R-A subunit expressed in *Xenopus laevis* oocytes. *Neuropharmacology* 33, 473–482.

Dwyer, T. M., Adams, D. J., and Hille, B. (1980). The permeability of the endplate channel to organic cations in frog muscle. *J. Gen. Physiol.* 75, 469–492.

Edwards, E., Harkins, K. L., Ashby, C. R., Jr., and Wang, R. Y. (1991). The effects of 5-HT_3 receptor agonists on phosphoinositide hydrolysis in the rat frontocingulate and entorhinal cortices. *J. Pharmacol. Exp. Ther.* 256, 1025–1032.

Eiselé, J.-L., Bertrand, S., Galzi, J.-L., Devillers-Thiéry, A., Changeux, J.-P., and Bertrand, D. (1993). Chimaeric nicotinic-serotonergic receptor combines distinct ligand binding and channel specificities. *Nature* 366, 479–483.

Ferrer-Montiel, A. V. and Montal, M. (1993). A negative charge in the M2 transmembrane segment of the neuronal α_7 acetylcholine receptor increases permeability to divalent cations. *FEBS Lett.* 324, 185–190.

Feuerstein, T.-J. and Hertting, G. (1986). Serotonin (5-HT) enhances hippocampal noradrenaline (NA) release: Evidence for facilitatory 5-HT receptors within the CNS. *Naunyn-Schmiedeberg's Arch. Pharmacol.* 333, 191–197.

Filatov, G. N., Aylwin, M. L., and White, M. M. (1993). Selective enhancement of the interaction of curare with the nicotinic acetylcholine receptor. *Mol. Pharmacol.* 44, 237–241.

Fozard, J. R. (1984). MDL 72222: A potent and highly selective antagonist at neuronal 5-hydroxytryptamine receptors. *Naunyn-Schmiedeberg's Arch. Pharmacol.* 326, 36–44.

Fozard, J. R. (1989). The development and early clinical evaluation of selective 5-HT_3 receptor antagonists. In *The Peripheral Actions of 5-Hydroxytryptamine.* Fozard, J. R., Ed., Oxford: Oxford University Press, pp. 354–376.

Fozard, J. R. (1990). Agonists and antagonists of 5-HT_3 receptors. In *Cardiovascular Pharmacology of 5-Hydroxytryptamine: Prospective Therapeutic Applications.* Saxena, P. R., Kluwer, D., Wallis, D. I., Wouters, W., and Bevan, P., Eds., Kluwer Academic Publishers, Dordrecht, The Netherlands, pp. 101–115.

Fozard, J. R. and Host, M. (1982). Selective antagonism of the Bezold-Jarisch effect of 5-HT in the rat by antagonists at neuronal 5-HT receptors. *Br. J. Pharmacol.* 77, 520P.

Fozard, J. R. and Mwaluko, G. M. P. (1976). Mechanisms of the indirect sympathomimetic effect of 5-hydroxytryptamine on the isolated heart of the rabbit. *Br. J. Pharmacol.* 57, 115–125.

Frazer, A., Maayani, S., and Wolfe, B. B. (1990). Subtypes of receptors for serotonin. *Annu. Rev. Pharmacol. Toxicol.* 30, 307–348.

Furakawa, K., Akaike, N., Onodera, H., and Kogure, K. (1992). Expression of 5-HT_3 receptors in PC12 cells treated with NGF and 8-Br-c-AMP. *J. Neurophysiol.* 67, 812–819.

Gaddum, J. H. and Picarelli, Z. P. (1957). Two kinds of tryptamine receptor. *Br. J. Pharmacol. Chemother.* 12, 323–328.

Galzi, J.-L., Revah, F., Bessis, A., and Changeux, J.-P. (1991). Functional architecture of the nicotinic acetylcholine receptor: From electric organ to brain. *Annu. Rev. Pharmacol.* 31, 37–72.

Galzi, J.-L., Devillers-Thiéry, A., Hussy, N., Bertrand, S., Changeux, J.-P., and Bertrand, D. (1992). Mutations in the channel domain of a neuronal nicotinic receptor convert ion selectivity from cationic to anionic. *Nature* 353, 846–849.

Gasic, G. P. and Heinemann, S. (1992). Determinants of the calcium permeation of ligand-gated ion channels. *Current Op. Cell Biol.* 4, 670–677.

Gill, C. H., Peters, J. A., Lambert, J. J., Hope, A. G., and Julius, D. (1993). Modulation by divalent cations of current responses mediated by a cloned murine 5-HT receptor (5-HT_3 R-A) expressed in HEK 293 cells. *Br. J. Pharmacol.* 109, 98P.

Goldfarb, J., Walcott, J., and Blandina, P. (1993). Serotonergic modulation of L-glutamic acid-evoked release of endogenous noradrenaline from the rat hypothalamus. *J. Pharmacol, Exp. Ther.* 267, 45–50.

Gordon, J. C., Sarbin, N. S., Barefoot, D. S., and Pinkus, L. M. (1990). Solubilization of a 5-HT_3 binding site from rabbit small bowel muscularis membranes. *Eur. J. Pharmacol. (Mol Pharmacol. Sect.)* 188, 313–319.

Gozlan, H., Schechter, L. E., Bolanos, F., Emerit, M. B., Miquel, M. C., Nielsen, M., and Hamon, M. (1989). Determination of the molecular size of the 5-HT_3 receptor binding site by radiation inactivation. *Eur. J. Pharmacol. Mol. Pharmacol. Sect.* 172, 497–500.

Greenshaw, A. J. (1993). Behavioural pharmacology of 5-HT_3 receptor antagonists: A critical update on therapeutic potential. *Trends Pharmacol. Sci.* 14, 265–270.

Hamon, M. (1992). *Central and peripheral 5-HT_3 Receptors.* Hamon, M., Ed., Academic Press, London.

Hargreaves, A. C., Lummis, S. C. R., and Taylor, C. W. (1994). Effects of the selective 5-HT_3 receptor agonist 1- (m-chlorophenyl) biguanide on undifferentiated intracellular free Ca^{2+} concentration in N1E-115 cells. *Br. J. Pharmacol.* 112, 99p

Higashi, H. and Nishi, S. (1982). 5-Hydroxytryptamine receptors on visceral primary afferent neurones of rabbit nodose ganglia. *J. Physiol.* 323, 543–567.

Hille, B. (1992). *Ionic Channels of Excitable Membranes.* Sinauer Associates, Sunderland, MA.

Hope, A. G. (1993). Cloning and functional expression of a murine 5-HT_3 receptor. Ph.D. thesis, University of Dundee, Scotland.

Hope, A. G., Downie, D. L., Sutherland, L., Lambert, J. J., Peters, J. A., and Burchell, B. (1993). Cloning and functional expression of an apparent splice variant of the murine 5-HT_3 receptor A subunit. *Eur. J. Pharmacol. (Mol. Pharmacol. Section)* 245, 187–192.

Humphrey, P. P. A., Hartig, P., and Hoyer, D. (1993). A proposed new nomenclature for 5-HT receptors. *Trends Pharmacol. Sci.* 14, 233–236.

Hussy, N. and Jones, K. (1993). Differences in single channel conductances of mouse 5-HT_3 receptors may indicate the presence of receptor subtypes. *Soc. Neurosci. Abstr.* 19, p282, 120. 18.

Imoto, K. (1993). Ion channels: Molecular basis of ion selectivity. *FEBS Lett.* 325, 100-103.

Imoto, K., Busch, C., Sakmann, B., Mishina, M., Konno, T., Nakai, J., Bujo, H., Mori, Y., Fukuda, K., and Numa, S. (1988). Rings of negatively charged amino acids determine the acetylcholine receptor channel conductance. *Nature* 335, 645–648.

Isenberg, K. E., Ukhun, I. A., Holstad, S. G., Jafri, S., Uchida, U., Zorumski, C. F., and Yang, J. (1993). Partial cDNA cloning and NGF regulation of a rat 5-HT_3 receptor subunit. *NeuroReport* 5, 121–124.

Jiang, L. H., Ashby, Jr., C. R., Kasser, R. J., and Wang, R. Y. (1990). The effect of intraventricular administration of the 5-HT_3 receptor agonist 2-methylserotonin on the release of dopamine in the nucleus accumbens: An *in vivo* chronocoulometric study. *Brain Res.* 513, 156–160.

Johnson, D. S. and Heinemann, S. F. (1992). Cloning and expression of the rat 5-HT_3 receptor reveals species-specific sensitivity of curare antagonism. *Soc. Neurosci. Abstr.* 18, 249.

Johnson, R. M., Inouye, G. T., Eglen, R. M., and Wong, E. H. F. (1993). 5-HT_3 receptors lack modulatory influence on acetylcholine release in rat entorhinal cortex. *Naunyn Schmiedeberg's Arch. Pharmacol.* 346, 1–7.

Karlin, A. (1993). Structure of nicotinic acetylcholine receptors. *Current Op. Neurobiol.* 3, 299–309.

Keller, B., Hollman, M., Heinemann, S., and Konnerth, A. (1992). Calcium influx through subunits GluR1/GluR3 of kainate/AMPA receptor channels is regulated by cAMP dependent protein kinase. *EMBO J.* 11, 891–896.

Kienker, P., Tomaselli, G., Jurman, M., and Yellen, G. (1994). Conductance mutations of the nicotinic acetylcholine receptor do not act by a simple electrostatic mechanism. *Biophys. J.* 66, 325–334.

King, F. D. (1994). Structure activity relationships of 5-HT_3 receptor antagonists. In *5-Hydroxytryptamine-3 Receptor Antagonists.* Jones, B. J., King, F., and Sanger, G. J., Eds., CRC Press, Boca Raton, FL, pp. 1–44.

Konno, T., Busch, C., Von Kitzing, E., Imoto, K., Wang, F., Nakai, J., Mishini, M., Numa, S., and Sakmann, B. (1991). Rings of anionic amino acids as structural determinants of ion selectivity in the acetylcholine receptor channel. *Proc. R. Soc. London B* 224, 69–79.

Kooyman, A. R. (1993). Modulation of the 5-hydroxytryptamine$_3$ receptor-operated ion current by differential occupation of agonist recognition sites. Ph.D. Thesis, University of Utrecht, Utrecht, The Netherlands.

Kooyman, A. R., Zwart, R., and Vijverberg, H. P. M. (1993a). Tetraethylammonium ions block 5-HT_3 receptor-operated ion current at the agonist recognition site and prevent desensitization in cultured mouse neuroblastoma cells. *Eur. J. Pharmacol. (Mol. Pharmacol. Section)* 246, 247–254.

Kooyman, A. R., Van Hooft, J. V., and Vijverberg, H. P. M. (1993b). 5-Hydroxyindole slows desensitization of the serotonin 5-HT_3 receptor-operated ion current in N1E-115 neuroblastoma cells. *Br. J. Pharmacol.* 108, 287–289.

Lambert, J. J., Peters, J. A., Hales, T. G., and Dempster, J. (1989). The properties of 5-HT_3 receptors in clonal cell lines studied by patch-clamp techniques. *Br. J. Pharmacol.* 97, 27–40.

Laporte, A. M., Kidd, E. J., Verge, D., Golzan, H., and Hamon, M. (1992). Autoradiographic mapping of central 5-HT_3 receptors. In *Central and Peripheral 5-HT_3 Receptors.* Hamon, M., Ed., Academic Press, London, pp. 157–187.

Lovinger, D. M. (1991). Inhibition of 5-HT_3 receptor-mediated ion current by divalent metal cations in NCB-20 neuroblastoma cells. *J. Neurophysiol.* 66, 1329–1337.

Lovinger, D. M. and White, G. (1991). Ethanol potentiation of 5-hydroxytryptamine$_3$ receptor-mediated ion current in neuroblastoma cells and isolated adult mammalian neurones. *Mol. Pharmacol.* 40, 263–270.

Lovinger, D. M. and Zhou, Q. (1993). Trichloroethanol potentiation of 5-hydroxytryptamine$_3$ receptor mediated ion current in nodose ganglion neurones from the adult rat. *J. Pharmacol. Exp. Ther.* 265, 771–777.

Luddens, H. and Wisden, W. (1991). Function and pharmacology of multiple $GABA_A$ receptor subunits. *Trends Pharmacol. Sci.* 12, 49–51.

Lummis, S. C. R. and Martin, I. L. (1992). Solubilization, purification and functional reconstitution of 5-hydroxytryptamine$_3$ receptors from N1E-115 neuroblastoma cells. *Mol. Pharmacol.* 41, 18–23.

Lummis, S. C. R., Nielsen, M., Kilpatrick, G. J., and Martin, I. L. (1990). Target size of 5-HT_3 receptors in N1E-115 neuroblastoma cells and rat brain. *Eur. J. Pharmacol. (Mol. Pharmacol. Sect.),* 189, 229–232.

Lummis, S. C. R., Baker, J., and Sepulveda, M.-I. (1993). Characterization of a radiolabelled 5-HT_3 receptor agonist, [^{3}H]m-chlorophenylbiguanide. *Br. J. Pharmacol.* 108, 104P.

Malone, H. M., Peters, J. A., and Lambert, J. J. (1991a). Physiological and pharmacological properties of $5HT_3$ receptors — a patch clamp study. *Neuropeptides* 19, 25–30.

Malone, H. M., Peters, J. A., and Lambert, J. J. (1991b). (+)-Tubocurarine and cocaine reveal species differences in the 5-HT_3 receptors of rabbit, mouse and guinea-pig nodose ganglion neurones. *Br. J. Pharmacol.* 104, 68P.

Malone, H. M., Peters, J. A., and Lambert, J. J. (1994). The permeability of 5-HT_3 receptors of rabbit nodose ganglion neurones to organic and monovalent cations. *J. Physiol.* 475. P, 151P.

Maricq, A. V., Peterson, A. S. Brake, A. J., Myers, R. M., and Julius, D. (1991). Primary structure and functional expression of the 5-HT_3 receptor, a serotonin-gated ion channel. *Science* 254, 432–437.

Martin, K. F., Hannon, S., Phillips, I., and Heal, D. J. (1992). Opposing roles for 5-HT_{1B} and 5-HT_3 receptors in the control of 5-HT release in rat hippocampus *in vivo*. *Br. J. Pharmacol.* 106, 139–142.

Maura, G., Andrioli, G. C., Cavazzani, P. and Raiteri, M. (1992). 5-Hydroxytryptamine$_3$ receptors sited on cholinergic axon terminals of human cerebral cortex mediate inhibition of acetylcholine release. *J. Neurochem.* 58, 2334–2337.

McKernan, R. M. (1992). Biochemical properties of the 5-HT_3 receptors. In *Central and Peripheral 5-HT_3 Receptors*. Hamon, M., Ed., Academic Press, London, pp. 90–102.

McKernan, R. M., Quirk, K., Jackson, R. G., and Ragan, C. I. (1990a). Solubilization of the 5-hydroxytryptamine$_3$ receptor from pooled rat cortical and hippocampal membranes. *J. Neurochem.* 54, 924–930.

McKernan, R. M., Biggs, C. S., Gillard, N. P., Quirk, K., and Ragan, C. I. (1990c). Molecular size of the 5-HT_3 receptor solubilized from NCB 20 cells. *Biochem. J.* 269, 623–628.

McKernan, R. M., Gillard, N. P., Quirk, K., Kneen, C. O., Stevenson, G. I., Swain, C. J., and Ragan, C. I. (1990b). Purification of the 5-hydroxytryptamine 5-HT_3 receptor from NCB 20 cells. *J. Biol. Chem.* 265, 13572-13577.

Miquel, M.-C., Emerit, M. B., Bolanos, F. J., Schechter, L. E., Gozlan, H., and Hamon, M. (1990). Physiochemical properties of serotonin 5-HT_3 binding sites solubilized from membranes of NG108-15 neuroblastoma-glioma cells. *J. Neurochem.* 55, 1526–1536.

Miquel, M.-C., Emerit, M. B., Gozlan, H., and Hamon, M. (1991). Involvement of tryptophan residue(s) in the specific binding of agonists/antagonists to 5-HT_3 receptors in NG 108-15 clonal cells. *Biochem. Pharmacol.* 42, 1453–1461.

Mulle, C., Choquet, D., Korn, H., and Changeux, J.-P. (1992). Calcium influx through nicotinic receptor in rat central neurones: Its relevance to cellular regulation. *Neuron* 8, 135–143.

Neijt, H. C., Te Duits, I. J., and Vijverberg, H. P. M. (1988). Pharmacological characterization of serotonin 5-HT_3 receptor-mediated electrical response in cultured mouse neuroblastoma cells. *Neuropharmacology* 27, 301–307.

Neijt, H. C., Plomp, J. J., and Vijverberg, H. P. M. (1989). Kinetics of the membrane current mediated by serotonin 5-HT_3 receptors in cultured mouse neuroblastoma cells. *J. Physiol.* 411, 257–269.

Paudice, P. and Raiteri, M. (1991). Cholecystokinin release mediated by 5-HT_3 receptors in rat cerebral cortex and nucleus accumbens. *Br. J. Pharmacol.* 103, 1790–1794.

Pedersen, S. E. and Cohen, J. B. (1990). d-Tubocurarine binding sites are located at α-γ and α-δ subunit interfaces of the nicotinic acetylcholine receptor. *Proc. Natl. Acad. Sci. U.S.A.* 87, 2785–2789.

Peters, J. A. and Lambert, J. J. (1989). Electrophysiology of 5-HT_3 receptors in neuronal cell lines. *Trends. Pharmacol. Sci.* 10, 172–175.

Peters, J. A. and Usherwood, P. N. R. (1983). 5-Hydroxytryptamine responses of murine neuroblastoma cells; ions and putative antagonists. *Br. J. Pharmacol.* 80, 523P.

Peters, J. A., Hales, T. G., and Lambert, J. J. (1988). Divalent cations modulate 5-HT_3 receptor-induced currents in N1E-115 neuroblastoma cells. *Eur. J. Pharmacol.* 151, 491–495.

Peters, J. A., Lambert, J. J., and Malone, H. M. (1994). Electrophysiological studies of 5-HT_3 receptors. In *5-HT_3 Receptor Antagonists*. Jones, B. J., King, F., and Sanger, G. J., Eds., CRC Press, Boca Raton, FL, pp. 115–153.

Peters, J. A., Malone, H. M., and Lambert, J. J. (1990). Antagonism of 5-HT_3 receptor mediated currents in murine N1E-115 neuroblastoma cells by (+)-tubocurarine. *Neurosci. Lett.* 10, 107–112.

Peters, J. A., Lambert, J. J., and Malone, H. M. (1991a). Physiological and pharmacological aspects of 5-HT_3 receptor function. In *Aspects of Synaptic Transmission, LTP, Galanin, Opioids, Autonomic and 5-HT.* Stone, T. W., Ed., Taylor and Francis, London, pp. 283–313.

Peters, J. A., Malone, H. M., and Lambert, J. J. (1991b). Ketamine potentiates 5-HT_3 receptor-mediated currents in rabbit nodose ganglion neurones. *Br. J. Pharmacol.* 103, 1623–1625.

Peters, J. A., Malone, H. M., and Lambert, J. J. (1991c). Characterization of 5-HT_3 receptor mediated electrical responses in nodose ganglion neurones and clonal neuroblastoma cells maintained in culture. In *Serotonin: Molecular Biology, Receptors and Functional Effects.* Fozard, J. R. and Saxena, P. R., Eds., Birkhauser, Basel, Switzerland, pp. 84–94.

Peters, J. A., Malone, H. M., and Lambert, J. J. (1992). Recent advances in the electrophysiological characterization of 5-HT_3 receptors. *Trends Pharmacol. Sci.* 13, 391–397.

Peters, J. A., Malone, H. M., and Lambert, J. J. (1993). An electrophysiological investigation of the properties of 5-HT_3 receptors of rabbit nodose ganglion neurones in culture. *Br. J. Pharmacol.* 110, 665–676.

Pless, D. D. and Lennarz, W. J. (1977). Enzymatic conversion of proteins to glycoproteins. *Proc. Natl. Acad. Sci. U.S.A.* 74, 134–138.

Pritchett, D. B. and Seeburg, P. H. (1991). γ-Aminobutyric acid type A receptor point mutation increases the affinity of compounds for the benzodiazepine site. *Proc. Natl. Acad. Sci. U.S.A.* 88, 1421–1425.

Raymond, L. A., Blackstone, C. D., and Huganir, R. L. (1993). Phosphorylation of amino acid neurotransmitter receptors in synaptic plasticity. *Trends Neurosci.* 16, 147–153.

Reiser, G. (1992). Biochemical responses to 5-HT_3-receptor stimulation. In *Central and Peripheral 5-HT_3 Receptors.* Hamon, M., Ed., Academic Press, London, pp. 129–156.

Revah, F., Bertrand, D., Galzi, J.-L., Devillers-Thiéry, A., Mulle, C., Hussy, N., Bertrand S., Ballivet, M., and Changeux, J.-P. (1991). Mutations in the channel domain alter desensitization of a neuronal nicotinic receptor. *Nature* 353, 846–849.

Richardson, B. P., Engel, G., Donatsch, P., and Stadler, P. A. (1985). Identification of serotonin M-receptor subtypes and their specific blockade by a new class of drugs. *Nature* 316, 126–131.

Robertson, B. and Bevan, S. (1991). Properties of 5-hydroxytryptamine$_3$ receptor-gated currents in adult rat dorsal root ganglion neurones. *Br. J. Pharmacol.* 102, 272–276.

Ropert, N. and Guy, N. (1991). Serotonin facilitates GABAergic transmission in the CA1 region of rat hippocampus *in vitro*. *J. Physiol.* 441, 121–136.

Russell, D. and Kenny, G. N. C. (1992). 5-HT_3 antagonists in post-operative nausea and vomiting. *Br. J. Anaesthesiol.* 69 (Suppl. 1), 63–68S.

Sanger, G. J. (1992). The pharmacology of anti-emetic agents. In *Emesis in Anti-Cancer Therapy.* Andrews, P. L. R. and Sanger, G. J., Eds., Chapman and Hall, London, pp. 179–210.

Sanger, G. J. and Nelson, D. R. (1989). Selective and functional 5-hydroxytryptamine$_3$ receptor antagonism by BRL 43694 (granisetron). *Eur. J. Pharmacol.* 159, 113–124.

Schmieden, V., Kuhse, J., and Betz, H. (1993). Mutation of glycine receptor subunit creates β-alanine receptor responsive to GABA. *Science* 262, 256–258.

Sepúlveda, M.-I. and Lummis, S. C. R (1994). 5-HT_3 agonists can distinguish cloned 5-HT_3 receptors from NCB20 and NG108-15 cell lines. *Br. J. Pharmacol.* 112, 316p.

Sepúlveda, M.-I., Lummis, S. C. R., and Martin, I. L. (1991). The agonist properties of m-chlorophenylbiguanide and 2-methyl-5-hydroxytryptamine on 5-HT_3 receptors in N1E-115 neuroblastoma cells. *Br. J. Pharmacol.* 104, 536–540.

Shao, X. M., Yakel, J. L., and Jackson, M. B. (1991). Differentiation of NG 108-15 cells alters channel conductance and desensitization kinetics of the 5-HT_3 receptor. *J. Neurophysiol.* 65, 630–638.

Sommer, B. and Seeburg, P. (1992). Glutamate receptor channels: Novel properties and new clones. *Trends Pharmacol. Sci.* 13, 291–296.

Strange, P. G. (1988). The structure and mechanism of the neurotransmitter receptors. Implications for the structure and function of the central nervous system. *Biochem. J.* 249, 309–318.

Sugita, S., Shen, K.-Z., and North, R. A. (1992). 5-Hydroxytryptamine is a fast excitatory transmitter at 5-HT_3 receptors in rat amygdala. *Neuron* 8, 199–203.

Surprenant, A. (1990). Whole cell and single channel currents produced by activation of 5-HT_3 receptors in enteric neurones. *Neurosci. Lett.* 38, S115.

Surprenant, S., Matsumoto, S., and Gerzanich, V. (1991). 5-HT_3 receptors in guinea-pig coeliac neurones. *Soc. Neurosci Abstr.* 17, 601.

Swope, S. L., Moss, S. J., Blackstone, C. D., and Huganir R. L. (1992). Phosphorylation of ligand-gated ion channels: A possible mode of synaptic plasticity. *FASEB J.* 6, 2514–2523.

Turton, S., Gillard, N. P., Stephenson, F. A., and McKernan, R. M. (1993). Antibodies against the 5-HT_3-A receptor identify a 54 kDa protein affinity-purified from NCB-20 cells. *Mol. Neuropharmacol.* 3, 167–171.

Vandenberg, R. J., Handford, C. A., and Schofield, P. R. (1992). Distinct agonist- and antagonist-binding sites on the glycine receptor. *Neuron* 9, 491–496.

Vernino, S., Amador, M., Luetje, C. W., Patrick, J., and Dani, J. A. (1992). Calcium modulation and high calcium permeability of neuronal nicotinic receptors. *Neuron* 8, 127–134.

Villarroel, A., Herlitze, S., Witzemann, V., Koenen, M., and Sakmann, B. (1992). Asymmetry of the rat acetylcholine receptor subunits in the narrow region of the pore. *Proc. R. Soc. London B* 249, 317–324.

Wallis, D. I. (1989). Interaction of 5-hydroxytryptamine with autonomic and sensory neurones. In *The Peripheral Actions of 5-Hydroxytryptamine.* Fozard, J. R., Ed., Oxford University Press, Oxford, pp. 220–246.

Werner, P., Humbert, Y., Boess, F. G., Reid, J., Jones, K., and Kawashima, E. (1993). Organisation of the murine 5-HT_3 receptor gene and investigation of its splice variants. *Soc. Neurosci. Abstr.* 19, 1164.

Wieland, H. A., Luddens, H., and Seeburg, P. H. (1992). A single histidine in $GABA_A$ receptors is essential for benzodiazepine agonist binding. *J. Biol. Chem.* 267, 1426–1429.

Yakel. J. L. (1992). 5-HT_3 receptors as cation channels. In *Central and Peripheral 5-HT_3 Receptors.* Hamon, M., Ed., Academic Press, London, pp. 103–128.

Yakel, J. L. and Jackson, M. B. (1988). 5-HT_3 receptors mediate rapid responses in cultured hippocampus and a clonal cell line. *Neuron* 1, 615–621.

Yakel, J. L., Trussel, L. O. and Jackson, M. B. (1988). Three serotonin responses in cultured mouse hippocampal and striatal neurones. *J. Neurosci.* 8, 1273–1285.

Yakel, J. L., Shao, X. M., and Jackson, M. B. (1990). The selectivity of the channel coupled to the 5-HT_3 receptor. *Brain Res.* 533, 46–52.

Yakel, J. L., Shao, X. M., Jackson, M. B. (1991). Activation and desensitization of the 5-HT_3 receptor in a rat glioma × mouse neuroblastoma hybrid cell. *J. Physiol.* 436, 293–308.

Yakel, J. L., Lagrutta, A., Adelman, J. P., and North, R. A. (1993). Single amino acid substitution affects desensitization of the 5-hydroxytryptamine type 3 receptor expressed in *Xenopus* oocytes. *Proc. Nat. Acad. Sci. U.S.A.* 90, 5030–5033.

Yang, J. (1990). Ion permeation through 5-HT-gated channels in neuroblastoma N18 cells. *J. Gen. Physiol.* 96, 1177–1198.

Yang, J., Mathie, A., and Hille, B. (1992). 5-HT_3 receptor channels in dissociated rat superior cervical ganglion neurones. *J. Physiol.* 448, 237–256.

6

Ionotropic Glutamate Receptors

Rolf Sprengel and Peter H. Seeburg

2.6.0 Introduction

In the central nervous system (CNS) of vertebrate and mammalian species ionotropic glutamate receptors (iGluRs) mediate the fast synaptic action of L-glutamate, the major excitatory neurotransmitter. iGluRs have traditionally been classified into three types, mainly based on pharmacological and biophysical properties: α-amino-3-hydroxy-5-methyl-4-isoxazole propionate (AMPA) receptors, high-affinity kainate receptors, and *N*-methyl-D-aspartate (NMDA) receptors (reviewed in Cotman et al., 1987; Mayer and Westbrook, 1987; Collingridge and Lester, 1989; Monaghan et al., 1989; McDonald and Johnston, 1990; Watkins et al., 1990). These cation-selective channels are essential elements in synaptic function and plasticity, and further mediate neuropathology in a number of brain diseases, including epilepsy (reviewed by Choi, 1988; Olney, 1989; Collingridge and Singer, 1990; Dingledine et al., 1990; Meldrum and Garthwaite, 1990).

Ionotropic GluRs are composed of subunits and for each receptor type several subtypes exist that differ in subunit composition and functional properties. Based on expression cloning of an AMPA receptor subunit (Hollmann et al., 1990) and an NMDA receptor subunit (Moriyoshi et al., 1991), the combined efforts of several laboratories have led to the characterization of 16 iGluR subunits for the three glutamate-gated channel types in rodent brain (reviewed by Gasic and Hollmann, 1992; Nakanishi, 1992; Seeburg, 1993; Sprengel and Seeburg, 1993; Hollmann and Heinemann, 1994) (Table 1).

Each subunit is encoded by its own gene, and many of the primary transcripts undergo alternative splicing, which increases the molecular complexity underlying channel formation. In addition, recoding through a process of RNA editing affects several positions in iGluR subunits with profound

0-8493-8322-6/95/$0.00+$.50

Table 1. Cloned Ionotropic Glutamate Receptor Subunits

Receptor type	Subunit	Alternative nomenclature	Species	References	GenBank accession no.
AMPA	GluR-A	h-KR4	Human	Potier et al. (1992)	X58633
		h-GluH1	Human	Puckett et al. (1991)	M64752
		h-HBGR1	Human	Sun et al. (1992)	M81886
		r-GluR-A	Rat	Keinänen et al. (1990)	M36418
		r-GluR-A	Rat	Sommer et al. (1990)	M38060
		r-GluR-K1	Rat	Hollmann et al. (1989)	X17184
		m-α1	Mouse	Sakimura et al. (1990)	X57497
AMPA	GluR-B	r-Glu-R-B	Rat	Keinänen et al. (1990)	M36419
		r-GluR-B	Rat	Sommer et al. (1990)	M38061
		r-GluR2	Rat	Boulter et al. (1990)	M85035
		r-GluR-K2	Rat	Nakanishi et al. (1990)	X54655
		m-α2	Mouse	Sakimura et al. (1990)	X57498
AMPA	GluR-C	r-GluR-C	Rat	Keinänen et al. (1990)	M36420
		r-GluR-C	Rat	Sommer et al. (1990)	M38062
		r-GluR3	Rat	Boulter et al. (1990)	M85036
		r-GluR-K3	Rat	Nakanishi et al. (1990)	X54656
AMPA	GluR-D	r-GluR-D	Rat	Keinänen et al. (1990)	M36421
		r-GluR-D	Rat	Sommer et al. (1990)	M38063
		r-GluR4	Rat	Bettler et al. (1990)	M85037
		r-GluR-4c	Rat	Gallo et al. (1992)	S94371
Kainate	GluR5	h-GluR5-1d	Human	Gregor et al. (1993a)	L19058
		r-GluR5-1	Rat	Bettler et al. (1990)	M83560
		r-GluR5-2	Rat	Bettler et al. (1990)	M83561
		r-GluR-5a	Rat	Sommer et al. (1992)	Z11712
		r-GluR-5b	Rat	Sommer et al. (1992)	Z11713
		r-GluR-5c	Rat	Sommer et al. (1992)	Z11714
		m-GluR5-3	Mouse	Gregor et al. (1993a)	X66118
Kainate	GluR6	r-GluR6	Rat	Egebjerg et al. (1991)	Z11548
		r-GluR-6	Rat	Lomeli et al. (1992)	Z11715
		m-β2	Mouse	Morita et al. (1992)	D10054
		m-GluR6-2	Mouse	Gregor et al. (1993a)	X66117
Kainate	GluR7	r-GluR7	Rat	Bettler et al. (1992)	M83552
		r-GluR-7	Rat	Lomeli et al. (1992)	Z11716
Kainate	KA-1	h-EAA1	Human	Kamboj et al. (1994)	
		r-KA-1	Rat	Werner et al. (1991)	X59996
	KA-2	h-EAA2	Human	Kamboj et al. (1992)	S40369
		r-KA-2	Rat	Herb et al. (1992)	Z11581
		m-γ2	Mouse	Sakimura et al. (1992)	D10011
Unknown	delta-1	r-delta-1	Rat	Lomeli et al. (1993)	Z17238
		m-δ1	Mouse	Yamazaki et al. (1992a)	D10171
	delta-2	r-delta-2	Rat	Lomeli et al. (1993)	Z17239
		m-δ2	Mouse	Araki et al. (1993)	D13266
Kainate-binding protein	KBP	Chick-KBP	Chick	Gregor et al. (1989)	X17700
		Frog-KBP	Frog	Wada et al. (1989)	X17314
NMDA	NMDAR1	h-NMDAR1	Human	Karp et al. (1993)	D13515
		h-NR1	Human	Planells et al. (1993)	L05666
		h-NR1-1	Human	Foldes et al. (1993)	L13266
		h-NR1-2	Human	Foldes et al. (1993)	L13267
		h-NR1-3	Human	Foldes et al. (1993)	L13268

Table 1. (Continued).

Receptor type	Subunit	Alternative nomenclature	Species	References	GenBank accession no.
		r-NMDAR1	Rat	Moriyoshi et al. (1991)	X63255
		r-NMDA-R-1A	Rat	Nakanishi et al. (1992)	S44964[a]
		r-NMDA-R-1B	Rat	Nakanishi et al. (1992)	S45121[a]
		r-NMDAR1b	Rat	Sugihara et al. (1992)	S39217[a]
		r-NMDAR1c	Rat	Sugihara et al. (1992)	S39218[a]
		r-NMDAR1d	Rat	Sugihara et al. (1992)	S39219[a]
		r-NMDAR1e	Rat	Sugihara et al. (1992)	S39217[a]
		r-NMDAR1f	Rat	Sugihara et al. (1992)	S39218[a]
		r-NMDAR1g	Rat	Sugihara et al. (1992)	S39219[a]
		r-NMDAR1tru.	Rat	Sugihara et al. (1992)	S39221[a]
		r-NMDAR1	Rat	Hollmann et al. (1993)	L08228
		r-R1-LL	Rat	Anantharam et al. (1992)	X65227
		m-ζ1	Mouse	Yamazaki et al. (1992b)	D10028
		m-ζ1-2	Mouse	Yamazaki et al. (1992b)	S37525
NMDA	NR2A	r-NR2A	Rat	Monyer et al. (1992)	M91561
		r-NMDAR2A	Rat	Ishii et al. (1993)	D13211
		m-ε1	Mouse	Meguro et al. (1992)	D10217
NMDA	NR2B	r-NR2B	Rat	Monyer et al. (1992)	M91562
		m-ε2	Mouse	Kutsuwada et al. (1992)	D10651
NMDA	NR2C	r-NR2C	Rat	Monyer et al. (1992)	M91563
		r-NMDAR2C	Rat	Ishii et al. (1993)	D13212
		m-ε3	Mouse	Kutsuwada et al. (1992)	D10694
NMDA	NR2D	r-NMDAR2D	Rat	Ishii et al. (1993)	D13213
		r-NMDAR2D-2	Rat	Ishii et al. (1993)	D13214
		r-NR2D	Rat	Monyer et al. (1994)	L31612
		m-ε4	Rat	Ikeda et al. (1992)	D12822
Invertebrate with unspecified subtype		DNMDAR-I	*Drosophila*	Ultsch et al. (1993)	X71790
		DGluR-I	*Drosophila*	Ultsch et al. (1992)	M97192
		DGluR-II	*Drosophila*	Schuster et al. (1991)	M73271
		LymGluR	Snail	Hutton et al. (1991)	X60086
		C-C06E1.4	*C. elegans*	Sulston et al. (1992)	L16560

[a] Accession number = sequence of alternative exon.

functional consequences. Subunits are glycosylated membrane-anchored polypeptides with an average length of 900 amino acids. The NR2 subunits of the NMDA receptors have larger sizes (>1300 amino acid residues) because of extended carboxy-termini. The shorter kainate-binding proteins (<500 amino acid residues) found in chick and frog but not in mammals lack a large part of the amino-terminal protein domains of other iGluR subunits. All subunits are synthesized as precursor proteins with an amino-terminal signal peptide of 15 to 22 amino acids in length. The hydrophobicity plots of the mature polypeptides predict the presence of four hydrophobic segments (M1 to M4), which determine the transmembrane topology, and which participate in ion channel formation (Figure 1). Four membrane spanning segments predict the topology of members of the nicotinic acetylcholine recep-

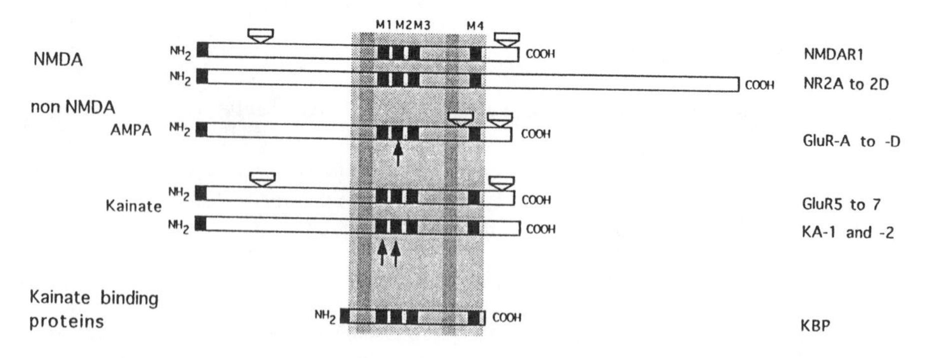

FIGURE 1. Schematic presentation of the ionotropic glutamate receptor subunits. The hydrophobic segments M1 to M4 and the amino-terminal signal peptide are indicated by black boxes. In subunits GluR-B, GluR5, and GluR6 selected amino acid residues in hydrophobic segments M1 and M2 are controlled by site-specific RNA editing (arrows). Areas within the polypeptide sequence that show sequence variability created by alternatively spliced transcripts are indicated by small superimposed boxes. The stippled area denotes sequences conserved among all subunits. The regions proximal to M1 and between M3 and M4 that are homologous to the *E. coli* glutamine binding protein are highlighted by darker shading.

tor/$GABA_A$ receptor superfamily with subunits having extracellularly located amino- and carboxy-termini (Unwin, 1993). However, experimental data suggest that iGluR subunits assume a different topology with amino- and carboxy-terminus on opposite sides of the cell membrane, the carboxy-terminus being intracellularly located. Posttranslational phosphorylation was observed at residues in the carboxy-terminal domain (Tingley et al., 1993) and in a sequence immediately following M3 (Raymond et al., 1993; Wang et al., 1993). Amino acids within M2 line the channel pore. Numerous studies give evidence that particular amino acids in this segment determine the ion selectivity of the AMPA and NMDA receptor channels (see below).

Alternative splicing generates subunit variants with different carboxy-termini and additional or substituted peptide sequences in the amino-terminal extracellular protein domain or in the segment between M3 and M4. In the AMPA receptor subunit GluR-B and in high-affinity kainate receptor subunits GluR5 and GluR6 selected residues in M1 and M2 are controlled by site-specific RNA editing (see below). The central portion from M1 to M4 ("core sequence") is well conserved in all subunits (Figure 1).

Significant similarity with a periplasmatic glutamine-binding protein of *Escherichia coli* was found in sequences preceding M1 and M4 (Nakanishi et al., 1990), suggesting that these sequences may form a two-domain agonist-binding site (O'Hara et al., 1993). The effect of mutations in the proximal domain supports this notion (Uchino et al., 1992). Thus, the basic structural design of ionotropic glutamate receptors and the sequence of their agonist-binding sites are ancient in evolutionary terms. iGluRs are found throughout the animal kingdom, including *Caenorhabditis, Lymnaea,* and *Drosophila.*

2.6.1 AMPA Receptor Subunits

Glutamate-gated receptor channels of the AMPA preferring type mediate the majority of all fast synaptic excitatory neurotransmission in the vertebrate CNS. When activated by L-glutamate or by AMPA these channels are characterized by rapid onset, offset, and desensitization kinetics and generally possess low divalent ion permeabilities. AMPA receptor channels can also be activated by neurotoxins such as kainate and domoate, and these agonists often produce large nondesensitizing current responses (see Table 2 for properties).

Four subunits exist for AMPA receptor assembly. These are termed GluR-A to -D (Keinänen et al., 1990) or GluR1 to 4 (Boulter et al., 1990) for rat and α1 to α4 for mouse (see Table 1 for nomenclature). The sequences of the rat, mouse, and — where known — human AMPA receptor subunits are approximately 70% identical. Each subunit occurs in one of two molecular forms, termed FLIP and FLOP, with respect to a 35 amino acid segment preceding M4 (Sommer et al., 1990). The FLIP versions of all subunits are expressed prominently before birth and remain largely invariant in their expression during postnatal development and in the adult. The expression of FLOP variants increases throughout the developing brain, and reaches adult levels by postnatal day 14 (Monyer et al., 1991). These splice forms affect the desensitization kinetics of AMPA receptor channels (Sommer et al., 1990).

Functional studies using recombinantly expressed AMPA receptor subunits have revealed that each subunit can form homooligomeric channels and can assemble with any of the others to produce heterooligomeric channels. The GluR-B subunit is functionally dominant in heterooligomeric receptors. AMPA receptor channels containing GluR-B have very low divalent ion permeability whereas channels lacking GluR-B possess high Ca^{2+} and Mg^{2+} permeabilities (Hume et al., 1991; Hollmann et al., 1991; Verdoorn et al., 1991; Burnashev et al., 1992a; Dingledine et al., 1992). Although neurons generally express AMPA receptors with low Ca^{2+} permeability, exceptions to this have been observed (Iino et al., 1990; Gilbertson et al., 1991). As documented by *in situ* hybridization and analysis of subunit expression in single cells in brain slices, neurons can in fact regulate the extent to which AMPA receptor channels are Ca^{2+} permeable by controlling the read-out of the GluR-B gene (Burnashev et al., 1992b; Jonas et al., 1994). A critical molecular determinant for the differential Ca^{2+} permeabilities resides in the channel-forming segment M2. GluR-B carries an arginine residue in a critical M2 position (Q/R site), whereas the other AMPA receptor subunits have a glutamine in their Q/R site (Figure 3b). The arginine residue in GluR-B M2 is not encoded by the GluR-B gene but is generated from a CAG (glutamine) to CGG (arginine) codon change through a process of RNA editing (Sommer et al., 1991; Higuchi et al., 1993).

2.6.1.1 GluR-A (GluR1)

Rat GluR1 was the first iGluR subunit obtained by cloning (Hollmann et al., 1989). At present, cDNAs encoding GluR-A have been characterized from rat, mouse, and human species (Keinänen et al., 1990; Boulter et al., 1990;

Table 2. Functional Characteristics and Chromosomal Localization of the Receptor Subunits

Subunits	Homomeric channels	Heteromeric channels	Functional characteristics	Chromosomal localization
GluR-A	+	With GluR-B, -C, and -D	Rapid kinetics; low Ca^{2+}	5q31.3–33.3
GluR-B	+	With GluR-A, -C, and -D	permeability; L-Glu activates	4q32–33
GluR-C	+	With GluR-A, -B, and -D	desensitizing currents;	Xq25–26
GluR-D	+	With GluR-A, -B, and -C	kainate activates non-desensitizing currents	11q22–23
GluR5	+	With KA-1 or KA-2	Rapid kinetics;	21q21.1–22.1
GluR6	+	With KA-1 or KA-2	partial Ca^{2+} permeability;	10 (mouse)
GluR7	–	Not known	L-Glu and kainate	4(mouse)
KA-1	–	With GluR5 or GluR6	activate desensitizing	Not known
KA-2	–	With GluR5 or GluR6	currents	Not known
delta-1	–	Not known	Not known	Not known
delta-2	–	Not known	Not known	Not known
NMDAR1	+	With NR2A to D	Slow kinetics, large conductance;	9q34.3
NR2A	–	With NMDAR1	little desensitization; high Ca^{2+}	16p13
NR2B	–	With NMDAR1	permeability; voltage-dependent	Not known
NR2C	–	With NMDAR1	Mg^{2+} block; glycine as	7q25
NR2D	–	With NMDAR1	coagonist	Not known
Chick KBP	–	Not known	Not known	Not known
Frog KBP	–	Not known	Not known	Not known
DNMDAR-I	–	Not known	Not known	3R-83AB *(Drosophila)*
DGluR-I	+	Not known	Kainate activates channel	3L-65C *(Drosophila)*
DGluR-II	+	Not known	L-Glu and L-Asp gated channel	2L-25F *(Drosophila)*
LymGluR	–	Not known	Not known	Not known
C-C06E1.4	Not known	Not known	Not known	3 CELCO6E1 *(C. elegans)*

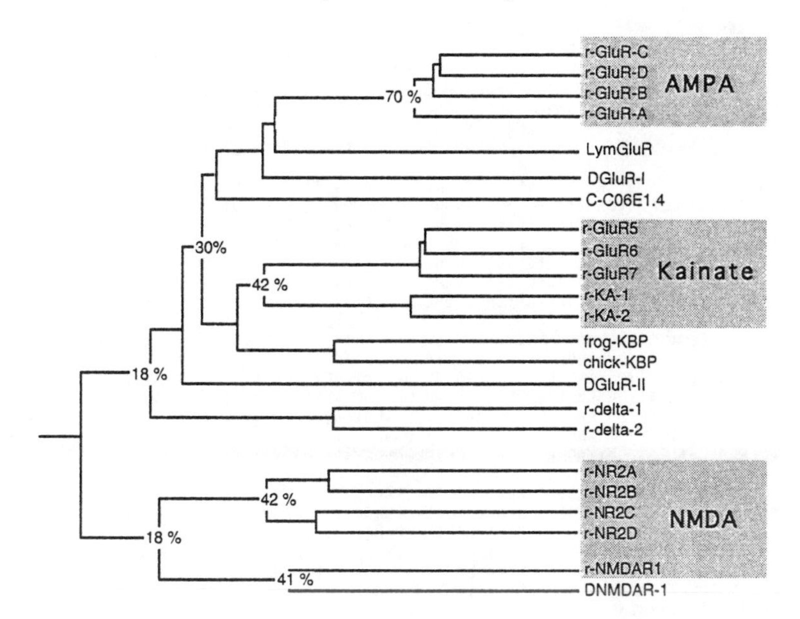

FIGURE 2. Phylogenetic tree of selected ionotropic glutamate receptor subunits. The tree was generated by comparing the entire polypeptide sequences of cloned receptor subunits. The software used is Megalign, commercially available by DNASTAR Ltd., London. The alignment was performed using the CLUSTAL program (Higgins and Sharp, 1988) and the default amino acid identity matrix table. The name and the appropriate reference for each subunit are listed in Table 1. The total length of the vertical lines between different receptors gives their putative phylogenetic distance in arbitrary units. The averaged amino acid identity between individual subunits or subunit families is indicated in percent at key branch points.

Nakanishi et al., 1990; Sakimura et al., 1990; Puckett et al., 1991; Sun et al., 1992; Potier et al., 1992) (Figure 3a). *In situ* hybridization (Keinänen et al., 1990; Monyer et al., 1991; Tölle et al., 1993) and immunocytochemistry (Blackstone et al., 1992a,b; Hampson et al., 1992; Craig et al., 1993; Molnar et al., 1993; Martin et al., 1993) show that GluR-A, like the other AMPA receptor subunits, is expressed very widely throughout the mammalian CNS during development and in the adult. Specific antibodies detect GluR-A on neuronal cell bodies and processes within most regions of the brain; within the cerebellum, however, GluR-A is localized on Bergmann glia. At the subcellular level GluR-A is localized in the postsynaptic membrane in dendrites (Hampson et al., 1992; Blackstone et al., 1992a).

The GluR-A subunit occurs in two splice forms, FLIP and FLOP, which differ in a 35 amino acid segment close to M4 (Sommer et al., 1990). Northern analysis indicates a size of 7 kb for the GluR-A transcript in rat brain (Hollmann et al., 1989; Partin et al., 1993). The human GluR-A gene has been mapped to chromosome 5q33 (McNamara et al., 1992; Puckett et al., 1991; Sun et al., 1992).

```
1   MQHIFAFFCTGFLGAVVGANFPNNIQIGGLFPNQQSQEHAAFRFALSQLTEPPKLLPQID h-KR4   (Potier)
1   MQHIFAFFCTGFLGAVVGANFPNNIQIGGLFPNQQSQEHAAFRFALSQLTEPPKLLPQID h-GluH1   (Puckett)
1   MQHIFAFFCTGFLGAVVGANFPNNIQIGGLFPNQQSQEHAAFRFALSQLTEPPKLLPQID h-HBGR1  (Sun)
1   MPYIFAFFCTGFLGAVVGANFPNNIQIGGLFPNQQSQEHAAFRFALSQLTEPPKLLPQID r-GluR-K1   (Hollmann)
1   MPYIFAFFCTGFLGAVVGANFPNNIQIGGLFPNQQSQEHAAFRFALSQLTEPPKLLPQID r-GluR-A   (Sommer)
1   MPYIFAFFCTGFLGAVVGANFPNNIQIGGLFPNQQSQEHAAFRFALSQLTEPPKLLPQID r-GluR-A   (Keinaenen)
1   MPYIFAFFCTGFLGAVVGANFPNNIQIGGLFPNQQSQEHAAFRFALSQLTEPPKLLPQID m-alpha1   (Sakimura)

61  IVNISDTFEMTYRFCSQFSKGVYAIFGFYERRTVNMLTSFCGALHVCFITPSFPVDTSNQ h-KR4   (Potier)
61  IVNISDSFEMTYRFCSQFSKGVYAIFGFYERRTVNMLTSFCGALHVCFITPSFPVDTSNQ h-GluH1   (Puckett)
61  IVNISDSFEMTYRFCSQFSKGVYAIFGFYERRTVNMLTSFCGALHVCFITPSFPVDTSNQ h-HBGR1  (Sun)
61  IVNISDTFEMTYRFCSQFSKGVYAIFGFYERRTVNMLTSFCGALHVCFITPSFPVDTSNQ r-GluR-K1   (Hollmann)
61  IVNISDTFEMTYRFCSQFSKGVYAIFGFYERRTVNMLTSFCGALHVCFITPSFPVDTSNQ r-GluR-A   (Sommer)
61  IVNISDTFEMTYRFCSQFSKGVYAIFGFYERRTVNMLTSFCGALHVCFITPSFPVDTSNQ r-GluR-A   (Keinaenen)
61  IVNISDSFEMTYRFCSQFSKGVYAIFGFYERRTVNMLTSFCGALHVCFITPSFPVDTSNQ m-alpha1   (Sakimura)

121 FVLQLRPELQDALISIIDHYKWQKFVYIYDADRGLSVLQKVLDTAAEKNWQVTDVNILTT h-KR4   (Potier)
121 FVLQLRPELQDALISIIDHYKWQKFVYIYDADRGLSVLQKVLDTAAEKNWQVTAVNILTT h-GluH1   (Puckett)
121 FVLQLRPELQDALISIIDHYKWQKFVYIYDADRGLSVLQKVLDTAAEKNWQVTAVNILTT h-HBGR1  (Sun)
121 FVLQLRPELQEALISIIDHYKWQTFVYIYDADRGLSVLQRVLDTAAEKNWQVTAVNILTT r-GluR-K1   (Hollmann)
121 FVLQLRPELQEALISIIDHYKWQTFVYIYDADRGLSVLQRVLDTAAEKNWQVTAVNILTT r-GluR-A   (Sommer)
121 FVLQLRPELQEALISIIDHYKWQTFVYIYDADRGLSVLQRVLDTAAEKNWQVTAVNILTT r-GluR-A   (Keinaenen)
121 FVLQLRPELQEALISIIDHYKWQTFVYIYDADRGLSVLQRVLDTAAEKNWQVTAVNILTT m-alpha1   (Sakimura)

181 TEEGYRMLFQDLEKKKERLVVVDCESERLNAILGQIIKLEKNGIGYHYILANLGFMDIDL h-KR4   (Potier)
181 TEEGYRMLFQDLEKKKERLVVVDCESERLNAILGQIIKLEKNGIGYHYILANLGFMDIDL h-GluH1   (Puckett)
181 TEEGYRMLFQDLEKKKERLVVVDCESERLNAILGQIIKLEKNGIGYHYILANLGFMDIDL h-HBGR1  (Sun)
181 TEEGYRMLFQDLEKKKERLVVVDCESERLNAILGQIVKLEKNGIGYHYILANLGFMDIDL r-GluR-K1   (Hollmann)
181 TEEGYRMLFQDLEKKKERLVVVDCESERLNAILGQIVKLEKNGIGYHYILANLGFMDIDL r-GluR-A   (Sommer)
181 TEEGYRMLFQDLEKKKERLVVVDCESERLNAILGQIVKLEKNGIGYHYILANLGFMDIDL r-GluR-A   (Keinaenen)
181 TEEGYRMLFQDLEKKKERLVVVDCESERLNAILGQIVKLEKNGIGYHYILANLGFMDIDL m-alpha1   (Sakimura)

241 NKFKESGANVTGFQLVNYTDTIPAKIMQQWKNSDARDHTRVDWKRPKYTSALTYDGVKVM h-KR4   (Potier)
241 NKFKESGANVTGFQLVNYTDTIPAKIMQQWKNSDARDHTRVDWKRPKYTSALTYDGVKVM h-GluH1   (Puckett)
241 NKFKESGANVTGFQLVNYTDTIPAKIMQQWKNSDARDHTRVDWKRPKYTSALTYDGVKVM h-HBGR1  (Sun)
241 NKFKESGRNVTGFQLVNYTDTIPARIMQQWRTSDSRDHTRVDWKRPKYTSALTYDGVKVM r-GluR-K1   (Hollmann)
241 NKFKESGRNVTGFQLVNYTDTIPARIMQQWRTSDSRDHTRVDWKRPKYTSALTYDGVKVM r-GluR-A   (Sommer)
241 NKFKESGRNVTGFQLVNYTDTIPARIMQQWRTSDSRDHTRVDWKRPKYTSALTYDGVKVM r-GluR-A   (Keinaenen)
241 NKFKESGANVTGFQLVNYTDTIPARIMQQWRTSDARDHTRVDWKRPKYTSALTYDGVKVM m-alpha1   (Sakimura)

301 AEAFQSLRRQRIDISRRGNAGDCLANPAVPWGQGIDIQRALQQVEFEGLTGNVQFNEKGR h-KR4   (Potier)
301 AEAFQSLRRQRIDISRRGNAGDCLANPAVPWGQGIDIQRALQQVRFEGLTGNVQFNEKGR h-GluH1   (Puckett)
301 AEAFQSLRRQRIDISRRGNAGDCLANPAVPWGQGIDIQRALQQVRFEGLTGNVQFNEKGR h-HBGR1  (Sun)
301 AEAFQSLRRQRIDISRRGNAGDCLANPAVPWGQGIDIQRALQQVRFEGLTGNVQFNEKGR r-GluR-K1   (Hollmann)
301 AEAFQSLRRQRIDISRRGNAGDCLANPAVPWGQGIDIQRALQQVRFEGLTGNVQFNEKGR r-GluR-A   (Sommer)
301 AEAFQSLRRQRIDISRRGNAGDCLANPAVPWGQGIDIQRALQQVRFEGLTGNVQFNEKGR r-GluR-A   (Keinaenen)
301 AEAFQSLRRQRIDISRRGNAGDCLANPAVPWGQGIDIQRALQQVRFEGLTGNVQFNEKGR m-alpha1   (Sakimura)

361 RTNYTIHVIEMKHDGIRKIGYWNEDDKFVPAATDEQAGGDNSSVQNRTYIVTTILEDPYV h-KR4   (Potier)
361 RTNYTLHVIEMKHDGIRKIGYWNEDDKFVPAATDAQAGGDNSSVQNRTYIVTTILEDPYV h-GluH1   (Puckett)
361 RTNYTLHVIEMKHDSIRKIGYWNEDDKFVPAATDAQAGGDNSSVQNRTYIVTTILEDPYV h-HBGR1  (Sun)
361 RTNYTLHVIEMKHDGIRKIGYWNEDDKFVPAATDAQAGGDNSSVQNRTYIVTTILEDPYV r-GluR-K1   (Hollmann)
361 RTNYTLHVIEMKHDGIRKIGYWNEDDKFVPAATDAQAGGDNSSVQNRTYIVTTILEDPYV r-GluR-A   (Sommer)
361 RTNYTLHVIEMKHDGIRKIGYWNEDDKFVPAATDAQAGGDNSSVQNRTYIVTTILEDPYV r-GluR-A   (Keinaenen)
361 RTNYTLHVIEMKHDGIRKIGYWNEDDKFVPAATDAQAGGDNSSVQNRTYIVTTILEDPYV m-alpha1   (Sakimura)

421 MLKKNANQFEGNDRYEGYCVELAAEIAKHVGYSYRLEIVSDGKYGARDPDTKAWNGMVGE h-KR4   (Potier)
421 MLKKNANQFEGNDRYEGYCVELAAEIAKHVGYSYRLEIVSDGKYGARDPDTKAWNGMVGE h-GluH1   (Puckett)
421 MLKKNANQFEGNDRYEGYCVELAAEIAKHVGYSYRLEIVSDGKYGARDPDTKAWNGMVGE h-HBGR1  (Sun)
421 MLKKNANQFEGNDRYEGYCVELAAEIAKHVGYSYRLEIVSDGKYGARDPDTKAWNGMVGE r-GluR-K1   (Hollmann)
421 MLKKNANQFEGNDRYEGYCVELAAEIAKHVGYSYRLEIVSDGKYGARDPDTKAWNGMVGE r-GluR-A   (Sommer)
421 MLKKNANQFEGNDRYEGYCVELAAEIAKHVGYSYRLEIVSDGKYGARDPDTKAWNGMVGE r-GluR-A   (Keinaenen)
421 MLKKNANQFEGNDRYEGYCVELAAEIAKHVGYSYRLEIVSDGKYGARDPDTKAWNGMVGE m-alpha1   (Sakimura)
```

FIGURE 3. Amino acid sequence alignment of published ionotropic glutamate receptor subunits. Alignments were performed using the CLUSTAL program (Higgins and Sharp, 1988). The four hydrophobic segments M1 to M4 are indicated by a vertical bar. The hydrophobic amino-terminal signal peptide sequences not present in the mature membrane-integrated subunits are displayed by a shaded box. The names and the appropriate references for the subunit sequences are listed in Table 1. The order of alignments is (a) GluR-A, (b) GluR-B, (c) GluR-C, (d) GluR-D, (e) GluR5, (f) GluR6, (g) GluR7, (h) KA-1, (i) KA-2, (j) delta-1, (k) delta-2, (l) NMDAR1, (m) NR2A, (n) NR2B, (o)

```
     ——————1——————                                   ——————2——
481 LVYGRADVAVAPLTITLVREEVIDFSKPFMSLGISIMIKKPQKSKPGVFSFLDPLAYEIW h-KR4    (Potier)
481 LVYGRADVAVAPLTITLVREEVIDFSKPFMSLGISIMIKKPQKSKPGVFSFLDPLAYEIW h-GluH1   (Puckett)
481 LVYGRADVAVAPLTITLVREEVIDFSKPFMSLGISIMIKKPQKSKPGVFSFLDPLAYEIW h-HBGR1   (Sun)
481 LVYGRADVAVAPLTITLVREEVIDFSKPFMSLGISIMIKKPQKSKPGVFSFLDPLAYEIW r-GluR-K1  (Hollmann)
481 LVYGRADVAVAPLTITLVREEVIDFSKPFMSLGISIMIKKPQKSKPGVFSFLDPLAYEIW r-GluR-A   (Sommer)
481 LVYGRADVAVAPLTITLVREEVIDFSKPFMSLGISIMIKKPQKSKPGVFSFLDPLAYEIW r-GluR-A   (Keinaenen)
481 LVYGRADVAVAPLTITLVREEVIDFSKPFMSLGISIMIKKPQKSKPGVFSFLDPLAYEIW m-alpha1   (Sakimura)

541 MCIVFAYIGVSVVLFLVSRFSPYEWHSEEFEEGRDQTTSDQSNEFGIFNSLWFSLGAFMQ h-KR4    (Potier)
541 MCIVFAYIGVSVVLFLVSRFSPYEWHSEEFEEGRDQTTSDQSNEFGIFNSLWFSLGAFMQ h-GluH1   (Puckett)
541 MCIVFAYIGVSVVLFLVSRFSPYEWHSEEFEEGRDQTTSDQSNEFGIFNSLWFSLGAFMQ h-HBGR1   (Sun)
541 MCIVFAYIGVSVVLFLVSRFSPYEWHSEEFEEGRDQTTSDQSNEFGIFNSLWFSLGAFMQ r-GluR-K1  (Hollmann)
541 MCIVFAYIGVSVVLFLVSRFSPYEWHSEEFEEGRDQTTSDQSNEFGIFNSLWFSLGAFMQ r-GluR-A   (Sommer)
541 MCIVFAYIGVSVVLFLVSRFSPYEWHSEEFEEGRDQTTSDQSNEFGIFNSLWFSLGAFMQ r-GluR-A   (Keinaenen)
541 MCIVFAYIGVSVVLFLVSRFSPYEWHSEEFEEGRDQTTSDQSNEFGIFNSLWFSLGAFMQ m-alpha1   (Sakimura)
    ——          ————————3————————
601 QGCDISPRSLSGRIVGGVWWFFTLIIISSYTANLAAFLTVERMVSPIESAEDLAKQTEIA h-KR4    (Potier)
601 QGCDISPRSLSGRIVGGVWWFFTLIIISSYTANLAAFLTVERMVSPIESAEDLAKQTEIA h-GluH1   (Puckett)
601 QGCDISPRSLSGRIVGGVWWFFTLIIISSYTANLAAFLTVERMVSPIESAEDLAKQTEIA h-HBGR1   (Sun)
601 QGCDISPRSLSGRIVGGVWWFFTLIIISSYTANLAAFLTVERMVSPIESAEDLAKQTEIA r-GluR-K1  (Hollmann)
601 QGCDISPRSLSGRIVGGVWWFFTLIIISSYTANLAAFLTVERMVSPIESAEDLAKQTEIA r-GluR-A   (Sommer)
601 QGCDISPRSLSGRIVGGVWWFFTLIIISSYTANLAAFLTVERMVSPIESAEDLAKQTEIA r-GluR-A   (Keinaenen)
601 QGCDISPRSLSGRIVGGVWWFFTLIIISSYTANLAAFLTVERMVSPIESAEDLAKQTEIA m-alpha1   (Sakimura)

661 YGTLEAGSTKEFFRRSKIAVFEKMWTYMKSAEPSVFVRTTEEGMIRVRKSKGKYAYLLES h-KR4    (Potier)
661 YGTLEAGSTKEFFRRSKIAVFEKMWTYMKSAEPSVFVRTTEEGMIRVRKSKGKYAYLLES h-GluH1   (Puckett)
661 YGTLEAGSTKEFFRRSKIAVFEKMWTYMKSAEPSVFVRTTEEGMIRVRKSKGKYAYLLES h-HBGR1   (Sun)
661 YGTLEAGSTKEFFRRSKIAVFEKMWTYMKSAEPSVFVRTTEEGMIRVRKSKGKYAYLLES r-GluR-K1  (Hollmann)
661 YGTLEAGSTKEFFRRSKIAVFEKMWTYMKSAEPSVFVLTTEEGMIRVRKTKGKYAYLLES r-GluR-A   (Sommer)
661 YGTLEAGSTKEFFRRSKIAVFEKMWTYMKSAEPSVFVLTTEEGMIRVRKTKGKYAYLLES r-GluR-A   (Keinaenen)
661 YGTLEAGSTKEFFRRSKIAVFEKMWTYMKSAEPSVFVRTTEEGMIRVRKSKGKYAYLLES m-alpha1   (Sakimura)
                                            .— — — — — — — — — —
721 TMNEYIEQRKPCDTMKVGGNLDSKGYGIATPKGSALRNPVNLAVLKLNEQGLLDKLKNKW h-KR4    (Potier)
721 TMNEYIEQRKPCDTMKVGGNLDSKGYGIATPKGSALRGPVNLAVLKLSEQGVLDKLKSKW h-GluH1   (Puckett)
721 TMNEYIEQRKPCDTMKVGGNLDSKGYGIATPKGSALRNPVNLAVLKLNEQGLLDKLKNKW h-HBGR1   (Sun)
721 TMNEYIEQRKPCDTMKVGGNLDSKGYGIATPKGSALRNPVNLAVLKLNEQGLLDKLKNKW r-GluR-K1  (Hollmann)
721 TMNEYIEQRKPCDTMKVGGNLDSKGYGIATPKGSALRGPVNLAVLKLSEQGVLDKLKSKW r-GluR-A   (Sommer)
721 TMNEYIEQRKPCDTMKVGGNLDSKGYGIATPKGSALRNPVNLAVLKLNEQGLLDKLKNKW r-GluR-A   (Keinaenen)
721 TMNEYIEQRKPCDTMKVGGNLDSKGYGIATPKGSALRGPVNLAVLKLSEQGVLDKLKSKW m-alpha1   (Sakimura)
    — — — — —                    —————————4—————————
781 WYDKGECGRGGGDSKDKTSALSLSNVAGVFYILIGGLGLAMLVALIEFCYKSRSESKRME h-KR4    (Potier)
781 WYDKGECGSKDSGSKDKTSALSLSNVAGVFYILIGGLGLAMLVALIEFCYKSRSESKRMK h-GluH1   (Puckett)
781 WYDKGECGSGGGDSKDKTSALSLSNVAGVFYILIGGLGLAMLVALIEFCYKSRSESKRMK h-HBGR1   (Sun)
781 WYDKGECGTGGGDSKDKTSALSLSNVAGVFYILIGGLGLAMLVALIEFCYKSRSESKRMK r-GluR-K1  (Hollmann)
781 WYDKGECGSKDSGSKDKTSALSLSNVAGVFYILIGGLGLAMLVALIEFCYKSRSESKRMK r-GluR-A   (Sommer)
781 WYDKGECGTGGGDSKDKTSALSLSNVAGVFYILIGGLGLAMLVALIEFCYKSRSESKRMK r-GluR-A   (Keinaenen)
781 WYDKGECGSKDSGSKDKTSALSLSNVAGVFYILIGGLGLAMLVALIEFCYKSRSESKRMK m-alpha1   (Sakimura)

841 GFCLIPQQSINEAIRTSTLPRNS-GAGASSGGSGENGRVVSHDFPKSMQSIPCMSHSSGM h-KR4    (Potier)
841 GFCLIPQQSINEAIRTSTLPRNSAGTAPSSGGSGENGRVVSHDFPKSMQSIPCMSHSSGM h-GluH1   (Puckett)
841 GFCLIPQQSINEAIRTSTLPRNS-GAGASSGGSGENGRVVSHDFPKSMQSIPCMSHSSGM h-HBGR1   (Sun)
841 GFCLIPQQSINEAIRTSTLPRNSGAGASGGGGSGENGRVVSQDFPKSMQSIPSMSHSSGM r-GluR-K1  (Hollmann)
841 GFCLIPQQSINEAIRTSTLPRNSGAGASGGGGSGENGRVVSQDFPKSMQSIPSMSHSSGM r-GluR-A   (Sommer)
841 GFCLIPQQSINEAIRTSTLPRNSGAGASGGGGSGENGRVVSQDFPKSMQSIPSMSHSSGM r-GluR-A   (Keinaenen)
841 GFCLIPQQSINEAIRTSTLPRNSGAGASGGSGSGENGRVVSQDFPKSMQSIPCMSHSSGM m-alpha1   (Sakimura)

900 PLGATGL                                                      h-KR4    (Potier)
901 PLGATGL                                                      h-GluH1   (Puckett)
900 PLGATGL                                                      h-HBGR1   (Sun)
901 PLGATGL                                                      r-GluR-K1  (Hollmann)
901 PLGATGL                                                      r-GluR-A   (Sommer)
901 PLGATGL                                                      r-GluR-A   (Keinaenen)
901 PLGATGL                                                      m-alpha1   (Sakimura)
```

FIGURE 3a(2).

NR2C, (p) NR2D, (q) KBP, (r) invertebrate subunits, and (s) DNMDAR1/NMDAR1. Amino acids that can be changed by RNA editing in GluR-B, GluR5, and GluR6 are indicated by dots. In alignments a to d, amino acids encoded by FLIP and FLOP exons are marked on top by a broken line. FLIP sequences are characterized by AKDSG, and FLOP sequences by SGGGD. In all alignments except r and s, residues differing between the aligned sequences are boxed. In r and s, identical residues are boxed.

```
1     MQKIMHISVLLSPVLWGLIFGVSSNSIQIGGLFPRGADQEYSAFRVGMVQFSTSEFRLTP r-GluR2   (Boulter)
1     MQKIMHISVLLSPVLWGLIFGVSSNSIQIGGLFPRGADQEYSAFRVGMVQFSTSEFRLTP r-GluR-B  (Keinaenen)
1     MQKIMHISVLLSPVLWGLIFGVSSNSIQIGGLFPRGADQEYSAFRVGMVQFSTSEFRLTP r-GluR-K2 (Nakanishi)
1     MQKIMHISVLLSPVLWGLIFGVSSNSIQIGGLFPRGADQEYSAFRVGMVQFSTSEFRLTP r-GluR-B  (Sommer)
1     MQKIMHISVLLSPVLWGLIFGVSSNSIQIGGLFPRGADQEYSAFRVGMVQFSTSEFRLTP m-alpha2  (Sakimura)

61    HIDNLEVANSFAVTNAFCSQFSRGVYAIFGFYDKKSVNTITSFCGTLHVSFITPSFPTDG r-GluR2   (Boulter)
61    HIDNLEVANSFAVTNAFCSQFSRGVYAIFGFYDKKSVNTITSFCGTLHVSFITPSFPTDG r-GluR-B  (Keinaenen)
61    HIDNLEVANSFAVTNAFCSQFSRGVYAIFGFYDKKSVNTITSFCGTLHVSFITPSFPTDG r-GluR-K2 (Nakanishi)
61    HIDNLEVANSFAVTNAFCSQFSRGVYAIFGFYDKKSVNTITSFCGTLHVSFITPSFPTDG r-GluR-B  (Sommer)
61    HIDNLEVANSFAVTNAFCSQFSRGVYAIFGFYDKKSVNTITSFCGTLHVSFITPSFPTDG m-alpha2  (Sakimura)

121   THPFVIQMRPDLKGALLSLIEYYQWDKFAYLYDSDRGLSTLQAVLDSAAEKKWQVTAINV r-GluR2   (Boulter)
121   THPFVIQMRPDLKGALLSLIEYYQWDKFAYLYDSDRGLSTLQAVLDSAAEKKWQVTAINV r-GluR-B  (Keinaenen)
121   THPFVIQMRPDLKGALLSLIEYYQWDKFAYLYDSDRGLSTLQAVLDSAAEKKWQVTAINV r-GluR-K2 (Nakanishi)
121   THPFVIQMRPDLKGALLSLIEYYQWDKFAYLYDSDRGLSTLQAVLDSAAEKKWQVTAINV r-GluR-B  (Sommer)
121   THPFVIQMRPDLKGALLSLIEYYQWDKFAYLYDSDRGLSTLQAVLDSAAEKKWQVTAINV m-alpha2  (Sakimura)

181   GNINNDKKDETYRSLFQDLELKKERRVILDCERDKVNDIVDQVITIGKHVKGYHYIIANL r-GluR2   (Boulter)
181   GNINNDKKDETYRSLFQDLELKKERRVILDCERDKVNDIVDQVITIGKHVKGYHYIIANL r-GluR-B  (Keinaenen)
181   GNINNDKKDETYRSLFQDLELKKERRVILDCERDKVNDIVDQVITIGKHVKGYHYIIANL r-GluR-K2 (Nakanishi)
181   GNINNDKKDETYRSLFQDLELKKERRVILDCERDKVNDIVDQVITIGKHVKGYHYIIANL r-GluR-B  (Sommer)
181   GNINNDKKDETYRSLFQDLELKKERRVILDCERDKVNDIVDQVITIGKHVKGYHYIIANL m-alpha2  (Sakimura)

241   GFTDGDLLKIQFGGANVSGFQIVDYDDSLVSKFIERWSTLEEKEYPGAHTATIKYTSALT r-GluR2   (Boulter)
241   GFTDGDLLKIQFGGANVSGFQIVDYDDSLVSKFIERWSTLEEKEYPGAHTATIKYTSALT r-GluR-B  (Keinaenen)
241   GFTDGDLLKIQFGGANVSGFQIVDYDDSLVSKFIERWSTLEEKEYPGAHTATIKYTSALT r-GluR-K2 (Nakanishi)
241   GFTDGDLLKIQFGGANVSGFQIVDYDDSLVSKFIERWSTLEEKEYPGAHTATIKYTSALT r-GluR-B  (Sommer)
241   GFTDGDLLKIQFGGANVSGFQIVVYDDSLASKFIERWSTLEGKEYPGAHTATIKYTSALT m-alpha2  (Sakimura)

301   YDAVQVMTEAFRNLRKQRIEISRRGNAGDCLANPAVPWGQGVEIERALKQVQVEGLSGNI r-GluR2   (Boulter)
301   YDAVQVMTEAFRNLRKQRIEISRRGNAGDCLANPAVPWGQGVEIERALKQVQVEGLSGNI r-GluR-B  (Keinaenen)
301   YDAVQVMTEAFRNLRKQRIEISRRGNAGDCLANPAVPWGQGVEIERALKQVQVEGLSGNI r-GluR-K2 (Nakanishi)
301   YDAVQVMTEAFRNLRKQRIEISRRGNAGDCLANPAVPWGQGVEIERALKQVQVEGLSGNI r-GluR-B  (Sommer)
301   YDAVQVMTEAFRNLRKQRIEISRRGNAGDCLANPAVPWGQGVEIERALKQVQVEGLSGNI m-alpha2  (Sakimura)

361   KFDQNGKRINYTINIMELKTNGPRKIGYWSEVDKMVVTLTELPSGNDTSGLENKTVVVTT r-GluR2   (Boulter)
361   KFDQNGKRINYTINIMELKTNGPRKIGYWSEVDKMVVTLTELPSGNDTSGLENKTVVVTT r-GluR-B  (Keinaenen)
361   KFDQNGKRINYTINIMELKTNGPRKIGYWSEVDKMVVTLTELPSGNDTSGLENKTVVVTT r-GluR-K2 (Nakanishi)
361   KFDQNGKRINYTINIMELKTNGPRKIGYWSEVDKMVVTLTELPSGNDTSGLENKTVVVTT r-GluR-B  (Sommer)
361   KFDQNGKRINYTINIMELKTNGPRKIGYWSEVDKMVVTLTELPSGNDTSGLENKTVVVTT m-alpha2  (Sakimura)

421   ILESPYVMMKKNHEMLEGNERYEGYCVDLAAEIAKHCGFKYKLTIVGDGKYGARDADTKI r-GluR2   (Boulter)
421   ILESPYVMMKKNHEMLEGNERYEGYCVDLAAEIAKHCGFKYKLTIVGDGKYGARDADTKI r-GluR-B  (Keinaenen)
421   ILESPYVMTKKNHEMLEGNERYEGYCVDLAAEIAKHCGFKYKLTIVGDGKYGARDADTKI r-GluR-K2 (Nakanishi)
421   ILESPYVMMKKNHEMLEGNERYEGYCVDLAAEIAKHCGFKYKLTIVGDGKYGARDADTKI r-GluR-B  (Sommer)
421   ILESPYVMMKKNHEMLEGNERYEGYCVDLAAEIAKHCGFKYKLTIVGDGKYGARDADTKI m-alpha2  (Sakimura)

481   WNGMVGELVYGKADIAIAPLTITLVREEVIDFSKPFMSLGISIMIKKPQKSKPGVFSFLD r-GluR2   (Boulter)
481   WNGMVGELVYGKADIAIAPLTITLVREEVIDFSKPFMSLGISIMIKKPQKSKPGVFSFLD r-GluR-B  (Keinaenen)
481   WNGMVGELVYGKADIAIAPLTITLVREEVIDFSKPFMSLGISIMIKKPQKSKPGVFSFLD r-GluR-K2 (Nakanishi)
481   WNGMVGELVYGKADIAIAPLTITLVREEVIDFSKPFMSLGISIMIKKPQKSKPGVFSFLD r-GluR-B  (Sommer)
481   WNGMVGELVYGKADIAIAPLTITLVREEVIDFSKPFMSLGISIMIKKPQKSKPGVFSFLD m-alpha2  (Sakimura)
                     1
541   PLAYEIWMCIVFAYIGVSVVLFLVSRFSPYEWHTEEFEDGRETQSSESTNEFGIFNSLWF r-GluR2   (Boulter)
541   PLAYEIWMCIVFAYIGVSVVLFLVSRFSPYEWHTEEFEDGRETQSSESTNEFGIFNSLWF r-GluR-B  (Keinaenen)
541   PLAYEIWMCIVFAYIGVSVVLFLVSRFSPYEWHTEEFEDGRETQSSESTNEFGIFNSLWF r-GluR-K2 (Nakanishi)
541   PLAYEIWMCIVFAYIGVSVVLFLVSRFSPYEWHTEEFEDGRETQSSESTNEFGIFNSLWF r-GluR-B  (Sommer)
541   PLAYEIWMCIVFAYIGVSVVLFLVSRFSPYEWHTEEFEDGRETQSSESTNEFGIFNSLWF m-alpha2  (Sakimura)
```

FIGURE 3b(1).

2.6.1.2 GluR-B (GluR2)

The rat and mouse GluR-B subunits have been characterized (Boulter et al., 1990, Keinänen et al., 1990; Nakanishi et al., 1990; Sakimura et al., 1990). GluR-B (Figure 3b) can be viewed as the principal AMPA receptor subunit because heterooligomeric receptor channels containing GluR-B have low divalent ion permeabilities and different ion conductance and gating proper-

```
    —2——•——            ————————3——————
601 SLGAFMRQGCDISPRSLSGRIVGGVWWFFTLIIISSYTANLAAFLTVERMVSPIESAEDL r-GluR2   (Boulter)
601 SLGAFMRQGCDISPRSLSGRIVGGVWWFFTLIIISSYTANLAAFLTVERMVSPIESAEDL r-GluR-B  (Keinaenen)
601 SLGAFMRQGCDISPRSLSGRIVGGVWWFFTLIIISSYTANLAAFLTVERMVSPIESAEDL r-GluR-K2 (Nakanishi)
601 SLGAFMRQGCDISPRSLSGRIVGGVWWFFTLIIISSYTANLAAFLTVERMVSPIESAEDL r-GluR-B  (Sommer)
601 SLGAFMRQGCDISPRSLSGRIVGGVWWFFTLIIISSYTANLAAFLTVERMVSPIESAEDL m-alpha2  (Sakimura)

661 SKQTEIAYGTLDSGSTKEFFRRSKIAVFDKMWTYMRSAEPSVFVRTTAEGVARVRKSKGK r-GluR2   (Boulter)
661 SKQTEIAYGTLDSGSTKEFFRRSKIAVFDKMWTYMRSAEPSVFVRTTAEGVARVRKSKGK r-GluR-B  (Keinaenen)
661 SKQTEIAYGTLDSGSTKEFFRRSKIAVFDKMWTYMRSAEPSVFVRTTAEGVARVRKSKGK r-GluR-K2 (Nakanishi)
661 SKQTEIAYGTLDSGSTKEFFRRSKIAVFDKMWTYMRSAEPSVFVRTTAEGVARVRKSKGK r-GluR-B  (Sommer)
661 SKQTEIAYGTLDSGSTKEFFRRSKIAVFDKMWTYMRSAEPSVFVRTTAEGVARVRKSKGK m-alpha2  (Sakimura)
                                                    .— — — — — —
721 YAYLLESTMNEYIEQRKPCDTMKVGGNLDSKGYGIATPKGSSLGNAVNLAVLKLNEQGLL r-GluR2   (Boulter)
721 YAYLLESTMNEYIEQRKPCDTMKVGGNLDSKGYGIATPKGSSLGNAVNLAVLKLNEQGLL r-GluR-B  (Keinaenen)
721 YAYLLESTMNEYIEQRKPCDTMKVGGNLDSKGYGIATPKGSSLGTPVNLAVLKLSEQGVL r-GluR-K2 (Nakanishi)
721 YAYLLESTMNEYIEQRKPCDTMKVGGNLDSKGYGIATPKGSSLGTPVNLAVLKLSEQGVL r-GluR-B  (Sommer)
721 YAYLLESTMNEYIEQRKPCDTMKVGGNLDSKGYGIATPKGSSLGNAVNLAVLKLNEQGLL m-alpha2  (Sakimura)
    — — — — — — — —                   ———————4————————
781 DKLKNKWWYDKGECGSGGGDSKEKTSALSLSNVAGVFYILVGGLGLAMLVALIEFCYKSR r-GluR2   (Boulter)
781 DKLKNKWWYDKGECGSGGGDSKEKTSALSLSNVAGVFYILVGGLGLAMLVALIEFCYKSR r-GluR-B  (Keinaenen)
781 DKLKNKWWYDKGECGAKDSGSKEKTSALSLSNVAGVFYILVGGLGLAMLVALIEFCYKSR r-GluR-K2 (Nakanishi)
781 DKLKNKWWYDKGECGAKDSGSKEKTSALSLSNVAGVFYILVGGLGLAMLVALIEFCYKSR r-GluR-B  (Sommer)
781 DKLKNKWWYDKGECGSGGGDSKEKTSALSLSNVAGVFYILVGGLGLAMLVALIEFCYKSR m-alpha2  (Sakimura)

841 AEAKRMKVAKNPQNINPSSSQNSQNFATYKEGYNVYGIESVKI                  r-GluR2   (Boulter)
841 AEAKRMKVAKNPQNINPSSSQNSQNFATYKEGYNVYGIESVKI                  r-GluR-B  (Keinaenen)
841 AEAKRMKVAKNPQNINPSSSQNSQNFATYKEGYNVYGIESVKI                  r-GluR-K2 (Nakanishi)
841 AEAKRMKVAKNPQNINPSSSQNSQNFATYKEGYNVYGIESVKI                  r-GluR-B  (Sommer)
841 AEAKRMKVAKNAQNINPSSSQNSQNFATYKEGYNVYGIESVKI                  m-alpha2  (Sakimura)
```

FIGURE 3b(2).

ties than receptors without GluR-B (Burnashev et al., 1992a). In the channel-forming segment M2 GluR-B carries a positively charged residue (arginine) in a position (Q/R site) where the other AMPA receptor subunits harbor a glutamine. The arginine is the result of recoding the primary GluR-B transcript by RNA editing (Sommer et al., 1991; Higuchi et al., 1993).

Alternative splicing generates variants of the GluR-B subunit, which occur in Flip and Flop forms (Sommer et al., 1990) and in two carboxy-terminal variants (Köhler et al., 1994). The functional properties of the carboxy-terminal variants are unknown. There are two mRNA sizes for GluR-B in the rodent brain, approximately 4 and 6 kb in length (Partin et al., 1993). The human GluR-B gene has been localized on chromosome 4q32–33 (McNamara et al., 1992; Sun et al., 1992), and the M1 and M2 segments are encoded on a single exon (Sommer et al., 1991).

2.6.1.3 GluR-C (GluR3)

The nucleotide sequence for the rat GluR-C coding region was described independently by three laboratories (Boulter et al., 1990, Keinänen et al., 1990, Nakanishi et al., 1990) (Figure 3c). Similar to GluR-A and GluR-B, the GluR-C subunits are prominently expressed in the CNS, and the expression of the GluR-C Flip and Flop subunit forms is developmentally regulated (see above). The GluR-C mRNA size is approximately 6 kb in rat brain (Partin et al., 1993). The M1–M2 exon organization is identical to that of the GluR-B gene (Sommer et al., 1991). The human gene is localized on chromosome Xq25–26 (McNamara et al., 1992).

```
1   MGQSVLRAVFFLVLGLLGHSHGGFPNTISIGGLFMRNTVQEHSAFRFAVQLYNTNQNTTE r-GluR3   (Boulter)
1   MGQSVLRAVFFLVLGLLGHSHGGFPNTISIGGLFMRNTVQEHSAFRFAVQLYNTNQNTTE r-GluR-C  (Keinaenen)
1   MGQSVLRAVFFLVLGLLGHSHGGFPNTISIGGLFMRNTVQEHSAFRFAVQLYNTNQNTTE r-GluR-C  (Sommer)
1   MGQSVLRAVFFLVLGLLGHSHGGFPNTISIGGLFMRNTVQEHSAFRFAVQLYNTNQNTTE r-GluR-K3 (Nakanishi)

61  KPFHLNYHVDHLDSSNSFSVTNAFCSQFSRGVYAIFGFYDQMSMNTLTSFCGALHTSFVT r-GluR3   (Boulter)
61  KPFHLNYHVDHLDSSNSFSVTNAFCSQFSRGVYAIFGFYDQMSMNTLTSFCGALHTSFVT r-GluR-C  (Keinaenen)
61  KPFHLNYHVDHLDSSNSFSVTNAFCSQFSRGVYAIFGFYDQMSMNTLTSFCGALHTSFVT r-GluR-C  (Sommer)
61  KPFHLNYHVDHLDSSNSFSVTNAFCSQFSRGVYAIFGFYDQMSMNTLTSFCGALHTSFVT r-GluR-K3 (Nakanishi)

121 PSFPTDADVQFVIQMRPALKGAILSLLSYYKWEKFVYLYDTERGFSVLQAIMEAAVQNNW r-GluR3   (Boulter)
121 PSFPTDADVQFVIQMRPALKGAILSLLSYYKWEKFVYLYDTERGFSVLQAIMEAAVQNNW r-GluR-C  (Keinaenen)
121 PSFPTDADVQFVIQMRPALKGAILSLLSYYKWEKFVYLYDTERGFSVLQAIMEAAVQNNW r-GluR-C  (Sommer)
121 PSFPTDADVQFVIQMRPALKGAILSLLSYYKWEKFVYLYDTERGFSVLQAIMEAAVQNNW r-GluR-K3 (Nakanishi)

181 QVTARSVGNIKDVQEFRRIIEEMDRRQEKRYLIDCEVERINTILEQVVILGKHSRGYHYM r-GluR3   (Boulter)
181 QVTARSVGNIKDVQEFRRIIEEMDRRQEKRYLIDCEVERINTILEQVVILGKHSRGYHYM r-GluR-C  (Keinaenen)
181 QVTARSVGNIKDVQEFRRIIEEMDRRQEKRYLIDCEVERINTILEQVVILGKHSRGYHYM r-GluR-C  (Sommer)
181 QVTARSVGNIKDVQEFRRIIEEMDRRQEKRYLIDCEVERINTILEQVVILGKHSRGYHYM r-GluR-K3 (Nakanishi)

241 LANLGFTDILLERVMHGGANITGFQIVNNENPMVQQFIQRWVRLDEREFPEAKSAPLKYT r-GluR3   (Boulter)
241 LANLGFTDILLERVMHGGANITGFQIVNNENPMVQQFIQRWVRLDEREFPEAKNAPLKYT r-GluR-C  (Keinaenen)
241 LANLGFTDILLERVMHGGANITGFQIVNNENPMVQQFIQRWVRLDEREFPEAKNAPLKYT r-GluR-C  (Sommer)
241 LANLGFTDILLERVMHGGANITGFQIVNNENPMVQQFIQRWVRLDEREFPEAKNAPLKYT r-GluR-K3 (Nakanishi)

301 SALTHDAILVIAEAFRYLRRQRVDVSRRGSAGDCLANPAVPWGQGIDIERALKMVQVQGM r-GluR3   (Boulter)
301 SALTHDAILVIAEAFRYLRRQRVDVSRRGSAGDCLANPAVPWSQGIDIERALKMVQVQGM r-GluR-C  (Keinaenen)
301 SALTHDAILVIAEAFRYLRRQRVDVSRRGSAGDCLANPAVPWSQGIDIERALKMVQVQGM r-GluR-C  (Sommer)
301 SALTHDAILVIAEAFRYLRRQRVDVSRRGSAGDCLANPAVPWSQGIDIERALKMVQVQGM r-GluR-K3 (Nakanishi)

361 TGNIQFDTYGRRTNYTIDVYEMKVSGSRKAGYWSEYERFVPFSDQQISNDSSSSENRTIV r-GluR3   (Boulter)
361 TGNIQFDTYGRRTNYTIDVYEMKVSGSRKAGYWNEYERFVPFSDQQISNDSSSSENRTIV r-GluR-C  (Keinaenen)
361 TGNIQFDTYGRRTNYTIDVYEMKVSGSRKAGYWNEYERFVPFSDQQISNDSSSSENRTIV r-GluR-C  (Sommer)
361 TGNIQFDTYGRRTNYTIDVYEMKVSGSRKAGYWNEYERFVPFSDQQISNDSSSSENRTIV r-GluR-K3 (Nakanishi)

421 VTTILESPYVVMYKKNHEQLEGNERYEGYCVDLAYEIAKHVRIKYKLSIVGDGKYGARDPE r-GluR3   (Boulter)
421 VTTILESPYVVMYKKNHEQLEGNERYEGYCVDLAYEIAKHVRIKYKLSIVGDGKYGARDPE r-GluR-C  (Keinaenen)
421 VTTILESPYVVMYKKNHEQLEGNERYEGYCVDLAYEIAKHVRIKYKLSIVGDGKYGARDPE r-GluR-C  (Sommer)
421 VTTILESPYVVMYKKNHEQLEGNERYEGYCVDLAYEIAKHVRIKYKLSIVGDGKYGARDPE r-GluR-K3 (Nakanishi)

481 TKIWNGMVGELVYGRADIAVAPLTITLVREEVIDFSNAFMSLGISIMIKKPQKSKPGVFS r-GluR3   (Boulter)
481 TKIWNGMVGELVYGRADIAVAPLTITLVREEVIDFSKPFMSLGISIMIKKPQKSKPGVFS r-GluR-C  (Keinaenen)
481 TKIWNGMVGELVYGRADIAVAPLTITLVREEVIDFSKPFMSLGISIMIKKPQKSKPGVFS r-GluR-C  (Sommer)
481 TKIWNGMVGELVYGRADIAVAPLTITLVREEVIDFSKPFMSLGISIMIKKPQKSKPGVFS r-GluR-K3 (Nakanishi)
               ----------1----------                    ----
541 FLDPLAYEIWMCIVFAYIGVSVVLFLVSRFSPYEWHLEDNNEEPRDPQSPPDPPNEFGIF r-GluR3   (Boulter)
541 FLDPLAYEIWMCIVFAYIGVSVVLFLVSRFSPYEWHLEDNNEEPRDPQSPPDPPNEFGIF r-GluR-C  (Keinaenen)
541 FLDPLAYEIWMCIVFAYIGVSVVLFLVSRFSPYEWHLEDNNEEPRDPQSPPDPPNEFGIF r-GluR-C  (Sommer)
541 FLDPLAYEIWMCIVFAYIGVSVVLFLVSRFSPYEWHLEDNNEEPRDPQSPPDPPNEFGIF r-GluR-K3 (Nakanishi)
    ----2----------             ----------3----------
601 NSLWFSLGAFMQQGCDISPRSLSGRIVGGVWWFFTLIIISSYTANLAAFLTVERMVSPIE r-GluR3   (Boulter)
601 NSLWFSLGAFMQQGCDISPRSLSGRIVGGVWWFFTLIIISSYTANLAAFLTVERMVSPIE r-GluR-C  (Keinaenen)
601 NSLWFSLGAFMQQGCDISPRSLSGRIVGGVWWFFTLIIISSYTANLAAFLTVERMVSPIE r-GluR-C  (Sommer)
601 NSLWFSLGAFMQQGCDISPRSLSGRIVGGVWWFFTLIIISSYTANLAAFLTVERMVSPIE r-GluR-K3 (Nakanishi)

661 SAEDLAKQTEIAYGTLDSGSTKEFFRRSKIAV-EKMWSYMKSAEPSVFTKTTADGVARVR r-GluR3   (Boulter)
661 SAEDLAKQTEIAYGTLDSGSTKEFFRRSKIAVYEKMWSYMKSAEPSVFTKTTADGVARVR r-GluR-C  (Keinaenen)
661 SAEDLAKQTEIAYGTLDSGSTKEFFRRSKIAVYEKMWSYMKSAEPSVFTKTTADGVARVR r-GluR-C  (Sommer)
661 SAEDLAKQTEIAYGTLDSGSTKEFFRRSKIAVYEKMWSYMKSAEPSVFTKTTADGVARVR r-GluR-K3 (Nakanishi)
```

FIGURE 3c(1).

2.6.1.4 GluR-D (GluR4)

For the GluR-D subunit only the rat cDNA has been isolated (Boulter et al., 1990; Keinänen et al., 1990; Gallo et al., 1992). It is the lowest expressed AMPA receptor subunit in the rodent brain. Highest expression levels are seen in the cerebellar granule cells. In addition to the Flip and Flop splice forms, GluR-D occurs with two different carboxy-termini (Figure 3d), also the products of differential splicing (Gallo et al., 1992). The functional consequence of this carboxy-terminal sequence difference is unknown. GluR-D mRNA sizes of

```
                                                                        — — — —
720 KSKGKFAFLLESTMNEYIEQRKPCDTMKVGGNLDSKGYGVATPKGSALGNAVNLAVLKLN r-GluR3    (Boulter)
721 KSKGKFAFLLESTMNEYIEQRKPCDTMKVGGNLDSKGYGVATPKGSALGNAVNLAVLKLN r-GluR-C   (Keinaenen)
721 KSKGKFAFLLESTMNEYIEQRKPCDTMKVGGNLDSKGYGVATPKGSALGTPVNLAVLKLS r-GluR-C   (Sommer)
721 KSKGKFAFLLESTMNEYIEQRKPCDTMKVGGNLDSKGYGVATPKGSALGTPVNLAVLKLS r-GluR-K3  (Nakanishi)
    — — — — — — — — — — — .                    ————————4————————
780 EQGLLDKLKNKW-YDKGECGSGGGDSKDKTSALSLSNVAGVFYILVGGLGLAMMVALIEF r-GluR3    (Boulter)
781 EQGLLDKLKNKWWYDKGECGSGGGDSKDKTSALSLSNVAGVFYILVGGLGLAMMVALIEF r-GluR-C   (Keinaenen)
781 EQGILDKLKNKWWYDKGECGAKDSGSKDKTSALSLSNVAGVFYILVGGLGLAMMVALIEF r-GluR-C   (Sommer)
781 EQGILDKLKNKWWYDKGECGAKDSGSKDKTSALSLSNVAGVFYILVGGLGLAMMVALIEF r-GluR-K3  (Nakanishi)

839 CYKSRAESKRMKLTKNTQNFKPAPATNTQNYATYREGYNVYGTESVKI             r-GluR3    (Boulter)
841 CYKSRAESKRMKLTKNTQNFKPAPATNTQNYATYREGYNVYGTESVKI             r-GluR-C   (Keinaenen)
841 CYKSRAESKRMKLTKNTQNFKPAPATNTQNYATYREGYNVYGTESVKI             r-GluR-C   (Sommer)
841 CYKSRAESKRMKLTKNTQNFKPAPATNTQNYATYREGYNVYGTESVKI             r-GluR-K3  (Nakanishi)
```

FIGURE 3c(2).

7, 4.5, and 3 kb have been reported in rat brain (Gallo et al., 1992). The M1–M2 exon organization is identical to that of the GluR-B gene (Sommer et al., 1991). The human GluR-D gene is on chromosome 11q22–23 (McNamara et al., 1992).

2.6.2 High-Affinity Kainate Receptor Subunits

The potent neurotoxins kainate and domoate bind to high-affinity sites in the mammalian brain (Monaghan and Cotman, 1982). These sites fall in two classes with respect to their affinity to kainate: low-affinity sites and high-affinity sites (Coyle, 1983; Hampson et al., 1987). Two subunit families from which these kainate receptors can be reconstituted in heterologous expression systems have been identified by molecular cloning. The subunits of one family — GluR5, GluR6, and GluR7 — show 70% sequence identity among each other and 30% to AMPA receptor subunits. Upon *in vitro* expression each subunit forms binding sites for [^{3}H]kainate with moderate to high affinities. Homooligomeric GluR5 and GluR6 channels can be activated by glutamate and kainate and both agonists elicit fast desensitizing currents (Table 2). The collective expression patterns of GluR5 to 7 in the rat brain largely mimic the distribution of high-affinity binding sites for [^{3}H]kainate (Monaghan and Cotman, 1982; Wisden and Seeburg, 1993).

Two other subunits (KA-1 and KA-2) comprise the second family. KA-1 and KA-2 differ from GluR5 to 7 by sequence and function. They show about 36% sequence similarity to GluR5, 6, and 7, and neither subunit forms glutamate-gated channels when expressed on its own. Both coassemble with GluR5 or GluR6 to form channels of heterooligomeric configurations (see Table 2 for functional properties of heterooligomeric channels). KA-1 and KA-2 gene expression does not appear to be subject to RNA editing, and no alternatively spliced forms have been reported.

The expression levels of the kainate receptor subunits are generally lower than levels observed for the AMPA receptor subunits. The subunits appear to be developmentally regulated, and are generally more restricted to specific brain areas (Bettler et al., 1990; Wisden and Seeburg, 1993;

```
1   MRIICRQIVLLFSGFWGLAMGAFPSSVQIGGLFIRNTDQEYTAFRLAIFLHNTSPNASEA r-GluR4  (Bettler)
1   MRIICRQIVLLFSGFWGLAMGAFPSSVQIGGLFIRNTDQEYTAFRLAIFLHNTSPNASEA r-GluR4c (Gallo)
1   MRIICRQIVLLFSGFWGLAMGAFPSSVQIGGLFIRNTDQEYTAFRLAIFLHNTSPNASEA r-GluR-D (Keinaenen)
1   MRIICRQIVLLFSGFWGLAMGAFPSSVQIGGLFIRNTDQEYTAFRLAIFLHNTSPNASEA r-GluR-D (Sommer)

61  PFNLVPHVDNIETANSFAVTNAFCSQYSRGVFAIFGLYDKRSVHTLTSFCRRLHISLITP r-GluR4  (Bettler)
61  PFNLVPHVDNIETANSFAVTNAFCSQYSRGVFAIFGLYDKRSVHTLTSFCSALHISLITP r-GluR4c (Gallo)
61  PFNLVPHVDNIETANSFAVTNAFCSQYSRGVFAIFGLYDKRSVHTLTSFCSALHISLITP r-GluR-D (Keinaenen)
61  PFNLVPHVDNIETANSFAVTNAFCSQYSRGVFAIFGLYDKRSVHTLTSFCSALHISLITP r-GluR-D (Sommer)

121 SFPTEGESQFVLQLRPSLRGALLSLLDHYEWNCFVFLYDTDRGYSILQAIMEKAGQNGWH r-GluR4  (Bettler)
121 SFPTEGESQFVLQLRPSLRGALLSLLDHYEWNCFVFLYDTDRGYSILQAIMEKAGQNGWH r-GluR4c (Gallo)
121 SFPTEGESQFVLQLRPSLRGALLSLLDHYEWNCFVFLYDTDRGYSILQAIMEKAGQNGWH r-GluR-D (Keinaenen)
121 SFPTEGESQFVLQLRPSLRGALLSLLDHYEWNCFVFLYDTDRGYSILQAIMEKAGQNGWH r-GluR-D (Sommer)

181 VSAICVENFNDVSYRQLLEELDRRQEKKFVIDCEIERLQNILEQIVSVGKHVKGYHYIIA r-GluR4  (Bettler)
181 VSAICVENFNDVSYRQLLEELDRRQEKKFVIDCEIERLQNILEQIVSVGKHVKGYHYIIA r-GluR4c (Gallo)
181 VSAICVENFNDVSYRQLLEELDRRQEKKFVIDCEIERLQNILEQIVSVGKHVKGYHYIIA r-GluR-D (Keinaenen)
181 VSAICVENFNDVSYRQLLEELDRRQEKKFVIDCEIERLQNILEQIVSVGKHVKGYHYIIA r-GluR-D (Sommer)

241 NLGFKDISLERFIHGGANVTGFQLVDFNTPMVTKLMDRWKKLDQREYPGSETPPKYTSAL r-GluR4  (Bettler)
241 NLGFKDISLERFIHGGANVTGFQLVDFNTPMVTKLMDRWKKLDQREYPGSETPPKYTSAL r-GluR4c (Gallo)
241 NLGFKDISLERFIHGGANVTGFQLVDFNTPMVTKLMDRWKKLDQREYPGSETPPKYTSAL r-GluR-D (Keinaenen)
241 NLGFKDISLERFIHGGANVTGFQLVDFNTPMVTKLMDRWKKLDQREYPGSETPPKYTSAL r-GluR-D (Sommer)

301 TYDGVLVMAETFRSLRRQKIDISRRGNAGDCLANPAAPWGQGIDMERTLKQVRIQGLTGN r-GluR4  (Bettler)
301 TYDGVLVMAETFRSLRRQKIDISRRGNAGDCLANPAAPWGQGIDMERTLKQVRIQGLTGN r-GluR4c (Gallo)
301 TYDGVLVMAETFRSLRRQKIDISRRGNAGDCLANPAAPWGQGIDMERTLKQVRIQGLTGN r-GluR-D (Keinaenen)
301 TYDGVLVMAETFRSLRRQKIDISRRGNAGDCLANPAAPWGQGIDMERTLKQVRIQGLTGN r-GluR-D (Sommer)

361 VQFDHYGRRVNYTMDVFELKSTGPRKVGYWNDMDKLVLIQDMPTLGNDTAAIENRTVVVT r-GluR4  (Bettler)
361 VQFDHYGRRVNYTMDVFELKSTGPRKVGYWNDMDKLVLIQDMPTLGNDTAAIENRTVVVT r-GluR4c (Gallo)
361 VQFDHYGRRVNYTMDVFELKSTGPRKVGYWNDMDKLVLIQDMPTLGNDTAAIENRTVVVT r-GluR-D (Keinaenen)
361 VQFDHYGRRVNYTMDVFELKSTGPRKVGYWNDMDKLVLIQDMPTLGNDTAAIENRTVVVT r-GluR-D (Sommer)

421 TIMESPYVMYKKNHEMFEGNDKYEGYCVDLASESAKHIGIKYKIAIVPDGKYGARDADTK r-GluR4  (Bettler)
421 TIMESPYVMYKKNHEMFEGNDKYEGYCVDLASEIAKHIGIKYKIAIVPDGKYGARDADTK r-GluR4c (Gallo)
421 TIMESPYVMYKKNHEMFEGNDKYEGYCVDLASEIAKHIGIKYKIAIVPDGKYGARDADTK r-GluR-D (Keinaenen)
421 TIMESPYVMYKKNHEMFEGNDKYEGYCVDLASEIAKHIGIKYKIAIVPDGKYGARDADTK r-GluR-D (Sommer)

481 IWNGMVGELVYGKAEIAIAPLTITLVREEVIDFSKPFMSLGISIMIKKPQKSKPGVFSFL r-GluR4  (Bettler)
481 IWNGMVGELVYGKAEIAIAPLTITLVREEVIDFSKPFMSLGISIMIKKPQKSKPGVFSFL r-GluR4c (Gallo)
481 IWNGMVGELVYGKAEIAIAPLTITLVREEVIDFSKPFMSLGISIMIKKPQKSKPGVFSFL r-GluR-D (Keinaenen)
481 IWNGMVGELVYGKAEIAIAPLTITLVREEVIDFSKPFMSLGISIMIKKPQKSKPGVFSFL r-GluR-D (Sommer)
           ______1________                          ________
541 DPLAYEIWMCIVFAYIGVSVVLFLVSRFSPYEWHTEEPEDGKEGPSDQPPNEFGIFNSLW r-GluR4  (Bettler)
541 DPLAYEIWMCIVFAYIGVSVVLFLVSRFSPYEWHTEEPEDGKEGPSDQPPNEFGIFNSLW r-GluR4c (Gallo)
541 DPLAYEIWMCIVFAYIGVSVVLFLVSRFSPYEWHTEEPEDGKEGPSDQPPNEFGIFNSLW r-GluR-D (Keinaenen)
541 DPLAYEIWMCIVFAYIGVSVVLFLVSRFSPYEWHTEEPEDGKEGPSDQPPNEFGIFNSLW r-GluR-D (Sommer)
    __2______            _______3________
601 FSLGAFMQQGCDISPRSLSGRIVGGVWWFFTLIIISSYTANLAAFLTVERMVSPIESAED r-GluR4  (Bettler)
601 FSLGAFMQQGCDISPRSLSGRIVGGVWWFFTLIIISSYTANLAAFLTVERMVSPIESAED r-GluR4c (Gallo)
601 FSLGAFMQQGCDISPRSLSGRIVGGVWWFFTLIIISSYTANLAAFLTVERMVSPIESAED r-GluR-D (Keinaenen)
601 FSLGAFMQQGCDISPRSLSGRIVGGVWWFFTLIIISSYTANLAAFLTVERMVSPIESAED r-GluR-D (Sommer)

661 LAKQTEIAYGTLDSGSTKEFFRRSKIAVYEKMWTYMRSAEPSVFTRTTAEGVARVRKSKG r-GluR4  (Bettler)
661 LAKQTEIAYGTLDSGSTKEFFRRSKIAVYEKMWTYMRSAEPSVFTRTTAEGVARVRKSKG r-GluR4c (Gallo)
661 LAKQTEIAYGTLDSGSTKEFFRRSKIAVYEKMWTYMRSAEPSVFTRTTAEGVARVRKSKG r-GluR-D (Keinaenen)
661 LAKQTEIAYGTLDSGSTKEFFRRSKIAVYEKMWTYMRSAEPSVFTRTTAEGVARVRKSKG r-GluR-D (Sommer)
```

FIGURE 3d(1).

Bahn et al., 1994). The physiological role of the high-affinity kainate receptors in central neurons has not been resolved. With the exception of peripheral neurons (Huettner, 1990; Partin et al., 1993) desensitizing responses to kainate — a hallmark of these receptor channels (Table 2) — have not been reported when patching CNS cells in slices, only in a subset of cultured hippocampal cells (Lerma et al., 1993). However, kainic acid depolarizes hippocampal CA3 neurons in a slice preparation (Robinson and Deadwyler, 1981).

```
721 KFAFLLESTMNEYIEQRKPCDTMKVGGNLDSKGYGVATPKGSSLRTPVNLAVLKLSEAGV r-GluR4   (Bettler)
721 KFAFLLESTMNEYIEQRKPCDTMKVGGNLDSKGYGVATPKGSSLGNAVNLAVLKLNEQGL r-GluR4c  (Gallo)
721 KFAFLLESTMNEYTEQRKPCDTMKVGGNLDSKGYGVATPKGSSLGNAVNLAVLKLNEQGL r-GluR-D  (Keinaenen)
721 KFAFLLESTMNEYTEQRKPCDTMKVGGNLDSKGYGVATPKGSSLRTPVNLAVLKLSEAGV r-GluR-D  (Sommer)
                                          4
781 LDKLKNKWWYDKGECGPKDSGSKDKTSALSLSNVAGVFYILVGGLGLAMLVALIEFCYKS r-GluR4   (Bettler)
781 LDKLKNKWWYDKGECGSGGGDSKDKTSALSLSNVAGVFYILVGGLGLAMLVALIEFCYKS r-GluR4c  (Gallo)
781 LDKLKNKWWYDKGECGSGGGDSKDKTSALSLSNVAGVFYILVGGLGLAMLVALIEFCYKS r-GluR-D  (Keinaenen)
781 LDKLKNKWWYDKGECGPKDSGSKDKTSALSLSNVAGVFYILVGGLGLAMLVALIEFCYKS r-GluR-D  (Sommer)

841 RAEAKRMKLTFSEAIRNKARLSITGSVGENGRVLTPDCPKAVHTGTAIRQSSGLAVIASD r-GluR4   (Bettler)
841 RAEAKRMKVAKSAQTFNPTSSQNTHNLATYREGYNVYGTESIKI                 r-GluR4c  (Gallo)
841 RAEAKRMKLTFSEATRNKARLSITGSVGENGRVLTPDCPKAVHTGTAIRQSSGLAVIASD r-GluR-D  (Keinaenen)
841 RAEAKRMKLTFSEATRNKARLSITGSVGENGRVLTPDCPKAVHTGTAIRQSSGLAVIASD r-GluR-D  (Sommer)

901 LP                                                           r-GluR4   (Bettler)

901 LP                                                           r-GluR-D  (Keinaenen)
901 LP                                                           r-GluR-D  (Sommer)
```

FIGURE 3d(2).

2.6.2.1 GluR5

The rat GluR5 subunit cDNA was the first member of this family to be cloned (Bettler et al., 1990; Sommer et al., 1992). Nucleotide sequences for the human (h-GluR5-1d) and a partial cDNA for the mouse GluR5 (m-GluR5-3) were published recently (Gregor et al., 1993a). This glutamate receptor subunit (Figure 3e) displays 32% amino acid identity with the kainate/AMPA receptor subunits GluR-A to -D, and forms homomeric ion channels that are weakly responsive to L-glutamate in *Xenopus* oocytes and in mammalian expression systems.

GluR5 receptors exhibit properties of a high-affinity domoate- (K_D ~2 n*M*) and kainate- (K_D ~70 n*M*) binding site. The homomeric channels are gated in decreasing order of sensitivity by domoate, kainate, glutamate, and AMPA. In contrast to recombinantly expressed GluR-A to -D channels, currents elicited at GluR5 channels desensitize to all agonists (Sommer et al., 1992). Several different isoforms of the GluR5 subunit can be engendered by posttranscriptional modification of the GluR5 mRNA. Alternative splicing generates GluR5 subunits (Sommer et al., 1992; Gregor et al., 1993a) that differ with respect to an insertion of 15 residues in the extracellularly located large amino-terminal region and with respect to four different carboxy-terminal domains following M4 (Figure 3e). The functional differences of these GluR5 isoforms have not been resolved. Similar to the GluR-B subunit, the GluR5 subunit mRNA is affected by RNA editing. Analysis revealed that in the adult rodent brain between 30 and 70% of the GluR5 RNA is edited in the Q/R site in M2. It has not been reported that homomeric GluR5(Q) and GluR5(R) forms have different Ca^{2+} permeabilities but an effect on the I–V relationship in heteromeric GluR5(Q)/GluR5(R) has been observed (Sommer et al., 1992).

In the rodent brain the principal GluR5 mRNA is approximately 4 kb in length (Gregor et al., 1993a; Partin et al., 1993). The GluR5 gene is expressed in subsets of neurons throughout the developing and adult central and peripheral nervous systems. During embryogenesis, GluR5 transcripts

```
1   MEHGTLLAQPGLWTRDTSWALLYFLCYILPQTAPQVLRIGGIFETVENEPVNVEELAFKF h-GluR5-1d  (Gregor)
1   MERSTVLIQPGLWTRDTSWTLLYFLCYILPQTSPQVLRIGGIFETVENEPVNVEELAFKF r-GluR5-2   (Bettler)
1   MERSTVLIQPGLWTRDTSWTLLYFLCYILPQTSPQVLRIGGIFETVENEPVNVEELAFKF r-GluR5-1   (Bettler)
1   MERSTVLIQPGLWTRDTSWTLLYFLCYILPQTSPQVLRIGGIFETVENEPVNVEELAFKF r-GluR-5a   (Sommer)
1   MERSTVLIQPGLWTRDTSWTLLYFLCYILPQTSPQVLRIGGIFETVENEPVNVEELAFKF r-GluR-5b   (Sommer)
1   MERSTVLIQPGLWTRDTSWTLLYFLCYILPQTSPQVLRIGGIFETVENEPVNVEELAFKF r-GluR-5c   (Sommer)
1   MERGTVLIQPGLWTRDTSWTLLYFLCYILPQTSPQVLRIGGIFETVENEPVNVEELAFKF m-GluR5-3   (Gregor)

61  AVTSINRNRTLMPNTTLTYDIQRINLFDSFEASRRACDQLALGVAALFGPSHSSSVSAVQ h-GluR5-1d  (Gregor)
61  AVTSINRNRTLMPNTTLTYDIQRINLFDSFEASRRACDQLALGVAALFGPSHSSSVSAVQ r-GluR5-2   (Bettler)
61  AVTSINRNRTLMPNTTLTYDIQRINLFDSFEASRRACDQLALGVAALFGPSHSSSVSAVQ r-GluR5-1   (Bettler)
61  AVTSINRNRTLMPNTTLTYDIQRINLFDSFEASRRACDQLALGVAALFGPSHSSSVSAVQ r-GluR-5a   (Sommer)
61  AVTSINRNRTLMPNTTLTYDIQRINLFDSFEASRRACDQLALGVAALFGPSHSSSVSAVQ r-GluR-5b   (Sommer)
61  AVTSINRNRTLMPNTTLTYDIQRINLFDSFEASRRACDQLALGVAALFGPSHSSSVSAVQ r-GluR-5c   (Sommer)
61  AVTSINRNRTLMPNTTLTYDIQRINLFDSFEASRRACDQLALGVAALFGPSHSSSVSAVQ m-GluR5-3   (Gregor)

121 SICNALEVPHIQTRWKHPSVDNKDLFYINLYPDYAAISRAILDLVLYYNWKTVTVVYEDS h-GluR5-1d  (Gregor)
121 SICNALEVPHIQTRWKHPSVDSRDLFYINLYPDYAAISRAVLDLVLYYNWKTVTVVYEDS r-GluR5-2   (Bettler)
121 SICNALEVPHIQTRWKHPSVDSRDLFYINLYPDYAAISRAVLDLVLYYNWKTVTVVYEDS r-GluR5-1   (Bettler)
121 SICNALEVPHIQTRWKHPSVDSRDLFYINLYPDYAAISRAVLDLVLYYNWKTVTVVYEDS r-GluR-5a   (Sommer)
121 SICNALEVPHIQTRWKHPSVDSRDLFYINLYPDYAAISRAVLDLVLYYNWKTVTVVYEDS r-GluR-5b   (Sommer)
121 SICNALEVPHIQTRWKHPSVDSRDLFYINLYPDYAAISRAVLDLVLYYNWKTVTVVYEDS r-GluR-5c   (Sommer)
121 SICNALEVPHIQTRWKHPSVDNRDLFYINLYPDYAAISRAVLDLVLYYNWKTVTVVYEDS m-GluR5-3   (Gregor)

181 TGLIRLQELIKAPSRYNIKIKIRQLPSGNKDAKPLLKEMKKGKEFYVIFDCSHETAAEIL h-GluR5-1d  (Gregor)
181 TGLIRLQELIKAPSRYNIKIKIRQLPPANKDAKPLLKEMKKSKEFYVIFDCSHETAAEIL r-GluR5-2   (Bettler)
181 TGLIRLQELIKAPSRYNIKIKIRQLPPANKDAKPLLKEMKKSKEFYVIFDCSHETAAEIL r-GluR5-1   (Bettler)
181 TGLIRLQELIKAPSRYNIKIKIRQLPPANKDAKPLLKEMKKSKEFYVIFDCSHETAAEIL r-GluR-5a   (Sommer)
181 TGLIRLQELIKAPSRYNIKIKIRQLPPANKDAKPLLKEMKKSKEFYVIFDCSHETAAEIL r-GluR-5b   (Sommer)
181 TGLIRLQELIKAPSRYNIKIKIRQLPPANKDAKPLLKEMKKSKEFYVIFDCSHETAAEIL r-GluR-5c   (Sommer)
181 TGLIRLQELIKAPSRYNIKIKIRQLPSGNKDAKPLLKEMKKGKEFYVIFDCSHETAAEIL m-GluR5-3   (Gregor)

241 KQILFMGMMTEYYHYFFTTLDLFALDLELYRYSGVNMTGFRLLNIDNPHVSSIIEKWSME h-GluR5-1d  (Gregor)
241 KQILFMGMMTEYYHYFFTTLDLFALDLELYRYSGVNMTGFRLLNIDNPHVSSIIEKWSME r-GluR5-2   (Bettler)
241 KQILFMGMMTEYYHYFFTTLDLFALDLELYRYSGVNMTGFRLLNIDNPHVSSIIEKWSME r-GluR5-1   (Bettler)
241 KQILFMGMMTEYYHYFFTTLDLFALDLELYRYSGVNMTGFRLLNIDNPHVSSIIEKWSME r-GluR-5a   (Sommer)
241 KQILFMGMMTEYYHYFFTTLDLFALDLELYRYSGVNMTGFRLLNIDNPHVSSIIEKWSME r-GluR-5b   (Sommer)
241 KQILFMGMMTEYYHYFFTTLDLFALDLELYRYSGVNMTGFRLLNIDNPHVSSIIEKWSME r-GluR-5c   (Sommer)
241 KQILFMGMMTEYYHYFFTTLDLFALDLELYRYSGVNMTGFRLLNIDNPHVSSIIEKWSME m-GluR5-3   (Gregor)

301 RLQAPPRPETGLLDGMMTTEAALMYDAVYMVAIASHRASQLTVSSLQCHRHKPWRLGPRF h-GluR5-1d  (Gregor)
301 RLQAPPRPETGLLDGMMTTEAALMYDAVYMVAIASHRASQLTVSSLQCHRHKPWRLGPRF r-GluR5-2   (Bettler)
301 RLQAPPRPETGLLDGMMTTEAALMYDAVYMVAIASHRASQLTVSSLQCHRHKPWRLGPRF r-GluR5-1   (Bettler)
301 RLQAPPRPETGLLDGMMTTEAALMYDAVYMVAIASHRASQLTVSSLQCHRHKPWRLGPRF r-GluR-5a   (Sommer)
301 RLQAPPRPETGLLDGMMTTEAALMYDAVYMVAIASHRASQLTVSSLQCHRHKPWRLGPRF r-GluR-5b   (Sommer)
301 RLQAPPRPETGLLDGMMTTEAALMYDAVYMVAIASHRASQLTVSSLQCHRHKPWRLGPRF r-GluR-5c   (Sommer)
301 RLQAPPRPETGLLDGVMTTEAALMYDAVYMVAIASHRASQLTVSSLQCHRHKPWRLGPRF m-GluR5-3   (Gregor)

361 MNLIKEARWDGLTGHITFNKTNGLRKDFDLDIISLKEEGTEKAAGEVSKHLYKVWKKIGI h-GluR5-1d  (Gregor)
361 MNLIKEARWDGLTGRITFNKTDGLRKDFDLDIISLKEEGTEK---------------IGI r-GluR5-2   (Bettler)
361 MNLIKEARWDGLTGRITFNKTDGLRKDFDLDIISLKEEGTEKASGEVSKHLYKVWKKIGI r-GluR5-1   (Bettler)
361 MNLIKEARWDGLTGRITFNKTDGLRKDFDLDIISLKEEGTEKASGEVSKHLYKVWKKIGI r-GluR-5a   (Sommer)
361 MNLIKEARWDGLTGRITFNKTDGLRKDFDLDIISLKEEGTEKASGEVSKHLYKVWKKIGI r-GluR-5b   (Sommer)
361 MNLIKEARWDGLTGRITFNKTDGLRKDFDLDIISLKEEGTEKASGEVSKHLYKVWKKIGI r-GluR-5c   (Sommer)
361 MNLIKEARWDGLTGRITFNKTDGLRKDFDLDIISLKEEGTEK---------------IGI m-GluR5-3   (Gregor)

421 WNSNSGLNMTDSNKDKSSNITDSLANRTLIVTTILEEPYVMYRKSDKPLYGNDRFEGYCL h-GluR5-1d  (Gregor)
406 WNSNSGLNMTDGNRDRSNNITDSLANRTLIVTTILEEPYVMYRKSDKPLYGNDRFEGYCL r-GluR5-2   (Bettler)
421 WNSNSGLNMTDGNRDRSNNITDSLANRTLIVTTILEEPYVMYRKSDKPLYGNDRFEGYCL r-GluR5-1   (Bettler)
421 WNSNSGLNMTDGNRDRSNNITDSLANRTLIVTTILEEPYVMYRKSDKPLYGNDRFEAYCL r-GluR-5a   (Sommer)
421 WNSNSGLNMTDGNRDRSNNITDSLANRTLIVTTILEEPYVMYRKSDKPLYGNDRFEAYCL r-GluR-5b   (Sommer)
421 WNSNSGLNMTDGNRDRSNNITDSLANRTLIVTTILEEPYVMYRKSDKPLYGNDRFEAYCL r-GluR-5c   (Sommer)
406 WNSNSGLNMTDGNRDRSNNITDSLANRTLIVTTILEEPYVMYRKSDKPLYGNDRFEGYCL m-GluR5-3   (Gregor)
```

FIGURE 3e(1).

are detected in areas of neuronal differentiation and synapse formation (Bettler et al., 1990; Bahn et al., 1994). GluR5 transcripts can be found by *in situ* hybridization in the cingulate and piriform cortex, the subiculum, lateral septal nuclei, anteroventral thalamus, suprachiasmatic nucleus, the tegmental nuclei, pontine nuclei, and Purkinje cells (Wisden and Seeburg,

```
481 DLLKELSNILGFIYDVKLVPDGKYGAQNDKGEWNGMVKELIDHRADLAVAPLTITYVREK h-GluR5-1d  (Gregor)
466 DLLKELSNILGFLYDVKLVPDGKYGAQNDKGEWNGMVKELIDHRADLAVAPLTITYVREK r-GluR5-2   (Bettler)
481 DLLKELSNILGFLYDVKLVPDGKYGAQNDKGEWNGMVKELIDHRADLAVAPLTITYVREK r-GluR5-1   (Bettler)
481 DLLKELSNILGFLYDVKLVPDGKYGAQNDKGEWNGMVKELIDHRADLAVAPLTITYVREK r-GluR-5a   (Sommer)
481 DLLKELSNILGFLYDVKLVPDGKYGAQNDKGEWNGMVKELIDHRADLAVAPLTITYVREK r-GluR-5b   (Sommer)
481 DLLKELSNILGFLYDVKLVPDGKYGAQNDKGEWNGMVKELIDHRADLAVAPLTITYVREK r-GluR-5c   (Sommer)
466 DLLKELSNILGFLYDVKLVPDGKYGAQNDKGEWNGMVKELIDHRADLAVAPLTITYVREK m-GluR5-3   (Gregor)
                                                           1
541 VIDFSKPFMTLGISILYRKPNGTNPGVFSFLNPLSPDIWMYVLLACLGVSCVLFVIARFT h-GluR5-1d  (Gregor)
526 VIDFSKPFMTLGISILYRKPNGTNPGVFSFLNPLSPDIWMYVLLACLGVSCVLFVIARFT r-GluR5-2   (Bettler)
541 VIDFSKPFMTLGISILYRKPNGTNPGVFSFLNPLSPDIWMYVLLACLGVSCVLFVIARFT r-GluR5-1   (Bettler)
541 VIDFSKPFMTLGISILYRKPNGTNPGVFSFLNPLSPDIWMYVLLACLGVSCVLFVIARFT r-GluR-5a   (Sommer)
541 VIDFSKPFMTLGISILYRKPNGTNPGVFSFLNPLSPDIWMYVLLACLGVSCVLFVIARFT r-GluR-5b   (Sommer)
541 VIDFSKPFMTLGISILYRKPNGTNPGVFSFLNPLSPDIWMYVLLACLGVSCVLFVIARFT r-GluR-5c   (Sommer)
526 VIDFSKPFMTLGISILYRKPNGTNPGVFSFLNPLSPDIWMYVLLACLGVSCVLFVIARFT m-GluR5-3   (Gregor)
                             2        •
601 PYEWYNPHPCNPDSDVVENNFTLLNSFWFGVGALMQQGSELMPKALSTRIVGGIWWFFTL h-GluR5-1d  (Gregor)
586 PYEWYNPHPCNPDSDVVENNFTLLNSFWFGVGALMQQGSELMPKALSTRIVGGIWWFFTL r-GluR5-2   (Bettler)
601 PYEWYNPHPCNPDSDVVENNFTLLNSFWFGVGALMQQGSELMPKALSTRIVGGIWWFFTL r-GluR5-1   (Bettler)
601 PYEWYNPHPCNPDSDVVENNFTLLNSFWFGVGALMQQGSELMPKALSTRIVGGIWWFFTL r-GluR-5a   (Sommer)
601 PYEWYNPHPCNPDSDVVENNFTLLNSFWFGVGALMQQGSELMPKALSTRIVGGIWWFFTL r-GluR-5b   (Sommer)
601 PYEWYNPHPCNPDSDVVENNFTLLNSFWFGVGALMQQGSELMPKALSTRIVGGIWWFFTL r-GluR-5c   (Sommer)
586 PYEWYNPHPCNPDSDVVENNFTLLNSFWFGVGALMRQGSELMPKALSTRIVGGIWWFFTL m-GluR5-3   (Gregor)
     3
661 IIISSYTANLAAFLTVERMESPIDSADDLAKQTKIEYGAVRDGSTMTFFKKSKISTYEKM h-GluR5-1d  (Gregor)
646 IIISSYTANLAAFLTVERMESPIDSADDLAKQTKIEYGAVRDGSTMTFFKKSKISTYEKM r-GluR5-2   (Bettler)
661 IIISSYTANLAAFLTVERMESPIDSADDLAKQTKIEYGAVRDGSTMTFFKKSKISTYEKM r-GluR5-1   (Bettler)
661 IIISSYTANLAAFLTVERMESPIDSADDLAKQTKIEYGAVRDGSTMTFFKKSKISTYEKM r-GluR-5a   (Sommer)
661 IIISSYTANLAAFLTVERMESPIDSADDLAKQTKIEYGAVRDGSTMTFFKKSKISTYEKM r-GluR-5b   (Sommer)
661 IIISSYTANLAAFLTVERMESPIDSADDLAKQTKIEYGAVRDGSTMTFFKKSKISTYEKM r-GluR-5c   (Sommer)
646 IIISSYTANLAAFLTVERMESPIDSADDLAKQTKIEYGAVRDGSTMTFFKKSKISTYEKM m-GluR5-3   (Gregor)

721 WAFMSSRQQTALVRNSDEGIQRVLTTDYALLMESTSIEYVTQRNCNLTQIGGLIDSKGYG h-GluR5-1d  (Gregor)
706 WAFMSSRQQSALVKNSDEGIQRVLTTDYALLMESTSIEYVTQRNCNLTQIGGLIDSKGYG r-GluR5-2   (Bettler)
721 WAFMSSRQQSALVKNSDEGIQRVLTTDYALLMESTSIEYVTQRNCNLTQIGGLIDSKGYG r-GluR5-1   (Bettler)
721 WAFMSSRQQSALVKNSDEGIQRVLTTDYALLMESTSIEYVTQRNCNLTQIGGLIDSKGYG r-GluR-5a   (Sommer)
721 WAFMSSRQQSALVKNSDEGIQRVLTTDYALLMESTSIEYVTQRNCNLTQIGGLIDSKGYG r-GluR-5b   (Sommer)
721 WAFMSSRQQSALVKNSDEGIQRVLTTDYALLMESTSIEYVTQRNCNLTQIGGLIDSKGYG r-GluR-5c   (Sommer)
706 WAFMSSRQPVL                                                  m-GluR5-3   (Gregor)

781 VGTPIGSPYRDKITIAILQLQEEGKLHMMKEKWWRGNGCPEEDNKEASALGVENIGGIFI h-GluR5-1d  (Gregor)
766 VGTPIGSPYRDKITIAILQLQEEGKLHMMKEKWWRGNGCPEEDSKEASALGVENIGGIFI r-GluR5-2   (Bettler)
781 VGTPIGSPYRDKITIAILQLQEEGKLHMMKEKWWRGNGCPEEDSKEASALGVENIGGIFI r-GluR5-1   (Bettler)
781 VGTPIGSPYRDKITIAILQLQEEGKLHMMKEKWWRGNGCPEEDSKEASALGVENIGGIFI r-GluR-5a   (Sommer)
781 VGTPIGSPYRDKITIAILQLQEEGKLHMMKEKWWRGNGCPEEDSKEASALGVENIGGIFI r-GluR-5b   (Sommer)
781 VGTPIGSPYRDKITIAILQLQEEGKLHMMKEKWWRGNGCPEEDSKEASALGVENIGGIFI r-GluR-5c   (Sommer)

       4
841 VLAAGLVLSVFVAIGEFIYKSRKNNDIEQAFCFFYGLQCKQTHPTNSTSGTTLSTDLECG h-GluR5-1d  (Gregor)
826 VLAAGLVLSVFVAIGEFLYKSRKNNDVEQ-----------------------------CL r-GluR5-2   (Bettler)
841 VLAAGLVLSVFVAIGEFLYKSRKNNDVEQ-----------------------------CL r-GluR5-1   (Bettler)
841 VLAAGLVLSVFVAIGEFLYKSRKNNDVEQ-----------------------------CL r-GluR-5a   (Sommer)
841 VLAAGLVLSVFVAIGEFLYKSRKNNDVEQHY                              r-GluR-5b   (Sommer)
841 VLAAGLVLSVFVAIGEFLYKSRKNNDVEQKGKSSRLRFYFRNKVRFHGSKKESLGVEKCL r-GluR-5c   (Sommer)

901 KLIREERGIRKQSSVHTV                                           h-GluR5-1d  (Gregor)
857 SFNAIMEELGISLKNQKKLKKKSRTKGKSSFTSILTCHQRRTQRKETVA            r-GluR5-2   (Bettler)
872 SFNAIMEELGISLKNQKKLKKKSRTKGKSSFTSILTCHQRRTQRKETVA            r-GluR5-1   (Bettler)
872 SFNAIMEELGISLKNQKKLKKKSRTKGKSSFTSILTCHQRRTQRKETVA            r-GluR-5a   (Sommer)
872                                                              r-GluR-5b   (Sommer)
901 SFNAIMEELGISLKNQKKLKKKSRTKGKSSFTSILTCHQRRTQRKETVA            r-GluR-5c   (Sommer)
```

FIGURE 3e(2).

1993). The rat GluR5 gene shows peaks of expression around birth in the sensory cortex, in CA1 interneurons in stratum oriens, and in the septum and thalamus (Bahn et al., 1994). The human gene for GluR5 has been mapped to chromosome 21q21.1–22.1 (Eubanks et al., 1993; Gregor et al., 1993b; Potier et al., 1993).

```
1   MKIISPVLSNLVFSRSIKVLLCLLWIGYSQGTTHVLRFGGIFEYVESGPMGAEELAFRFA r-GluR6   (Egebjerg)
1   MKIISPVLSNLVFSRSIKVLLCLLWIGYSQGTTHVLRFGGIFEYVESGPMGAEELAFRFA r-GluR-6  (Lomeli)
1   MKIISPVLSNLVFSRSIKVLLCLLWIGYSQGTTHVLRFGGIFEYVESGPMGAEELAFRFA m-beta2   (Morita)
1   MKIISPVLSNLVFSRSIKVLLCLLWIGYSQGTTHVLRFGGIFEYVESGPMGAEELAFRFA m-GluR6-2 (Gregor)

61  VNTINRNRTLLPNTTLTYDTQKINLYDSFEASKKACDQLSLGVAAIFGPSHSSSANAVQS r-GluR6   (Egebjerg)
61  VNTINRNRTLLPNTTLTYDTQKINLYDSFEASKKACDQLSLGVAAIFGPSHSSSANAVQS r-GluR-6  (Lomeli)
61  VNTINRNRTLLPNTTLTYDTQKINLYDSFEASKKACDQLSLGVAAIFGPSHSSSANAVQS m-beta2   (Morita)
61  VNTINRNRTLLPNTTLTYDTQKINLYDSFEASKKACDQLSLGVAAIFGPSHSSSANAVQS m-GluR6-2 (Gregor)

121 ICNALGVPHIQTRWKHQVSDNKDSFYVSLYPDFSSLSRAILDLVQFFKWKTVTVVYDDST r-GluR6   (Egebjerg)
121 ICNALGVPHIQTRWKHQVSDNKDSFYVSLYPDFSSLSRAILDLVQFFKWKTVTVVYDDST r-GluR-6  (Lomeli)
121 ICNALGVPHIQTRWKHQVSDNKDSFYVSLYPDFSSLSRAILDLVQFFKWKTVTVVYDDST m-beta2   (Morita)
121 ICNALGVPHIQTRWKHQVSDNKDSFYVSLYPDFSSLSRAILDLVQFFKWKTVTVVYDDST m-GluR6-2 (Gregor)

181 GLIRLQELIKAPSRYNLRLKIRQLPADTKDAKPLLKEMKRGKEFHVIFDCSHEMAAGILK r-GluR6   (Egebjerg)
181 GLIRLQELIKAPSRYNLRLKIRQLPADTKDAKPLLKEMKRGKEFHVIFDCSHEMAAGILK r-GluR-6  (Lomeli)
181 GLIRLQELIKAPSRYNLRLKIRQLPADTKDAKPLLKEMKRGKEFHVIFDCSHEMAAGILK m-beta2   (Morita)
181 GLIRLQELIKAPSRYNLRLKIRQLPADTKDAKPLLKEMKRGKEFHVIFDCSHEMAAGILK m-GluR6-2 (Gregor)

241 QALAMGMMTEYYHYI--TLDLFALDVEPYRYSGVNMTGFRILNTENTQVSSIIEKWSMER r-GluR6   (Egebjerg)
241 QALAMGMMTEYYHYIFTTLDLFALDVEPYRYSGVNMTGFRILNTENTQVSSIIEKWSMER r-GluR-6  (Lomeli)
241 QALAMGMMTEYYHYI--TLDLFALDVEPYRYSGVNMTGFRILNTENTQVSSIIEKWSMER m-beta2   (Morita)
241 QALAMGMMTEYYHYI--TLDLFALDVEPYRYSGVNMTGFRILNTENTQVSSIIEKWSMER m-GluR6-2 (Gregor)

299 LQAPPKPDSGLLDGFMTTDAALMYDAVHVVSVAVQQFPQMTVSSLQCNRHKPWRFGTRFM r-GluR6   (Egebjerg)
301 LQAPPKPDSGLLDGFMTTDAALMYDAVHVVSVAVQQFPQMTVSSLQCNRHKPWRFGTRFM r-GluR-6  (Lomeli)
299 LQAPPKPDSGLLDGFMTTDAALMYDAVHVVSVAVQQFPQMTVSSLQCNRHKPWRFGTRFM m-beta2   (Morita)
299 LQAPPKPDSGLLDGFMTTDAALMYDAVHVVSVAVQQFPQMTVSSLQCNRHKPWRFGTRFM m-GluR6-2 (Gregor)

359 SLIKEAHWEGLTGRITFNKTNGLRTDFDLDVISLKEEGLEKIGTWDPASGLNMTESQKGK r-GluR6   (Egebjerg)
361 SLIKEAHWEGLTGRITFNKTNGLRTDFDLDVISLKEEGLEKIGTWDPASGLNMTESQKGK r-GluR-6  (Lomeli)
359 SLIKEAHWEGLTGRITFNKTNGLRTDFDLDVISLKEEGLEKIGTWDPSSGLNMTESQKGK m-beta2   (Morita)
359 SLIKEAHWEGLTGRITFNKTNGLRTDFDLDVISLKEEGLEKIGTWDPSSGLNMTESQKGK m-GluR6-2 (Gregor)

419 PANITDSLSNRSLIVTTILEEPYVLFKKSDKPLYGNDRFEGYCIDLLRELSTILGFTYEI r-GluR6   (Egebjerg)
421 PANITDSLSNRSLIVTTILEEPYVLFKKSDKPLYGNDRFEGYCIDLLRELSTILGFTYEI r-GluR-6  (Lomeli)
419 PANITDSLSNRSLIVTTILEEPYVLFKKSDKPLYGNDRFEGYCIDLLRELSTILGFTYEI m-beta2   (Morita)
419 PANITDSLSNRSLIVTTILEEPYVLFKKSDKPLYGNDRFEGYCIDLLRELSTILGFTYEI m-GluR6-2 (Gregor)

479 RLVEDGKYGAQDDVNGQWNGMVRELIDHKADLAVAPLAITYVREKVIDFSKPFMTLGISI r-GluR6   (Egebjerg)
481 RLVEDGKYGAQDDVNGQWNGMVRELIDHKADLAVAPLAITYVREKVIDFSKPFMTLGISI r-GluR-6  (Lomeli)
479 RLVEDGKYGAQDDVNGQWNGMVRELIDHKADLAVAPLAITYVREKVIDFSKPFMTLGISI m-beta2   (Morita)
479 RLVEDGKYGAQDDVNGQWNGMVRELIDHKADLAVAPLAITYVREKVIDFSKPFMTLGISI m-GluR6-2 (Gregor)
                                   ——●——●–1————
539 LYRKPNGTNPGVFSFLNPLSPDIWMYVLLACLGVSCVLFVIARFSPYEWYNPHPCNPDSD r-GluR6   (Egebjerg)
541 LYRKPNGTNPGVFSFLNPLSPDIWMYVLLACLGVSCVLFVIARFSPYEWYNPHPCNPDSD r-GluR-6  (Lomeli)
539 LYRKPNGTNPGVFSFLNPLSPDIWMYVLLACLGVSCVLFVIARFSPYEWYNPHPCNPDSD m-beta2   (Morita)
539 LYRKPNGTNPGVFSFLNPLSPDIWMYVLLACLGVSCVLFVIARFSPYEWYNPHPCNPDSD m-GluR6-2 (Gregor)
                ————2————●——       ————3————
599 VVENNFTLLNSFWFGVGALMRQGSELMPKALSTRIVGGIWWFFTLIIISSYTANLAAFLT r-GluR6   (Egebjerg)
601 VVENNFTLLNSFWFGVGALMQQGSELMPKALSTRIVGGIWWFFTLIIISSYTANLAAFLT r-GluR-6  (Lomeli)
599 VVENNFTLLNSFWFGVGALMRQGSELMPKALSTRIVGGIWWFFTLIIISSYTANLAAFLT m-beta2   (Morita)
599 VVENNFTLLNGFWFGVGALMRQGSELMPKALSTRIVGGIWWFFTLIIISSYTANLAAFLT m-GluR6-2 (Gregor)

659 VERMESPIDSADDLAKQTKIEYGAVEDGATMTFFKKSKISTYDKMWAFMSSRRQSVLVKS r-GluR6   (Egebjerg)
661 VERMESPIDSADDLAKQTKIEYGAVEDGATMTFFKKSKISTYDKMWAFMSSRRQSVLVKS r-GluR-6  (Lomeli)
659 VERMESPIDSADDLAKQTKIEYGAVEDGATMTFFKKSKISTYDKMWAFMSSRRQSVLVKS m-beta2   (Morita)
659 VERMESPIDSADDLAKQTKIEYGAVEDGATMTFFKKSKISTYDKMWAFMSSRRQSVLVKS m-GluR6-2 (Gregor)
```

FIGURE 3f(1).

2.6.2.2 GluR6

The GluR6 subunit has been sequence characterized from rat and mouse species (Egebjerg et al., 1991; Lomeli et al., 1992; Morita et al., 1992; Gregor et al., 1993a). Similar to GluR5, GluR6 homomeric channels are sensitive to kainate, quisqualate, and L-glutamate. The GluR6 subunit occurs in several molecular forms (Figure 3f). Two splice variants, r-GluR6 and m-GluR6-2, have been described that differ in carboxy-terminal sequences (Gregor et al.,

```
719 NEEGIQRVLTSDYAFLMESTTIEFVTQRNCNLTQIGGLIDSKGYGVGTPMGSPYRDKITI r-GluR6   (Egebjerg)
721 NEEGIQRVLTSDYAFLMESTTIEFVTQRNCNLTQIGGLIDSKGYGVGTPMGSPYRDKITI r-GluR-6   (Lomeli)
719 NEEGIQRVLTSDYAFLMESTTIEFVTQRNCNLTQIGGLIDSKGYGVGTPMGSPYRDKITI m-beta2   (Morita)
719 NEEGIQRVLTSDYAFLMESTTIEFVTQRNCNLTQIGGLIDSKGYGVGTPMGSPYRDKITI m-GluR6c   (Gregor)
                                                   ______4______
779 AILQLQEEGKLHMMKEKWWRGNGCPEEESKEASALGVQNIGGIFIVLAAGLVLSVFVAVG r-GluR6   (Egebjerg)
781 AILQLQEEGKLHMMKEKWWRGNGCPEEESKEASALGVQNIGGIFIVLAAGLVLSVFVAVG r-GluR-6   (Lomeli)
779 AILQLQEEGKLHMMKEKWWRGNGCPEEESKEASALGVQNIGGIFIVLAAGLVLSVFVAVG m-beta2   (Morita)
779 AILQLQEEGKLHMMKEKWWRGNGCPEEESKEASALGVQNIGGIFIVLAAGLVLSVFVAVG m-GluR6-2   (Gregor)

839 EFLYKSKKNAQLEKRSFCSAMVEELRMSLKCQRRLKHKPQAPVIVKTEEVINMHTFNDRR r-GluR6   (Egebjerg)
841 EFLYKSKKNAQLEKRSFCSAMVEELRMSLKCQRRLKHKPQAPVIVKTEEVINMHTFNDRR r-GluR-6   (Lomeli)
839 EFLYKSKK[TLNWKRGPSVAPWWKN]                                   m-beta2   (Morita)
839 EFLYKSKKNAQLEK[ESSIWLVPPYHPDTV]                              m-GluR6-2   (Gregor)

899 LPGKETMA                                                     r-GluR6   (Egebjerg)
901 LPGKETMA                                                     r-GluR-6   (Lomeli)
                                                                 m-beta2   (Morita)
                                                                 m-GluR6-2   (Gregor)
```

FIGURE 3f(2).

1993a). The carboxy-terminal variation reported for the mouse homolog, β2 (Morita et al., 1992), is due to a sequence error in the cDNA. GluR6 M2 is target for Q/R site specific RNA editing (Sommer et al., 1991). For GluR6, two additional editing sites are located in M1 (Figure 3f). There, isoleucine (I) and tyrosine (Y) codons of the gene are edited to a valine (V) and cysteine (C) codon, respectively. Thus eight different GluR6 forms can be expressed with respect to different editing positions alone (Köhler et al., 1993). The functional consequences of the edited positions concern the Ca^{2+} permeability of the agonist-activated channels. Several of these variants, when expressed in a homomeric configuration, show different Ca^{2+}/Cs^{+} reversal potentials for glutamate-evoked currents (Egebjerg and Heinemann, 1993; Köhler et al., 1993). A phosphorylation site for cAMP-dependent protein kinase has been characterized in molecular and functional terms (Raymond et al., 1993; Wang et al., 1993).

The GluR6 subunit is synthesized from a 6-kb mRNA (Gregor et al., 1993a; Partin et al., 1993) and is most abundant in cerebellar granule cells, with lower levels in caudate putamen and the pyramidal cell layers and dentate granule cells of hippocampus (Wisden and Seeburg, 1993). In the rat, GluR6 shows a prenatal expression peak in the neocortex (Bahn et al., 1994).

The gene for the GluR6 subunit was mapped on mouse chromosome 10 near Waltzer and Jackson circler mutant phenotypes (Gregor et al., 1993b).

2.6.2.3 GluR7

A third member of this glutamate receptor subfamily (Figure 3g), which shows approximately 70% amino acid identity to GluR5 and GluR6 subunits, was described by two groups (Bettler et al., 1992; Lomeli et al., 1992). Similar to GluR5 and GluR6, the GluR7 homomeric receptors have a rank order of agonist affinity (domoate > kainate >> L-glutamate, quisqualate >> AMPA, NMDA) and a dissociation constant for kainate (95 and 77 n*M*, respectively), characteristic of the low-affinity population of the high-affinity kainate-binding sites described in brain. However, no agonist-activated cur-

```
1                                   MPHVIRIGGIFEYADGPNAQVMNAEEHAF r-GluR7   (Bettler)
1   MTAPWRRLRSLVWEYWAGFLVCAFWIPDSRGMPHVIRIGGIFEYADGPNAQVMNAEEHAF r-GluR-7  (Lomeli)

30  RFSANIINRNRTLLPNTTLTYDIQRIHFHDSFEATKKACDQLALGVVAIFGPSQGSCINA r-GluR7   (Bettler)
61  RFSANIINRNRTLLPNTTLTYDIQRIHFHDSFEATKKACDQLALGVVAIFGPSQGSCTNA r-GluR-7  (Lomeli)

90  VQSICNALEVPHIQLRWKHHPLDNKDTFYVNLYPDYASLSHAILDLVQSLKWRSATVVYD r-GluR7   (Bettler)
121 VQSICNALEVPHIQLRWKHHPLDNKDTFYVNLYPDYASLSHAILDLVQSLKWRSATVVYD r-GluR-7  (Lomeli)

150 DSTGLIRLQELIMAPSRYNIRLKIRQLPIDSDDSRPLLKEMKRGREFRIIFDCSHTMAAQ r-GluR7   (Bettler)
181 DSTGLIRLQELIMAPSRYNIRLKIRQLPIDSDDSRPLLKEMKRGREFRIIFDCSHTMAAQ r-GluR-7  (Lomeli)

210 ILKQAMAMGMMTEYYHFIFTTLDLYALDLEPYRYSGVNLTGFRILNVDNPHVSAIVEKWS r-GluR7   (Bettler)
241 ILKQAMAMGMMTEYYHFIFTTLDLYALDLEPYRYSGVNLTGFRILNVDNAHVSAIVEKWS r-GluR-7  (Lomeli)

270 MERLQAAPRAESGLLDGVMMTDAALLYDAVHIVSVCYQRAPQMTVNSLQCHRHKAWRFGG r-GluR7   (Bettler)
301 MERLQAAPRAESGLLDGVMMTDAALLYDAVHIVSVCYQRASQMTVNSLQCHRHKPWRFGG r-GluR-7  (Lomeli)

330 RFMNFIKEAQWEGLTGRIVFNKTSGLRTDFDLDIISLKEDGLEKVGVWSPADGLNITEVA r-GluR7   (Bettler)
361 RFMNFIKEAQWEGLTGRIVFNKTSGLRTDFDLDIISLKEDGLEKVGVWSPADGLNITEVA r-GluR-7  (Lomeli)

390 KGRGPNVTDSLTNRSLIVTTLLEEPFVMFRKSDRTLYGNDRFEGYCIDLLKELAHILGFS r-GluR7   (Bettler)
421 KGRGPNVTDSLTNRSLIVTTLLEEPFVMFRKSDRTLYGNDRFEGYCIDLLKELAHILGFS r-GluR-7  (Lomeli)

450 YEIRLVEDGKYGAQDDKGQWNGMVKELIDHKADLAVAPLTITHVREKAIDFSKPFMTLGV r-GluR7   (Bettler)
481 YEIRLVEDGKYGAQDDKGQWNGMVKELIDHKADLAVAPLTITHVREKAIDFSKPFMTLGV r-GluR-7  (Lomeli)
                                 ------1------
510 SILYRKPNGTNPSVFSFLNPLSPDIWMYVLLAYLGVSCVLFVIARFSPYEWYDAHPCNPG r-GluR7   (Bettler)
541 SILYRKPNGTNPSVFSFLNPLSPDIWMYVLLAYLGVSCVLFVIARFSPYEWYDAHPCNPG r-GluR-7  (Lomeli)
          ------2------                   ------3------
570 SEVVENNFTLLNSFWFGMGSLMQQGSELMPKALSTRIIGGIWWFFTLIIISSYTANLAAF r-GluR7   (Bettler)
601 SEVVENNFTLLNSFWFGMGSLMQQGSELMPKALSTRIIGGIWWFFTLIIISSYTANLAAF r-GluR-7  (Lomeli)

630 LTVERMESPIDSADDLAKQTKIEYGAVKDGATMTFFKKSKISTFEKMWAFMSSKPSALVK r-GluR7   (Bettler)
661 LTVERMESPIDSADDLAKQTKIEYGAVKDGATMTFFKKSKISTFEKMWAFMSSKPSALVK r-GluR-7  (Lomeli)

690 NNEEGIQRTLTADYALLMESTTIEYITQRNCNLTQIGGLIDSKGYGIGTPMGSPYRDKIT r-GluR7   (Bettler)
721 NNEEGIQRTLTADYALLMESTTIEYITQRNCNLTQIGGLIDSKGYGIGTPMGSPYRDKIT r-GluR-7  (Lomeli)
                                                ------4------
750 IAILQLQEEDKLHIMKEKWWRGSGCPEEENKEASALGIQKIGGIFIVLAAGLVLSVLVAV r-GluR7   (Bettler)
781 IAILQLQEEDKLHIMKEKWWRGSGCPEEENKEASALGIQKIGGIFIVLAAGLVLSVLVAV r-GluR-7  (Lomeli)

810 GEFIYKLRKTAEREQRSFCSTVADEIRFSLTCQRRLKHKPQPPMMVKTDAVINMHTFNDR r-GluR7   (Bettler)
841 GEFIYKLRKTAEREQRSFCSTVADEIRFSLTCQRRLKHKPQPPMMVKTDAVINMHTFNDR r-GluR-7  (Lomeli)

870 RLPGKDSMSCSTSLAPVFP                                          r-GluR7   (Bettler)
901 RLPGKDSMSCSTSLAPVFP                                          r-GluR-7  (Lomeli)
```

FIGURE 3g.

rents can be detected with GluR7 transiently expressed in *Xenopus* oocytes or cultured mammalian cells. In addition, coexpression with other subunits of the iGluR family failed to produce any detectable GluR7-specific effects. It is not known why GluR7 does not form gateable channels.

GluR7 gene transcripts are prominently expressed in the inner neocortical layers and some cells in layer II, subiculum, caudate putamen, reticular thalamus, ventral medial hypothalamic nucleus, pontine nuclei, and in putative stellate/basket cells in the cerebellum (Wisden and Seeburg, 1993). As determined by haplotype analysis, the mouse gene for GluR7 maps close to the clasper neurological mutant on chromosome 4 (Gregor et al., 1993b).

2.6.2.4 KA-1

A cDNA species from rat brain encoding the glutamate receptor subunit KA-1 has 30% sequence similarity with the AMPA receptor subunits GluR-

```
1   MPRVSAPLVLLPAWLLMVACSPHSLRIAAILDDPMECSRGERLSITLAKNRINRAPERLGKAKVEVDIFE r-KA-1  (Herb)
1   MPRVSAPLVLLPAWLVMVACSPHSLRIAAILDDPMECSRGERLSITLAKNRINRAPERLGKAKVEVDIFE h-EAA1  (Kamboj)

71  LLRDSEYETAETMCQILPKGVVAVLGPSSSPASSSIISNICGEKEVPHFKVAPEEFVRFQLQRFTTLNLH r-KA-1  (Herb)
71  LLRDSEYETAETMCQILPKGVVAVLGPSSSPASSSIISNICGEKEVPHFKVAPEEFVKFQLQRFTTLNLH h-EAA1  (Kamboj)

141 PSNTDISVAVAGILNFFNCTTACLICAKAECLLNLEKLLRQFLISKDTLSVRMLDDTRDPTPLLKEIRDD r-KA-1  (Herb)
141 PSNTDISVAVAGILNFFNCTTACLICAKAECLLNLEKLLRQFLISKDTLSVRMLDDTRDPTPLLKEIRDD h-EAA1  (Kamboj)

211 KTATIIIHANASMSHTILLKAAELGMVSAYYTYIFTNLEFSLQRMDSLVDDRVNILGFSIFNQSHAFFQE r-KA-1  (Herb)
211 KTATIIIHANASMSHTILLKAAELGMVSAYYTYIFTNLEFSLQRTDSLVDDRVNILGFSIFNQSHAFFQE h-EAA1  (Kamboj)

281 FSQSLNQSWQENCDHVPFTGPALSSALLFDAVYAVVTAVQELNRSQEIGVKPLSCGSAQIWQHGTSLMNY r-KA-1  (Herb)
281 FAQSLNQSWQENCDHVPFTGPALSSALLFDAVYAVVTAVQELNRSQEIGVKPLSCGSAQIWQHGTSLMNY h-EAA1  (Kamboj)

351 LRMVELEGLTGHIEFNSKGQRSNYALKILQFTRNGFRQIGQWHVAEGLSMDSRLYASNISDSLFNTTLVV r-KA-1  (Herb)
351 LRMVELEGLTGHIEFNSKGQRSNYALKILQFTRNGFRQIGQWHVAEGLSMDSHLYASNISDTLFNTTLVV h-EAA1  (Kamboj)

421 TTILENPYLMLKGNHQDMEGNDRYEGFCVDMLKELAEILRFNYKIRLVGDGVYGVPEANGTWTGMVGELI r-KA-1  (Herb)
421 TTILENPYLMLKGNHQEMEGNDRYEGFCVDMLKELAEILRFNYKIRLVGDGVYGVPEANGTWTGMVGELI h-EAA1  (Kamboj)
                                                                  1
491 ARKADLAVAGLTITAEREKVIDFSKPFMTLGISILYRVHMGRRPGYFSSLDPFSPGVWLFMLLAYLAVSC r-KA-1  (Herb)
491 ARKADLAVAGLTITAEREKVIDFSKPFMTLGISILYRIHMGRRPGYFSSLDPFSPGVWLFMLLAYLAVSC h-EAA1  (Kamboj)
                                         2
561 VLFLVARLTPYEWYSPHPCAQGRCNLLVNQYSLGNSLWFPVGGFMQQGSTIAPRALSTRCVSGVWWAFTL r-KA-1  (Herb)
561 VLFLVARLTPYEWYSPHPCAQGRCNLLVNQYSLGNSLWFPVGGFMQQGSTIAPRALSTRCVSGVWWAFTL h-EAA1  (Kamboj)
     3
631 IIISSYTANLAAFLTVQRMEVPIESVDDLADQTAIEYGTIHGGSSMTFFQNSRYQTYQRMWNYMYSKQPS r-KA-1  (Herb)
631 IIISSYTANLAAFLTVQRMDVPIESVDDLADQTAIEYGTIHGGSSMTFFQNSRYQTYQRMWNYMYSKQPS h-EAA1  (Kamboj)

701 VFVKSTEEGIARVLNSNYAFLLESTMNEYYRQRNCNLTQIGGLLDTKGYGIGMPVGSVFRDEFDLAILQL r-KA-1  (Herb)
701 VFVKSTEEGIARVLNSNYAFLLESTMNEYYRQRNCNLTQIGGLLDTKGYGIGMPVGSVFRDEFDLAILQL h-EAA1  (Kamboj)
                                             4
771 QENNRLEILKRKWWEGGKCPKEEDHRAKGLGMENIGGIFVVLICGLIVAIFMAMLEFLWTLRHSEASEVS r-KA-1  (Herb)
771 QENNRLEILKRKWWEGGKCPKEEDHRAKGLGMENIGGIFVVLICGLIVAIFMAMLEFLWTLRHSEASEVS h-EAA1  (Kamboj)

841 VCQEMMTELRSIILCQDNIHPRRRRSGGLPPQPPVLEERRPRGTATLSNGKLCGAGEPDQLAQRLAQEAA r-KA-1  (Herb)
841 VCQEMVTELRSIILCQDSIHPRRRRAAVPPPRPPIPEERRPRGTATLSNGKLCGAGEPDQLAQRLAQEAA h-EAA1  (Kamboj)

911 LVARGCTHIRVCPECRRFQGLRARPSPARSEESLEWDKTTNSSEPE                         r-KA-1  (Herb)
911 LVARGCTHIRVCPECRRFQGLRARPSPARSEESLEWEKTTNSSEPE                         h-EAA1  (Kamboj)
```

FIGURE 3h.

A to -D (Werner et al., 1991). The human KA-1 homolog of this subunit, humEAA-1, was recently cloned (Kamboj et al., 1994) (Figure 3h). The pharmacological profiles of recombinant KA-1 and humEAA-1 determined in binding experiments with [^{3}H]kainate were similar (kainate > quisqualate > glutamate >> AMPA) with a dissociation constant of about 5 n*M* for kainate. The affinity for kainate is significantly higher than that found for the GluR5 to GluR7 subunits, suggesting that the KA-1 subunit is associated with the high-affinity kainate-binding sites that are seen in [^{3}H]kainate-binding studies in the brain and that display a similar rank order of potency (Werner et al., 1991; Herb et al., 1992). For the KA-1 subunit no different splice forms or edited versions have been reported. The carboxy-terminal sequence divergence found between rat KA-1 and humEAA-1 (Kamboj et al., 1994) is due to the use of a frame-shifted rat KA-1 sequence. The rat KA-1 subunit was first reported with a frameshift in the carboxy-terminal sequence (Werner et al., 1991). The correct KA-1 sequence (Herb et al., 1992) is in the Genbank/EMBL database (Table 1).

Homooligomeric expression of KA-1 in cultured cells does not elicit ligand-gated ion channel activity, but coexpression with GluR5 and GluR6 produces functional channels.

The KA-1 subunit is expressed at low levels in the rat CNS but shows high expression in the CA3 region of the hippocampus and dentate gyrus (Werner et al., 1991; Wisden and Seeburg, 1993). KA-1 mRNA of 4.5 kb size has been reported (Partin et al., 1993). The genomic locus for the KA-1 subunit has not been determined.

2.6.2.5 KA-2

The cloning of the KA-2 subunit cDNA has been described for the rat (Herb et al., 1992), the human (Kamboj et al., 1992), and the mouse (termed $\gamma 2$; Sakimura et al., 1992) (Figure 3i). Like KA-1, the KA-2 subunit when expressed *in vitro* exhibits high affinity for [^{3}H]kainate (K_D approximately 15 n*M*). Homomeric KA-2 expression does not generate agonist-sensitive channels and channel activity is observed only when KA-2 is coexpressed with GluR5 or GluR6 subunits. Specifically, the combination GluR5(R)/KA-2 produces channels, and currents through heteromeric GluR5(Q)/KA-2 channels show increased desensitization and different current–voltage relations when compared with homomeric GluR5(Q) currents. GluR6/KA-2 channels are gated by AMPA, which fails to activate homomeric GluR6 receptor channels. The KA-2 subunit provides a nice example for the fact that subunit combinations can produce new emergent properties, including the formation of agonist-sensitive channels (Herb et al., 1992).

KA-2 mRNA is widely expressed in embryonic and adult rat brain, including layers II–VI of neocortex, hippocampal pyramidal (CA1-CA3), and dentate granule cells (Wisden and Seeburg, 1993). KA-2 mRNA is approximately 5 kb in size (Partin et al., 1994). The KA-2 gene has not been mapped.

2.6.3 Orphan Receptor Subunits

A small family of iGluR subunits, delta-1 and delta-2, has been isolated by homology screening from rodent brain cDNA libraries. These subunits show 22% sequence similarity with GluR5, 6, and 7. Their expression does not generate binding sites for known excitatory amino acid receptor ligands nor channel activity elicited by glutamate, kainate, or domoate in HEK 293 cells and *Xenopus* oocytes expressing the delta subunits. Thus, delta-1 and delta-2 currently remain orphan subunits in search of a function. These subunits may participate in the assembly of particular AMPA or kainate receptor channels although no functional evidence for this has been forthcoming.

2.6.3.1 Delta-1

The delta-1 subunit was isolated from rat and mouse brain cDNA libraries (Yamazaki et al., 1992a; Lomeli et al., 1993). The cDNAs encode putative iGluR subunits with low sequence identity with other such subunits (Figure 3j). Despite the low sequence similarity the major structural features — the

```
1   MPAELLLLLIVAFANPSCQVLSSLRMAAILDDQTVCGRGERLALALAREQINGIIEVPAK r-KA-2    (Herb)
1   MPAELLLLLIVAFASPSCQVLSSLRMAAILDDQTVCGRGERLALALAREQINGIIEVPAK h-EAA2    (Kamboj)
1   MPAELLLLLIVAFANPSCQVLSSLRMAAILDDQTVCGRGERLALALAREQINGIIEVPAK m-gamma2  (Sakimura)

61  ARVEVDIFELQRDSQYETTDTMCQILPKGVVSVLGPSSSPASASTVSHICGEKEIPHIKV r-KA-2    (Herb)
61  ARVEVDIFELQRDSQYETTDTMCQILPKGVVSVLGPSSSPASASTVSHICGEKEIPHIKV h-EAA2    (Kamboj)
61  ARVEVDIFELQRDSQYETTDTMCQILPKGVVSVLGPSSSPASASTVSHICGEKEIPHIKV m-gamma2  (Sakimura)

121 GPEETPRLQYLRFASVSLYPSNEDVSLAVSRILKSFNYPSASLICAKAECLLRLEELVRG r-KA-2    (Herb)
121 GPEETPRLQYLRFASVSLYPSNEDVSLAVSRILKSFNYPSASLICAKAECLLRLEELVRG h-EAA2    (Kamboj)
121 GPEETPRLQYLRFASVSLYPSNEDVSLAVSRILKSFNYPSASLICAKAECLLRLEELVRG m-gamma2  (Sakimura)

181 FLISKETLSVRMLDDSRDPTPLLKEIRDDKVSTIIIDANASISHLVLRKASELGMTSAFY r-KA-2    (Herb)
181 FLISKETLSVRMLDDSRDPTPLLKEIRDDKVSTIIIDANASISHLILRKASELGMTSAFY h-EAA2    (Kamboj)
181 FLISKETLSVRMLDDSRDPTPLLKEIRDDKVSTIIIDANASISHLVLRKASELGMTSAFY m-gamma2  (Sakimura)

241 KYILTTMDFPILHLDGIVEDSSNILGFSMFNTSHPFYPEFVRSLNMSWRENCEASTYPGP r-KA-2    (Herb)
241 KYILTTMDFPILHLDGIVEDSSNILGFSMFNTSHPFYPEFVRSLNMSWRENCEASTYLGP h-EAA2    (Kamboj)
241 KYILTTMDFPILHLDGIVEDSSNILGFSMFNTSHPFYPEFVRSLNMSWRENCEASTYPGP m-gamma2  (Sakimura)

301 ALSAALMFDAVHVVVSAVRELNRSQEIGVKPLACTSANIWPHGTSLMNYLRMVEYDGLTG r-KA-2    (Herb)
301 ALSAALMFDAVHVVVSAVRELNRSQEIGVKPLACTSANIWPHGTSLMNYLRMVEYDGLTG h-EAA2    (Kamboj)
301 ALSAALMFDAVHVVVSAVRELNRSQEIGVKPLACTSANIWPHGTSLMNYLRMVEYDGLTG m-gamma2  (Sakimura)

361 RVEFNSKGQRTNYTLRILEKSRQGHREIGVWYSNRTLAMNATTLDINLSQTLANKTLVVT r-KA-2    (Herb)
361 RVEFNSKGQRTNYTLRILEKSRQGHREIGVWYSNRTLAMNATTLDINLSQTLANKTLVVT h-EAA2    (Kamboj)
361 RVEFNSKGQRTNYTLRILEKSRQGHREIGVWYSNRTLAMNATTLDINLSQTLANKTLVVT m-gamma2  (Sakimura)

421 TILENPYVMRRPNFQALSGNERFEGFCVDMLRELAELLRFRYRLRLVEDGLYGAPEPNGS r-KA-2    (Herb)
421 TILENPYVMRRPNFQGLSGNERFEGFCVDMLRELAELLPFPYRLRLVEDGLYGAPEPNGS h-EAA2    (Kamboj)
421 TILENPYVMRRPNFQALSGNERFEGFCVDMLRELAELLRFRYRLRLVEDGLYGAPEPNGS m-gamma2  (Sakimura)

481 WTGMVGELINRKADLAVAAFTITAEREKVIDFSKPFMTLGISILYRVHMGRKPGYFSFLD r-KA-2    (Herb)
481 WTGMVGELINRKADLAVAAFTITAEREKVIDFSKPFMTLGISILYRVHMGRKPGYFSFLD h-EAA2    (Kamboj)
481 WTGMVGELINRKADLAVAAFTITAEREKVIDFSKPFMTLGISILYRVHMGRKPGYFSFLD m-gamma2  (Sakimura)
          ———————1———————                            ———————2——
541 PFSPAVWLFMLLAYLAVSCVLFLAARLSPYEWYNPHPCLRARPHILENQYTLGNSLWFPV r-KA-2    (Herb)
541 PFSPAVWLFMLLAYLAVSCVLFLAARLSPYEWYNPHPCLRARPHILENQYTLGNSLWFPV h-EAA2    (Kamboj)
541 PFSPAVWLFMLLAYLAVSCVLFLAARLSPYEWYNPHPCLRARPHILENQYTLGNSLWFPV m-gamma2  (Sakimura)
    —2————            ————————3————————
601 GGFMQQGSEIMPRALSTRCVSGVWWAFTLIIISSYTANLAAFLTVQRMEVPVESADDLAD r-KA-2    (Herb)
601 GGFMQQGSEIMPRALSTRCVSGVWWAFTLIIISSYTANLAAFLTVQRMEVPVESADDLAD h-EAA2    (Kamboj)
601 GGFMQQGSEVMPRALSTRCVSGVWWAFTLIIISSYTANLAAFLTVQRMEVPVESADDLAD m-gamma2  (Sakimura)

661 QTNIEYGTIHAGSTMTFFQNSRYQTYQRMWNYMQSKQPSVFVKSTEEGIARVLNSRYAFL r-KA-2    (Herb)
661 QTNIEYGTIHAGSTMTFFQNSRYQTYQRMWNYMQSKQPSVFVKSTEEGIAAVLNSRYAFL h-EAA2    (Kamboj)
661 QTNIEYGTIHAGSTMTFFQNSRYQTYQRMWNYMQSKQPSVFVKSTEEGIARVLNSRYAFL m-gamma2  (Sakimura)

721 LESTMNEYHRRLNCNLTQIGGLLDTKGYGIGMPLGSPFRDEITLAILQLQENNRLEILKR r-KA-2    (Herb)
721 LESTMNEYHRRLNCNLTQIGGLLDTKGYGIGMPLGSPFRDEITLAILQLQENNRLEILKR h-EAA2    (Kamboj)
721 LESTMNEYHRRLNCNLTQIGGLLDTKGYGIGMPLGSPFRDEITLAILQLQENNRLEILKR m-gamma2  (Sakimura)
                                  ————————4————————
781 KWWEGGRCPKEEDHRAKGLGMENIGGIFVVLICGLIIAVFVAVMEFIWSTRRSAESEEVS r-KA-2    (Herb)
781 KWWEGGRCPKEEDHRAKGLGMENIGGIFIVLICGLIIAVFVAVMEFIWSTRRSAESEEVS h-EAA2    (Kamboj)
781 KWWEGGRCPKEEDHRAKGLGMENIGGIFVVLICGLIIAVFVAVMEFIWSTRRSAESEEVS m-gamma2  (Sakimura)

841 VCQEMLQELRHAVSCRKTSRSRRRRRPGGPSRALLSLRAVREMRLSNGKLYSAGAGGDAG r-KA-2    (Herb)
841 VCQEMLQELRHAVSCRKTSRSRRRRRPGGPSRALLSLRAVREMRLSNGKLYSAGAGGDAG h-EAA2    (Kamboj)
841 VCQEMLQELRHAVSCRKTSRSRRRRRPGGPSRALLSLRAVREMRLSNGKLYSAGAGGDAG m-gamma2  (Sakimura)

901 -AHGGPQRLLDDPGPPGGPRPQAPTPCTHVRVCQECRRIQALRASGAGAPPRGLGTPAEA r-KA-2    (Herb)
901 SAHGGPQRLLDDPGPPSGARPAAPTPCTHVRVCQECRRIQALRASGAGAPPRGLGVPAEA h-EAA2    (Kamboj)
901 -AHGGPQRLLDDPGPPGGPRPQAPTPCTHVRVCQECRRIQALRASGAGAPPRGLGTPAEA m-gamma2  (Sakimura)

960 TSPPRPRPGPTGPRELTEHE                                         r-KA-2    (Herb)
961 TSPPRPRPGPAGPRELAEHE                                         h-EAA2    (Kamboj)
960 TSPPRPRPGPTGPRELTEHE                                         m-gamma2  (Sakimura)
```

FIGURE 3i.

four hydrophobic segments and the putative agonist binding site in front of M1 and in the M3–M4—loop are present in the mouse and rat delta-1 polypeptides (Figure 3j). The overall expression of the delta-1 subunit in adult brain is low. In younger animals expression is higher and is seen in the

```
1    MEALTLWLLPWICQCVTVRADSIIHIGAIFEENAAKDDRVFQLAVSDLSLNDDILQSEKI r-delta-1  (Lomeli)
1    MEALTLWLLPWICQCVTVRADSIIHIGAIFEENAAKDDRVFQLAVSDLSLNDDILQSEKI m-delta1   (Yamazaki)

61   TYSIKVIEANNPFQAVQEACDLMTQGILALVTSTGCASANALQSLTDAMHIPHLFVQRNP r-delta-1  (Lomeli)
61   TYSIKVIEANNPFQAVQEACDLMTQGILALVTSTGCASANALQSLTDAMHIPHLFVQRNP m-delta1   (Yamazaki)

121  GGSPRTACHLNPSPDGEAYTLASRPPVRLNDVMLRLVTELRWQKFVMFYDSEYDIRGLQS r-delta-1  (Lomeli)
121  GGSPRTACHLNPSPDGEAYTLASRPPVRLNDVMLRLVTELRWQKFVMFYDSEYDIRGFQS m-delta1   (Yamazaki)

181  FLDQASRLGLDVSLQKVDKNISHVFTSLFTTMKTEELNRYRDTLRRAILLLSPQGAHSFI r-delta-1  (Lomeli)
181  FLDQASRLGLDVSLQKVDKNISHVFTSLFTTMKTEELNRYRDTLRRAILLLSPQGAHSFI m-delta1   (Yamazaki)

241  NEAVETNLASKDSHWVFVNEEISDPEILDLVHSALGRMTVVRQIFPSAKDNQKCMRNNHR r-delta-1  (Lomeli)
241  NEAVETNLASKDSHWVFVNEEISDPEILDLVHSALGRMTVVRQIFPSAKDNQKCMRNNHR m-delta1   (Yamazaki)

301  ISSLLCDPQEGYLQMLQISNLYLYDSVLMLANAFHRKLEDRKWHSMASLNCIRKSTKPWN r-delta-1  (Lomeli)
301  ISSLLCDPQEGYLQMLQISNLYLYDSVLMLANAFHRKLEDRKWHSMASLNCIRKSTKPWN m-delta1   (Yamazaki)

361  GGRSMLDTIKKGHITGLTGVMEFREDSSNPYVQFEILGTTYSETFGKDMRKLATWDSEKG r-delta-1  (Lomeli)
361  GGRSMLDTIKKGHITGLTGVMEFREDSSNPYVQFEILGTTYSETFGKDMRKLATWDSEKG m-delta1   (Yamazaki)

421  LNGSLQERPMGSRLQGLTLKVVTVLEEPFVMVAENILGQPKRYKGFSIDVLDALAKALGF r-delta-1  (Lomeli)
421  LNGSLQERPMGSRLQGLTLKVVTVLEEPFVMVAENILGQPKRYKGFSIDVLDALAKALGF m-delta1   (Yamazaki)

481  KYEIYQAPDGRYGHQLHNTSWNGMIGELISKRADLAISAITITPERESVVDFSKRYMDYS r-delta-1  (Lomeli)
481  KYEIYQAPDGRYGHQLHNTSWNGMIGELISKRADLAISAITITPERESVVDFSKRYMDYS m-delta1   (Yamazaki)
                                   ______1______
541  VGILIKKPEEKISIFSLFAPFDFAVWACIAAAIPVVGVLIFVLNRIQAVRSQSATQPRPS r-delta-1  (Lomeli)
541  VGILIKKPEEKISIFSLFAPFDFAVWACIAAAIPVVGVLIFVLNRIQAVRSQSATQPRPS m-delta1   (Yamazaki)
    ______2______                  _____3_____
601  ASATLHSAIWIVYGAFVQQGGESSVNSVAMRIVMGSWWLFTLIVCSSYTANLAAFLTVSR r-delta-1  (Lomeli)
601  ASATLHSAIWIVYGAFVQQGGESSVNSVAMRIVMGSWWLFTLIVCSSYTANLAAFLTVSR m-delta1   (Yamazaki)

661  MDSPVRTFQDLSKQLEMSYGTVRDSAVYEYFRAKGTNPLEQDSTFAELWRTISKNGGADN r-delta-1  (Lomeli)
661  MDNPIRTFQDLSKQLEMSYGTVRDSAVYEYFRAKGTNPLEQDSTFAELWRTISKNGGADN m-delta1   (Yamazaki)

721  CVSNPSEGIRKAKKGNYAFLWDVAVVEYAALTDDDCSVTVIGNSISSKGYGIALQHGSPY r-delta-1  (Lomeli)
721  CVSNPSEGIRKAKKGNYAFLWDVAVVEYAALTDDDCSVTVIGNSISSKGYGIALQHGSPY m-delta1   (Yamazaki)
                                                        ________
781  RDLFSQRILELQDTGDLDVLKQKWWPHTGRCDLTSHSSAQTDGKSLKLHSFAGVFCILAI r-delta-1  (Lomeli)
781  RDLFSQRILELQDTGDLDVLKQKWWPHTGRCDLTSHSSTQTEGKSLKLHSFAGVFCILAI m-delta1   (Yamazaki)
     __4_____
841  GLLLACLVAALELWWNSNRCHQETPKEDKEVNLEQVHRRINSLMDEDIAHKQISPASIEL r-delta-1  (Lomeli)
841  GLLLACLVAALELWWNSNRCHQETPKEDKEVNLEQVHRRINSLMDEDIAHKQISPASIEL m-delta1   (Yamazaki)

901  SALEMGGLAPSQALEPSREYQNTQLSVSTFLPEQSSHGTSRTLSSGPSSNLPLPLSSSAT r-delta-1  (Lomeli)
901  SALEMGGLAPSQALEPTREYQNTQLSVSTFLPEQSSHGTSRTLSSGPSSNLPLPLSSSAT m-delta1   (Yamazaki)

961  MPSIQCKHRSPNGGLFRQSPVKTPIPMSFQPVPGGVLPEALDTSHGTSI            r-delta-1  (Lomeli)
961  MPSIQCKHRSPNGGLFRQSPVKTPIPMSFQPVPGGVLPEALDTSHGTSI            m-delta1   (Yamazaki)
```

FIGURE 3j.

caudate putamen and anterioventral thalamic nucleus. The timing of the delta-1 expression may indicate a role in development such as neurite extension and differentiation (Lomeli et al., 1993).

2.6.3.2 Delta-2

A second member of the delta subunit family has been characterized (Lomeli et al., 1993; Araki et al., 1993). The delta-2 subunit (Figure 3k) exhibits 56% amino acid sequence identity to the delta-1 subunit. In contrast to delta-1 the delta-2 subunit is expressed at high levels in the Purkinje cell of the cerebellum. In mouse a similar expression pattern has been reported for delta-2. There, a delta-2 mRNA of 6400 nucleotides and a delta-2 specific protein of about 110 kDa can be found (Araki et al., 1993).

```
1    MEVFPLLFFLSFWWSRTWDLATSDSIIHIGAIFDESAKKDDEVFRTAVGDLNQNEEILQT r-delta-2  (Lomeli)
1    MEVFPLLLFLSFCWSRTWDLATADSIIHIGAIFDESAKKDDEVFRTAVGDLNQNEEILQT m-delta2  (Araki)

61   EKITFSVTFVDGNNPFQAVQEACELMNQGILALVSSIGCTSAGSLQSLGRLTMHIPHLFI r-delta-2  (Lomeli)
61   EKITFSVTFVDGNNPFQAVQEACELMNQGILALVSSIGCTSAGSLQSLAD-AMHIPHLFI m-delta2  (Araki)

121  QRSTAGTPRSGCGLTRSNRNDDYTLSVRPPVYLNEVILRVVTEYAWQKFIIFYDSEYDIR r-delta-2  (Lomeli)
120  QRSTAGTPRSGCGLTRSNRNDDYTLSVRPPVYLNEVILRVVTEYAWQKFIIFYDSEYDIR m-delta2  (Araki)

181  GIQEFLDKVSQQGMDVALQKVENNINKMITTLFDTMRIEELNRYRDTLRRAILVMNPATA r-delta-2  (Lomeli)
180  GIQEFLDKVSQQGMDVALQKVENNINKMITTLFDTMRIEELNRYRDTLRRAILVMNPATA m-delta2  (Araki)

241  KSFISEVVETNLVAFDCHWIIINEEINDVDVQELVRRSIGRLTIIRQTFPVPQNISQRCF r-delta-2  (Lomeli)
240  KSFISEVVETNLVAFDCHWIIINEEINDVDVQELVRRSIGRLTIIRQTFPVPQNISQRCF m-delta2  (Araki)

301  RGNHRISSTLCDPKDPFAQNMEISHLCIYDAVLLLANAFHKKLEDRKWHSMASLSCIRKN r-delta-2  (Lomeli)
300  RGNHRISSSLCDPKDPFAQNMEISNLYIYDTVLLLANAFHKKLEDRKWHSMASLSCIRKN m-delta2  (Araki)

361  SKPWQGGRSMLETIKKGGVNGLTGDLEFGENGGNPNVHFEILGTNYGEELGRGVRKLGCW r-delta-2  (Lomeli)
360  SKPWQGGRSMLETIKKGGVNGLTGDLEFGENGGNPNVHFEILGTNYGEELGRGVRKLGCW m-delta2  (Araki)

421  NPVTGLNGSLTDKKLENNMRGVVLRVVTVLEEPFVMVSENVLGKPKKYQGFSIDVLDALS r-delta-2  (Lomeli)
420  NPVTGLNGSLTDKKLENNMRGVVLRVVTVLEEPFVMVSENVLGKPKKYQGFSIDVLDALS m-delta2  (Araki)

481  NYLGFNYEIYVAPDHKYGSPQEDGTWNGLVGELVFKRADIGISALTITPDRENVVDFTTR r-delta-2  (Lomeli)
480  NYLGFNYEIYVAPDHKYGSPQEDGTWNGLVGELVFKRADIGISALTITPDRENVVDFTTR m-delta2  (Araki)
                                             ______1______
541  YMDYSVGVLLRRAEKTVDMFACLAPFDLSLWACIAGTVLLVGLLVYLLNWLNPPRLQMGS r-delta-2  (Lomeli)
540  YMDYSVGVLLRRAEKTVDMFACLAPFDLSLWACIAGTVLLVGLLVYLLNWLNPPRLQMGS m-delta2  (Araki)
        ______2______                ______3______
601  MTSTTLYNSMWFVYGSFVQQGGEVPYTTLATRMMMGAWWLFALIVISSYTANLAAFLTIT r-delta-2  (Lomeli)
600  MTSTTLYNSMWFVYGSFVQQGGEVPYTTLATRMMMGAWWLFALIVISSYTANLAAFLTIT m-delta2  (Araki)

661  RIESSIQSLQDLSKQTDIPYGTVLDSAVYQHVRMKGLNPFERDSMYSQMWRMINRSNGSE r-delta-2  (Lomeli)
660  RIESSIQSLQDLSKQTDIPYGTVLDSAVYQHVRMKGLNPFERDSMYSQMWRMINRSNGSE m-delta2  (Araki)

721  NNVLESQAGIQKVKYGNYAFVWDAAVLEYVAINDPDCSFYTVGNTVADRGYGIALQHGSP r-delta-2  (Lomeli)
720  NNVLESQAGIQKVKYGNYAFVWDAAVLEYVAINDPDCSFYTVGNTVADRGYGIALQHGSP m-delta2  (Araki)
                                                         ________
781  YRDVFSQRILELQQSGDMDILKHKWWPKNGQCDLYSSVDAKQKGGALDIKSLAGVFCILA r-delta-2  (Lomeli)
780  YRDVFSQRILELQQSGDMDILKHKWWPKNGQCDLYSSVDAKQKGGALDIKSLAGVFCILA m-delta2  (Araki)
     _4_______
841  AGIVLSCLIAVLETWWSRRKGSRVPSKEDDKEIDLEHLHRRVNSLCTDDDSPHKQFSTSS r-delta-2  (Lomeli)
840  AGIVLSCLIAVLETWWSRRKGSRVPSKEDDKEIDLEHLHRRVNSLCTDDDSPHKQFSTSS m-delta2  (Araki)

901  IDLTPLDIDTLPTRQALEQISDFRNTHITTTTFIPEQIQTLSRTLSAKAASGFTFGSVPE r-delta-2  (Lomeli)
900  IDLTPLDIDTLPTRQALEQISDFRNTHITTTTFIPEQIQTLSRTLSAKAASGFAFGSVPE m-delta2  (Araki)

961  HRTGPFRHRAPNGGFFRSPIKTMSSIPYQPTPTLGLNLGNDPDRGTSI             r-delta-2  (Lomeli)
960  HRTGPFRHRAPNGGFFRSPIKTMSSIPYQPTPTLGLNLGNDPDRGTSI             m-delta2  (Araki)
```

FIGURE 3k.

2.6.4 NMDA Receptor Subunits

This iGluR is characterized by high Ca^{2+} permeability, voltage-dependent Mg^{2+} block, high single channel conductance, and relatively slow gating kinetics. Furthermore, glycine is required as coagonist for channel activation. The conditional activation and the high Ca^{2+} permeability render this channel an important element in activity-dependent synaptic plasticity (reviewed by Collingridge and Singer, 1991; Bliss and Collingridge, 1993; Bourne and Nicoll, 1993). The brain expresses several genes whose products can assemble to generate different NMDA receptor subtypes. These subunits belong to two sequence families, the NR1 and the NR2 family.

2.6.4.1 NMDAR1 (NR1)

NR1 is the major NMDA receptor subunit, first characterized by expression cloning (Moriyoshi et al., 1991). Several cDNA sequences for rat, mouse, and human NR1 subunits have been published (Nakanishi et al., 1992; Yamazaki et al., 1992b; Karp et al., 1993; Foldes et al., 1993; Planells et al., 1993) (Figure 3k). The NR1 gene is prominently expressed in all neurons and the NR1 subunit seems to be the principal constituent of NMDA receptors. NR1 transcripts can be detected by *in situ* hybridization in virtually every neuron in the rodent brain, making the NR1 gene the highest expressed gene of all iGluR subunits (Moriyoshi et al., 1991; Watanabe et al., 1993; Tölle et al., 1993; Monyer et al., 1992, 1994; Laurie and Seeburg, 1994). Light and electron microscopic studies with a selective antibody confirm this distribution (Petralia et al., 1994).

When expressed *in vitro,* homooligomeric NR1 channels form that can be gated by the concerted action of L-glutamate and glycine, and the channels are blocked by Mg^{2+} ions in a voltage-dependent manner. However, homooligomeric NR1 assembly is inefficient and such channels may not play a functional role in the CNS. The particular functional properties of NMDA receptors — especially the Ca^{2+} permeability — have been traced, in part, to the asparagine residue in the channel forming segment M2 located in a position homologous to the Q/R site in AMPA/kainate receptor channels (Burnashev et al., 1992c; Sakurada et al., 1993).

The NR1 gene gives rise to nine alternatively spliced transcripts. These differ in carboxy-terminal sequences and in the presence of a 26 amino acid long segment in the amino-terminal region, and one form is truncated within the amino-terminal domain (Anantharam et al., 1992; Sugihara et al., 1992; Nakanishi et al., 1992; Hollmann et al., 1993) (Figure 3k). The splice forms to which no major functional differences have been assigned show differential expression patterns in the developing and the adult rodent brain (Laurie and Seeburg, 1994). One of the alternative carboxy-terminal sequences contains a serine phosphorylation site for protein kinase C (Tingley et al., 1993). The mRNA for the rat and mouse NR1 is approximately 4 kb in length (Moriyoshi et al., 1991; Meguro et al., 1992). The NR1 gene with 22 exons spans over 25 kb (Hollmann et al., 1993), and two major transcriptional start sites were identified at nucleotides –276 and –238 relative to the translational start codon AUG (Bai and Kusiak, 1993). The human gene is on chromosome 9q34.3 (Collins et al., 1993; Takano et al., 1993).

2.6.4.2 NMDAR2 (NR2)

The family of NMDAR2 subunits comprises four highly sequence-related members termed NR2A to NR2D for the rat and ε1 to ε4 for mouse (Ikeda et al., 1992; Kutsuwada et al., 1992; Monyer et al., 1992; Ishii et al., 1993). These subunits have unusually long carboxy-terminal extensions of >500 amino acids (Figures 1 and 3l–o). As is the case for NR1, the NR2 subunits carry an asparagine residue in their M2 Q/R/N site. This residue is a determinant for the

```
1   MSTMRLLTLALLFSCSVARAACDPKIVNIGAVLSTRKHEQMFREAVNQANKRHGSWKIQL h-NMDAR1   (Karp)
1   MSTMRLLTLALLFSCSVARAACDPKIVNIGAVLSTRKHEQMFREAVNQANKRHGSWKIQL h-NR1   (Planells)
1   MSTMRLLTLALLFSCSVARAACDPKIVNIGAVLSTRKHEQMFREAVNQANKRHGSWKIQL h-NR1-1   (Foldes)
1                                           MFREAVNQANKRHGSWKIQL h-NR1-3   (Foldes)
1   MSTMHLLTFALLFSCSFARAACDPKIVNIGAVLSTRKHEQMFREAVNQANKRHGSWKIQL r-NMDAR1   (Miriyoshi)
1   MSTMHLLTFALLFSCSFARAACDPKIVNIGAVLSTRKHEQMFREAVNQANKRHGSWKIQL r-NMDA-R1A   (Nakanishi)
1   MSTMHLLTFALLFSCSFARAACDPKIVNIGAVLSTRKHEQMFREAVNQANKRHGSWKIQL r-NMDA-R1B   (Nakanishi)
1   MSTMHLLTFALLFSCSFARAACDPKIVNIGAVLSTRKHEQMFREAVNQANKRHGSWKIQL r-R1-LL   (Bayley)
1   MSTMHLLTFALLFSCSFARAACDPKIVNIGAVLSTRKHEQMFREAVNQANKRHGSWKIQL r-NMDAR1b(Sugihara)
1   MSTMHLLTFALLFSCSFARAACDPKIVNIGAVLSTRKHEQMFREAVNQANKRHGSWKIQL r-NMDAR1c   (Sugihara)
1   MSTMHLLTFALLFSCSFARAACDPKIVNIGAVLSTRKHEQMFREAVNQANKRHGSWKIQL r-NMDAR1d   (Sugihara)
1   MSTMHLLTFALLFSCSFARAACDPKIVNIGAVLSTRKHEQMFREAVNQANKRHGSWKIQL r-NMDAR1e   (Sugihara)
1   MSTMHLLTFALLFSCSFARAACDPKIVNIGAVLSTRKHEQMFREAVNQANKRHGSWKIQL r-NMDAR1f   (Sugihara)
1   MSTMHLLTFALLFSCSFARAACDPKIVNIGAVLSTRKHEQMFREAVNQANKRHGSWKIQL r-NMDAR1g   (Sugihara)
1   MSTMHLLTFALLFSCSFARAACDPKIVNIGAVLSTRKHEQMFREAVNQANKRHGSWKIQL m-zeta1   (Yamazsaki)
1   MSTMHLLTFALLFSCSFARAACDPKIVNIGAVLSTRKHEQMFREAVNQANKRHGSWKIQL m-zeta1-2   (Yamazaki)
1   MSTMHLLTFALLFSCSFARAACDPKIVNIGAVLSTRKHEQMFREAVNQANKRHGSWKIQL r-NMDAR1tru   (Sugihara)

61  NATSVTHKPNAIQMALSVCEDLISSQVYAILVSHPPTPNDHFTPTPVSYTAGFYRIPVLG h-NMDAR1   (Karp)
61  NATSVTHKPNAIQMALSVCEDLISSQVYAILVSHPPTPNDHFTPTPVSYTAGFYRIPVLG h-NR1   (Planells)
61  NATSVTHKPNAIQMALSVCEDLISSQVYAILVSHPPTPNDHFTPTPVSYTAGFYRIPVLG h-NR1-1   (Foldes)
21  NATSVTHKPNAIQMALSVCEDLISSQVYAILVSHPPTPNDHFTPTPVSYTAGFYRIPVLG h-NR1-3   (Foldes)
61  NATSVTHKPNAIQMALSVCEDLISSQVYAILVSHPPTPNDHFTPTPVSYTAGFYRIPVLG r-NMDAR1   (Miriyoshi)
61  NATSVTHKPNAIQMALSVCEDLISSQVYAILVSHPPTPNDHFTPTPVSYTAGFYRIPVLG r-NMDA-R1A   (Nakanishi)
61  NATSVTHKPNAIQMALSVCEDLISSQVYAILVSHPPTPNDHFTPTPVSYTAGFYRIPVLG r-NMDA-R1B   (Nakanishi)
61  NATSVTHKPNAIQMALSVCEDLISSQVYAILVSHPPTPNDHFTPTPVSYTAGFYRIPVLG r-R1-LL   (Bayley)
61  NATSVTHKPNAIQMALSVCEDLISSQVYAILVSHPPTPNDHFTPTPVSYTAGFYRIPVLG r-NMDAR1b   (Sugihara)
61  NATSVTHKPNAIQMALSVCEDLISSQVYAILVSHPPTPNDHFTPTPVSYTAGFYRIPVLG r-NMDAR1c   (Sugihara)
61  NATSVTHKPNAIQMALSVCEDLISSQVYAILVSHPPTPNDHFTPTPVSYTAGFYRIPVLG r-NMDAR1d   (Sugihara)
61  NATSVTHKPNAIQMALSVCEDLISSQVYAILVSHPPTPNDHFTPTPVSYTAGFYRIPVLG r-NMDAR1e   (Sugihara)
61  NATSVTHKPNAIQMALSVCEDLISSQVYAILVSHPPTPNDHFTPTPVSYTAGFYRIPVLG r-NMDAR1f   (Sugihara)
61  NATSVTHKPNAIQMALSVCEDLISSQVYAILVSHPPTPNDHFTPTPVSYTAGFYRIPVLG r-NMDAR1g(Sugihara)
61  NATSVTHKPNAIQMALSVCEDLISSQVYAILVSHPPTPNDHFTPTPVSYTAGFYRIPVLG m-zeta1   (Yamazsaki)
61  NATSVTHKPNAIQMALSVCEDLISSQVYAILVSHPPTPNDHFTPTPVSYTAGFYRIPVLG m-zeta1-2   (Yamazaki)
61  NATSVTHKPNAIQMALSVCEDLISSQVYAILVSHPPTPNDHFTPTPVSYTAGFYRIPVLG r-NMDAR1tru   (Sugihara)

121 LTTRMSIYSDKSIHLSFLRTVPPYSHQSSVWFEMMRVYSWNHIILLVSDDHEGRAAQKRL h-NMDAR1   (Karp)
121 LTTRMSIYSDKSIHLSFLRTVPPYSHQSSVWFEMMRVYSWNHIILLVSDDHEGRAAQKRL h-NR1   (Planells)
121 LTTRMSIYSDKSIHLSFLRTVPPYSHQSSVWFEMMRVYSWNHIILLVSDDHEGRAAQKRL h-NR1-1   (Foldes)
81  LTTRMSIYSDKSIHLSFLRTVPPYSHQSSVWFEMMRVYSWNHIILLVSDDHEGRAAQKRL h-NR1-3   (Foldes)
121 LTTRMSIYSDKSIHLSFLRTVPPYSHQSSVWFEMMRVYNWNHIILLVSDDHEGRAAQKRL r-NMDAR1   (Miriyoshi)
121 LTTRMSIYSDKSIHLSFLRTVPPYSHQSSVWFEMMRVYNWNHIILLVSDDHEGRAAQKRL r-NMDA-R1A   (Nakanishi)
121 LTTRMSIYSDKSIHLSFLRTVPPYSHQSSVWFEMMRVYNWNHIILLVSDDHEGRAAQKRL r-NMDA-R1B   (Nakanishi)
121 LTTRMSIYSDKSIHLSFLRTVPPYSHQSSVWFEMMRVYNWNHIILLVSDDHEGRAAQKRL r-R1-LL   (Bayley)
121 LTTRMSIYSDKSIHLSFLRTVPPYSHQSSVWFEMMRVYNWNHIILLVSDDHEGRAAQKRL r-NMDAR1b   (Sugihara)
121 LTTRMSIYSDKSIHLSFLRTVPPYSHQSSVWFEMMRVYNWNHIILLVSDDHEGRAAQKRL r-NMDAR1c   (Sugihara)
121 LTTRMSIYSDKSIHLSFLRTVPPYSHQSSVWFEMMRVYNWNHIILLVSDDHEGRAAQKRL r-NMDAR1d   (Sugihara)
121 LTTRMSIYSDKSIHLSFLRTVPPYSHQSSVWFEMMRVYNWNHIILLVSDDHEGRAAQKRL r-NMDAR1e   (Sugihara)
121 LTTRMSIYSDKSIHLSFLRTVPPYSHQSSVWFEMMRVYNWNHIILLVSDDHEGRAAQKRL r-NMDAR1f   (Sugihara)
121 LTTRMSIYSDKSIHLSFLRTVPPYSHQSSVWFEMMRVYNWNHIILLVSDDHEGRAAQKRL r-NMDAR1g   (Sugihara)
121 LTTRMSIYSDKSIHLSFLRTVPPYSHQSSVWFEMMRVYNWNHIILLVSDDHEGRAAQKRL m-zeta1   (Yamazsaki)
121 LTTRMSIYSDKSIHLSFLRTVPPYSHQSSVWFEMMRVYNWNHIILLVSDDHEGRAAQKRL m-zeta1-2   (Yamazaki)
121 LTTRMSIYSDKPSVHHTGRKEQARSTKCNTSCLQPVAEAVNILSLQCTAPDHQEGTGKKL r-NMDAR1tru.   (Sugihara)
```

FIGURE 3I(1).

voltage-dependent Mg^{2+} block in heterooligomeric NMDA receptors (Burnashev et al., 1992c; Mori et al., 1992). NR2 subunits form channels only when coexpressed with NR1. However, heteromeric NR1–NR2 channels have distinct properties that depend on which NR2 subunit participates in receptor assembly. These properties include different time courses of deactivation, different strengths of Mg^{2+} block and different glycine sensitivities (Kutsuwada et al., 1992; Ishii et al., 1993; Wafford et al., 1993; Monyer et al., 1992, 1994). The expression of NR2 genes is highly regulated both with respect to developmental periods and brain regions (Watanabe et al., 1993; Monyer et al., 1992, 1994). The stoichiometry and composition of native NMDA receptors is unknown. Heterologous expression and immunoprecipitation studies suggest that

```
181 ETLLEERESK---------------------AEKVLQFDPGTKNVTALLMEAKELEARVI h-NMDAR1   (Karp)
181 ETLLEERESK---------------------AEKVLQFDPGTKNVTALLMEAKELEARVI h-NR1   (Planells)
181 ETLLEERESK---------------------AEKVLQFDPGTKNVTALLMEAKELEARVI h-NR1-1   (Foldes)
141 ETLLEERESK---------------------AEKVLQFDPGTKNVTALLMEAKELEARVI h-NR1-3   (Foldes)
181 ETLLEERESK---------------------AEKVLQFDPGTKNVTALLMEARELEARVI r-NMDAR1   (Miriyoshi)
181 ETLLEERESK---------------------AEKVLQFDPGTKNVTALLMEARELEARVI r-NMDA-R1A   (Nakanishi)
181 ETLLEERESKSKKRNYENLDQLSYDNKRGPKAEKVLQFDPGTKNVTALLMEARELEARVI r-NMDA-R1B   (Nakanishi)
181 ETLLEERESKSKKRNYENLDQLSYDNKRGPKAEKVLQFDPGTKNVTALLMEARELEARVI r-R1-LL   (Bayley)
181 ETLLEERESKSKKRNYENLDQLSYDNKRGPKAEKVLQFDPGTKNVTALLMEARELEARVI r-NMDAR1b   (Sugihara)
181 ETLLEERESK---------------------AEKVLQFDPGTKNVTALLMEARELEARVI r-NMDAR1c   (Sugihara)
181 ETLLEERESK---------------------AEKVLQFDPGTKNVTALLMEARELEARVI r-NMDAR1d   (Sugihara)
181 ETLLEERESK---------------------AEKVLQFDPGTKNVTALLMEARELEARVI r-NMDAR1e   (Sugihara)
181 ETLLEERESKSKKRNYENLDQLSYDNKRGPKAEKVLQFDPGTKNVTALLMEARELEARVI r-NMDAR1f   (Sugihara)
181 ETLLEERESKSKKRNYENLDQLSYDNKRGPKAEKVLQFDPGTKNVTALLMEARELEARVI r-NMDAR1g   (Sugihara)
181 ETLLEERESK---------------------AEKVLQFDPGTKNVTALLMEARDLEARVI m-zeta1   (Yamazsaki)
181 ETLLEERESK---------------------AEKVLQFDPGTKNVTALLMEARDLEARVI m-zeta1-2   (Yamazaki)
181 T                                                            r-NMDAR1tru.   (Sugihara)

220 ILSASEDDAATVYRAAAMLNMTGSGYVWLVGEREISGNALRYAPDGILGLQLINGKNESA h-NMDAR1   (Karp)
220 ILSASEDDAATVYRAAAMLNMTGSGYVWLVGEREISGNALRYAPDGILGLQLINGKNESA h-NR1   (Planells)
220 ILSASEDDAATVYRAAAMLNMTGSGYVWLVGEREISGNALRYAPDGILGLQLINGKNESA h-NR1-1   (Foldes)
180 ILSASEDDAATVYRAAAMLNMTGSGYVWLVGEREISGNALRYAPDGILGLQLINGKNESA h-NR1-3   (Foldes)
220 ILSASEDDAATVYRAAAMLNMTGSGYVWLVGEREISGNALRYAPDGIIGLQLINGKNESA r-NMDAR1   (Miriyoshi)
220 ILSASEDDAATVYRAAAMLNMTGSGYVWLVGEREISGNALRYAPDGIIGLQLINGKNESA r-NMDA-R1A   (Nakanishi)
241 ILSASEDDAATVYRAAAMLNMTGSGYVWLVGEREISGNALRYAPDGIIGLQLINGKNESA r-NMDA-R1B   (Nakanishi)
241 ILSASEDDAATVYRAAAMLNMTGSGYVWLVGEREISGNALRYAPDGIIGLQLINGKNESA r-R1-LL   (Bayley)
241 ILSASEDDAATVYRAAAMLNMTGSGYVWLVGEREISGNALRYAPDGIIGLQLINGKNESA r-NMDAR1b   (Sugihara)
220 ILSASEDDAATVYRAAAMLNMTGSGYVWLVGEREISGNALRYAPDGIIGLQLINGKNESA r-NMDAR1c   (Sugihara)
220 ILSASEDDAATVYRAAAMLNMTGSGYVWLVGEREISGNALRYAPDGIIGLQLINGKNESA r-NMDAR1d   (Sugihara)
220 ILSASEDDAATVYRAAAMLNMTGSGYVWLVGEREISGNALRYAPDGIIGLQLINGKNESA r-NMDAR1e(Sugihara)
241 ILSASEDDAATVYRAAAMLNMTGSGYVWLVGEREISGNALRYAPDGIIGLQLINGKNESA r-NMDAR1f   (Sugihara)
241 ILSASEDDAATVYRAAAMLNMTGSGYVWLVGEREISGNALRYAPDGIIGLQLINGKNESA r-NMDAR1g   (Sugihara)
220 ILSASEDDAATVYRAAAMLNMTGSGYVWLVGEREISGNALRYAPDGIIGLQLINGKNESA m-zeta1   (Yamazsaki)
220 ILSASEDDAATVYRAAAMLNMTGSGYVWLVGEREISGNALRYAPDGIIGLQLINGKNESA m-zeta1-2   (Yamazaki)

280 HISDAVGVVAQAVHELLEKENITDPPRGCVGNTNIWKTGPLFKRVLMSSKYADGVTGRVE h-NMDAR1   (Karp)
280 HISDAVGVVAQAVHELLEKENITDPPRGCVGNTNIWKTGPLFKRVLMSSKYADGVTGRVE h-NR1   (Planells)
280 HISDAVGVVAQAVHELLEKENITDPPRGCVGNTNIWKTGPLFKRVLMSSKYADGVTGRVE h-NR1-1   (Foldes)
240 HISDAVGVVAQAVHELLEKENITDPPRGCVGNTNIWKTGPLFKRVLMSSKYADGVTGRVE h-NR1-3   (Foldes)
280 HISDAVGVVAQAVHELLEKENITDPPRGCVGNTNIWKTGPLFKRVLMSSKYADGVTGRVE r-NMDAR1   (Miriyoshi)
280 HISDAVGVVAQAVHELLEKENITDPPRGCVGNTNIWKTGPLFKRVLMSSKYADGVTGRVE r-NMDA-R1A   (Nakanishi)
301 HISDAVGVVAQAVHELLEKENITDPPRGCVGNTNIWKTGPLFKRVLMSSKYADGVTGRVE r-NMDA-R1B   (Nakanishi)
301 HISDAVGVVAQAVHELLEKENITDPPRGCVGNTNIWKTGPLFKRVLMSSKYADGVTGRVE r-R1-LL   (Bayley)
301 HISDAVGVVAQAVHELLEKENITDPPRGCVGNTNIWKTGPLFKRVLMSSKYADGVTGRVE r-NMDAR1b   (Sugihara)
280 HISDAVGVVAQAVHELLEKENITDPPRGCVGNTNIWKTGPLFKRVLMSSKYADGVTGRVE r-NMDAR1c(Sugihara)
280 HISDAVGVVAQAVHELLEKENITDPPRGCVGNTNIWKTGPLFKRVLMSSKYADGVTGRVE r-NMDAR1d   (Sugihara)
280 HISDAVGVVAQAVHELLEKENITDPPRGCVGNTNIWKTGPLFKRVLMSSKYADGVTGRVE r-NMDAR1e   (Sugihara)
301 HISDAVGVVAQAVHELLEKENITDPPRGCVGNTNIWKTGPLFKRVLMSSKYADGVTGRVE r-NMDAR1f   (Sugihara)
301 HISDAVGVVAQAVHELLEKENITDPPRGCVGNTNIWKTGPLFKRVLMSSKYADGVTGRVE r-NMDAR1g   (Sugihara)
280 HISDAVGVVAQAVHELLEKENITDPPRGCVGNTNIWKTGPLFKRVLMSSKYADGVTGRVE m-zeta1   (Yamazsaki)
280 HISDAVGVVAQAVHELLEKENITDPPRGCVGNTNIWKTGPLFKRVLMSSKYADGVTGRVE m-zeta1-2   (Yamazaki)
                                                  MSSKYADGVTGRVE h-NR1-2   (Foldes)
```

FIGURE 3l(2).

one or two different NR2 subunits can assemble with NR1 (Monyer et al., 1992, 1994; Meguro et al., 1992; Ishii et al., 1993; Wafford et al., 1993; Sheng et al., 1994).

2.6.4.2.1 NMDAR2A (NR2A)

Using the sequence information of NR1 the NR2A subunit cDNA was isolated from rat and mouse brain RNA by PCR-mediated homology screens (Meguro et al., 1992; Monyer et al., 1992; Ishii et al., 1993). NR2A is 1455 amino acids long, and is characterized by an extended carboxy-terminal sequence (Figure 3l). NR2A splice variants have not been reported. NR2A

```
340 FNEDGDRKFANYSIMNLQNRKLVQVGIYNGTHVIPNDRKIIWPGGETEKPRGYQMSTRLK h-NMDAR1   (Karp)
340 FNEDGDRKFANYSIMNLQNRKLVQVGIYNGTHVIPNDRKIIWPGGETEKPRGYQMSTRLK h-NR1   (Planells)
340 FNEDGDRKFANYSIMNLQNRKLVQVGIYNGTHVIPNDRKIIWPGGETEKPRGYQMSTRLK h-NR1-1   (Foldes)
300 FNEDGDRKFANYSIMNLQNRKLVQVGIYNGTHVIPNDRKIIWPGGETEKPRGYQMSTRLK h-NR1-3   (Foldes)
340 FNEDGDRKFANYSIMNLQNRKLVQVGIYNGTHVIPNDRKIIWPGGETEKPRGYQMSTRLK r-NMDAR1   (Miriyoshi)
340 FNEDGDRKFANYSIMNLQNRKLVQVGIYNGTHVIPNDRKIIWPGGETEKPRGYQMSTRLK r-NMDA-R1A   (Nakanishi)
361 FNEDGDRKFANYSIMNLQNRKLVQVGIYNGTHVIPNDRKIIWPGGETEKPRGYQMSTRLK r-NMDA-R1B   (Nakanishi)
361 FNEDGDRKFANYSIMNLQNRKLVQVGIYNGTHVIPNDRKIIWPGGETEKPRGYQMSTRLK r-R1-LL   (Bayley)
361 FNEDGDRKFANYSIMNLQNRKLVQVGIYNGTHVIPNDRKIIWPGGETEKPRGYQMSTRLK r-NMDAR1b   (Sugihara)
340 FNEDGDRKFANYSIMNLQNRKLVQVGIYNGTHVIPNDRKIIWPGGETEKPRGYQMSTRLK r-NMDAR1c   (Sugihara)
340 FNEDGDRKFANYSIMNLQNRKLVQVGIYNGTHVIPNDRKIIWPGGETEKPRGYQMSTRLK r-NMDAR1d   (Sugihara)
340 FNEDGDRKFANYSIMNLQNRKLVQVGIYNGTHVIPNDRKIIWPGGETEKPRGYQMSTRLK r-NMDAR1e   (Sugihara)
361 FNEDGDRKFANYSIMNLQNRKLVQVGIYNGTHVIPNDRKIIWPGGETEKPRGYQMSTRLK r-NMDAR1f   (Sugihara)
361 FNEDGDRKFANYSIMNLQNRKLVQVGIYNGTHVIPNDRKIIWPGGETEKPRGYQMSTRLK r-NMDAR1g   (Sugihara)
340 FNEDGDRKFANYSIMNLQNRKLVQVGIYNGTHVIPNDRKIIWPGGETEKPRGYQMSTRLK m-zeta1   (Yamazsaki)
340 FNEDGDRKFANYSIMNLQNRKLVQVGIYNGTHVIPNDRKIIWPGGETEKPRGYQMSTRLK m-zeta1-2   (Yamazaki)
15  FNEDGDRKFANYSIMNLQNRKLVQVGIYNGTHVIPNDRKIIWPGGETEKPRGYQMSTRLK h-NR1-2   (Foldes)

400 IVTIHQEPFVYVKPTLSDGTCKEEFTVNGDPVKKVICTGPNDTSPGSPRHTVPQCCYGFC h-NMDAR1   (Karp)
400 IVTIHQEPFVYVKPTLSDGTCKEEFTVNGDPVKKVICTGPNDTSPGSPRHTVPQCCYGFC h-NR1   (Planells)
400 IVTIHQEPFVYVKPTLSDGTCKEEFTVNGDPVKKVICTGPNDTSPGSPRHTVPQCCYGFC h-NR1-1   (Foldes)
360 IVTIHQEPFVYVKPTLSDGTCKEEFTVNGDPVKKVICTGPNDTSPGSPRHTVPQCCYGFC h-NR1-3   (Foldes)
400 IVTIHQEPFVYVKPTMSDGTCKEEFTVNGDPVKKVICTGPNDTSPGSPRHTVPQCCYGFC r-NMDAR1   (Miriyoshi)
400 IVTIHQEPFVYVKPTMSDGTCKEEFTVNGDPVKKVICTGPNDTSPGSPRHTVPQCCYGFC r-NMDA-R1A   (Nakanishi)
421 IVTIHQEPFVYVKPTMSDGTCKEEFTVNGDPVKKVICTGPNDTSPGSPRHTVPQCCYGFC r-NMDA-R1B   (Nakanishi)
421 IVTIHQEPFVYVKPTMSDGTCKEEFTVNGDPVKKVICTGPNDTSPGSPRHTVPQCCYGFC r-R1-LL   (Bayley)
421 IVTIHQEPFVYVKPTMSDGTCKEEFTVNGDPVKKVICTGPNDTSPGSPRHTVPQCCYGFC r-NMDAR1b   (Sugihara)
400 IVTIHQEPFVYVKPTMSDGTCKEEFTVNGDPVKKVICTGPNDTSPGSPRHTVPQCCYGFC r-NMDAR1c   (Sugihara)
400 IVTIHQEPFVYVKPTMSDGTCKEEFTVNGDPVKKVICTGPNDTSPGSPRHTVPQCCYGFC r-NMDAR1d   (Sugihara)
400 IVTIHQEPFVYVKPTMSDGTCKEEFTVNGDPVKKVICTGPNDTSPGSPRHTVPQCCYGFC r-NMDAR1e   (Sugihara)
421 IVTIHQEPFVYVKPTMSDGTCKEEFTVNGDPVKKVICTGPNDTSPGSPRHTVPQCCYGFC r-NMDAR1f   (Sugihara)
421 IVTIHQEPFVYVKPTMSDGTCKEEFTVNGDPVKKVICTGPNDTSPGSPRHTVPQCCYGFC r-NMDAR1g   (Sugihara)
400 IVTIHQEPFVYVKPTMSDGTCKEEFTVNGDPVKKVICTGPNDTSPGSPRHTVPQCCYGFC m-zeta1   (Yamazsaki)
400 IVTIHQEPFVYVKPTMSDGTCKEEFTVNGDPVKKVICTGPNDTSPGSPRHTVPQCCYGFC m-zeta1-2   (Yamazaki)
75  IVTIHQEPFVYVKPTLSDGTCKEEFTVNGDPVKKVICTGPNDTSPGSPRHTVPQCCYGFC h-NR1-2   (Foldes)

460 IDLLIKLARTMNFTYEVHLVADGKFGTQERVNNSNKKEWNGMMGELLSGQADMIVAPLTI h-NMDAR1   (Karp)
460 IDLLIKLARTMNFTYEVHLVADGKFGTQERVNNSNKKEWNGMMGELLSGQADMIVAPLTI h-NR1   (Planells)
460 IDLLIKLARTMNFTYEVHLVADGKFGTQERVNNSNKKEWNGMMGELLSGQADMIVAPLTI h-NR1-1   (Foldes)
420 IDLLIKLARTMNFTYEVHLVADGKFGTQERVNNSNKKEWNGMMGELLSGQADMIVAPLTI h-NR1-3   (Foldes)
460 IDLLIKLARTMNFTYEVHLVADGKFGTQERVNNSNKKEWNGMMGELLSGQADMIVAPLTI r-NMDAR1   (Miriyoshi)
460 IDLLIKLARTMNFTYEVHLVADGKFGTQERVNNSNKKEWNGMMGELLSGQADMIVAPLTI r-NMDA-R1A   (Nakanishi)
481 IDLLIKLARTMNFTYEVHLVADGKFGTQERVNNSNKKEWNGMMGELLSGQADMIVAPLTI r-NMDA-R1B   (Nakanishi)
481 IDLLIKLARTMNFTYEVHLVADGKFGTQERVNNSNKKEWNGMMGELLSGQADMIVAPLTI r-R1-LL   (Bayley)
481 IDLLIKLARTMNFTYEVHLVADGKFGTQERVNNSNKKEWNGMMGELLSGQADMIVAPLTI r-NMDAR1b   (Sugihara)
460 IDLLIKLARTMNFTYEVHLVADGKFGTQERVNNSNKKEWNGMMGELLSGQADMIVAPLTI r-NMDAR1c   (Sugihara)
460 IDLLIKLARTMNFTYEVHLVADGKFGTQERVNNSNKKEWNGMMGELLSGQADMIVAPLTI r-NMDAR1d   (Sugihara)
460 IDLLIKLARTMNFTYEVHLVADGKFGTQERVNNSNKKEWNGMMGELLSGQADMIVAPLTI r-NMDAR1e   (Sugihara)
481 IDLLIKLARTMNFTYEVHLVADGKFGTQERVNNSNKKEWNGMMGELLSGQADMIVAPLTI r-NMDAR1f   (Sugihara)
481 IDLLIKLARTMNFTYEVHLVADGKFGTQERVNNSNKKEWNGMMGELLSGQADMIVAPLTI r-NMDAR1g   (Sugihara)
460 VDLLIKLARTMNFTYEVHLVADGKFGTQERVNNSNKKEWNGMMGELLSGQADMIVAPLTI m-zeta1   (Yamazsaki)
460 VDLLIKLARTMNFTYEVHLVADGKFGTQERVNNSNKKEWNGMMGELLSGQADMIVAPLTI m-zeta1-2   (Yamazaki)
135 IDLLIKLARTMNFTYEVHLVADGKFGTQERVNNSNKKEWNGMMGELLSGQADMIVAPLTI h-NR1-2   (Foldes)
```

FIGURE 3I(3).

does not form homooligomeric receptors but assembles with NR1 to form heterooligomeric receptor channels. When expressed *in vitro* these show desensitization in the presence of agonists (L-glutamate, NMDA), and display a characteristic time-course of deactivation that is shorter than for the other binary NR1–NR2 receptor channels.

The NR2A subunit is expressed in the CNS mainly during postnatal and adult stages (Watanabe et al., 1993; Ishii et al., 1993; Monyer et al., 1994). The mRNA has been estimated to be 12 kb in rat and 20 kb in mouse (Ishii et al., 1993; Meguro et al., 1992). Human NR2A sequences have not been reported, but the human NR2A gene was mapped to 16p13 (Takano et al., 1993).

```
                                                          ———1———
520 NNERAQYIEFSKPFKYQGLTILVKKEIPRSTLDSFMQPFQSTLWLLVGLSVHVVAVMLYL h-NMDAR1 (Karp)
520 NNERAQYIEFSKPFKYQGLTILVKKEIPRSTLDSFMQPFQSTLWLLVGLSVHVVAVMLYL h-NR1 (Planells)
520 NNERAQYIEFSKPFKYQGLTILVKKEIPRSTLDSFMQPFQSTLWLLVGLSVHVVAVMLYL h-NR1-1 (Foldes)
480 NNERAQYIEFSKPFKYQGLTILVKKEIPRSTLDSFMQPFQSTLWLLVGLSVHVVAVMLYL h-NR1-3 (Foldes)
520 NNERAQYIEFSKPFKYQGLTILVKKEIPRSTLDSFMQPFQSTLWLLVGLSVHVVAVMLYL r-NMDAR1 (Miriyoshi)
520 NNERAQYIEFSKPFKYQGLTILVKKEIPRSTLDSFMQPFQSTLWLLVGLSVHVVAVMLYL r-NMDA-R1A (Nakanishi)
541 NNERAQYIEFSKPFKYQGLTILVKKEIPRSTLDSFMQPFQSTLWLLVGLSVHVVAVMLYL r-NMDA-R1B (Nakanishi)
541 NNERAQYIEFSKPFKYQGLTILVKKEIPRSTLDSFMQPFQSTLWLLVGLSVHVVAVMLYL r-R1-LL (Bayley)
541 NNERAQYIEFSKPFKYQGLTILVKKEIPRSTLDSFMQPFQSTLWLLVGLSVHVVAVMLYL r-NMDAR1b (Sugihara)
520 NNERAQYIEFSKPFKYQGLTILVKKEIPRSTLDSFMQPFQSTLWLLVGLSVHVVAVMLYL r-NMDAR1c (Sugihara)
520 NNERAQYIEFSKPFKYQGLTILVKKEIPRSTLDSFMQPFQSTLWLLVGLSVHVVAVMLYL r-NMDAR1d (Sugihara)
520 NNERAQYIEFSKPFKYQGLTILVKKEIPRSTLDSFMQPFQSTLWLLVGLSVHVVAVMLYL r-NMDAR1e (Sugihara)
541 NNERAQYIEFSKPFKYQGLTILVKKEIPRSTLDSFMQPFQSTLWLLVGLSVHVVAVMLYL r-NMDAR1f (Sugihara)
541 NNERAQYIEFSKPFKYQGLTILVKKEIPRSTLDSFMQPFQSTLWLLVGLSVHVVAVMLYL r-NMDAR1g (Sugihara)
520 NNERAQYIEFSKPFKYQGLTILVKKEIPRSTLDSFMQPFQSTLWLLVGLSVHVVAVMLYL m-zeta1 (Yamazsaki)
520 NNERAQYIEFSKPFKYQGLTILVKKEIPRSTLDSFMQPFQSTLWLLVGLSVHVVAVMLYL m-zeta1-2 (Yamazaki)
195 NNERAQYIEFSKPFKYQGLTILVKKEIPRSTLDSFMQPFQSTLWLLVGLSVHVVAVMLYL h-NR1-2 (Foldes)

                                     ————2————                ————
580 LDRFSPFGRFKVNSEEEEEDALTLSSAMWFSWGVLLNSGIGEGAPRSFSARILGMVWAGF h-NMDAR1 (Karp)
580 LDRFSPFGRFKVNSEEEEEDALTLSSAMWFSWGVLLNSGIGEGAPRSFSARILGMVWAGF h-NR1 (Planells)
580 LDRFSPFGRFKVNSEEEEEDALTLSSAMWFSWGVLLNSGIGEGAPRSFSARILGMVWAGF h-NR1-1 (Foldes)
540 LDRFSPFGRFKVNSEEEEEDALTLSSAMWFSWGVLLNSGIGEGAPRSFSARILGMVWAGF h-NR1-3 (Foldes)
580 LDRFSPFGRFKVNSEEEEEDALTLSSAMWFSWGVLLNSGIGEGAPRSFSARILGMVWAGF r-NMDAR1 (Miriyoshi)
580 LDRFSPFGRFKVNSEEEEEDALTLSSAMWFSWGVLLNSGIGEGAPRSFSARILGMVWAGF r-NMDA-R1A (Nakanishi)
601 LDRFSPFGRFKVNSEEEEEDALTLSSAMWFSWGVLLNSGIGEGAPRSFSARILGMVWAGF r-NMDA-R1B (Nakanishi)
601 LDRFSPFGRFKVNSEEEEEDALTLSSAMWFSWGVLLNSGIGEGAPRSFSARILGMVWAGF r-R1-LL (Bayley)
601 LDRFSPFGRFKVNSEEEEEDALTLSSAMWFSWGVLLNSGIGEGAPRSFSARILGMVWAGF r-NMDAR1b (Sugihara)
580 LDRFSPFGRFKVNSEEEEEDALTLSSAMWFSWGVLLNSGIGEGAPRSFSARILGMVWAGF r-NMDAR1c (Sugihara)
580 LDRFSPFGRFKVNSEEEEEDALTLSSAMWFSWGVLLNSGIGEGAPRSFSARILGMVWAGF r-NMDAR1d (Sugihara)
580 LDRFSPFGRFKVNSEEEEEDALTLSSAMWFSWGVLLNSGIGEGAPRSFSARILGMVWAGF r-NMDAR1e (Sugihara)
601 LDRFSPFGRFKVNSEEEEEDALTLSSAMWFSWGVLLNSGIGEGAPRSFSARILGMVWAGF r-NMDAR1f (Sugihara)
601 LDRFSPFGRFKVNSEEEEEDALTLSSAMWFSWGVLLNSGIGEGAPRSFSARILGMVWAGF r-NMDAR1g (Sugihara)
580 LDRFSPFGRFKVNSEEEEEDALTLSSAMWFSWGVLLNSGIGEGAPRSFSARILGMVWAGF m-zeta1 (Yamazsaki)
580 LDRFSPFGRFKVNSEEEEEDALTLSSAMWFSWGVLLNSGIGEGAPRSFSARILGMVWAGF m-zeta1-2 (Yamazaki)
255 LDRFSPFGRFKVNSEEEEEDALTLSSAMWFSWGVLLNSGIGEGAPRSFSARILGMVWAGF h-NR1-2 (Foldes)

    —3———
640 AMIIVASYTANLAAFLVLDRPEERITGINDPRLRNPSDKFIYATVKQSSVDIYFRRQVEL h-NMDAR1 (Karp)
640 AMIIVASYTANLAAFLVLDRPEERITGINDPRLRNPSDKFIYATVKQSSVDIYFRRQVEL h-NR1 (Planells)
640 AMIIVASYTANLAAFLVLDRPEERITGINDPRLRNPSDKFIYATVKQSSVDIYFRRQVEL h-NR1-1 (Foldes)
600 AMIIVASYTANLAAFLVLDRPEERITGINDPRLRNPSDKFIYATVKQSSVDIYFRRQVEL h-NR1-3 (Foldes)
640 AMIIVASYTANLAAFLVLDRPEERITGINDPRLRNPSDKFIYATVKQSSVDIYFRRQVEL r-NMDAR1 (Miriyoshi)
640 AMIIVASYTANLAAFLVLDRPEERITGINDPRLRNPSDKFIYATVKQSSVDIYFRRQVEL r-NMDA-R1A (Nakanishi)
661 AMIIVASYTANLAAFLVLDRPEERITGINDPRLRNPSDKFIYATVKQSSVDIYFRRQVEL r-NMDA-R1B (Nakanishi)
661 AMIIVASYTANLAAFLVLDRPEERITGINDPRLRNPSDKFIYATVKQSSVDIYFRRQVEL r-R1-LL (Bayley)
661 AMIIVASYTANLAAFLVLDRPEERITGINDPRLRNPSDKFIYATVKQSSVDIYFRRQVEL r-NMDAR1b (Sugihara)
640 AMIIVASYTANLAAFLVLDRPEERITGINDPRLRNPSDKFIYATVKQSSVDIYFRRQVEL r-NMDAR1c (Sugihara)
640 AMIIVASYTANLAAFLVLDRPEERITGINDPRLRNPSDKFIYATVKQSSVDIYFRRQVEL r-NMDAR1d (Sugihara)
640 AMIIVASYTANLAAFLVLDRPEERITGINDPRLRNPSDKFIYATVKQSSVDIYFRRQVEL r-NMDAR1e (Sugihara)
661 AMIIVASYTANLAAFLVLDRPEERITGINDPRLRNPSDKFIYATVKQSSVDIYFRRQVEL r-NMDAR1f (Sugihara)
661 AMIIVASYTANLAAFLVLDRPEERITGINDPRLRNPSDKFIYATVKQSSVDIYFRRQVEL r-NMDAR1g (Sugihara)
640 AMIIVASYTANLAAFLVLDRPEERITGINDPRLRNPSDKFIYATVKQSSVDIYFRRQVEL m-zeta1 (Yamazsaki)
640 AMIIVASYTANLAAFLVLDRPEERITGINDPRLRNPSDKFIYATVKQSSVDIYFRRQVEL m-zeta1-2 (Yamazaki)
315 AMIIVASYTANLAAFLVLDRPEERITGINDPRLRNPSDKFIYATVKQSSVDIYFRRQVEL h-NR1-2 (Foldes)
```

FIGURE 3l(4).

2.6.4.2.2 NMDAR2B (NR2B)

The rat and mouse NR2B subunit cDNAs have been isolated (Monyer et al., 1992; Kutsuwada et al., 1992). The NR2B subunit is 1482 residues long (Figure 3m) and can be coexpressed with NR1 *in vitro* to form a glutamate-gated channel with distinct deactivation time-courses (Monyer et al., 1992, 1994), strength of Mg^{2+} block (Kutsuwada et al., 1992; Ishii et al., 1993; Monyer et al., 1992, 1994), and glycine sensitivity (Kutsuwada et al., 1992; Ishii et al., 1993). The mouse and rat sequences show an amino acid variability at the very end of the carboxy-terminus, probably reflecting an error in the mouse sequence.

```
700 STMYRHMEKHNYESAAEAIQAVRDNKLHAFIWDSAVLEFEASQKCDLVTTGELFFRSGFG h-NMDAR1   (Karp)
700 STMYRHMEKHNYESAAEAIQAVRDNKLHAFIWDSAVLEFEASQKCDLVTTGELFFRSGFG h-NR1   (Planells)
700 STMYRHMEKHNYESAAEAIQAVRDNKLHAFIWDSAVLEFEASQKCDLVTTGELFFRSGFG h-NR1-1   (Foldes)
660 STMYRHMEKHNYESAAEAIQAVRDNKLHAFIWDSAVLEFEASQKCDLVTTGELFFRSGFG h-NR1-3   (Foldes)
700 STMYRHMEKHNYESAAEAIQAVRDNKLHAFIWDSAVLEFEASQKCDLVTTGELFFRSGFG r-NMDAR1   (Miriyoshi)
700 STMYRHMEKHNYESAAEAIQAVRDNKLHAFIWDSAVLEFEASQKCDLVTTGELFFRSGFG r-NMDA-R1A   (Nakanishi)
721 STMYRHMEKHNYESAAEAIQAVRDNKLHAFIWDSAVLEFEASQKCDLVTTGELFFRSGFG r-NMDA-R1B   (Nakanishi)
721 STMYRHMEKHNYESAAEAIQAVRDNKLHAFIWDSAVLEFEASQKCDLVTTGELFFRSGFG r-R1-LL   (Bayley)
721 STMYRHMEKHNYESAAEAIQAVRDNKLHAFIWDSAVLEFEASQKCDLVTTGELFFRSGFG r-NMDAR1b   (Sugihara)
700 STMYRHMEKHNYESAAEAIQAVRDNKLHAFIWDSAVLEFEASQKCDLVTTGELFFRSGFG r-NMDAR1c   (Sugihara)
700 STMYRHMEKHNYESAAEAIQAVRDNKLHAFIWDSAVLEFEASQKCDLVTTGELFFRSGFG r-NMDAR1d   (Sugihara)
700 STMYRHMEKHNYESAAEAIQAVRDNKLHAFIWDSAVLEFEASQKCDLVTTGELFFRSGFG r-NMDAR1e(Sugihara)
721 STMYRHMEKHNYESAAEAIQAVRDNKLHAFIWDSAVLEFEASQKCDLVTTGELFFRSGFG r-NMDAR1f   (Sugihara)
721 STMYRHMEKHNYESAAEAIQAVRDNKLHAFIWDSAVLEFEASQKCDLVTTGELFFRSGFG r-NMDAR1g   (Sugihara)
700 STMYRHMEKHNYESAAEAIQAVRDNKLHAFIWDSAVLEFEASQKCDLVTTGELFFRSGFG m-zeta1   (Yamazsaki)
700 STMYRHMEKHNYESAAEAIQAVRDNKLHAFIWDSAVLEFEASQKCDLVTTGELFFRSGFG m-zeta1-2   (Yamazaki)
375 STMYRHMEKHNYESAAEAIQAVRDNKLHAFIWDSAVLEFEASQKCDLVTTGELFFRSGFG h-NR1-2   (Foldes)

760 IGMRKDSPWKQNVSLSILKSHENGFMEDLDKTWVRYQECDSRSNAPATLTFENMAGVFML h-NMDAR1   (Karp)
760 IGMRKDSPWKQNVSLSILKSHENGFMEDLDKTWVRYQECDSRSNAPATLTFENMAGVFML h-NR1   (Planells)
760 IGMRKDSPWKQNVSLSILKSHENGFMEDLDKTWVRYQECDSRSNAPATLTFENMAGVFML h-NR1-1   (Foldes)
720 IGMRKDSPWKQNVSLSILKSHENGFMEDLDKTWVRYQECDSRSNAPATLTFENMAGVFML h-NR1-3   (Foldes)
760 IGMRKDSPWKQNVSLSILKSHENGFMEDLDKTWVRYQECDSRSNAPATLTFENMAGVFML r-NMDAR1   (Miriyoshi)
760 IGMRKDSPWKQNVSLSILKSHENGFMEDLDKTWVRYQECDSRSNAPATLTFENMAGVFML r-NMDA-R1A   (Nakanishi)
781 IGMRKDSPWKQNVSLSILKSHENGFMEDLDKTWVRYQECDSRSNAPATLTFENMAGVFML r-NMDA-R1B   (Nakanishi)
781 IGMRKDSPWKQNVSLSILKSHENGFMEDLDKTWVRYQECDSRSNAPATLTFENMAGVFML r-R1-LL   (Bayley)
781 IGMRKDSPWKQNVSLSILKSHENGFMEDLDKTWVRYQECDSRSNAPATLTFENMAGVFML r-NMDAR1b   (Sugihara)
760 IGMRKDSPWKQNVSLSILKSHENGFMEDLDKTWVRYQECDSRSNAPATLTFENMAGVFML r-NMDAR1c   (Sugihara)
760 IGMRKDSPWKQNVSLSILKSHENGFMEDLDKTWVRYQECDSRSNAPATLTFENMAGVFML r-NMDAR1d   (Sugihara)
760 IGMRKDSPWKQNVSLSILKSHENGFMEDLDKTWVRYQECDSRSNAPATLTFENMAGVFML r-NMDAR1e   (Sugihara)
781 IGMRKDSPWKQNVSLSILKSHENGFMEDLDKTWVRYQECDSRSNAPATLTFENMAGVFML r-NMDAR1f   (Sugihara)
781 IGMRKDSPWKQNVSLSILKSHENGFMEDLDKTWVRYQECDSRSNAPATLTFENMAGVFML r-NMDAR1g   (Sugihara)
760 IGMRKDSPWKQNVSLSILKSHENGFMEDLDKTWVRYQECDSRSNAPATLTFENMAGVFML m-zeta1   (Yamazsaki)
760 IGMRKDSPWKQNVSLSILKSHENGFMEDLDKTWVRYQECDSRSNAPATLTFENMAGVFML m-zeta1-2   (Yamazaki)
435 IGMRKDSPWKQNVSLSILKSHENGFMEDLDKTWVRYQECDSRSNAPATLTFENMAGVFML h-NR1-2   (Foldes)
      ----4------
820 VAGGIVAGIFLIFIEIAYKRHKDARRKQMQLAFAAVNVWRKNLQDRKSGRAEPDPKKKAT h-NMDAR1   (Karp)
820 VAGGIVAGIFLIFIEIAYKRHKDARRKQMQLAFAAVNVWRKNLQQYHPTDITGPLNLSDP h-NR1   (Planells)
820 VAGGIVAGIFLIFIEIAYKRHKDARRKQMQLAFAAVNVWRKNLQQYHPTDITGPLNLSDP h-NR1-1   (Foldes)
780 VAGGIVAGIFLIFIEIAYKRHKDARRKQMQLAFAAVNVWRKNLQDRKSGRAEPDPKKKAT h-NR1-3   (Foldes)
820 VAGGIVAGIFLIFIEIAYKRHKDARRKQMQLAFAAVNVWRKNLQDRKSGRAEPDPKKKAT r-NMDAR1   (Miriyoshi)
820 VAGGIVAGIFLIFIEIAYKRHKDARRKQMQLAFAAVNVWRKNLQDRKSGRAEPDPKKKAT r-NMDA-R1A   (Nakanishi)
841 VAGGIVAGIFLIFIEIAYKRHKDARRKQMQLAFAAVNVWRKNLQDRKSGRAEPDPKKKAT r-NMDA-R1B   (Nakanishi)
841 VAGGIVAGIFLIFIEIAYKRHKDARRKQMQLAFAAVNVWRKNLQDRKSGRAEPDPKKKAT r-R1-LL   (Bayley)
841 VAGGIVAGIFLIFIEIAYKRHKDARRKQMQLAFAAVNVWRKNLQDRKSGRAEPDPKKKAT r-NMDAR1b   (Sugihara)
820 VAGGIVAGIFLIFIEIAYKRHKDARRKQMQLAFAAVNVWRKNLQSTGGGRGALQNQKDTV r-NMDAR1c   (Sugihara)
820 VAGGIVAGIFLIFIEIAYKRHKDARRKQMQLAFAAVNVWRKNLQDRKSGRAEPDPKKKAT r-NMDAR1d   (Sugihara)
820 VAGGIVAGIFLIFIEIAYKRHKDARRKQMQLAFAAVNVWRKNLQQYHPTDITGPLNLSDP r-NMDAR1e   (Sugihara)
841 VAGGIVAGIFLIFIEIAYKRHKDARRKQMQLAFAAVNVWRKNLQSTGGGRGALQNQKDTV r-NMDAR1f   (Sugihara)
841 VAGGIVAGIFLIFIEIAYKRHKDARRKQMQLAFAAVNVWRKNLQQYHPTDITGPLNLSDP r-NMDAR1g   (Sugihara)
820 VAGGIVAGIFLIFIEIAYKRHKDARRKQMQLAFAAVNVWRKNLQDRKSGRAEPDPKKKAT m-zeta1   (Yamazsaki)
820 VAGGIVAGIFLIFIEIAYKRHKDARRKQMQLAFAAVNVWRKNLQSTGGGRGALQNQKDTV m-zeta1-2   (Yamazaki)
495 VAGGIVAGIFLIFIEIAYKRHKDARRKQMQLAFAAVNVWRKNLQSTGGGRGALQNQKDTV h-NR1-2   (Foldes)
```

FIGURE 3l(5).

NR2B is expressed in the embryonal and in the adult brain and its mRNA size has been determined for rat (15 kb) and mouse (20 kb) (Ishii et al., 1993; Kutsuwada et al., 1992). Splice variants and chromosomal localization have not been reported.

2.6.4.2.3 NMDAR2C (NR2C)

The NR2C subunit was isolated from rat and mouse brain by PCR-mediated homology screens (Kutsuwada et al., 1992; Monyer et al., 1992; Ishii et al., 1993). The NR2C subunit is 1250 amino acids in length (Figure 3n). In the cloned rat NR2C cDNA sequence there are two possible translational initiation

```
880 FRAITSTLASSFKRRRSSKDTSTGGGRGALQNQKDTVLPRRAIEREEGQLQLCSRHRES  h-NMDAR1   (Karp)
880 SVSTVV                                                      h-NR1   (Planells)
880 SVSTVV                                                      h-NR1-1   (Foldes)
840 FRAITSTLASSFKRRRSSKDTSTGGGRGALQNQKDTVLPRRAIEREEGQLQLCSRHTES  h-NR1-3   (Foldes)
880 FRAITSTLASSFKRRRSSKDTSTGGGRGALQNQKDTVLPRRAIEREEGQLQLCSRHRES  r-NMDAR1   (Miriyoshi)
880 FRAITSTLASSFKRRRSSKDTSTGGGRGALQNQKDTVLPRRAIEREEGQLQLCSRHRES  r-NMDA-R1A   (Nakanishi)
901 FRAITSTLASSFKRRRSSKDTSTGGGRGALQNQKDTVLPRRAIEREEGQLQLCSRHRES  r-NMDA-R1B   (Nakanishi)
901 FRAITSTLASSFKRRRSSKDTSTGGGRGALQNQKDTVLPRRAIEREEGQLQLCSRHRES  r-R1-LL   (Bayley)
901 FRAITSTLASSFKRRRSSKDTSTGGGRGALQNQKDTVLPRRAIEREEGQLQLCSRHRES  r-NMDAR1b   (Sugihara)
880 LPRRAIEREEGQLQLCSRHRES                                      r-NMDAR1c   (Sugihara)
880 FRAITSTLASSFKRRRSSKDTQYHPTDITGPLNLSDPSVSTVV                 r-NMDAR1d   (Sugihara)
880 SVSTVV                                                      r-NMDAR1e   (Sugihara)
901 LPRRAIEREEGQLQLCSRHRES                                      r-NMDAR1f   (Sugihara)
901 SVSTVV                                                      r-NMDAR1g   (Sugihara)
880 FRAITSTLASSFKRRRSSKDTSTGGGRGALQNQKDTVLPRRAIEREEGQLQLCSRHRES  m-zeta1   (Yamazsaki)
880 LPRRAIEREEGQLQLCSRHRES                                      m-zeta1-2   (Yamazaki)
555 LPRRAIEREEGQLQLCSRHRES                                      h-NR1-2   (Foldes)
```

FIGURE 3l(6).

codons but comparison with the mouse sequence identifies the second one as the correct start codon. The rat NR2C sequence was originally reported with a frame-shifted carboxy-terminus due to a single nucleotide deletion in the cloned cDNA (Monyer et al., 1992). This error has been corrected (Burnashev et al., 1992c), and the correct sequence is incorporated in the protein alignment of Figure 3n. The remaining sequence variability in the carboxy-terminal domain between the rat and the mouse NR2C subunits seems to reflect an error in the mouse sequence.

NR1–NR2C channels have slower deactivation kinetics than NR1–NR2A and NR1–NR2B channels and show an increased block by Ca^{2+} ions of their monovalent conductances (Stern et al., 1992). NR2C expression appears to be confined to the postnatal and adult brain, and is particularly prominent in cerebellar granule cells (Watanabe et al., 1992; Monyer et al., 1992, 1994). NR2C mRNA sizes have been published for mouse (4.5 and 12 kb) and rat (6 kb). Splice variants have not been reported. The human NR2C gene was mapped to 7q25 (Takano et al., 1993).

2.6.4.2.4 NMDAR2D (NR2D)

The NR2D subunit was isolated by a PCR-mediated homology screen (Ikeda et al., 1992; Ishii et al., 1993; Monyer et al., 1994). Expression is prominent in prenatal brain structures but is low in the adult brain where NR2D is mainly found in midline structures (Monyer et al., 1994). NR2D is 1356 amino acids in length (Figure 3o), and an NR2D mRNA size of 7 kb has been reported for rat (Ishii et al., 1993). The 30 residue mismatch in the carboxy-terminal region between the rat and mouse NR2D subunit is likely to reflect an error in the sequence of the cloned mouse cDNA. NR1–NR2D channels show the slowest deactivation kinetics of all NR1–NR2 channels; in Mg^{2+} and Ca^{2+} block these channels resemble the NR1–NR2C channel (Monyer et al., 1994). NR1–NR2D channels have been difficult to obtain by *in vitro* expression of the native mouse and rat NR2D cDNAs, most likely due to the high G+C content of the mRNAs, particularly around the start of translation. Improved expression levels of NR2D have been obtained by substituting part of the native nucleotide sequence

```
1    MGRLGYWTLLVLPALLVWRDPAQNAAAEKGPPALNIAVLLGHSHDVTERELRNLWGPEQA r-NMDAR2A  (Ishii)
1    MGRLGYWTLLVLPALLVWRDPAQNAAAEKGPPALNIAVLLGHSHDVTERELRNLWGPEQA r-NR2A  (Monyer)
1    MGRLGYWTLLVLPALLVWHGPAQNAAAEKGTPALNIAVLLGHSHDVTERELRNLWGPEQA m-epsilon1  (Meguro)

61   TGLPLDVNVVALLMNRTDPKSLITHVCDLMSGARIHGLVFGDDTDQEAVAQMLDFISSQT r-NMDAR2A  (Ishii)
61   TGLPLDVNVVALLMNRTDPKSLITHVCDLMSGARIHGLVFGDDTDQEAVAQMLDFISSQT r-NR2A  (Monyer)
61   TGLPLDVNVVALLMNRTDPKSLITHVCDLMSGARIHGLVFGDDTDQEAVAQMLDFISSQT m-epsilon1  (Meguro)

121  FIPILGIHGGASMIMADKDPTSTFFQFGASIQQQATVMLKIMQDYDWHVFSLVTTIFPGY r-NMDAR2A  (Ishii)
121  FIPILGIHGGASMIMADKDPTSTFFQFGASIQQQATVMLKIMQDYDWHVFSLVTTIFPGY r-NR2A  (Monyer)
121  FIPILGIHGGASMIMADKDPTSTFFQFGASIQQQATVMLKIMQDYDWHVFSLVTTIFPGY m-epsilon1  (Meguro)

181  RDFISFIKTTVDNSFVGWDMQNVITLDTSFEDAKTQVQLKKIHSSVILLYCSKDEAVLIL r-NMDAR2A  (Ishii)
181  RDFISFIKTTVDNSFVGWDMQNVITLDTSFEDAKTQVQLKKIHSSVILLYCSKDEAVLIL r-NR2A  (Monyer)
181  RDFISFIKTTVDNSFVGWDMQNVITLDTSFEDAKTQVQLKKIHSSVILLYCSKDEAVLIL m-epsilon1  (Meguro)

241  SEARSFGLTGYDFFWIVPSLVSGNTELIPKEFPSGLISVSYDDWDYSLEARVRDGLGILT r-NMDAR2A  (Ishii)
241  SEARSLGLTGYDFFWIVPSLVSGNTELIPKEFPSGLISVSYDDWDYSLEARVRDGLGILT r-NR2A  (Monyer)
241  SEARSLGLTGYDFFWIVPSLVSGNTELIPKEFPSGLISVSYDDWDYSLEARVRDGLGILT m-epsilon1  (Meguro)

301  TAASSMLEKFSYIPEAKASCYGQAEKPETPLHTLHQFMVNVTWDGKDLSFTEEGYQVHPR r-NMDAR2A  (Ishii)
301  TAASSMLEKFSYIPEAKASCYGQAEKPETPLHTLHQFMVNVTWDGKDLSFTEEGYQVHPR r-NR2A  (Monyer)
301  TAASSMLEKFSYIPEAKASCYGQTEKPETPLHTLHQFMVNVTWDGKDLSFTEEGYQVHPR m-epsilon1  (Meguro)

361  LVVIVLNKDREWEKVGKWENQTLSLRHAVWPRYKSFSDCEPDDNHLSIVTLEEAPFVIVE r-NMDAR2A  (Ishii)
361  LVVIVLNKDREWEKVGKWENQTLSLRHAVWPRYKSFSDCEPDDNHLSIVTLEEAPFVIVE r-NR2A  (Monyer)
361  LVVIVLNKDREWEKVGKWENQTLRLRHAVWPRYKSFSDCEPDDNHLSIVTLEEAPFVIVE m-epsilon1  (Meguro)

421  DIDPLTETCVRNTVPCRKFVKINNSTNEGMNVKKCCKGFCIDILKKLSRTVKFTYDLYLV r-NMDAR2A  (Ishii)
421  DIDPLTETCVRNTVPCRKFVKINNSTNEGMNVKKCCKGFCIDILKKLSRTVKFTYDLYLV r-NR2A  (Monyer)
421  DIDPLTETCVRNTVPCRKFVKINNSTNEGMNVKKCCKGFCIDILKKLSRTVKFTYDLYLV m-epsilon1  (Meguro)

481  TNGKHGKKVNNVWNGMIGEVVYQRAVMAVGSLTINEERSEVVDFSVPFVETGISVMVSRS r-NMDAR2A  (Ishii)
481  TNGKHGKKVNNVWNGMIGEVVYQRAVMAVGSLTINEERSEVVDFSVPFVETGISVMVSRS r-NR2A  (Monyer)
481  TNGKHGKKVNNVWNGMIGEVVYQRAVMAVGSLTINEERSEVVDFSVPFVETGISVMVSRS m-epsilon1  (Meguro)
                             ----------1----------
541  NGTVSPSAFLEPFSASVWVMMFVMLLIVSAIAVFVFEYFSPVGYNRNLAKGKAPHGPSFT r-NMDAR2A  (Ishii)
541  NGTVSPSAFLEPFSASVWVMMFVMLLIVSAIAVFVFEYFSPVGYNRNLAKGKAPHGPSFT r-NR2A  (Monyer)
541  NGTVSPSAFLEPFSASVWVMMFVMLLIVSAIAVFVFEYFSPVGYNRNLAKGKAPHGPSFT m-epsilon1  (Meguro)
     ---------2-----------           ---------3----------
601  IGKAIWLLWGLVFNNSVPVQNPKGTTSKIMVSVWAFFAVIFLASYTANLAAFMIQEEFVD r-NMDAR2A  (Ishii)
601  IGKAIWLLWGLVFNNSVPVQNPKGTTSKIMVSVWAFFAVIFLASYTANLAAFMIQEEFVD r-NR2A  (Monyer)
601  IGKAIWLLWGLVFNNSVPVQNPKGTTSKIMVSVWAFFAVIFLASYTANLAAFMIQEEFVD m-epsilon1  (Meguro)

661  QVTGLSDKKFQRPHDYSPPFRFGTVPNGSTERNIRNNYPYMHQYMTRFNQRGVEDALVSL r-NMDAR2A  (Ishii)
661  QVTGLSDKKFQRPHDYSPPFRFGTVPNGSTERNIRNNYPYMHQYMTRFNQRGVEDALVSL r-NR2A  (Monyer)
661  QVTGLSDKKFQRPHDYSPPFRFGTVPNGSTERNIRNNYPYMHQYMTKFNQRGVEDALVSL m-epsilon1  (Meguro)

721  KTGKLDAFIYDAAVLNYKAGRDEGCKLVTIGSGYIFATTGYGIALQKGSPWKRQIDLALL r-NMDAR2A  (Ishii)
721  KTGKLDAFIYDAAVLNYKAGRDEGCKLVTIGSGYIFATTGYGIALQKGSPWKRQIDLALL r-NR2A  (Monyer)
721  KTGKLDAFIYDAAVLNYKAGRDEGCKLVTIGSGYIFATTGYGIALQKGSPWKRQIDLALL m-epsilon1  (Meguro)
                                        ---------4----------
781  QFVGDGEMEELETLWLTGICHNEKNEVMSSQLDIDNMAGVFYMLAAAMALSLITFIWEHL r-NMDAR2A  (Ishii)
781  QFVGDGEMEELETLWLTGICHNEKNEVMSSQLDIDNMAGVFYMLAAAMALSLITFIWEHL r-NR2A  (Monyer)
781  QFVGDGEMEELETLWLTGICHNEKNEVMSSQLDIDNMAGVFYMLAAAMALSLITFIWEHL m-epsilon1  (Meguro)

841  FYWKLRFCFTGVCSDRPGLLFSISRGIYSCIHGVHIEEKKKSPDFNLTGSQSNMLKLLRS r-NMDAR2A  (Ishii)
841  FYWKLRFCFTGVCSDRPGLLFSISRGIYSCIHGVHIEEKKKSPDFNLTGSQSNMLKLLRS r-NR2A  (Monyer)
841  FYWKLRFCFTGVCSDRPGLLFSISRGIYSCIHGVHIEEKKKSPDFNLTGSQSNMLKLLRS m-epsilon1  (Meguro)

901  AKNISNMSNMNSSRMDSPKRATDFIQRGSLIVDMVSDKGNLIYSDNRSFQGKDSIFGDNM r-NMDAR2A  (Ishii)
901  AKNISNMSNMNSSRMDSPKRATDFIQRGSLIVDMVSDKGNLIYSDNRSFQGKDSIFGDNM r-NR2A  (Monyer)
901  AKNISNMSNMNSSRMDSPKRAADFIQRGSLIVDMVSDKGNLIYSDNRSFQGKDSIFGENM m-epsilon1  (Meguro)
```

FIGURE 3m(1).

with a synthetic sequence having a reduced G+C content (Monyer et al., 1994).

An NR2D cDNA encoding a different carboxy-terminus was reported (Ishii et al., 1993). The *in vivo* expression of this variant has yet to be demonstrated. The chromosomal localization for NR2D is not known.

```
961  NELQTFVANRHKDNLSNYVFQGQHPLTLNESNPNTVEVAVSTESKGNSRPRQLWKKSMES r-NMDAR2A   (Ishii)
961  -ELQTFVANRHKDNLSNYVFQGQHPLTLNESNPNTVEVAVSTESKGNSRPRQLWKKSMES r-NR2A  (Monyer)
961  NELQTFVANRHKDSLSNYVFQGQHPLTLNESNPNTVEVAVSTESKGNSRPRQLWKKSMES m-epsilon1   (Meguro)

1021 LRQDSLNQNPVSQRDEKTAENRTHSLKSPRYLPEEVAHSDISETSSRATCHREPDNNKNH r-NMDAR2A   (Ishii)
1020 LRQDSLNQNPVSQRDEKTAENRTHSLKSPRYLPEEVAHSDISETSSRATCHREPDNNKNH r-NR2A  (Monyer)
1021 LRQDSLNQNPVSQRDEKTAENRTHSLKSPRYLPEEVAHSDISETSSRATCHREPDNNKNH m-epsilon 1  (Meguro)

1081 KTKDNFKRSMASKYPKDCSDVDRTYMKTKASSPRDKIYTIDGEKEPSFHLDPPQFVENIT r-NMDAR2A   (Ishii)
1080 KTKDNFKRSMASKYPKDCSDVDRTYMKTKASSPRDKIYTIDGEKEPSFHLDPPQFVENIT r-NR2A  (Monyer)
1081 KTKDNFKRSMASKYPKDCSEVERTYVKTKASSPRDKIYTIDGEKEPSFHLDPPQFIENIV m-epsilon1   (Meguro)

1141 LPENVGFPDTYQDHNENFRKGDSTLPMNRNPLHNEDGLPNNDQYKLYAKHFTLKDKGSPH r-NMDAR2A   (Ishii)
1140 LPENVGFPDTYQDHNENFRKGDSTLPMNRNPLHNEDGLPNNDQYKLYAKHFTLKDKGSPH r-NR2A  (Monyer)
1141 LPENVDFPDTYQDHNENFRKGDSTLPMNRNPLHNEDGLPNNDQYKLYAKHFTLKDKGSPH m-epsilon1   (Meguro)

1201 SEGSDRYRQNSTHCRSCLSNLPTYSGHFTMRSPFKCDACLRMGNLYDIDEDQMLQETGNP r-NMDAR2A   (Ishii)
1200 SEGSDRYRQNSTHCRSCLSNLPTYSGHFTMRSPFKCDACLRMGNLYDIDEDQMLQETGNP r-NR2A  (Monyer)
1201 SEGSDRYRQNSTHCRSCLSNLPTYSGHFTMRSPFKCDACLRMGNLYDIDEDQMLQETGNP m-epsilon1   (Meguro)

1261 ATREEVYQQDWSQNNALQFQKNKLRINRQHSYDNILDKPREIDLSRPSRSISLKDRERLL r-NMDAR2A   (Ishii)
1260 ATREEVYQQDWSQNNALQFQKNKLRINRQHSYDNILDKPREIDLSRPSRSISLKDRERLL r-NR2A  (Monyer)
1261 ATREEAYQQDWSQNNALQFQKNKLKINRQHSYDNILDKPREIDLSRPSRSISLKDRERLL m-epsilon1   (Meguro)

1321 EGNLYGSLFSVPSSKLLGNKSSLFPQGLEDSKRSKSLLPDHASDNPFLHTYGDDQRLVIG r-NMDAR2A   (Ishii)
1320 EGNLYGSLFSVPSSKLLGNKSSLFPQGLEDSKRSKSLLPDHASDNPFLHTYGDDQRLVIG r-NR2A  (Monyer)
1321 EGNLYGSLFSVPSSKLLGNKSSLFPQGLEDSKRSKSLLPDHTSDNPFLHTYGDDQRLVIG m-epsilon1   (Meguro)

1381 RCPSDPYKHSLPSQAVNDSYLRSSLRSTASYCSRDSRGHSDVYISEHVMPYAANKNTMYS r-NMDAR2A   (Ishii)
1380 RCPSDPYKHSLPSQAVNDSYLRSSLRSTASYCSRDSRGHSDVYISEHVMPYAANKNTMYS r-NR2A  (Monyer)
1381 RCPSDPYKHSLPSQAVNDSYLRSSLRSTASYCSRDSRGHSDVYISEHVMPYAANKNNMYS m-epsilon1   (Meguro)

1441 TPRVLNSCSNRRVYKKMPSIESDV                                     r-NMDAR2A   (Ishii)
1440 TPRVLNSCSNRRVYKKMPSIESDV                                     r-NR2A  (Monyer)
1441 TPRVLNSCSNRRVYKKMPSIESDV                                     m-epsilon1   (Meguro)
```

FIGURE 3m(2).

2.6.5 Kainate-Binding Proteins (KBPs)

Simultaneous to the description of the first iGluR subunit GluR1 (Hollmann et al., 1989) the sequence of the kainate binding proteins of chick (Gregor et al., 1989) and frog (Wada et al., 1989) was reported. Both proteins (Figure 3q) are sequence-related to the subunits of the AMPA/kainate receptor, and share approximately 35% amino acid residues with subunits of the high-affinity kainate receptor. In contrast to these and to the other glutamate receptors the KBPs contain a short amino-terminus and are only 464 and 470 amino acids in length, respectively (Figures 1 and 3p). Although the KBPs are structurally related to iGluRs it remains to be elucidated if these subunits form functional ion channels or if other subunits are needed for channel activity. Alternatively, KBPs may have entirely different functions. No mammalian homologs have been found.

2.6.5.1 Chick KBP

For the cloning of the chick KBP, an abundant oligomeric protein with high [^{3}H]kainate-binding activity was isolated from chick cerebellum (Gregor et al., 1988). The purified protein displayed a pharmacological profile similar to that of a kainate receptor and represented a single polypeptide of M_r 49 kDa. The localization of this protein is exclusively on Bergmann glial membranes of cerebellum, in close proximity to established glutamatergic synapses.

```
1    MKPSAECCSPKFWLVLAVLAVSGSKARSQKSAPSIGIAVILVGTSDEVAIKDAHEKDDFH m-epsilon2 (Kutsuwada)
1    MKPSAECCSPKFWLVLAVLAVSGSKARSQKSPPSIGIAVILVGTSDEVAIKDAHEKDDFH r-NR2B (Monyer)

61   HLSVVPRVELVAMNETDPKSIITRICDLMSDRKIQGVVLADDTDQEAIAQILDFISAQTL m-epsilon2 (Kutsuwada)
61   HLSVVPRVELVAMNETDPKSIITRICDLMSDRKIQGVVFADDTDQEAIAQILDFISAQTL r-NR2B (Monyer)

121  TPILGIHGGSSMIMADKDESSMFFQFGPSIEQQASVMLNIMEEYDWYIFSIVTTYFPGYQ m-epsilon2 (Kutsuwada)
121  TPILGIHGGSSMIMADKDESSMFFQFGPSIEQQASVMLNIMEEYDWYIFSIVTTYFPGYQ r-NR2B (Monyer)

181  DFVNKIRSTIENSFVGWELEEVLLLDMSLDDGDSKIQNQLKKLQSPIILLYCTKEEATYI m-epsilon2 (Kutsuwada)
181  DFVNKIRSTIENSFVGWELEEVLLLDMSLDDGDSKIQNQLKKLQSPIILLYCTKEEATYI r-NR2B (Monyer)

241  FEVANSVGLTGYGYTWIVPSLVAGDTDTVPSEFPTGLISVSYDEWDYGLPARVRDGIAII m-epsilon2 (Kutsuwada)
241  FEVANSVGLTGYGYTWIVPSLVAGDTDTVPSEFPTGLISVSYDEWDYGLPARVRDGIAII r-NR2B (Monyer)

301  TTAASDMLSEHSFIPEPKSSCYNTHEKRIYQSNMLNRYLINVTFEGRNLSFSEDGYQMHP m-epsilon2 (Kutsuwada)
301  TTAASDMLSEHSFIPEPKSSCYNTHEKRIYQSNMLNRYLINVTFEGRNLSFSEDGYQMHP r-NR2B (Monyer)

361  KLVIILLNKERKWERVGKWKDKSLQMKYYVWPRMCPETEEQEDDHLSIVTLEEAPFVIVE m-epsilon2 (Kutsuwada)
361  KLVIILLNKERKWERVGKWKDKSLQMKYYVWPRMCPETEEQEDDHLSIVTLEEAPFVIVE r-NR2B (Monyer)

421  SVDPLSGTCMRNTVPCQKRIISENKTDEEPGYIKKCCKGFCIDILKKISKSVKFTYDLYL m-epsilon2 (Kutsuwada)
421  SVDPLSGTCMRNTVPCQKRIISENKTDEEPGYIKKCCKGFCIDILKKISKSVKFTYDLYL r-NR2B (Monyer)

481  VTNGKHGKKINGTWNGMIGEVVMKRAYMAVGSLTINEERSEVVDFSVPFIETGISVMVSR m-epsilon2 (Kutsuwada)
481  VTNGKHGKKINGTWNGMIGEVVMKRAYMAVGSLTINEERSEVVDFSVPFIETGISVMVSR r-NR2B (Monyer)

                                         1
541  SNGTVSPSAFLEPFSADVWVMMFVMLLIVSAVAVFVFEYFSPVGYNRCLADGREPGGPSF m-epsilon2 (Kutsuwada)
541  SNGTVSPSAFLEPFSADVWVMMFVMLLIVSAVAVFVFEYFSPVGYNRCLADGREPGGPSF r-NR2B (Monyer)

            2                             3
601  TIGKAIWLLWGLVFNNSVPVQNPKGTTSKIMVSVWAFFAVIFLASYTANLAAFMIQEEYV m-epsilon2 (Kutsuwada)
601  TIGKAIWLLWGLVFNNSVPVQNPKGTTSKIMVSVWAFFAVIFLASYTANLAAFMIQEEYV r-NR2B (Monyer)

661  DQVSGLSDKKFQRPNDFSPPFRFGTVPNGSTERNIRNNYAEMHAYMGKFNQRGVDDALLS m-epsilon2 (Kutsuwada)
661  DQVSGLSDKKFQRPNDFSPPFRFGTVPNGSTERNIRNNYAEMHAYMGKFNQRGVDDALLS r-NR2B (Monyer)

721  LKTGKLDAFIYDAAVLNYMAGRDEGCKLVTIGSGKVFASTGYGIAIQKDSGWKRQVDLAI m-epsilon2 (Kutsuwada)
721  LKTGKLDAFIYDAAVLNYMAGRDEGCKLVTIGSGKVFASTGYGIAIQKDSGWKRQVDLAI r-NR2B (Monyer)

                                                 4
781  LQLFGDGEMEELEALWLTGICHNEKNEVMSSQLDIDNMAGVFYMLGAAMALSLITFICEH m-epsilon2 (Kutsuwada)
781  LQLFGDGEMEELEALWLTGICHNEKNEVMSSQLDIDNMAGVFYMLGAAMALSLITFICEH r-NR2B (Monyer)

841  LFYWQFRHCFMGVCSGKPGMVFSISRGIYSCIHGVAIEERQSVMNSPTATMNNTHSNILR m-epsilon2 (Kutsuwada)
841  LFYWQFRHCFMGVCSGKPGMVFSISRGIYSCIHGVAIEERQSVMNSPTATMNNTHSNILR r-NR2B (Monyer)

901  LLRTAKNMANLSGVNGSPQSALDFIRRESSVYDISEHRRSFTHSDCKSYNNPPCEENLFS m-epsilon2 (Kutsuwada)
901  LLRTAKNMANLSGVNGSPQSALDFIRRESSVYDISEHRRSFTHSDCKSYNNPPCEENLFS r-NR2B (Monyer)

961  DYISEVERTFGNLQLKDSNVYQDHYHHHHRPHSIGSTSSIDGLYDCDNPPFTTQPRSISK m-epsilon2 (Kutsuwada)
961  DYISEVERTFGNLQLKDSNVYQDHYHHHHRPHSIGSTSSIDGLYDCDNPPFTTQPRSISK r-NR2B (Monyer)

1021 KPLDIGLPSSKHSQLSDLYGKFSFKSDRYSGHDDLIRSDVSDISTHTVTYGNIEGNAAKR m-epsilon2 (Kutsuwada)
1021 KPLDIGLPSSKHSQLSDLYGKFSFKSDRYSGHDDLIRSDVSDISTHTVTYGNIEGNAAKR r-NR2B (Monyer)

1081 RKQQYKDSLKKRPASAKSRREFDEIELAYRRRPPRSPDHKRYFRDKEGLRDFYLDQFRTK m-epsilon2 (Kutsuwada)
1081 RKQQYKDSLKKRPASAKSRREFDEIELAYRRRPPRSPDHKRYFRDKEGLRDFYLDQFRTK r-NR2B (Monyer)

1141 ENSPHWEHVDLTDIYKERSDDFKRDSVSGGGPCTNRSHLKHGTGDKHGVVGGVPAPWEKN m-epsilon2 (Kutsuwada)
1141 ENSPHWEHVDLTDIYKERSDDFKRDSVSGGGPCTNRSHLKHGTGEKHGVVGGVPAPWEKN r-NR2B (Monyer)

1201 LTNVDWEDRSGGNFCRSCPSKLHNYSSTVAGQNSGRQACIRCEACKKAGNLYDISEDNSL m-epsilon2 (Kutsuwada)
1201 LTNVDWEDRSGGNFCRSCPSKLHNYSSTVAGQNSGRQACIRCEACKKAGNLYDISEDNSL r-NR2B (Monyer)

1261 QELDQPAAPVAVSSNASTTKYPQSPTNSKAQKKNRNKLRRQHSYDTFVDLQKEEAALAPR m-epsilon2 (Kutsuwada)
1261 QELDQPAAPVAVTSNASSTKYPQSPTNSKAQKKNRNKLRRQHSYDTFVDLQKEEAALAPR r-NR2B (Monyer)

1321 SVSLKDKGRFMDGSPYAHMFEMPAGESSFANKSSVTTAGHHHNNPGSGYMLSKSLYPDRV m-epsilon2 (Kutsuwada)
1321 SVSLKDKGRFMDGSPYAHMFEMPAGESSFANKSSVPTAGHHHNNPGSGYMLSKSLYPDRV r-NR2B (Monyer)

1381 TQNPFIPLLGMISACFTAANPTSSGSPRWQGRRKQGRTSGPLSPISQWCRPFHGAVPGRF m-epsilon2 (Kutsuwada)
1381 TQNPFIPTFGDDQCLLHGSKSYFFRQPTVAGASKTRPDFRALVTNKPVVVTLHGAVPGRF r-NR2B (Monyer)

1441 QKDICIGNQSNPCVPNNKNPRAFNGSSNGHVYEKLSSIESDV                   m-epsilon2 (Kutsuwada)
1441 QKDICIGNQSNPCVPNNKNPRAFNGSSNGHVYEKLSSIESDV                   r-NR2B (Monyer)
```

FIGURE 3n.

Amino acid sequence information from fragments of this purified protein was used to isolate the cDNA for this protein from a cDNA library. The cDNA sequence revealed that the overall structure of this protein (Figure 3p) is well related to iGluR subunits, with an amino-terminal domain followed by four hydrophobic segments, each about 20 amino acids in length. As mentioned

```
1                MGGALGPALLLTSLLGAWARLGAGQGEQAVTVAVVFGSSGPLQTQAR r-NR2C (Monyer)
1   MPALVSLQDPPVDMGGALGPALLLTSLLGAWARLGAGQGEQAVTVAVVFGSSGPLQTQAR r-NMDAR2C (Ishii)
1                MGGALGPALLLTSLLGAWAGLGAGQGEQAVTVAVVFGSSGPLQAQAR m-epsilon3 (Kutsuwada)

48  TRLTSQNFLDLPLEIQPLTVGVNNTNPSSILTQICGLLGAARVHGIVFEDNVDTEAVAQL r-NR2C (Monyer)
61  TRLTSQNFLDLPLEIQPLTVGVNNTNPSSILTQICGLLGAARVHGIVFEDNVDTEAVAQL r-NMDAR2C (Ishii)
48  TRLTPQNFLDLPLEIQPLTIGVNNTNPSSILTQICGLLGAARVHGIVFEDNVDTEAVAQL m-epsilon3 (Kutsuwada)

108 LDFVSSQTHVPILSISGGSAVVLTPKEPGSAFLQLGVSLEQQLQVLFKVLEEYDWSAFAV r-NR2C (Monyer)
121 LDFVSSQTHVPILSISGGSAVVLTPKEPGSAFLQLGVSLEQQLQVLFKVLEEYDWSAFAV r-NMDAR2C (Ishii)
108 LDFVSSQTHVPILSISGGSAVVLTPKEPGSAFLQLGVSLEQQLQVLFKVLEEYDWSAFAV m-epsilon3 (Kutsuwada)

168 ITSLHPGHALFLEGVRAVADASYLSWRLLDVLTLELGPGGPRARTQRLLRQVDAPVLVAY r-NR2C (Monyer)
181 ITSLHPGHALFLEGVRAVADASYLSWRLLDVLTLELGPGGPRARTQRILRQVDAPVLVAY r-NMDAR2C (Ishii)
168 ITSLHPGHALFLEGVRAVADASYLSWRLLDVLTLELGPGGPRARTQRLLRQVDAPVLVAY m-epsilon3 (Kutsuwada)

228 CSREEAEVLFAEAAQAGLVGPGHVWLVPNLALGSTDAPPAAFPVGLISVVTESWRLSLRQ r-NR2C (Monyer)
241 CSREEAEVLFAEAAQAGLVGPGHVWLVPNLALGSTDAPPAAFPVGLISVVTESWRLSLRQ r-NMDAR2C (Ishii)
228 CSREEAEVLFAEAAQAGLVGPGHVWLVPNLALGSTDAPPAAFPVGLISVVTESWRLSLRQ m-epsilon3 (Kutsuwada)

288 KVRDGVAILALGAHSYRRQYGTLPAPAGDCRSHPGPVSPAREAFYRHLLNVTWEGRDFSF r-NR2C (Monyer)
301 KVRDGVAILALGAHSYRRQYGTLPAPAGDCRSHPGPVSPAREAFYRHLLNVTWEGRDFSF r-NMDAR2C (Ishii)
288 KVRDGVAILALGAHSYRRQYGTLPAPAGDCRSHPGPVSPAREAFYRHLLNVTWEGRDFSF m-epsilon3 (Kutsuwada)

348 SPGGYLVRPTMVVIALNRHRLWEMVGRWDHGVLYMKYPVWPRYSTSLQPVVDSRHLTVAT r-NR2C (Monyer)
361 SPGGYLVRPTMVVIALNRHRLWEMVGRWDHGVLYMKYPVWPRYSTSLQPVVDSRHLTVAT r-NMDAR2C (Ishii)
348 SPGGYLVQPTMVVIALNRHRLWEMVGRWDHGVLYMKYPVWPRYSTSLQPVVDSRHLTVAT m-epsilon3 (Kutsuwada)

408 LEERPFVIVESPDPGTGGCVPNTVPCRRQSNHTFSSGDLTPYTKLCCKGFCIDILKKLAK r-NR2C (Monyer)
421 LEERPFVIVESPDPGTGGCVPNTVPCRRQSNHTFSSGDLTPYTKLCCKGFCIDILKKLAK r-NMDAR2C (Ishii)
408 LEERPFVIVESPDPGTGGCVPNTVPCRRQSNHTFSSGDITPYTKLCCKGFCIDILKKLAK m-epsilon3 (Kutsuwada)

468 VVKFSYDLYLVTNGKHGKRVRGVWNGMIGEVYYKRADMAIGSLTINEERSEIIDFSVPFV r-NR2C (Monyer)
481 VVKFSYDLYLVTNGKHGKRVRGVWNGMIGEVYYKRADMAIGSLTINEERSEIIDFSVPFV r-NMDAR2C (Ishii)
468 VVKFSYDLYLVTNGKHGKRVRGVWNGMIGEVYYKRADMAIGSLTINEERSEIIDFSVPFV m-epsilon3 (Kutsuwada)
                                      ________1________
528 ETGISVMVSRSNGTVSPSAFLEPYSPAVWVMMFVMCLTVVAITVFMFEYFSPVSYNQNLT r-NR2C (Monyer)
541 ETGISVMVSRSNGTVSPSAFLEPYSPAVWVMMFVMCLTVVAITVFMFEYFSPVSYNQNLT r-NMDAR2C (Ishii)
528 ETGISVMVARSNGTVSPSAFLEPYSPAVWVMMFVMCLTVVAITVFMFEYFSPVSYNQNLT m-epsilon3 (Kutsuwada)
             ________2________          ________3________
588 KGKKPGGPSFTIGKSVWLLWALVFNNSVPIENPRGTTSKIMVLVWAFFAVIFLASYTANL r-NR2C (Monyer)
601 KGKKPGGPSFTIGKSVWLLWALVFNNSVPIENPRGTTSKIMVLVWAFFAVIFLASYTANL r-NMDAR2C (Ishii)
588 KGKKSGGPSFTIGKSVWLLWALVFNNSVPIENPRGTTSKIMVLVWAFFAVIFLASYTANL m-epsilon3 (Kutsuwada)

648 AAFMIQEQYIDTVSGLSDKKFQRPQDQYPPFRFGTVPNGSTERNIRSNYRDMHTHMVKFN r-NR2C (Monyer)
661 AAFMIQEQYIDTVSGLSDKKFQRPQDQYPPFRFGTVPNGSTERNIRSNYRDMHTHMVKFN r-NMDAR2C (Ishii)
648 AAFMIQEQYIDTVSGLSDKKFQRPQDQYPPFRFGTVPNGSTERNIRSNYRDMHTHMVKFN m-epsilon3 (Kutsuwada)

708 QRSVEDALTSLKMGKLDAFIYDAAVLNYMAGKDEGCKLVTIGSGKVFATTGYGIAMQKDS r-NR2C (Monyer)
721 QRSVEDALTSLKMGKLDAFIYDAAVLNYMAGKDEGCKLVTIGSGKVFATTGYGIAMQKDS r-NMDAR2C (Ishii)
708 QRSVEDALTSLKMGKLDAFIYDAAVLNYMAGKDEGCKLVTIGSGKVFATTGYGIAMQKDS m-epsilon3 (Kutsuwada)
                                                  _______4_______
768 HWKRAIDLALLQLLGDGETQKLETVWLSGICQNEKNEVMSSKLDIDNMAGVFYMLLVAMG r-NR2C (Monyer)
781 HWKRAIDLALLQLLGDGETQKLETVWLSGICQNEKNEVMSSKLDIDNMAGVFYMLLVAMG r-NMDAR2C (Ishii)
768 HWKRAIDLALLQELGDGETQKLETVWLSGICHNEKNEVMSSKLDIDNMAGVFYMLLVAMG m-epsilon3 (Kutsuwada)
    ______
828 LALLVFAWEHLVYWKLRHSVPNSSQLDFLLAFSRGIYSCFNGVQSLPSPARPPSPDLTAD r-NR2C (Monyer)
841 LALLVFAWEHLVYWKLRHSVPNSSQLDFLLAFSRGIYSCFNGVQSLPSPARPPSPDLTAD r-NMDAR2C (Ishii)
828 LALLVFAWEHLVYWKLRHSVPSSSQLDFLLAFSRGIYSCFNGVQSLPSPARPPSPDLTAG m-epsilon3 (Kutsuwada)
```

FIGURE 3o(1).

above the major difference is the short amino-terminal domain that extends for only 147–148 amino acids proximal of M1.

When expressed in transfected cells a K_D value similar to that of the native cerebellar kainate binding protein of 560 nM for [^{3}H]kainate was obtained. In the same cell culture system, and in *Xenopus* oocytes, no channel activity was observed for the chick KBP (Gregor et al., 1989, 1992).

The biological function of the chick KBP remains to be elucidated. It might be involved in coordinating granule cell migration in the developing

```
888  SAQANVLKMLQAARDMVNTADVSSSLDRATRTIENWGNNRRVPAPTASGPRSSTPGPPGQ r-NR2C    (Monyer)
901  SAQANVLKMLQAARDMVNTADVSSSLDRATRTIENWGNNRRVPAPTASGPRSSTPGPPGQ r-NMDAR2C   (Ishii)
888  SAQANVLKMLQAARDMVSTADVSGSLDRATRTIENWGNNRRAPAPTTSGPRSCTPGPPGQ m-epsilon3   (Kutsuwada)

948  PSPSGWGPPGGGRTPLARRAPQPPARPA-TCGPPLPDVSRPSCRHASDARWPVRVGHQGP r-NR2C    (Monyer)
961  PSPSGWGPPGGGRTPLARRAPQPPARPA-TCGPPLPDVSRPSCRHASDARWPVRVGHQGP r-NMDAR2C   (Ishii)
948  PSPSGWRPPGGGRTPLARRAPQPPARPGPAQGRLSPTCPEHPAGTLGMRGGQCESGIRDR m-epsilon3   (Kutsuwada)

1007 HVSASERRALPERSLLPAHCHYSSFPRAERSGRPYLPLFPEPPEPDDLPLLGPEQLARRE r-NR2C    (Monyer)
1020 HVSASERRALPERSLLPAHCHYSSFPRAERSGRPYLPLFPEPPEPDDLPLLGPEQLARRE r-NMDAR2C   (Ishii)
1008 TSRPPERRALPERSLLHAHCHYSSFPRAERSGRPFLPLFPEPPEPDDLPLLGPEQLARRE m-epsilon3   (Kutsuwada)

1067 AMLRAAWARGPRPRHASLPSSVAEAFTRSNPLPARCTGHACACPCPQSRPSCRHLAQAQS r-NR2C    (Monyer)
1080 AMLRAAWARGPRPRHASLPSSVAEAFTRSNPLPARCTGHACACPCPQSRPSCRHLAQAQS r-NMDAR2C   (Ishii)
1068 ALLRAAWARGPRPRHASLPSSVAEAFTRSNPLPARCTGHACACPCPQSRPSCRHVAQTQS m-epsilon 3  (Kutsuwada)

1127 LRLPSYPEACVEGVPAGVA-TWQPRQHVCLHAHTRLPFCWGTVCRHPPPCTSHSPWLIGT r-NR2C    (Monyer)
1140 LRLPSYPEACVEGVPAGVA-TWQPRQHVCLHAHTRLPFCWGTVCRHPPPCTSHSPWLIGT r-NMDAR2C   (Ishii)
1128 LRLPSYREACVEGVPAGVAATWQPRQHVCLHTHTHLPFCWGTVCRHPPPCSSHSPWLIGT m-epsilon3   (Kutsuwada)

1186 WEPPAHRVRTLGLGTGYRDSGVLEEVSREACGTQGFPRSCTWRRVSSLESEV         r-NR2C    (Monyer)
1199 WEPPAHRVRTLGLGTGYRDSGVLEEVSREACGTQGFPRSCTWRRVSSLESEV         r-NMDAR2C   (Ishii)
1188 WEPPSHRGRTLGLGTGYRDSGVLEEVSREACGTQGFPRSCTWRRISSLESEV         m-epsilon3   (Kutsuwada)
```

FIGURE 3o(2).

cerebellum, given that it is expressed a few days after Bergmann glial cells develop and around the time when granule cells begin migration (Ortega et al., 1991).

The gene for the chick KBP covers at least 13,000 bp and contains 11 exons. The major transcriptional initiation site was identified 117 bases upstream from the translational start site, and the putative promoter region displays high GC content and contains TATA, CAATA and AP1 consensus sequences (Eshhar et al., 1992; Gregor et al., 1992).

2.6.5.2 Frog KBP

The frog KBP was molecularly characterized at the same time as the chick KBP (Wada et al., 1989). Wada and colleagues purified a 48-kDa kainate-binding protein from frog *(Rana pipiens berlandieri)* brain by the high-affinity kainate analog domoic acid. These researchers used the amino acid sequence information of the protein to isolate a cDNA that expresses a high-affinity kainate-binding protein in COS-7 cells. As for the chick KBP no electrophysiological responses were observed after kainate superfusion of *Xenopus* oocytes expressing the cloned frog KBP. The pharmacological properties of the transiently expressed frog KBP are similar to those of the biochemically purified KBP. The [^{3}H]kainate K_D value (5.5 n*M)* for the purified and the cloned frog KBP is 100 times higher than for the chick KBP, and the rank order of potencies based on K_D values is DOM > KA > GLU > CNQX.

The mRNA encoding frog KBP is seen by *in situ* hybridization in the optic tectum, telencephalon, and in cells lining the lateral ventricles (Wada et al., 1989). Electron microscopic immunocytochemical studies show that frog KBP has synaptic and extrasynaptic aspects of localization in the optic tectum, with most labeling being extrasynaptic (Wenthold et al., 1990).

```
1    MRGAGGPRGPRGPAKMLLLLALACASPFPEEVPGPGAAGGGTGGARPLNVALVFSGPAYA m-epsilon4   (Ikeda)
1    MRGAGGPRGPRGPAKMLLLLALACASPFPEEVPGPGAVGGGTGGARPLNVALVFSGPAYA r-NMDAR2D-1  (Ishii)
1    MRGAGGPRGPRGPAKMLLLLALACASPFPEEVPGPGAVGGGTGGARPLNVALVFSGPAYA r-NMDAR2D-2  (Ishii)
1    MRGAGGPRGPRGPAKMLLLLALACASPFPEEVPGPGAVGGGTGGARPLNVALVFSGPAYA r-NR2D  (Monyer)

61   AEAARLGPAVAAAVRSPGLDVRPVALVLNGSDPRSLVLQLCDLLSGLRVHGVVFEDDSRA m-epsilon4   (Ikeda)
61   AEAARLGPAVAAAVRSPGLDVRPVALVLNGSDPRSLVLQLCDLLSGLRVHGVVFEDDSRA r-NMDAR2D-1  (Ishii)
61   AEAARLGPAVAAAVRSPGLDVRPVALVLNGSDPRSLVLQLCDLLSGLRVHGVVFEDDSRA r-NMDAR2D-2  (Ishii)
61   AEAARLGPAVAAAVRSPGLDVRPVALVLNGSDPPSLVLQLCDLLSGLRVHGVVFEDDSRA r-NR2D  (Monyer)

121  PAVAPILDFLSAQTSLPIVAVHGGAALVLTPKEKGSTFLQLGSSTEQQLQVIFEVLEEYD m-epsilon4   (Ikeda)
121  PAVAPILDFLSAQTSLPIVAVHGGAALVLTPKEKGSTFLQLGSSTEQQLQVIFEVLEEYD r-NMDAR2D-1  (Ishii)
121  PAVAPILDFLSAQTSLPIVAVHGGAALVLTPKEKGSTFLQLGSSTEQQLQVIFEVLEEYD r-NMDAR2D-2  (Ishii)
121  PAVAPILDFLSAQTSLPIVAVHGGAALVLTPKEKGSTFLQLGSSTEQQLQVIFEVLEEYD r-NR2D  (Monyer)

181  WTSFVAVTTRAPGHRAFLSYIEVLTDGSLVGWEHRGALTLDPGAGEAVLGAQLRSVSAQI m-epsilon4   (Ikeda)
181  WTSFVAVTTRAPGHRAFLSYIEVLTDGSLVGWEHRGALTLDPGAGEAVLGAQLRSVSAQI r-NMDAR2D-1  (Ishii)
181  WTSFVAVTTRAPGHRAFLSYIEVLTDGSLVGWEHRGALTLDPGAGEAVLGAQLRSVSAQI r-NMDAR2D-2  (Ishii)
181  WTSFVAVTTRAPGHRAFLSYIEVLTDGSLVGWEHRGALTLDPGAGEAVLGAQLRSVSAQI r-NR2D  (Monyer)

241  RLLFCAREEAEPVFRAAEEAGLTGPGYVWFMVGPQLAGGGGSGVPGEPLLLPGGAPLPAG m-epsilon4   (Ikeda)
241  RLLFCAREEAEPVFRAAEEAGLTGPGYVWFMVGPQLAGGGGSGVPGEPLLLPGGSPLPAG r-NMDAR2D-1  (Ishii)
241  RLLFCAREEAEPVFRAAEEAGLTGPGYVWFMVGPQLAGGGGSGVPGEPLLLPGGSPLPAG r-NMDAR2D-2  (Ishii)
241  RLLFCAREEAEPVFRAAEEAGLTGPGYVWFMVGPQLAGGGGSGVPGEPLLLPGGSPLPAG r-NR2D  (Monyer)

301  LFAVRSAGWRDDLARRVAAGVAVVARGAQALLRDYGFLPELGHDCRAQNRTHRGESLHRY m-epsilon4   (Ikeda)
301  LFAVRSAGWRDDLARRVAAGVAVVARGAQALLRDYGFLPELGHDCRTQNRTHRGESLHRY r-NMDAR2D-1  (Ishii)
301  LFAVRSAGWRDDLARRVAAGVAVVARGAQALLRDYGFLPELGHDCRTQNRTHRGESLHRY r-NMDAR2D-2  (Ishii)
301  LFAVASAGWRDDLARRVAAGVAVVARGAQALLRDYGFLPELGHDCRTQNRTHRGESLHRY r-NR2D  (Monyer)

361  FMNITWDNRDYSFNEDGFLVNPSLVVISLTRDRTWEVVGSWEQQTLRLKYPLWSRYGRFL m-epsilon4   (Ikeda)
361  FMNITWDNRDYSFNEDGFLVNPSLVVISLTRDRTWEVVGSWEQQTLRLKYPLWSRYGRFL r-NMDAR2D-1  (Ishii)
361  FMNITWDNRDYSFNEDGFLVNPSLVVISLTRDRTWEVVGSWEQQTLRLKYPLWSRYGRFL r-NMDAR2D-2  (Ishii)
361  FMNITWDNRDYSFNEDGFLVNPSLVVISLTRDRTWEVVGSWEQQTLRLKYPLWSRYGRFL r-NR2D  (Monyer)

421  QPVDDTQHLTVATLEERPFVIVEPADPISGTCIRDSVPCRSQLNRTHSPPPDAPRPEKRC m-epsilon4   (Ikeda)
421  QPVDDTQHLTVATLEERPFVIVEPADPISGTCIRDSVPCRSQLNRTHSPPPDAPRPEKRC r-NMDAR2D-1  (Ishii)
421  QPVDDTQHLTVATLEERPFVIVEPADPISGTCIRDSVPCRSQLNRTHSPPPDAPRPEKRC r-NMDAR2D-2  (Ishii)
421  QPVDDTQHLTVATLEERPFVIVEPADPISGTCIRDSVPCRSQLNRTHSPPPDAPRPEKRC r-NR2D  (Monyer)

481  CKGFCIDILKRLAHTIGFSYDLYLVTNGKHGKKIDGVWNGMIGEVFYQRADMAIGSLTIN m-epsilon4   (Ikeda)
481  CKGFCIDILKRLAHTIGFSYDLYLVTNGKHGKKIDGVWNGMIGEVFYQRADMAIGSLTIN r-NMDAR2D-1  (Ishii)
481  CKGFCIDILKRLAHTIGFSYDLYLVTNGKHGKKIDGVWNGMIGEVFYQRADMAIGSLTIN r-NMDAR2D-2  (Ishii)
481  CKGFCIDILKRLAHTIGFSYDLYLVTNGKHGKKIDGVWNGMIGEVFYQRADMAIGSLTIN r-NR2D  (Monyer)
                                                   ________1______
541  EERSEIVDFSVPFVETGISVMVARSNGTVSPSAFLEPYSPAVWVMMFVMCLTVVAVTVFI m-epsilon4   (Ikeda)
541  EERSEIVDFSVPFVETGISVMVARSNGTVSPSAFLEPYSPAVWVMMFVMCLTVVAVTVFI r-NMDAR2D-1  (Ishii)
541  EERSEIVDFSVPFVETGISVMVARSNGTVSPSAFLEPYSPAVWVMMFVMCLTVVAVTVFI r-NMDAR2D-2  (Ishii)
541  EERSEIVDFSVPFVETGISVMVARSNGTVSPSAFLEPYSPAVWVMMFVMCLTVVAVTVFI r-NR2D  (Monyer)
                               __________2__________       _____
601  FEYLSPVGYNRSLATGKRPGGSTFTIGKSIWLLWALVFNNSVPVENPRGTTSKIMVLVWA m-epsilon4   (Ikeda)
601  FEYLSPVGYNRSLATGKRPGGSTFTIGKSIWLLWALVFNNSVPVENPRGTTSKIMVLVWA r-NMDAR2D-1  (Ishii)
601  FEYLSPVGYNRSLATGKRPGGSTFTIGKSIWLLWALVFNNSVPVENPRGTTSKIMVLVWA r-NMDAR2D-2  (Ishii)
601  FEYLSPVGYNRSLATGKRPGGSTFTIGKSIWLLWALVFNNSVPVENPRGTTSKIMVLVWA r-NR2D  (Monyer)
     __3________
661  FFAVIFLASYTANLAAFMIQEEYVDTVSGLSDRKFQRPQEQYPPLKFGTVPNGSTEKNIR m-epsilon4   (Ikeda)
661  FFAVIFLASYTANLAAFMIQEEYVDTVSGLSDRKFQRPQEQYPPLKFGTVPNGSTEKNIR r-NMDAR2D-1  (Ishii)
661  FFAVIFLASYTANLAAFMIQEEYVDTVSGLSDRKFQRPQEQYPPLKFGTVPNGSTEKNIR r-NMDAR2D-2  (Ishii)
661  FFAVIFLASYTANLAAFMIQEEYVDTVSGLSDRKFQRPQEQYPPLKFGTVPNGSTEKNIR r-NR2D  (Monyer)
```

FIGURE 3p(1).

2.6.6 Invertebrate Glutamate Receptor Subunits

In insects and crustacea glutamate acts as a neurotransmitter in the central nervous system and at the neuromuscular junction. There the receptors are expected to be equivalent to the nicotinic acetylcholine receptor, and cationic excitatory (D-type) and anionic inhibitory (H-type) glutamate-gated channels were described by patch-clamp techniques (Fraser et al., 1990).

```
721  SNYPDMHSYMVRYNQPRVEEALTQLKAGKLDAFIYDAAVLNYMARKDEGCKLVTIGSGKV m-epsilon4   (Ikeda)
721  SNYPDMHSYMVRYNQPRVEEALTQLKAGKLDAFIYDAAVLNYMARKDEGCKLVTIGSGKV r-NMDAR2D-1  (Ishii)
721  SNYPDMHSYMVRYNQPRVEEALTQLKAGKLDAFIYDAAVLNYMARKDEGCKLVTIGSGKV r-NMDAR2D-2  (Ishii)
721  SNYPDMHSYMVRYNQPRVEEALTQLKAGKLDAFIYDAAVLNYMARKDEGCKLVTIGSGKV r-NR2D  (Monyer)

781  FATTGYGIALHKGSRWKRPIDLALLQFLGDDEIEMLERLWLSGICHNDKIEVMSSKLDID m-epsilon4   (Ikeda)
781  FATTGYGIALHKGSRWKRPIDLALLQFLGDDEIEMLERLWLSGICHNDKIEVMSSKLDID r-NMDAR2D-1  (Ishii)
781  FATTGYGIALHKGSRWKRPIDLALLQFLGDDEIEMLERLWLSGICHNDKIEVMSSKLDID r-NMDAR2D-2  (Ishii)
781  FATTGYGIALHKGSRWKRPIDLALLQFLGDDEIEMLERLWLSGICHNDKIEVMSSKLDID r-NR2D  (Monyer)
     __________4__________
841  NMAGVFYMLLVAMGLSLLVFAWEHLVYWRLRHCLGPTHRMDFLLAFSRGMYSCCSAEAAP m-epsilon4   (Ikeda)
841  NMAGVFYMLLVAMGLSLLVFAWEHLVYWRLRHCLGPTHRMDFLLAFSRGMYSCCSAEAAP r-NMDAR2D-1  (Ishii)
841  NMAGVFYMLLVAMGLSLLVFAWEHLVYWRLRHCLGPTHRMDFLLAFSRGMYSCCSAEAAP r-NMDAR2D-2  (Ishii)
841  NMAGVFYMLLVAMGLSLLVFAWEHLVYWRLRHCLGPTHRMDFLLAFSRGMYSCCSAEAAP r-NR2D  (Monyer)

901  PPAKPPPPPQPLPSPAYPAARPPPGPAPFVPRERAAADRWRRAKGTGPPGGAALADGFHR m-epsilon4   (Ikeda)
901  PPAKPPPPPQPLPSPAYPAARPPPGPAPFVPRERAAADRWRRAKGTGPPGGAAIADGFHR r-NMDAR2D-1  (Ishii)
901  PPAKPPPPPQPLPSPAYPAARPPPGPAPFVPRERAAADRWRRAKGTGPPGGAAIADGFHR r-NMDAR2D-2  (Ishii)
901  PPAKPPPPPQPLPSPAYPAARPPPGPAPFVPRERAAADRWRRAKGTGPPGGAAIADGFHR r-NR2D  (Monyer)

961  YYGPIEPQGLGLGEARAAPRGAAGRPLSPPTTQPPQKPPPSYFAIVREQEPAEPPAGAFP m-epsilon4   (Ikeda)
961  YYGPIEPQGLGLGEARAAPRGAAGRPLSPPTTQPPQKPPPSYFAIVREQEPTEPPAGAFP r-NMDAR2D-1  (Ishii)
961  YYGPIEPQGLGLGEARAAPRGAAGRPLSPPTTQPPQKPPPSYFAIVREQEPTEPPAGAFP r-NMDAR2D-2  (Ishii)
961  YYGPIEPQGLGLGEARAAPRGAAGRPLSPPTTQPPQKPPPSYFAIVREQEPTEPPAGAFP r-NR2D  (Monyer)

1021 GFPSPPAPPAAAAAAVGPPLCRLAFEDESPPAPSAGRVLTPRASRCWVGARAARALGPRP m-epsilon4   (Ikeda)
1021 GFPSPPAPPAAAAAAVGPPLCRLAFEDESPPAPSRWPRSDPESQPLLGGGAGGPSAGAPT r-NMDAR2D-1  (Ishii)
1021 GFPSPPAPPAAAAAAVGPPLCRLAFEDESPPAPSRWPRSDPESQPLLGGGAGGPSAGAPT r-NMDAR2D-2  (Ishii)
1021 GFPSPPAPPAAAAAAVGPPLCRLAFEDESPPAPSRWPRSDPESQPLLGGGAGGPSAGAPT r-NR2D  (Monyer)

1081 HHRRVRTAPPPCAYLDLEPSPSDSEDSESLGGASLGGLEPWWFADFPYPYAERLGPPPGR m-epsilon4   (Ikeda)
1081 APPPRRAAPPPCAYLDLEPSPSDSEDSESLGGASLGGLEPWWFADFPYPYAERLGPPPGR r-NMDAR2D-1  (Ishii)
1081 APPPRRAAPPPCAYLDLEPSPSDSEDSESLGGASLGGLEPWWFADFPYPYAERLGPPPGR r-NMDAR2D-2  (Ishii)
1081 APPPRRAAPPPCAYLDLEPSPSDSEDSESLGGASLGGLEPWWFADFPYPYAERLGPPPGR r-NR2D  (Monyer)

1141 YWSVDKLGGWRAGSWDYLPPRGGPAWHCRHCASLELLPPPRHLSCSHDGLDGGWWAPPPP m-epsilon4   (Ikeda)
1141 YWSVDKLGGWRAGSWDYLPPRGGPAWHCRHCASLELLPPPRHLSCSHDGLDGGWWAPPPP r-NMDAR2D-1  (Ishii)
1141 YWSVDKLGGWRAGSWDYLPPRGGPAWHCRHCASLELLPPPRHLSCSHDGLDGGWWAPPPP r-NMDAR2D-2  (Ishii)
1141 YWSVDKLGGWRAGSWDYLPPRGGPAWHCRHCASLELLPPPRHLSCSHDGLDGGWWAPPPP r-NR2D  (Monyer)

1201 PWAAGPPAPRRARCGCPRPHPHRPRASHRAPAAAPHHHRHRRAAGGWDLPPPAPTSRSLE m-epsilon4   (Ikeda)
1201 PWAAGPPPRRRARCGCPRPHPHRPRASHRAPAAAPHHHRHRRAAGGWDFPPPAPTSRSLE r-NMDAR2D-1  (Ishii)
1201 PWAAGPPPRRRARCGCPRPHPHRPRASHRAPAAAPHHHRHRRAAGGWDFPPPAPTSRSLE r-NMDAR2D-2  (Ishii)
1201 PWAAGPPPRRRARCGCPRPHPHRPRASHRAPAAAPHHHRHRRAAGGWDFPPPAPTSRSLE r-NR2D  (Monyer)

1261 DLSSCPRAAPTRRLTGPSRHARRCPHAAHWGPPLPTASHRRHRGGDLGTRRGSAHFSSLE m-epsilon4   (Ikeda)
1261 DLSSRPCPPHRTGDTGAGTWAHAGALRISPAWSPRYDAAPAPTPTPAAPSVSAGHGPRGR r-NMDAR2D-1  (Ishii)
1261 DLSSCPRAAPTRRLTGPSRHARRCPHAAHWGPPLPTASHRRHRGGDLGTRRGSAHFSSLE r-NMDAR2D-2  (Ishii)
1261 DLSSCPRAAPTRRLTGPSRHARRCPHAAHWGPPLPTASHRRHRGGDLGTRRGSAHFSSLE r-NR2D  (Monyer)

1321 SEV                                                          m-epsilon4   (Ikeda)
1321 AKWTGPSWVGKDRNGPGRTPPGAASCAPTPFALGEL                         r-NMDAR2D-1  (Ishii)
1321 SEV                                                          r-NMDAR2D-2  (Ishii)
1321 SEV                                                          r-NR2D  (Monyer)
```

FIGURE 3p(2).

Up to date five subunits of these invertebrate receptors have been characterized, three from *Drosophila* (Schuster et al., 1991; Ultsch et al., 1992), one from the pond snail *Lymnaea stagnalis* (Hutton et al., 1991), and one from the nematode *C. elegans* (Sulston et al., 1992). All these subunits show the structural features characteristic of the iGluR family (Figures 1 and 3q).

2.6.6.1 DGluR-I

PCR-mediated homology screen on genomic *Drosophila* DNA led to isolation of a gene encoding a *Drosophila* kainate-selective glutamate receptor (DGluR-I) with significant homology to mammalian iGluR subunits (Ultsch

```
1   MDKGLHFIFCVVTAVLLLRESSQTGAMRNDDAMIKPNDLRGPEENLPSLTVTTILEDPYV chick-KBP (Gregor)
1           MEKALMLFLAVSLLSLGHTDGKENAETLLKERTKRQIPKTLTVTTILEKPFA frog-KBP  (Wada)

61  MVRSAE-LEGYCIDLLKALASMLHFSYKVKVVGDGKYGAISPSGNWTGMIGEILRQEADI chick-KBP (Gregor)
53  MKTESDALEGYAIDLLSELTQSLGFNYTLHIVKDGKYGSKDQEGNWSGMVGEIIRKEADL frog-KBP  (Wada)
                                                         ______1_
120 AVAPLTVTSAREEVVSFTTPFLQTGIGILLRKETISQEMSFFHFLAPFSKETWTGLLFAY chick-KBP (Gregor)
113 AIAPLTITSVRENAISFTKPFMQTGIGILLKKDTAAESSYMFGFLNPFSKELWIGIIISY frog-KBP  (Wada)
    __________                          _____2_______          ____
180 VLTCVCLFLVARLSPCEWNEPKNEENHFTFLNSLWFGAGALTLQGVTPRPKAFSVRVIAA chick-KBP (Gregor)
173 VITSLCLFLVGRLSPCEWTEPASEQNQFTLLNSLWYGVGALTLQGAEPQPKALSARIIAV frog-KBP  (Wada)
    _3________
240 IWWLFTIALLAAYIAN--ALLSSGSEQLS-IQTFEDLVKQRKLEFGTLDGSSTFYFFKNS chick-KBP (Gregor)
233 IWWVFSITLLAAYIGSFASYINSNTNQTPNIQSVEDLLKQDKLDFGTLSNSSTLNFFKNS frog-KBP  (Wada)

297 KNPIHRMVYEYMDKRRDHVLVKTYQEAVQRVMESNYAFIGESISQDLAAARHCNLIRAPE chick-KBP (Gregor)
293 KNPTFQMIYEYMDKRKDRVLVKTFSEGVQRVRESNYAFLGESISQDFVVAKHCDLIRAPE frog-KBP  (Wada)

357 VIGARGFGIATAQASPWTKKLSVAVLKLRETGDLDYLRNKWWESSCLHKSREGWSPLQPQ chick-KBP (Gregor)
353 MIGGRGYGIAAELDSPLIRPLTIAILELFESGKLEYLRQKWWENTCSTQDQTGWVPVQPH frog-KBP  (Wada)
      _______4_______
417 ALGGLFLTLAIGLALGVIAAMVELSNKSRHAAGHIKKSCCSIFTEEMCTRLRIKENTRQT chick-KBP (Gregor)
413 TLGGIFLILGIGLALGLIVSFMELMCKSRSNAEQQKKSCCSAFSEEIAQRFGKTQNQEGL frog-KBP  (Wada)

477 QETSGRANA                                                    chick-KBP (Gregor)
473 EKKSPTSNSCDEVKA                                              frog-KBP  (Wada)
```

FIGURE 3q.

et al., 1992) (Figure 3q). This receptor subunit is differentially expressed during *Drosophila* development, and accumulates in late embryogenesis in the CNS.

The expression of recombinant DGluRI in *Xenopus* oocytes generates kainate-operated ion channels blocked by the selective non-N-methyl-D-aspartate receptor antagonist 6-cyano-7-nitro-quinoxaline-2,3-dione (CNQX) (Watkins et al., 1990) and philanthotoxin. The DGluR-I gene maps to position 65C of *Drosophila* chromosome 3L.

2.6.6.2 DGluR-II

Using a similar approach a second clone (DGluR-II) was isolated (Schuster et al., 1991). This polypeptide is less related to the AMPA/kainate receptor family, and shows about 25% sequence similarity to the AMPA/kainate receptor and 14% to the NMDA receptor family (Figures 2 and 3q). In contrast to DGluR-I this clone is expressed in somatic muscle tissue of *Drosophila* embryos. Electrophysiological recordings of DGluR-II expressed in *Xenopus* oocytes revealed depolarizing responses to L-glutamate and L-aspartate but low sensitivity to quisqualate, AMPA, and kainate. The DGluR-II gene maps to chromosome 2L, position 25F, a region that is characterized by several lethal mutations.

2.6.6.3 DNMDAR-I

A *Drosophila* NMDA receptor subunit (DNMDAR-I) has been characterized that shows 46% amino acid identity to the rat NR1 polypeptide (Figure 3r) (Ultsch et al., 1993). The high sequence similarity to the NR1 subunit makes the identity as an NMDA receptor subtype very likely even though the

```
1   MKMC-----CPPIQVSLPVYGWGTLIGHFCLRLFFSYTIQ--M-LRESFGNNEEVSRVAL C-C06E1.4  (Sulston)
1   MHSRLKFLAYLHFICASSIF-WPEFSSAQQQQQTVSLTEK--IPLGAIF---EQGTDDVQ DGluR-I   (Ultsch)
1   MRLC-----------PVVIYAFIIIIGFLEGIIALGGDDRNEITVGAIFYENEK---EIE DGluR-II  (Schuster)
1   MDTCV-----FPLV----VL-WISM------RITSTLDE---VPIGGIF---DSRSVQAL LymGluR   (Hutton)

53  KAMEYTSDHINSRDDVPFKLAFDHRVVEEGAAVSWNMVNAVCDELKEGAMALLSSVDGKG C-C06E1.4  (Sulston)
55  SAFKYAL-LNHNLNVSSRRFELQAYVDVINTADAFKLSRLICNQFSRGVYSMLGAVSPDS DGluR-I   (Ultsch)
47  LSFDQAFREVNNMKFSELRFV--TIKRYMPTNDSFLLQQITCELISNGVAAIFGPSSKAA DGluR-II  (Schuster)
39  TAFRHEIHMFNRAYSHVYRYKLKNDTTILDVTDSFAVSNALCHHLSRGDLAIFGVSNASS LymGluR   (Hutton)

113 REGIRGVSDALEMPLVSLTALSNDDHQQQFGNL-FEVSVR-PPISELLADFIV-HKGWG C-C06E1.4  (Sulston)
114 FDTLHSYSNTFQMPFV--TPWFPEKVLAPSSGLLDFAISMR-PDYHQAIIDTIQ-YYGWQ DGluR-I   (Ultsch)
105 SDIVAQIANATGIPHIEYDLKLEATRQEQLNHQMSINVAPSLSVLSRAYFEIIKSNYEWR DGluR-II  (Schuster)
99  LATIQSYTDTFNVPFV--TISMAQN--NSHNG--SYQIYMR-PMYINALVDVIV-HYRWE LymGluR   (Hutton)

170 EVLVLIDPVHASLHLPSLWRHLRTRTNTSVKASMFDLPADEKQFEAYLMQFNMMRNNETN C-C06E1.4  (Sulston)
170 SIIYLYDSHDGLLRLQQIYQELKPGNETFRVQMVKRIANVTMAIE-FLHTLEDLGRFSKK DGluR-I   (Ultsch)
165 TFTLIYETPEGLARLQDL-MNIQALNSDYVK--LRNLADYADDYRILWKETD--ETFHEQ DGluR-II  (Schuster)
151 KVAFYYDSDEGLVRLQQLFQATNKYDKMIISIDTKRITSVENGYH-MLKELHLMDPEMEH LymGluR   (Hutton)

230 RILIDCASPKRLKKLLINIRSAQFNQANYHYVLANYDFLPY-DQEMFQNG-NINISGFNI C-C06E1.4  (Sulston)
229 RIVLDCPAEMAKEIIVQHVRDIKLGRRTYHYLLSGLVMDNHWPSDVVEFG-AINITGFRI DGluR-I   (Ultsch)
220 RIILDCE-PKTLKELLKVSIDFKLQGPFRNWFLTHLDTHNSGLRDIYNEDFKANITSVRL DGluR-II  (Schuster)
210 RVLLDVRTDKAEQIILKVMNDSKINNAKFHFLLGDLGMLEI-NTTHFKIG-GVNITGFQL LymGluR   (Hutton)

288 IN---KDGREYWSLKKHL----KTSSSLGGGDD----VSVEAAVGHDAMLVTWHGFAKCL C-C06E1.4  (Sulston)
288 VDSNRRAVRDFHDNRKRLEPSAKAKARTQGGPNSLPPISAQAALMYDAVFVLVEAFNRIL DGluR-I   (Ultsch)
279 K---VVDANPFERKKTRLTKV----------DQILGNQTMLPILIYDAVVLFASSARNVI DGluR-II  (Schuster)
268 VDPFNSTSELFISTWSSLDPVYWPGAGTN-------HVNYEAALAADSVRLFKSAFGSIL LymGluR   (Hutton)

337 QANDSLFHGTFRHRRFFNRGFPGIYCDPLSDRSHPNRPF---------SSFEHGKTIGVA C-C06E1.4  (Sulston)
348 RKK----PDQFRSNHLQRRSHGGSSSSSATGTNESSALLDCNTSKGWVTPWEQGEKISRV DGluR-I   (Ultsch)
326 RAMQP-FHPPNRH--------------------------CGSS----SPWMLGAFIVNE DGluR-II  (Schuster)
321 QKD----PN------FLRRSRSGTAGKS----------MKCTDDSEIKT--GHGQMILEE LymGluR   (Hutton)

388 FRNMKIGHKEGTL-TGNIEFDRFGNRKNFDVSIVDLVSNTKATFNSKEVLAWRQGVGFFS C-C06E1.4  (Sulston)
404 LRKVEI----DGL-SGEIRFDEDGRRINYTLHVVEMSVNSTLQ----QVAEWRDDAGLLP DGluR-I   (Ultsch)
354 MKTISEDDVEPHFKTENMKLDEYGQRIHFNLEIYKPTVNEP-------MMVWTPDNGI-- DGluR-II  (Schuster)
359 MKKVKF----EGV-TGHVAFNEQGHRKDFTLGVYNVAMTRGTA----KIGYWNEREGKLH LymGluR   (Hutton)

447 --NRTVAQHSRKSQND-----HKDNQVIVLTNLVAPFVMIKRECLEMANLTECQGNNKFE C-C06E1.4  (Sulston)
455 LHSHNYASSSRSASASIGDY-DRNHTYIVSSLLEEPYLSLKQYTY-GESLV---GNDRFE DGluR-I   (Ultsch)
405 --KKRLLNLELESAGTTQDFSEQRKVYTVVTHYEEPYFMMKE---DHENF---RGREKYE DGluR-II  (Schuster)
410 AHNPRLFQNNSS---------DMNRTRIVTTIIKEPYVMVNNVIRDGKPLV---GNEPVE LymGluR   (Hutton)

500 GFC-IDLLKLLADKIEEFNYEIKLGTKASEKILRIVFYKLFQAGSKQADGSWDGMIGELL C-C06E1.4  (Sulston)
510 GYC-KDLADMLAAQLG-IKYEIRLVQDGN-----------YGAENQYAPGGWDGMVGELI DGluR-I   (Ultsch)
457 GYAWISLASFPSSWSSITEFMI---VNGNGK---------YNPETKQ----WDGIIRKLI DGluR-II  (Schuster)
458 GFC-IDLTKAVAEKVG-FDFVIQFVKDGS-----------YGS--VLSNGTWDGIVGELI LymGluR   (Hutton)

559 SGRAHAVVASLTINQERERVVDFSKPFMTTGISIMIKKPDKQEFSVFSFMQPLSTEIWMY C-C06E1.4  (Sulston)
557 RKEADIAISAMTITAERERVIDFSKPFMTLGISIMIKKPVKQTPGVFSFLNPLSQEIWIS DGluR-I   (Ultsch)
501 DHHAQIGVCDLTITQMRRSVVDFTVPFMQLGISILHYKSPPEPKNQFAFLEPFAVEVWIY DGluR-II  (Schuster)
503 RHEADMAIAPFTITADRSRVIDFTKPFMSLGISIMIKRPQPAGKHFFSFMEPLSSEIWMC LymGluR   (Hutton)
                       1
619 IIFAYIGVSVVIFLVSRFSPYEWRVEETSRGGFT-----------------------ISN C-C06E1.4  (Sulston)
617 VILSYVGVSFVLYFVTRFPPYEWRIVRRPQADSTAQQPPGIIGGATLSEPQAHVPPVPPN DGluR-I   (Ultsch)
561 MIFAQLIMTLAFVFIARLSYREWL-------------PP---------NPAIQDPDELEN DGluR-II  (Schuster)
563 IVFAYIGVSVVLFLVSRFSPNEWH---------------------LSEAH---HSYIAN LymGluR   (Hutton)
                   2                                  3
656 DFSVYNCLWFTLAAFMQQGTDILPRSISGRIASSAWWFFTMIIVSSYTANLAAFLTLEKM C-C06E1.4  (Sulston)
677 EFTMLNSFWYSLAAFMQQGCDIIPPSIAGRIAAAVWWFFTIILISSYTANLAAFLTVERM D-GluR-I   (Ultsch)
599 IWNVNNSTWLMVGSIMQQGCDILPRGPHMRILTSMWWFFALMMLSLYTANLA-FLTSNKW D-GluR-II  (Schuster)
598 DFSISNSLWFSLGAFMQQGCDISPRSMSGRIVGSVWWFFTLIIISSYTANLAAFLTVERM LymGluR   (Hutton)
```

FIGURE 3r(1).

authors did not demonstrate NMDA or L-glutamate-induced activation of the *Xenopus* oocyte-expressed subunit. As for the mammalian NMDA receptors, coexpression with other subunits might be needed to obtain active channels. The DNMDAR-I subunit is highly expressed in the head of adult flies. The corresponding gene maps to position 83AB of chromosome 3R.

```
716 QAPIESVEDLAKQSKIKYGIQGGGSTASFFKYSSVQIYQRMWRYMESQVPPVFVASYAEG C-C06E1.4   (Sulston)
737 VAPIKTPEDLAMQTDVNYGTLLHGSTWEFFRRSQIGLHNKMWEYMNANQHHS-VHTYDEG DGluR-I   (Ultsch)
658 QSSIKSPQDLIEQDKVHFGSMRGGSTSLFFSESNDTDYQRAWNQMKDFNPSAFTSTNKEG DGluR-II   (Schuster)
658 LTPIDSAEDLARQTEIQYGTIMSGSTKAFFKNSQFQTYQRMWAYMTSAQPSVFVKTHEEG LymGluR   (Hutton)

776 IERVRSHKGRYAFLLEATANEYENTRKPCDTMKVGANLNSIGYGIATPFGSDWKDHINLA C-C06E1.4   (Sulston)
796 IRRVRQSKGKYALLVESPKNEYVNARPPCDTMKVGRNIDTKGFGVATPIGSPLRKRLNEA D-GluR-I   (Ultsch)
718 VARVRKEKGGYAFLMETTSLTYNIERN-CDLTQIGEQIGEKHYGLAVPLGSDYRTNLSVS D-GluR-II   (Schuster)
718 IQRVRQSNGKYAYLTESSTIDYVSNRKPCDTLKVGSNLNSDGFGIGTPVGSDLRDKLNFS LymGluR   (Hutton)
                                                   ______4______
836 ILALQERGELKKLENKWWYDRG-QCDAGITVDGSSAS-LNLSKVAGIFYILMGGMVISML C-C06E1.4   (Sulston)
856 VLTLKENGELLRIRNKWWFDKT-EC--NLDQETSTPNELSLSNVAGIYYILIGGLLLAVI DGluR-I   (Ultsch)
777 ILQLSERGELQKMKNKWWKNHNVTCDSYHEVDGD---ELSIIELGGVFLVLAGGVLIGVI DGluR-II   (Schuster)
778 VLELRENGDLAK-WEKIWFDRG-ECPQHSSNKEGAQSALTLANVAGIFYILIGGLVVAVL LymGluR   (Hutton)
    ___
894 AALGEFLYRSRIEARKSNSNSMVANFAKNLKSALSSQLRLSVEGGAVAQPGSQSHNAIRR C-C06E1.4   (Sulston)
913 VAIVEFFCRNKTPQLKSPG---------SNGSAGGVPGMLGSSTY----QRDSLSDAIMH DGluR-I   (Ultsch)
834 LGIFEFLWNVQNVAVEERVTPW-----QAFKAELIFALKFWVRKKPMRISSSSDKSSSRR DGluR-II   (Schuster)
836 SAAFEFLYKSRMDSRKSRM---------SFGSALRTKARLSFKGHIDSEQKTTGNGTRRR LymGluR   (Hutton)

954 QQVAAFLPANEK--EAFNNVDRPANTLYNTAV                             C-C06E1.4   (Sulston)
960 SQAKLAMQASSEYDERLVGVELASNVRYQYSM                             DGluR-I   (Ultsch)
889 SSGSR--------------RSSKEKSRSKTVS                             DGluR-II   (Schuster)
887 SHNSVTYTYTGP-TNVMGGSHAFEDSNTHTEV                             LymGluR   (Hutton)
```

FIGURE 3r(2).

2.6.6.4 LymGluR

A full-length cDNA encoding the putative receptor polypeptide LymGluR was isolated from the pond snail *Lymnaea stagnalis* (Hutton et al., 1991) (Figure 3q). A sequence comparison shows approximately 40% amino acid identity to the rat AMPA receptor subunits, GluR-A to GluR-D. When expressed in *Xenopus oocytes* no functional channels could be detected, and ligand binding has not been reported for this subunit.

2.6.6.5 C-C06E1.4

A gene encoding a protein (Figure 3q) of approximately 35% sequence similarity to the mammalian AMPA receptors was identified on a genomic fragment of the *Caenorhabditis elegans* genome. It was found on a cosmid sequenced in the initial phase of the *C. elegans* genome sequencing project (Sulston et al., 1992; Wilson et al., 1994). The gene contains 13 exons and 12 small introns. The hydrophobic transmembrane segments M2 and M3 are encoded by separate exons. In contrast to the gene organization of GluR-B, GluR5, and GluR6 the coding region for segment M1 is on two exons. Nothing is known about expression, function, and pharmacology of this subunit.

2.6.7 Concluding Remarks

Ingenious expression cloning strategies (Hollmann et al., 1989; Moriyoshi et al., 1991) have led to the isolation of cDNAs encoding key subunits of the iGluR superfamily. Based on sequence and structural features of these subunits, the different members of the iGluR superfamily could be characterized

```
1   MAMAEFVFCRPLFGLAIVLLVAPIDAAQRHTASDNPSTYNIGGVLSNSDSEEHFSTTIKH d-DNMDAR-I  (Ultsch)
1               MSTMHLLTFALLFSCSFARAACDPKIVNIGAVLSTRKHEQMFREAVNQ r-NMDAR1    (Miriyoshi)

61  LNFDQQYVPRKVTYYDKTIRMDKNPIKTVFNVCDKLIENRVYAVVVSHEQTSGD-LSPAA d-DNMDAR-I  (Ultsch)
49  AN--KRHGSWKIQLNATSVTHKPNAIQMALSVCEDLISSQVYAILVSHPPTPNDHFTPTP r-NMDAR1    (Miriyoshi)

120 VSYTSGFYSIPVIGISSRDAAFSDKNIHVSFLRTVPPYYHQADVWLEMLSHFAYTKVIII d-DNMDAR-I  (Ultsch)
107 VSYTAGFYRIPVLGLTTRMSIYSDKSIHLSFLRTVPPYSHQSSVWFEMMRVYNWNHIILL r-NMDAR1    (Miriyoshi)

180 HSSDTDGRAILGRFQTTSQTYYDDVDVRATVELIVEFEPKLESFTEHLIDMKTAQSRVYL d-DNMDAR-I  (Ultsch)
167 VSDDHEGRAAQKRL----ETLLEERESKA--EKVLQFDPGTKNVTALLMEARELEARVII r-NMDAR1    (Miriyoshi)

240 MYASTEDAQVIFRDAGEYNMTGEGHVWIVTEQALFSNN---TPDGVLGLQLEHAHSDKGH d-DNMDAR-I  (Ultsch)
221 LSASEDDAATVYRAAAMLNMTGSGYVWLVGEREISGNALRYAPDGIIGLQLINGKNESAH r-NMDAR1    (Miriyoshi)

297 IRDSVYVLASAIKEMISNETIAEAPKDCGDSAVNWESGKRLFQ--YLKSRNITGETGQVA d-DNMDAR-I  (Ultsch)
281 ISDAVGVVAQAVHELLEKENITDPPRGCVGNTNIWKTG-PLFKRVLMSSKYADGVTGRVE r-NMDAR1    (Miriyoshi)

355 FDDNGDRIYAGYDVINIREQQKKHVVGKFSYDSMRAKMRMRINDSEIIWPGKQRRKPEGI d-DNMDAR-I  (Ultsch)
340 FNEDGDRKFANYSIMNL---QNRKLVQVGIYNGT----HVIPNDRKIIWPGGETEKPRGY r-NMDAR1    (Miriyoshi)

415 MIPTHLRLLTIEEKPFVYVRR-MGD----DEFRCEPDERP---CPLFNNSDATANEF--- d-DNMDAR-I  (Ultsch)
393 QMSTRLKIVTIHQEPFVYVKPTMSDGTCKEEFTVNGDPVKKVICTGPNDTSPGSPRHTVP r-NMDAR1    (Miriyoshi)

464 -CCRGYCIDLLIELSKRINFTYDLALSPDGQFGHYILRNNTGAMTLRKEWTGLIGELVNE d-DNMDAR-I  (Ultsch)
453 QCCYGFCIDLLIKLARTMNFTYEVHLVADGKFGT----QERVNNSNKKEWNGMMGELLSG r-NMDAR1    (Miriyoshi)

523 RADMIVAPLTINPERAEYIEFSKPFKYQGITILEKKPSRSSTLVSFLQPFSNTLWILVMV d-DNMDAR-I  (Ultsch)
509 QADMIVAPLTINNERAQYIEFSKPFKYQGLTILVKKEIPRSTLDSFMQPFQSTLWLLVGL r-NMDAR1    (Miriyoshi)
          1                                 2
583 SVHVVALVLYLLDRFSPFGRFKLSHSDSNEEKALNLSSAVWFAWGVLLNSGIGEGTPRSF d-DNMDAR-I  (Ultsch)
569 SVHVVAVMLYLLDRFSPFGRFKV-NSEEEEEDALTLSSAMWFSWGVLLNSGIGEGAPRSF r-NMDAR1    (Miriyoshi)
                 3
643 SARVLGMVWAGFAMIIVASYTANLAAFLVLERPKTKLSGINDARLRNTMENLTCATVKGS d-DNMDAR-I  (Ultsch)
628 SARILGMVWAGFAMIIVASYTANLAAFLVLDRPEERITGINDPRLRNPSDKFIYATVKQS r-NMDAR1    (Miriyoshi)

703 SVDMYFRRQVELSNMYRTMEANNYATAEQAIQDVKKGKLMAFIWDSSRLEYEASKDCELV d-DNMDAR-I  (Ultsch)
688 SVDIYFRRQVELSTMYRHMEKHNYESAAEAIQAVRDNKLHAFIWDSAVLEFEASQKCDLV r-NMDAR1    (Miriyoshi)

763 TAGELFGRSGYGIGLQKGSPWTDAVTLAILEFHESGFMEKLDKQWIFHGHVQQNCELFEK d-DNMDAR-I  (Ultsch)
748 TTGELFFRSGFGIGMRKDSPWKQNVSLSILKSHENGFMEDLDKTWVRY----QECDSRSN r-NMDAR1    (Miriyoshi)
                    4
823 TPNTLGLKNMAGVFILVGVGIAGGVGLIIIEVIYKKHQVKKQKRLDIARHAADKWRGTIE d-DNMDAR-I  (Ultsch)
804 APATLTFENMAGVFMLVAGGIVAGIFLIFIEIAYKRHKDARRKQMQLAFAAVNVWRKNLQ r-NMDAR1    (Miriyoshi)

883 KRKTIRASLAMQRQYNVGLNSTHAPGTISLAVDKRRYPRLGQRLGPERAWPGDAADVLRI d-DNMDAR-I  (Ultsch)
864 DRKSGRAEPDPKK---------------------KATFRAITSTL----------ASSFKRR r-NMDAR1    (Miriyoshi)

943 RRPYELGNPGQSPKVMAANQPGMPMPMLGKTRPQQSVLPPRYSPGYTSDVSHLVV      d-DNMDAR-I  (Ultsch)
895 RSSKDTSTGGGRGALQNQKDTVLPRRAIEREEGQLQ-LCSRHRES                r-NMDAR1    (Miriyoshi)
```

FIGURE 3s.

from rodent species and, in part, from human. Although all subunits in this superfamily are structurally related, distinct subfamilies can be identified characterized by different functional signatures. A significant finding was that all subunits share structural homology with prokaryotic periplasmic proteins, indicating an ancient evolutionary origin of iGluR proteins. Accordingly, iGluR proteins are found throughout the animal kingdom.

The existence of L-glutamate-gated cation channels in the CNS that are structurally unrelated to members of the iGluR superfamily has been reported (Kumar et al., 1991; Smirnova et al., 1993). Further studies are required to assess whether the proteins encoded by the cloned cDNAs indeed possess the purported physiological functions.

A characteristic feature of the iGluR genes is that many of them give rise to different molecular subunit variants by transcript splicing or by RNA editing. The full spectrum of functional consequences has not been determined but will undoubtedly yield new insights into synaptic function. Novel methodologies such as combined electrophysiology and molecular biology on single cells (Eberwine et al., 1992; Lambolez et al., 1992; Bochet et al., 1994; Jonas et al., 1994) will serve as indispensable tools.

Acknowledgments

We thank Jutta Rami for secretarial assistance. We apologize to any colleague whose work pertinent to this chapter we failed to cite. Supported by the BMFT, DFG, and funds from the German Chemical Industry.

REFERENCES

Anantharam, V., Panchal, R. G., Wilson, A., Kolchine, V. V., Treistman, S. N., and Bayley, H. (1992). Combinatorial RNA splicing alters the surface charge on the NMDA receptor. *FEBS Lett.* 305, 27–30.

Araki, K., Meguro, H., Kushiya, E., Takayama, C., Inoue, Y., and Mishina, M. (1993). Selective expression of the glutamate receptor channel delta 2 subunit in cerebellar Purkinje cells. *Biochem. Biophys. Res. Commun.* 197, 1267–1276.

Bahn, S., Volk, B., and Wisden, W. (1994). Kainate receptor gene expression in the developing rat brain. *J. Neurosci.* 14, August issue.

Bai, G. and Kusiak, J. W. (1993). Cloning and analysis of the 5′ flanking sequence of the rat N-methyl-D-aspartate-receptor 1 (NMDAR1) gene. *Biochim. Biophys. Acta* 1152, 197–200.

Bettler, B., Boulter, J., Hermans-Borgmeyer, I., O'Shea-Greenfield, A., Deneris, E. S., Moll, C., Borgmeyer, U., Hollmann, M., and Heinemann, S. (1990). Cloning of a novel glutamate receptor subunit, GluR5: Expression in the nervous system during development. *Neuron* 5, 583–595.

Bettler, B., Egebjerg, J., Sharma, G., Pecht, G., Hermans-Borgmeyer, I., Moll, C., Stevens, C. F., and Heinemann, S. (1992). Cloning of a putative glutamate receptor: A low affinity kainate-binding subunit. *Neuron* 8, 257–265.

Blackstone, C. D., Moss, S. J., Martin, L. J., Levey, A. I., Price, D. L., and Huganir, R. L. (1992a). Biochemical characterization and localization of a non-N-methyl-D-aspartate glutamate receptor in rat brain. *J. Neurochem.* 58, 1118–1126.

Blackstone, C. D., Levey, A. I., Martin, L. J., Price, D. L., and Huganir, R. L. (1992b). Immunological detection of glutamate receptor subtypes in human central nervous system. *Ann. Neurol.* 31, 680–683.

Bliss, T. V. and Collingridge, G. L. (1993). A synaptic model of memory: Long-term potentiation in the hippocampus. *Nature (London)* 361, 31–39.

Bochet, P., Audinat, E., Lambolez, B., Crépel, F., Rossier, J., Iino, M., Tsuzuki, K., and Ozawa, S. (1994). Subunit composition at the single-cell level explains functional properties of a glutamate-gated channel. *Neuron* 12, 383–388.

Boulter, J., Hollmann, M., O'Shea-Greenfield, A., Hartley, M., Deneris, E., Maron, C., and Heinemann, S. (1990). Molecular cloning and functional expression of glutamate receptor subunit genes. *Science* 249, 1033–1037.

Bourne, H. R. and Nicoll, R. (1993). Molecular machines integrate coincident synaptic signals. *Cell/Neuron Rev. Suppl. Cell 72/Neuron* 10, 65–76.

Burnashev, N., Monyer, H., Seeburg, P. H., and Sakmann, B. (1992a). Divalent ion permeability of AMPA receptor channels is dominated by the edited form of a single subunit. *Neuron* 8, 189–198.

Burnashev, N., Khodorova, A., Jonas, P., Helm, P. J., Wisden, W., Monyer, H., Seeburg, P. H., and Sakmann, B. (1992b). Calcium-permeable AMPA-kainate receptors in fusiform cerebellar glial cells. *Science* 256, 1566–1570.

Burnashev, N., Schoepfer, R., Monyer, H., Ruppersberg, J. P., Gunther, W., Seeburg, P. H., and Sakmann, B. (1992c). Control by asparagine residues of calcium permeability and magnesium blockade in the NMDA receptor. *Science* 257, 1415–1419.

Choi, D. W. (1988). Glutamate neurotoxicity and diseases of the nervous system. *Neuron* 1, 623–634.

Collingridge, G. L. and Lester, R. A. (1989). Excitatory amino acid receptors in the vertebrate central nervous system. *Pharmacol. Rev.* 41, 143–210.

Collingridge, G. L. and Singer, W. (1990). Excitatory amino acid receptors and synaptic plasticity. *Trends Pharmacol. Sci.* 11, 290–296.

Collins, C., Duff, C., Duncan, A. M., Planells, C. R., Sun, W., Norremolle, A., Michaelis, E., Montal, M., Worton, R., and Hayden, M. R. (1993). Mapping of the human NMDA receptor subunit (NR1) and the proposed NMDA receptor glutamate-binding subunit (NMDARA1) to chromosomes 9q34. 3 and chromosome 8, respectively. *Genomics* 17, 237–239.

Cotman, C. W., Monaghan, D. T., Ottersen, O. P., and Storm-Mathisen, J. (1987). Anatomical organization of excitatory amino acid receptors and their pathways. *Trends Neurosci.* 10, 273–279.

Coyle, J. T. (1983). Neurotoxic action of kainic acid. *J. Neurochem.* 41, 1–11.

Craig, A. M., Blackstone, C. D., Huganir, R. L., and Banker, G. (1993). The distribution of glutamate receptors in cultured hippocampal neurons: Postsynaptic clustering of AMPA-selective subunits. *Neuron* 10, 1055–1068.

Dingledine, R., McBain, C. J., and McNamara, J. O. (1990). Excitatory amino acid receptors in epilepsy. *Trends Pharmacol. Sci.* 11, 334–338.

Dingledine, R., Hume, R. I., and Heinemann, S. F. (1992). Structural determinants of barium permeation and rectification in non-NMDA glutamate receptor channels. *J. Neurosci.* 12, 4080–4087.

Eberwine, J., Yeh, H. Miyashiro, K., Cao, Y., Nair, S., Finnell, R., Zettel, M., and Coleman, P. (1992). Analysis of gene expression in single live neurons. *Proc. Natl. Acad. Sci. U.S.A.* 89, 3010–3014.

Egebjerg, J. and Heinemann, S. F. (1993). Ca^{2+} permeability of unedited and edited versions of the kainate selective glutamate receptor GluR6. *Proc. Natl. Acad. Sci. U.S.A.* 90, 755–759.

Egebjerg, J., Bettler, B., Hermans-Borgmeyer, I., and Heinemann, S. (1991). Cloning of a cDNA for a glutamate receptor subunit activated by kainate but not AMPA. *Nature* 351, 745–748.

Eshhar, N., Hunter, C., Wenthold, R. J., and Wada, K. (1992). Structural characterization and expression of a brain specific gene encoding chick kainate binding protein. *FEBS Lett.* 297, 257–262.

Eubanks, J. H., Puranam, R. S., Kleckner, N. W., Bettler, B., Heinemann, S. F., and McNamara, J. O. (1993). The gene encoding the glutamate receptor subunit GluR5 is located on human chromosome 21q21.1–22.1 in the vicinity of the gene for familial amyotrophic lateral sclerosis. *Proc. Natl. Acad. Sci. U.S.A.* 90, 178–182.

Foldes, R. L., Rampersad, V., and Kamboj, R. K. (1993). Cloning and sequence analysis of cDNAs encoding human hippocampus N-methyl-D-aspartate receptor subunits: Evidence for alternative RNA splicing. *Gene* 131, 293–298.

Fraser, S. P., Djamgoz, M. B., Usherwood, P. N., O'Brien, J., Darlison, M. G., and Barnard, E. A. (1990). Amino acid receptors from insect muscle: Electrophysiological characterization in Xenopus oocytes following expression by injection of mRNA. *Mol. Brain Res.* 8, 331–341.

Gallo, V., Upson, L. M., Hayes, W. P., Vyklicky, L. J., Winters, C. A., and Buonanno, A. (1992). Molecular cloning and development analysis of a new glutamate receptor subunit isoform in cerebellum. *J. Neurosci.* 12, 1010–1023.

Gasic, G. P. and Hollmann, M. (1992). Molecular neurobiology of glutamate receptors. *Annu. Rev. Physiol.* 54, 507–536.

Gilbertson, T. A., Scobey, R., and Wilson, M. (1991). Permeation of calcium ions through non-NMDA glutamate channels in retinal bipolar cells. *Science* 251, 1613–1615.

Gregor, P., Eshhar, N., Ortega, A., and Teichberg, V. I. (1988). Isolation, immunochemical characterization and localization of the kainate sub-class of glutamate receptor from chick cerebellum. *EMBO J.* 7, 2673–2679.

Gregor, P., Mano, I., Maoz, I., McKeown, M., and Teichberg, V. I. (1989). Molecular structure of the chick cerebellar kainate-binding subunit of a putative glutamate receptor. *Nature* 342, 689–692.

Gregor, P., Yang, X., Mano, I., Takemura, M., Teichberg, V. I., and Uhl, G. R. (1992). Organization and expression of the gene encoding chick kainate binding protein, a member of the glutamate receptor family. *Mol. Brain Res.* 16, 179–186.

Gregor, P., O'Hara, B. F., Yang, X., and Uhl, G. R. (1993a). Expression and novel subunit isoforms of glutamate receptor genes GluR5 and GluR6. *NeuroReport* 4, 1343–1346.

Gregor, P., Reeves, R. H., Jabs, E. W., Yang, X., Dackowski, W., Rochelle, J. M., Brown, R. J., Haines, J. L., O'Hara, B. F., Uhl, G. R., and Seldin, M. F. (1993b). Chromosomal localization of glutamate receptor genes: Relationship to familial amyotrophic lateral sclerosis and other neurological disorders of mice and humans. *Proc. Natl. Acad. Sci. U.S.A.* 90, 3053–3057.

Hampson, D. R., Huie, D., and Wenthold, R. J. (1987). Solubilization of kainic acid binding sites from rat brain. *J. Neurochem.* 49, 1209–1215.

Hampson, D. R., Huang, X. P., Oberdorfer, M. D., Goh, J. W., Auyeung, A., and Wenthold, R. J. (1992). Localization of AMPA receptors in the hippocampus and cerebellum of the rat using an anti-receptor monoclonal antibody. *Neuroscience* 50, 11–22.

Herb, A., Burnashev, N., Werner, P., Sakmann, B., Wisden, W., and Seeburg, P. H. (1992). The KA-2 subunit of excitatory amino acid receptors shows widespread expression in brain and forms ion channels with distantly related subunits. *Neuron* 8, 775–785.

Higgins, D. G. and Sharp, P. M. (1988). Clustal: A package for performing multiple sequence alignment on a microcomputer. *Gene* 73, 237–244.

Higuchi, M., Single, F. N., Köhler, M., Sommer, B., Sprengel, R., and Seeburg, P. H. (1993). RNA editing of AMPA receptor subunit GluR-B: A base-paired intron-exon structure determines position and efficiency. *Cell* 75, 1361–1370.

Hollmann, M., O'Shea-Greenfield, A., Rogers, S. W., and Heinemann, S. (1989). Cloning by functional expression of a member of the glutamate receptor family. *Nature* 342, 643–648.

Hollmann, M. and Heinemann, S. (1994). Cloned glutamate receptors. *Ann. Rev. Neurosci.* 17, 31–108.

Hollmann, M., Hartley, M., and Heinemann, S. (1991). Ca^{2+} permeability of KA-AMPA gated glutamate receptor channels depends on subunit composition. *Science* 252, 851–853.

Hollmann, M., Boulter, J., Maron, C., Beasley, L., Sullivan, J., Pecht, G., and Heinemann, S. (1993). Zinc potentiates agonist-induced currents at certain splice variants of the NMDA receptor. *Neuron* 10, 943–954.

Huettner, J. E. (1990). Glutamate receptor channels in rat DRG neurons: Activation by kainate and quisqualate and blockade of desensitization by Con A. *Neuron* 5, 255–266.

Hume, R. I., Dingledine, R., and Heinemann, S. F. (1991). Identification of a site in glutamate receptor subunits that controls calcium permeability. *Science* 253, 1028–1031.

Hutton, M. L., Harvey, R. J., Barnard, E. A., and Darlison, M. G. (1991). Cloning of a cDNA that encodes an invertebrate glutamate receptor subunit. *FEBS Lett.* 292, 111–114.

Iino, M., Ozawa, S., and Tsuzuki, K. (1990). Permeation of calcium through excitatory amino acid receptor channels in cultured rat hippocampal neurones. *J. Physiol. (London)* 424, 151–165.

Ikeda, K., Nagasawa, M., Mori, H., Araki, K., Sakimura, K., Watanabe, M., Inoue, Y., and Mishina, M. (1992). Cloning and expression of the epsilon 4 subunit of the NMDA receptor channel. *FEBS Lett.* 313, 34–38.

Ishii, T., Moriyoshi, K., Sugihara, H., Sakurada, K., Kadotani, H., Yokoi, M., Akazawa, C., Shigemoto, R., Mizuno, N., Masu, M., and Nakanishi, S. (1993). Molecular characterization of the family of the N-methyl-D-aspartate receptor subunits. *J. Biol. Chem.* 268, 2836–2843.

Jonas, P., Racca, C., Sakmann, B., Seeburg, P. H., and Monyer, H. (1994). Differences in Ca^{2+} permeability of AMPA-type glutamate receptor channels in neocortical neurons caused by differential expression of the GluR-B subunit. *Neuron* 12, 1281–1289.

Kamboj, R. K., Schoepp, D. D., Nutt, S., Shekter, L., Korczak, B., True, R. A., Zimmerman, D. M., and Wosnick, M. A. (1992). Molecular structure and pharmacological characterization of humEAA2, a novel human kainate receptor subunit. *Mol. Pharmacol.* 42, 10–15.

Kamboj, R. K., Schoepp, D. D., Nutt, S., Shekter, L., Korczak, B., True, R. A., Rampersad, V., Zimmerman, D. M., and Wosnick, M. A. (1994). Molecular cloning, expression, and pharmacological characterization of HumEAA1, a human kainate receptor subunit. *J. Neurochem.* 62, 1–9.

Karp, S. J., Masu, M., Eki, T., Ozawa, K., and Nakanishi, S. (1993). Molecular cloning and chromosomal localization of the key subunit of the human N-methyl-D-aspartate receptor. *J. Biol. Chem.* 268, 3728–3733.

Keinänen, K., Wisden, W., Sommer, B., Werner, P., Herb, A., Verdoorn, T. A., Sakmann, B., and Seeburg, P. H. (1990). A family of AMPA-selective glutamate receptors. *Science* 249, 556–560.

Köhler, M., Burnashev, N., Sakmann, B., and Seeburg, P. H. (1993). Determinants of Ca^{2+} permeability in both TM1 and TM2 of high affinity kainate receptor channels: Diversity by RNA editing. *Neuron* 10, 491–500.

Köhler, M., Kornau, H. C., and Seeburg, P. H. (1994). The organization of the gene for the functionally dominant alpha-amino-3-hydroxy-5-methylisoxazole-4-propionic acid receptor subunit GluR-B, *J. Biol. Chem.*, 269, 17367–17370.

Kumar, K. N., Tilakaratne, N., Johnson, P. S., Allen, A. E., and Michaelis, E. K. (1991). Cloning of cDNA for the glutamate-binding subunit of an NMDA receptor complex. *Nature* 354, 70–73.

Kutsuwada, T., Kashiwabuchi, N., Mori, H., Sakimura, K., Kushiya, E., Araki, K., Meguro, H., Masaki, H., Kumanishi, T., Arakawa, M., and Mishina, M. (1992). Molecular diversity of the NMDA receptor channel. *Nature* 358, 36–41.

Lambolez, B., Audinat, E., Bochet, P., Crepel, F., and Rossier, J. (1992). AMPA receptor subunits expressed by single Purkinje cells. *Neuron* 9, 247–258.

Laurie, D. J. and Seeburg, P. H. (1994). Regional and developmental heterogeneity in splicing of the rat brain NMDAR1 mRNA. *J. Neurosci.* 14, 3180–3194.

Lerma, J., Paternain, A. V., Naranjo, J. R., and Mellström, B. (1993). Functional kainate-selective glutamate receptors in cultured hippocampal neurons. *Proc. Natl. Acad. Sci. U.S.A.* 90, 11688–11692.

Lomeli, H., Wisden, W., Köhler, M., Keinänen, K., Sommer, B., and Seeburg, P. H. (1992). High-affinity kainate and domoate receptors in rat brain. *FEBS Lett.* 307, 139–143.

Lomeli, H., Sprengel, R., Laurie, D. J., Köhr, G., Herb, A., Seeburg, P. H., and Wisden, W. (1993). The rat delta-1 and delta-2 subunits extend the excitatory amino acid receptor family. *FEBS Lett.* 315, 318–322.

Martin, L. J., Blackstone, C. D., Huganir, R. L., and Price, D. L. (1993). The striatal mosaic in primates: Striosomes and matrix are differentially enriched in ionotropic glutamate receptor subunits. *J. Neurosci.* 13, 782–792.

Mayer, M. L. and Westbrook, G. L. (1987). The physiology of excitatory amino acids in the vertebrate central nervous system. *Prog. Neurobiol.* 28, 197–276.

McDonald, J. W. and Johnston, M. V. (1990). Physiological and pathophysiological roles of excitatory amino acids during central nervous system development. *Brain. Res. Rev.* 15, 41–70.

McNamara, J. O., Eubanks, J. H., McPherson, J. D., Wasmuth, J. J., Evans, G. A., and Heinemann, S. F. (1992). Chromosomal localization of human glutamate receptor genes. *J. Neurosci.* 12, 2555–2562.

Meguro, H., Mori, H., Araki, K., Kushiya, E., Kutsuwada, T., Yamazaki, M., Kumanishi, T., Arakawa, M., Sakimura, K., and Mishina, M. (1992). Functional characterization of a heteromeric NMDA receptor channel expressed from cloned cDNAs. *Nature* 357, 70–74.

Meldrum, B. and Garthwaite, J. (1990). Excitatory amino acid neurotoxicity and neurodegenerative disease. *Trends Pharmacol. Sci.* 11, 379–387.

Molnar, E., Baude, A., Richmond, S. A., Patel, P. B., Somogyi, P., and McIlhinney, R. A. (1993). Biochemical and immunocytochemical characterization of antipeptide antibodies to a cloned GluR1 glutamate receptor subunit: Cellular and subcellular distribution in the rat forebrain. *Neuroscience* 53, 307–326.

Monaghan, D. T. and Cotman, C. W. (1982). The distribution of [^{3}H]kainic acid binding sites in rat CNS as determined by autoradiography. *Brain Res.* 252, 91–100.

Monaghan, D. T., Bridges, R. J., and Cotman, C. W. (1989). The excitatory amino acid receptors: Their classes, pharmacology, and distinct properties in the function of the central nervous system. *Annu. Rev. Pharmacol. Toxicol.* 29, 365–402.

Monyer, H., Seeburg, P. H., and Wisden, W. (1991). Glutamate-operated channels: Developmentally early and mature forms arise by alternative splicing. *Neuron* 6, 799–810.

Monyer, H., Sprengel, R., Schoepfer, R., Herb, A., Higuchi, M., Lomeli, H., Burnashev, N., Sakmann, B., and Seeburg, P. H. (1992). Heteromeric NMDA receptors: Molecular and functional distinction of subtypes. *Science* 256, 1217–1221.

Monyer, H., Burnashev, N., Laurie, D. J., Sakmann, B., and Seeburg, P. H. (1994). Developmental and regional expression in the rat brain and functional properties of new NMDA receptors. *Neuron* 12, 529–540.

Mori, H., Masaki, H., Yamakura, T., and Mishina, M. (1992). Identification by mutagenesis of a Mg($^{2+}$)-block site of the NMDA receptor channel. *Nature* 358, 673–675.

Morita, T., Sakimura, K., Kushiya, E., Yamazaki, M., Meguro, H., Araki, K., Abe, T., Mori, K. J., and Mishina, M. (1992). Cloning and functional expression of a cDNA encoding the mouse beta 2 subunit of the kainate-selective glutamate receptor channel. *Mol. Brain Res.* 14, 143–146.

Moriyoshi, K., Masu, M., Ishii, T., Shigemoto, R., Mizuno, N., and Nakanishi, S. (1991). Molecular cloning and characterization of the rat NMDA receptor. *Nature* 354, 31–37.

Nakanishi, N., Axel, R., and Shneider, N. A. (1992). Alternative splicing generates functionally distinct N-methyl-D-aspartate receptors. *Proc. Natl. Acad. Sci. U.S.A.* 89, 8552–8556.

Nakanishi, N., Shneider, N. A., and Axel, R. (1990). A family of glutamate receptor genes: Evidence for the formation of heteromultimeric receptors with distinct channel properties. *Neuron* 5, 569–581.

Nakanishi, S. (1992). Molecular diversity of glutamate receptors and implications for brain function. *Science* 258, 597–603.

O'Hara, P. J., Sheppard, P. O., Thogersen, H., Venezia, D., Haldeman, B. A., McGrane, V., Houamed, K. M., Thomsen, C., Gilbert, T. L., and Mulvihill, E. R. (1993). The ligand-binding domain in metabotropic glutamate receptors is related to bacterial periplasmic binding proteins. Neuron 11, 41–52.

Olney, J. S. (1989). Excitotoxicity and N-methyl-D-aspartate receptors. *Drug Dev. Res.* 17, 299.

Ortega, A., Eshhar, N., and Teichberg, V. I. (1991). Properties of kainate receptor/channels on cultured Bergmann glia. *Neuroscience* 41, 335–349.

Partin, K. M., Patneau, D. K., Winters, C. A., Mayer, M. L., and Buonanno, A. (1993). Selective modulation of desensitization at AMPA versus kainate receptors by cyclothiazide and concanavalin A. *Neuron* 11, 1069–1982.

Petralia, R. S., Yokotani, N., and Wenthold, R. J. (1994). Light and electron microscope distribution of the NMDA receptor subunit NR1 in the rat nervous system using a selective anti-peptide antibody. *J. Neurosci.* 14, 667–696.

Planells, C. R., Sun, W., Ferrer, M. A., and Montal, M. (1993). Molecular cloning, functional expression, and pharmacological characterization of an N-methyl-D-aspartate receptor subunit from human brain. *Proc. Natl. Acad. Sci. U.S.A.* 90, 5057–5061.

Potier, M. C., Spillantini, M. G., and Carter, N. P. (1992). The human glutamate receptor cDNA GluR1: Cloning, sequencing, expression and localization to chromosome 5. *DNA Seq.* 2, 211–218.

Potier, M. C., Dutriaux, A., Lambolez, B., Bochet, P., and Rossier, J. (1993). Assignment of the human glutamate receptor gene GLUR5 to 21q22 by screening a chromosome 21 YAC library. *Genomics* 15, 696–697.

Puckett, C., Gomez, C. M., Korenberg, J. R., Tung, H., Meier, T. J., Chen, X. N., and Hood, L. (1991). Molecular cloning and chromosomal localization of one of the human glutamate receptor genes. *Proc. Natl. Acad. Sci. U.S.A.* 88, 7557–7561.

Raymond, L. A., Blackstone, C. D., and Huganir, R. L. (1993). Phosphorylation and modulation of recombinant GluR6 glutamate receptors by cAMP-dependent protein kinase. *Nature* 361, 637–641.

Robinson, J. H. and Deadwyler, S. A. (1981). Kainic acid produces depolarization of CA3 pyramidal cells in the in vitro hippocampal slice. *Brain Res.* 221, 117–127.

Sakimura, K., Bujo, H., Kushiya, E., Araki, K., Yamazaki, M., Yamazaki, M., Meguro, H., Warashina, A., Numa, S., and Mishina, M. (1990). Functional expression from cloned cDNAs of glutamate receptor species responsive to kainate and quisqualate. *FEBS Lett.* 272, 73–80.

Sakimura, K., Morita, T., Kushiya, E., and Mishina, M. (1992). Primary structure and expression of the gamma 2 subunit of the glutamate receptor channel selective for kainate. *Neuron* 8, 267–274.

Sakurada, K., Masu, M., and Nakanishi, S. (1993). Alteration of Ca^{2+} permeability and sensitivity to Mg^{2+} and channel blockers by a single amino acid substitution in the N-methyl-D-aspartate receptor. *J. Biol. Chem.* 268, 410–415.

Schuster, C. M., Ultsch, A., Schloss, P., Cox, J. A., Schmitt, B., and Betz, H. (1991). Molecular cloning of an invertebrate glutamate receptor subunit expressed in Drosophila muscle. *Science* 254, 112–114.

Seeburg, P. H. (1993). The TINS/TIPS Lecture — The molecular biology of mammalian glutamate receptor channels. *Trends Neurosci.* 16, 359–365.

Sheng, M., Cummings, J., Roldan, L. A., Jan, Y. M., and Jan, L. Y. (1994). Changing subunit composition of heteromeric NMDA receptors during development of rat cortex. *Nature* 368, 144–147.

Smirnova, T., Stinnakre, J., and Mallet, J. (1993). Characterization of a presynaptic glutamate receptor. *Science* 262, 430–433.

Sommer, B., Keinanen, K., Verdoorn, T. A., Wisden, W., Burnashev, N., Herb, A., Köhler, M., Takagi, T., Sakmann, B., and Seeburg, P. H. (1990). Flip and flop: A cell-specific functional switch in glutamate-operated channels of the CNS. *Science* 249, 1580–1585.

Sommer, B., Köhler, M., Sprengel, R., and Seeburg, P. H. (1991). RNA editing in brain controls a determinant of ion flow in glutamate-gated channels. *Cell* 67, 11–19.

Sommer, B., Burnashev, N., Verdoorn, T. A., Keinänen, K., Sakmann, B., and Seeburg, P. H. (1992). A glutamate receptor channel with high affinity for domoate and kainate. *EMBO J.* 11, 1651–1656.

Sprengel, R. and Seeburg, P. H. (1993). The unique properties of glutamate receptor channels. *FEBS Lett.* 325, 90–94.

Stern, P., Behe, P., Schoepfer, R., and Colquhoun, D. (1992). Single-channel conductances of NMDA receptors expressed from cloned cDNAs: Comparison with native receptors. *Proc. R. Soc. London Biol.* 250, 271–277.

Sugihara, H., Moriyoshi, K., Ishii, T., Masu, M., and Nakanishi, S. (1992). Structures and properties of seven isoforms of the NMDA receptor generated by alternative splicing. *Biochem. Biophys. Res. Commun.* 185, 826–832.

Sulston, J., Du, Z., Thomas, K., Wilson, R., Hillier, L., Staden, R., Halloran, N., Green, P., Thierry, M. J., Qiu, L., and et al. (1992). The C. elegans genome sequencing project: A beginning. *Nature* 356, 37–41.

Sun, W., Ferre-Montiel, A. V., Schinder, A. F., McPherson, J. P., Evans, G. A., and Montal, M. (1992). Molecular cloning, chromosomal mapping, and functional expression of human brain glutamate receptors. *Proc. Natl. Acad. Sci. U.S.A.* 89, 1443–1447.

Takano, H., Onodera, O., Tanaka, H., Mori, H., Sakimura, K., Hori, T., Kobayashi, H., Mishina, M., and Tsuji, S. (1993). Chromosomal localization of the epsilon 1, epsilon 3, and zeta 1 subunit genes of the human NMDA receptor channel. *Biochem. Biophys. Res. Commun.* 197, 922–926.

Tingley, W. G., Roche, K. W., Thompson, A. K., and Huganir, R. L. (1993). Regulation of NMDA receptor phosphorylation by alternative splicing of the C-terminal domain. *Nature* 364, 70–73.

Tölle, T. R., Berthele, A., Zieglgansberger, W., Seeburg, P. H., and Wisden, W. (1993). The differential expression of 16 NMDA and non-NMDA receptor subunits in the rat spinal cord and in periaqueductal gray. *J. Neurosci.* 13, 5009–5028.

Uchino, S., Sakimura, K., Nagahari, K., and Mishina, M. (1992). Mutations in a putative agonist binding region of the AMPA-selective glutamate receptor channel. *FEBS Lett.* 308, 253–257.

Ultsch, A., Schuster, C. M., Laube, B., Schloss, P., Schmitt, B., and Betz, H. (1992). Glutamate receptors of Drosophila melanogaster: Cloning of a kainate-selective subunit expressed in the central nervous system. *Proc. Natl. Acad. Sci. U.S.A.* 89, 10484–10488.

Ultsch, A., Schuster, C. M., Laube, B., Betz, H., and Schmitt, B. (1993). Glutamate receptors of Drosophila melanogaster. Primary structure of a putative NMDA receptor protein expressed in the head of the adult fly. *FEBS Lett.* 324, 171–177.

Unwin, N. (1993). Nicotinic acetylcholine receptor at 9 A resolution. *J. Mol. Biol.* 229, 1101–1124.

Verdoorn, T. A., Burnashev, N., Monyer, H., Seeburg, P. H., and Sakmann, B. (1991). Structural determinants of ion flow through recombinant glutamate receptor channels. *Science* 252, 1715–1718.

Wada, K., Dechesne, C. J., Shimasaki, S., King, R. G., Kusano, K., Buonanno, A., Hampson, D. R., Banner, C., Wenthold, R. J., and Nakatani, Y. (1989). Sequence and expression of a frog brain complementary DNA encoding a kainate-binding protein. *Nature* 342, 684–689.

Wafford, K. A., Bain, C. J., LeBourdelles, B., Whiting, P. J., and Kemp, J. A. (1993). Preferential co-assembly of recombinant NMDA receptors composed of three different subunits. *NeuroReport* 4, 1347–1349.

Wang, L. Y., Taverna, F. A., Huang, X. P., MacDonald, J. F., and Hampson, D. R. (1993). Phosphorylation and modulation of a kainate receptor (GluR6) by cAMP-dependent protein kinase. *Science* 259, 1173–1175.

Watanabe, M., Inoue, Y., Sakimura, K., and Mishina, M. (1993). Distinct distributions of five N-methyl-D-aspartate receptor channel subunit messenger RNAs in the forebrain. *J. Comp. Neurol.* 338, 377–390.

Watkins, J. C., Krogsgaard, L. P., and Honore, T. (1990). Structure-activity relationships in the development of excitatory amino acid receptor agonists and competitive antagonists. *Trends Pharmacol. Sci.* 11, 25–33.

Wenthold, R. J., Hampson, D. R., Wada, K., Hunter, C., Oberdorfer, M. D., and Dechesne, C. (1990). Isolation, localization, and cloning of a kainic acid binding protein from frog brain. *J. Histochem. Cytochem.* 38, 1717–1723.

Werner, P., Voigt, M., Keinänen, K., Wisden, W., and Seeburg, P. H. (1991). Cloning of a putative high-affinity kainate receptor expressed predominantly in hippocampal CA3 cells. *Nature* 351, 742–744.

Wilson, R., Ainscough, R., Anderson, K., Baynes, C., Berks, M. et al. (1994). 2.2 Mb of contiguous nucleotide sequence from chromosome III of *C. elegans*. *Nature* 368, 32–38.

Wisden, W. and Seeburg, P. H. (1993). A complex mosaic of high-affinity kainate receptors in rat brain. *J. Neurosci.* 13, 3582–3598.

Yamazaki, M., Araki, K., Shibata, A., and Mishina, M. (1992a). Molecular cloning of a cDNA encoding a novel member of the mouse glutamate receptor channel family. *Biochem. Biophys. Res. Commun.* 183, 886–892.

Yamazaki, M., Mori, H., Araki, K., Mori, K. J., and Mishina, M. (1992b). Cloning, expression and modulation of a mouse NMDA receptor subunit. *FEBS Lett.* 300, 39–45.

7

GABA$_A$ Receptors

Rachel F. Tyndale, Richard W. Olsen, and Allan J. Tobin

2.7.0 Introduction

The predominant inhibitory neurotransmitter of the adult central nervous system (CNS), γ-aminobutyric acid (GABA), acts principally via the GABA$_A$ receptor (Olsen and Tobin, 1990). This receptor is a member of a superfamily of ligand-gated ion channels that also includes the nicotinic acetylcholine, glycine, and 5-HT$_3$ receptors (Barnard et al., 1987; Deneris et al., 1991; Sargent, 1993). Members of this superfamily are heterooligomeric glycoprotein receptors [(probably pentameric (Barnard et al., 1987; Anand et al., 1991) (Figure 1)] that span the membrane to form an ion channel (Vernallis et al., 1993; Macdonald and Twyman, 1991). Superfamily members share significant sequence identity and similarity, particularly in the membrane-spanning domains, suggesting that they are homologous (i.e., derived from a common ancestor by gene duplication and divergent evolution).

GABA$_A$ receptors are thought to be the major sites of action of a large number of clinically useful neuroactive drugs (Figure 1). (Olsen et al., 1991b; Ticku, 1991). Aberrations in GABA$_A$ reception may underlie many human neurological and psychiatric disorders such as epilepsy (Houser, 1991; Snodgrass, 1992; DeLorey and Olsen, 1992). Many pharmacological agents act on the GABA$_A$ receptor, including benzodiazepines, barbiturates, and other general anesthetics, some neurosteroids, convulsants such as picrotoxin, bicuculline, penicillin G, and β-carbolines, GABA analogues such as muscimol and possibly ethanol (Barnard et al., 1987; Ticku, 1991; Burt and Kamatchi, 1991). GABA$_A$ receptor ligands include an abundance of pharmacological efficacies from full to partial agonists, antagonists, as well as partial to full inverse agonists (for the benzodiazepine binding site) (Macdonald and Twyman, 1991; DeLorey and Olsen, 1992; Burt and Kamatchi, 1991; Olsen, 1991).

GABA binds to GABA$_A$ receptors allowing, a passive flow of negatively charged chloride ions down their electrochemical gradient into neurons, thus

0-8493-8322-6/95/$0.00+$.50

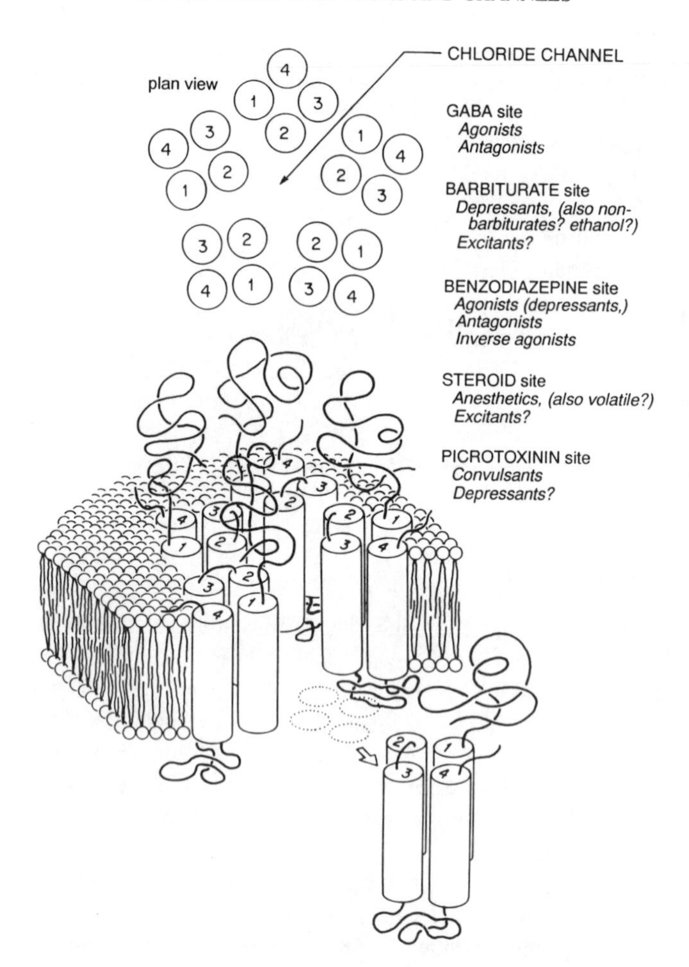

FIGURE 1. A model of the structure of the $GABA_A$ receptor chloride channel complex. The ligand-gated ion channel is thought to be pentameric, composed of members from the five subunit subfamilies (α, β, γ, δ, ρ). The second transmembrane domain is thought to be part of the channel core. The postulated binding sites are also indicated.

changing the membrane potential (Macdonald et al., 1991; Macdonald and Olsen, 1994). Agents that enhance the current may act by increasing (1) the channel conductance, (2) the frequency of channel openings (benzodiazepines), and/or (3) channel open times (barbiturates). Agents that reduce the current may act by decreasing the channel conductance, the channel open and burst frequencies (bicuculline, β-carbolines), and/or the channel open and burst durations (bicuculline, picrotoxin) (Macdonald and Twyman, 1991; Cherubini et al., 1991).

While activation of the adult CNS $GABA_A$ receptor almost always causes membrane hyperpolarization, in neonatal neurons, activation of the $GABA_A$ receptor leads to membrane depolarization. This effect is probably due to these cells having a different chloride gradient from mature neurons (for review see Cherubini et al., 1991).

Prior to the introduction of molecular biological techniques, the taxonomy of GABA$_A$ receptors was based chiefly on their pharmacological properties (Burt and Kamatchi, 1991). As knowledge of the receptor structure increased and as multiple GABA$_A$ receptor subunits were isolated and sequenced, it became clear that classifications based solely on pharmacological properties were inadequate.

Recent molecular biological studies have provided unequivocal evidence for the existence of multiple GABA$_A$ receptors distinct from the GABA$_B$ G protein-coupled receptor. A substantial number of GABA$_A$ receptor subunits [(13 in the rat: α1 (Lolait et al., 1989; Khrestchatisky et al., 1989), α2 (Khrestchatisky et al., 1991), α3 (Malherbe et al., 1990), α4 (Wisden et al., 1991), α5 (originally described as α4 by Khrestchatisky et al., 1989; Malherbe et al., 1990; Pritchett and Seeburg, 1990), α6 (Lueddens et al., 1990), β1 (Ymer et al., 1989), β2 (Ymer et al., 1989), β3 (Lolait et al., 1989; Ymer et al., 1989a), γ1 (Ymer et al., 1990), γ2 (Shivers et al., 1989), γ3 (Knoflach et al., 1991; Herb et al., 1992), δ (Shivers et al., 1989; Zhao et al., 1990); seven in the mouse: α1 (Keir et al., 1991), α2 (Wang et al., 1992), α3 (Wang et al., 1992), α6 (Kato, 1990), γ2 (Kofuji et al., 1991), γ3 (Wilson-Shaw et al., 1991) δ (Sommer et al., 1990); eight in the human: α1 (Schofield et al., 1989), α5 (Knoll et al., 1993), β1 (Schofield et al., 1989), β3 (Wagstaff et al., 1991a), γ1 (Ymer et al., 1990), γ2 (Pritchett et al., 1989), ρ1 (Cutting et al., 1991), ρ2 (Cutting et al., 1991); and 10 in the cow: α1 (Schofield et al., 1987), α2 (Levitan et al., 1988), α3 (Levitan et al., 1988), α4 (Ymer et al., 1989b), α6 (Lueddens et al., 1990), β1 (Schofield et al., 1987), β2 (Ymer et al., 1989), β3 (Ymer et al., 1989), γ1 (Ymer et al., 1990), γ2 (Whiting et al., 1990)] have been cataloged, and publications identifying additional subunits continue to emerge. In addition to mammals, GABA$_A$ receptor subunit mRNAs have been identified in species as diverse as chickens [(α1 (Bateson et al., 1990a), β3 (Bateson et al., 1990b), β4 (Bateson et al., 1991), γ2 (Glencourse et al., 1990)], the snail *Lymnaea stagnalis* (Harvey et al., 1991), and the fruit fly *Drosophila melanogaster (Rdl)* (French-Constant et al., 1991). Other lines of evidence (Harvey et al., 1991; French-Constant et al., 1991; Deng et al., 1991) also suggest that GABA$_A$ receptors exist in a multitude of vertebrates and invertebrates (Lunt, 1991; Darlison, 1992; Hebebrand et al., 1987; Trembley et al., 1988). This would suggest a major role for the GABA$_A$ receptor throughout evolutionary development.

2.7.1 GABA$_A$ Receptor Subunit Polypeptide Structures

Analysis of the amino acid sequences of all known GABA$_A$ receptor subunits suggests that each contains (1) a large extracellular N-terminus containing a cysteine bridge, (2) four transmembrane domains (M1–M4), (3) a variable (length and composition) intracellular loop (between M3 and M4), and (4) a short extracellular C-terminus. The extracellular N-terminus region contains consensus sites for N-linked glycosylation while the intracellular loop of some subunits contains the consensus sequence for phosphorylation by protein kinase A and C and tyrosine protein kinase (for review see Leidenheimer et al., 1991). All members of the $GABA_A$ receptor family (as well as the other members of the superfamily) contain an invariant proline residue in M1, while M2 is composed of several hydrophilic residues that may form the ion channel (Figure 2) (Barnard et al., 1987; Salpeter and Loring, 1985).

Differences in amino acid sequences are not evenly distributed throughout the polypeptides. This is depicted in Figures 2 and 3, which demonstrate the conserved and variant amino acid residues among the published rat $GABA_A$ receptor polypeptides. Presumed transmembrane domains and their immediate flanking regions are highly conserved. The least conserved predicted amino acid sequences of the subunits are believed to be concentrated in the cytoplasmic loop between the membrane-spanning domains M3 and M4 as well as in the extracellular N-terminal domain (Figure 2) (Persohn et al., 1992). Variation in the N-terminal extracellular region may allow for diversity of ligand binding while variation in the intracellular loop may allow for separate regulation via second messenger systems and phosphorylation (Leidenheimer et al., 1991; Moss et al., 1992a).

2.7.2 Multiple $GABA_A$ Receptor Subunits

Despite information available from the vast array of $GABA_A$ receptor ligands, the multiplicity of the individual $GABA_A$ receptor subunits and their potential myriad of combinations has exceeded most estimates based on pharmacological properties. For example, assuming (1) a pentameric structure for the receptor and (2) that each subunit is equivalent and (3) uniquely placed, one can estimate that over 3000 potential subunit combinations could exist. If, however, one assumes that two αs, two βs, and a γ or δ are utilized, then approximately 300 combinations are possible. From these estimates it is clear that vast numbers of $GABA_A$ receptor subunit combinations are potentially available.

$GABA_A$ receptors consist of at least 15 different receptor subunits, which can be divided into five subfamilies (α, β, γ, δ, and ρ) based on sequence similarity. Within and between subfamily similarities can be demonstrated for the rat sequences by comparing nucleotide identity or amino acid identity and similarity (Tables 1 and 2). Each member of a subfamily shares approximately 60–70% nucleotide sequence identity. At the amino acid level, each subfamily shares more than 70% amino acid similarity (Table 2).

The β subfamily appears to contain the highest degree of similarity, with greater that 88% amino acid identity between members and less than 63% with members of the other subfamilies. The α subfamily demonstrates the most diversity containing members with as little as 71% amino acid similarity while also having as much as 68% amino acid similarity to members of another subunit subfamily (Tables 1 and 2).

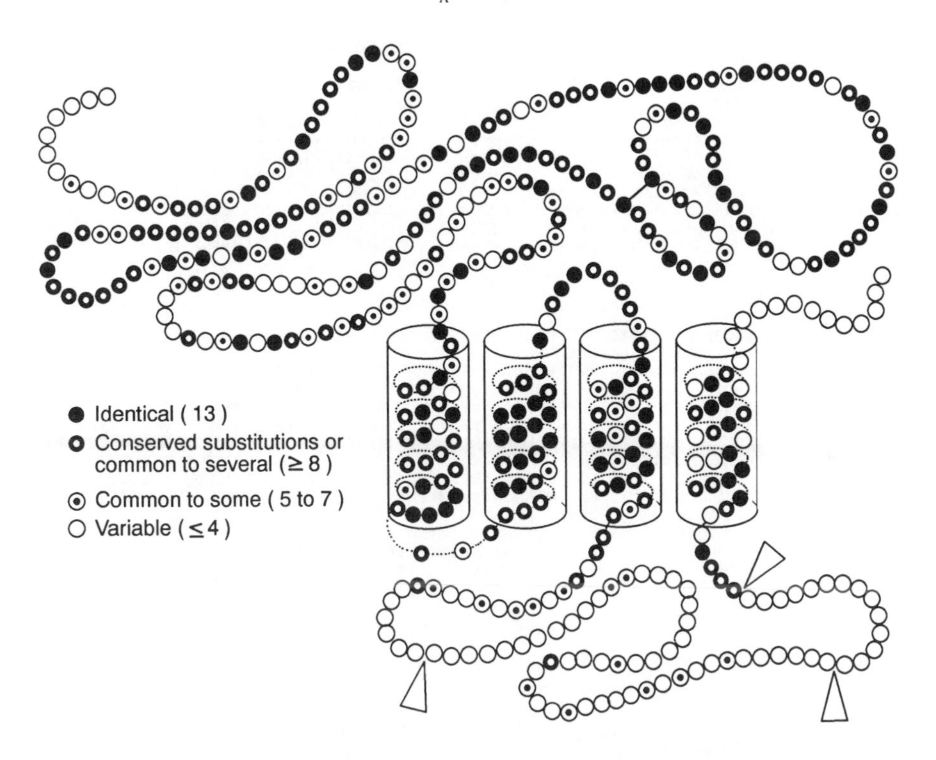

FIGURE 2. A generic GABA$_A$ receptor and proposed topographic structure. This figure is based on the amino acid (aa) alignment found in Figure 3 and starts with the fourth aa of the mature α1 sequence (Q). The NH_2-terminal half of the sequence is presumed to be extracellular and contains sites of potential glycosylation, which are not conserved between subunits, and a conserved cysteine bridge. The four cylinders represent putative α-helical membrane-spanning regions (M). The extremely variable intracellular loop found between M3 and M4 conforms to the α1-subunit with respect to size, and the composition is derived from the positions where α1 residues are found in the alignment (Figure 3, triangles indicate the location of large inserts in non-α1-subunits). The intracellular loop contains putative phosphorylation sites on many subunits. The extracellular loop following M4 is also variable in length and composition.

2.7.2.1 The GABA$_A$ Receptor α-Subunits

The first α GABA$_A$ receptor subunit cDNA was cloned in 1987 from the cow (Schofield et al., 1987). α-Subunits range in length from 428 to 521 amino acids, have mRNAs between 2.8 and 4.2 kb, and contain two to four N-linked glycosylation sites (Burt and Kamatchi, 1991).

In binding assays the type of α-subunit affects both the sensitivity of the receptor to GABA and the affinity of different ligands for the benzodiazepine binding site (Pritchett and Seeburg, 1990, 1991; Weiland et al., 1992), but not for the barbiturate or picrotoxin binding site (DeLorey and Olsen, 1992; Tobin

```
α1  Q----------------------PS----QDELKDNTTVFTRILDRLLDGYDNRLRPGLGERVTEVKTDIFVTSFGPVS
α2  N----------------------IQ----EDEAKNNITIFTRILDRLLDGYDNRLRPGLGDSITEVFTNIYVTSFGPVS
α3  QGESRRQEP-GDFVKQDIGGLSPKHAPDIPDDSTDNITIFTRILDRLLDGYDNRLRPGLGDAVTEVKTDIYVTSFGPVS
α4  QNS----KD-EKLCPEN----------------------FTRILDSLLDGYDNRLRPGFGGPVTEVKTDIYVTSFGPVS
α5  Q----------------------MPTSSVQDETNDNITIFTRILDGLLDGYDNRLRPGLGERITQVRTDIYVTSFGPVS
α6  QLE----DE-GNFYSEN----------------------VSRILDNLLEGYDNRLRPGFGGAVTEVKTDIYVTSFGPVS
β1  HSS----NEPSNMSYVKE------------T------------VDRLLKGYDIRLRPDFGGPPVDVGMRIDVASIDMVS
β2  QSV----NDPSNMSLVKE------------T------------VDRLLKGYDIRLRPDFGGPPVAVGMNIDIASIDMVS
β3  QSV----NDPGNMSFVKE------------T------------VDKLLKGYDIRLRPDFGGPPVCVGMNIDIASIDMVS
γ1  IDKA-DDEDDEDLTMNKTWVLAP----------KIHEGDITQILNSLLQGYDNKLRPDIGVRPTVIETDVYVNSIGPVD
γ2  --QK-SDDDYEDYASNKTWVLTP----------KVPEGDVTVILNNLLEGYDNKLRPDIGVKPTLIHTDMYVNSIGPVN
γ3  RSRRVEEDDSEDSPSNQKWVLAP----------KSQDTDVTLIILNKLLREYDKKLRPDIGIKPTVIDVDIYVNSIGPVS
δ   QPH----HGARAMNDIGDYVGSNLEISWLPN------------LDGLMEGYARNFRPGIGGPPVNVALALEVASIDHIS
                                               .. *. .*. ..**..*  . .   . ..*.. ..

α1  DHDMEYTIDVFFRQSWKDERLKFKGPMTVLRLNNLMASKIWTPDTFFHNGKKSVAHNMTMPNKLLRITEDGTLLYTMRL
α2  DTDMEYTIDVFFRQKWKDERLKFKGPMNILRLNNSMASKIWTPDTFFHNGKKSVAHNMTMPNKLLRIQDDGTLLYTMRL
α3  DTDMEYTIDVFFRQTWHDERLKFDGPMKILPLNNLLASKIWTPDTFFHNGKKSVAHNMTTPNKLLRLVDNGTLLYTMRL
α4  DVEMEYTMDVFFRQTWIDKRLKYDGPIEILRLNNMMVTKVWTPDTFFRNGKKSVSHNMTAPNKLFRIMRNGTILYTMRL
α5  DTEMEYTIDVFFRQSWKDERLRFKGPMQRLPLNNLLASKIWTPDTFFHNGKKSIAHNMTTPNKLLRLEDDGTLLYTMRL
α6  DVEMEYTMDVFFRQTWTDERLKFKGPAEILSLNNLMVSKIWTPDTFFRNGKKSIAHNMTTPNKLFRLMHNGTILYTMRL
β1  EVNMDYTLTMYFQQSWKDKRLSYSGIPLNLTLDNRVADQLWVPDTYFLNDKKSFVHGVTVKNRMIRLHPDGTVLYGLRI
β2  EVNMDYTLTMYFQQAWRDKRLSYNVIPLNLTLDNRVADQLWVPDTYFLNDKKSFVHGVTVKNRMIRLHPDGTVLYGLRI
β3  EVNMDYTLTMYFQQYWRDKRLAYSGIPLNLTLDNRVADQLWVPDTYFLNDKKSFVHGVTVKNRMIRLHPDGTVLYGLRI
γ1  PINMEYTIDIIFAQTWFDSRLKFNSTMKVLMLNSNMVGKIWIPDTFFRNSRKSDAHWITTPNRLLRIWSDGRVLYTLRL
γ2  AINMEYTIDIFFAQTWYDRRLKFNSTIKVLRLNSNMVGKIWIPDTFFRNSKKADAHWITTPNRMLRIWNDGRVLYTLRL
γ3  SINMEYQIDIFFAQTWTDSRLRFNSTMKILTLNSNMVGLIWIPDTIFRNSKTAEAHWITTPNQLLRIWNDGKILYTLRL
δ   EANMEYTMTVFLHQSWRDSRLSYNHTNETLGLDSRFVDKLWLPDTFIVNAKSAWFHDVTVENKLIRLQPDGVILYSIRI
      .*.* ... . * * * ** ..     * *..  .. .* *** . *....  * .*  *...*.  .* .**..*.
```

FIGURE 3. Alignments of the deduced amino acid sequences of the 13 rat $GABA_A$ receptor subunits. Asterisks below the alignment represent a perfect match among all 13 subunits; dots signify a conservative replacement. The alignment was performed using the Clustal Multiple Alignment Program described by Higgins and Sharp (1989), gap fixed = 10 and gap vary = 10. Putative transmembrane domains are indicated by double lines above the alignment.

```
α1  TVRAECPMHLEDFPMDAHACPLKFGSYAYTRAEVVYEWTREPARSVVVAEDGS-RLNQYDLLGQTVDSGIVQ-SSTGEYVVM
α2  TVQAECPMHLEDFPMDAHSCPLKFGSYAYTTSEVTYIWTYNPSDSVQVAPDGS-RLNQYDLLGQSIGKETIK-SSTGEYTVM
α3  TIHAECPMHLEDFPMDVHACPLKFGSYAYTKAEVIYSWTLGKNKSVEVAQDGS-RLNQYDLLGHVVGTEIIR-SSTGEYVVM
α4  TISAECPMRLVDFPMDGHACPLKFGSYAYPKSEMIYTWTKGPEKSVEVPKESS-SLVQYDLIGQTVSSETIK-SITGEYIVM
α5  TISAECPMQLEDFPMDAHACPLKFGSYAYPNSEVVYVWTNGSTKSVVVAEDGS-RLNQYHLMGQTVGTENIS-TSTGEYTIM
α6  TINADCPMRLVNFPMDGHACPLKFGSYAYPKSEIIYTWKKGPLYSVEVPEESS-SLLQYDLIGQTVSSETIK-SNTGEYVIM
β1  TTTAACMMDLRRYPLDEQNCTLEIESYGYTTDDIEFYWNGGEGAVTGVNKI---ELPQFSIVDYKMVSKKVEF-TTGAYPRL
β2  TTTAACMMDLRRYPLDEQNCTLEIESYGYTTDDIEFYWRGDDNAVTGVTKI---ELPQFSIVDYKLITKKVVF-STGSYPRL
β3  TTTAACMMDLRRYPLDEQNCTLEIESYGYTTDDIEFYWRGGDKAVTGVERI---ELPQFSIVEHRLVSRNVVF-ATGAYPRL
γ1  TINAECYLQLHNFPMDEHSCPLEFSSYGYPKNEIEYKWKKP---SVEVADPKYWRLYQFAFVGLRNSTEISH-TISGDYIIM
γ2  TIDAECQLQLHNFPMDEHSCPLEFSSYGYPREEIVYQWKRS---SVEVGDTRSWRLYQFSFVGLRNTTEVVK-TTSGDYVVM
γ3  TINAECQLQLHNFPMDAHACPLTFSSYGYPKEEMIYRWRKN---SVEAADQKSWRLYQFDFMGLRNTTEIVT-TSAGDYVVM
δ   TSTVACDMDLAKYPMDEQECMLDLESYGYSSEDIVYYWSENQEQIHGLDRL---QLAQFTITSYRFTTELMNFKSAGQFPRL
    *  ..* . *  .*.* ..* *...**.*. ... . *                * *. . .    .      .* .  .

                 =======================   =====================           ==========
α1  TTHFHLKRKIGYFVIQTYLPCIMTVILSQVSFWLNRESVPARTVFGVTTVLTMTTLSISARNSLPKVAYATAMDWFIAVCYA
α2  TAHFHLKRKIGYFVIQTYLPCIMTVILSQVSFWLNRESVPARTVFGVTTVLTMTTLSISARNSLPKVAYATAMDWFIAVCYA
α3  TTHFHLKRKIGYFVIQTYLPCIMTVILSQVSFWLNRESVPARTVFGVTTVLTMTTLSISARNSLPKVAYATAMDWFMAVCYA
α4  TVYFHLRRKMGYFMIQTYIPCIMTVILSQVSFWINKESVPARTVFGITTVLTMTTLSISARHSLPKVSYATAMDWFIAVCFA
α5  TAHFHLKRKIGYFVIQTYLPCIMTVILSQVSFWLNRESVPARTVFGVTTVLTMTTLSISARNSLPKVAYATAMDWFIAVCYA
α6  TVYFHLQRKMGYFMIQIYTPCIMTVILSQVSFWINKESVPARTVFGITTVLTMTTLSISARHSLPKVSYATAMDWFIAVCFA
β1  SLSFRLKRNIGYFILQTYMPSTLITILSWVSFWINYDASAARVALGITTVLTMTTISTHLRETLPKIPYVKAIDIYLMGCFV
β2  SLSFKLKRNIGYFILQTYMPSILITILSWVSFWINYDASAARVALGITTVLTMTTINTHLRETLPKIPYVKAIDMYLMGCFV
β3  SLSFRLKRNIGYFILQTYMPSIMITILSWVSFWINYDASAARVALGITTVLTMTTINTHLRETLPKIPYVKAIDMYLMGCFV
γ1  TIFFDLSRRMGYFTIQTYIPCILTVVLSWVSFWINKDAVPARTSLGITTVLTMTTLSTIARKSLPKVSYVTAMDLFVSVCFI
γ2  SVYFDLSRRMGYFTIQTYIPCTLIVVLSWVSFWINKDAVPARTSLGITTVLTMTTLSTIARKSLPKVSSVTAMDLFVSVCFI
γ3  TIYFELSRRMGYFTIQTYIPCILTVVLSWVSFWIKKDATPARTTLGITTVLTMTTLSTIARKSLPRVSYVTAMDLFVTVCFL
δ   SLHFQLRRNRGVYIIQSYMPSVLLVAMSWVSFWISQAAVPARVSLGITTVLTMTTLMVSARSSLPRASAIKALDVYFWICYV
    .  * * *. * . .* * *... . .* ****.. .. .**. .*.********. .  * .**. .  .*.* .  *.
```

FIGURE 3(2).

```
     =============
α1   FVFSALIEFATVNY-FTKRGYAWDGKSVV-PEK----------------PKKVDKPLIKKN--NTYAPTATSYTPNLARGDP
α2   FVFSALIEFATVNY-FTKRGWAWDGKSVV-NDK-----------------KKEKGSVMIQN--NAYAVAVANYAPNLSK-DP
α3   FVFSALIEFATVNY-FTKRSWAWEGKKVPEALE----------------MKKKTPAAPTKKTSTTFNIVGTTYPINLAL-DT
α4   FVFSALIEFAAVNY-FTNIQMQKAKKKISKPPPEVPAAPVLKEKHTETSLQNTHANLNMRKRTNALVHSESDVNSRTEVGNH
α5   FVFSALIEFATVNY-FTKRGWAWDGKKALEAAK----------------IKKKERELILNKSTNAFTTGKLTHPPNIPKEQ-
α6   FVFSALIEFAAVNY-FTNLQSQKAERQ----AQTAAKPPVAKSKTTE-SLE-----------AEIVVHSDSKYHLKKRISSL
β1   FVFLALLEYAFVNYIFFGKGPQ--KKGASKQDQS-------------------------ANEKNKLEMNKVQVDAHGNILLS
β2   FVFMALLEYALVNYIFFGRGPQRQKKAAEKAANA-------------------------NNEKMRLDVNKM--DPHENILLS
β3   FVFLALLEYAFVNYIFFGRGPQRQKKLAEKTAKA-------------------------KNDRSKSEINRV--DAHGNILLA
γ1   FVFAALMEYGTLHY-FTSNNKGKTTRD--------------RKLKSKTSVS-----------------PGLHAGST-LIPMN
γ2   FVFSALVEYGTLHY-FVSNRKPSKDKD--------------KK-KKNPAPT-----------------IDIRPRSA-TIQMN
γ3   FVFAALMEYATLNY-YSSCRKPTIRKK--------------KTSLLHPDSTRWIPDRISLQAPSNYSLLDMRPPPPVMITLN
δ    FVFAALVEYAFAHF----NADYRKKR----------------------------------KAKVKVTKPRAEMDVRNAIVLF
     *** **.*..  ..

α1   GLATIAKSATIEPKEVKP---------------E----------TKPPEPKKTFNSV-------------------------
α2   VLSTISKSATTPEPNKKP---------------E----------NKPAEAKKTFNSV-------------------------
α3   EFSTISKAAAAPSASSTP---------------TVIASPKTTYVQDSPAETKTYNSV-------------------------
α4   SSKTTAAQESSETTPKAHLASSPNPFSRANAAETISAAARGLSSAASPSPHGTLQPAPLRSASARPAFGARLGRIKTTVNTT
α5   -LPGGTGNAVGTASIRAS---------------E----------EKTSESKKTYNSI-------------------------
α6   TLPIVPSSEASKVLSRTPI----------------------------------LPSTPVTPPLLLPAIG-------------
β1   TLEIRNETSGSEVLTGVSDPKATMY-----SYDSASIQYRKPLSSREGFGR-GLDRHGVPGKGRIRR---RASQLKVKI---
β2   TLEIKNEMATSEAVMGLGDPRSTML-----AYDASSIQYRKAGLPRHSFGRNALERHVAQKKSRLRR---RASQLKITI---
β3   PMDVHNEM--NEVAGSVGDTRNSAI-----SFDNSGIQYRKQSMPKEGHGRYMGDRSIPHKKTHLRR---RSSQLKIKI---
γ1   NIS-MPQGEDDYGYQCLEGKDCATFFC---CFEDCRTGSW------------------------------REGRIHIRI---
γ2   NATHLTERDEEYGYECLDGKDCASFFC---CFEDCRTGAW------------------------------RHGRIHIRI---
γ3   NSMYWQEFEDTCVYECLDGKDCQSFFC---CYEECKSGSW------------------------------RRGRIHIDV---
δ    SLSAAGVSQELAISRRQGRVPGNLM----GSYRSVEVEAKKEGGSRPG------------GPGGIRS---RLKPI-------
```

FIGURE 3(3).

```
                                 =======================
α1   ----------------------SKIDRLSRIAFPLLFGIFNLVYWATYLNREPQLKAPTPHQ
α2   ----------------------SKIDRMSRIVFPVLFGTFNLVYWATYLNREPVLGVSP---
α3   ----------------------SKVDKISRIIFPVLFAIFNLVYWATYVNRESAIKGMIRKQ
α4   GVPGNVSATPPPSAPPPSGSGTSKIDKYARILFPVTFGAFNMVYWVVYLSKDT-MEKSESLM
α5   ----------------------SKIDKMSRIVFPILFGTFNLVYWATYLNREPVIKGATSPK
α6   --------------------GTSKIDQYSRILFPVAFAGFNLVYWIVYLSKDT-MEVSSTVE
β1   ----------------PDLTDVNSIDKWSRMFFPITFSLFNVVYWLYYVH------------
β2   ----------------PDLTDVNAIDRWSRIFFPVVFSFFNIVYWLYYVN------------
β3   ----------------PDLTDVNAIDRWSRIVFPFTFSLFNLVYWLYYVN------------
γ1   ----------------------AKIDSYSRIFFPTAFALFNLVYWVGYLY-----------L
γ2   ----------------------AKMDSYARIFFPTAFCLFNLVYWVSYLY-----------L
γ3   ----------------------SELDSYSRVFFPTSFLLFNLVYWVGYLY-----------L
δ    --------------------DADTIDIYARAVFPAAFAAVNIIYWAAYTM------------
                          . .*  .*  **  *   *..**  *
```

FIGURE 3(4).

et al., 1991). These benzodiazepine binding affinities are exquisitely sensitive to changes in single amino acid residues (Weiland et al., 1992; Pritchett and Seeburg, 1991; Korpi et al., 1993) or to modifications of the disulfide bond (Otero de Bengtsson et al., 1993). In addition, the α-subunits show distinctive tissue localizations (Khrestchatisky et al., 1991; Persohn et al., 1992; Tobin et al., 1991; Wisden et al., 1992; Laurie et al., 1992a; Zhang et al., 1992), developmental patterns (Tobin et al., 1991; Laurie et al., 1992b), and apparent molecular weights (DeLorey and Olsen, 1992; Olsen et al., 1991a).

2.7.2.2 The GABA$_A$ Receptor β-Subunits

The first GABA$_A$ receptor β-subunit was cloned from the cow in 1987 (Schofield et al., 1987). β-Subunits range in length from 449 to 450 amino

Table 1. Percent Nucleotide Identity of Rat $GABA_A$ Receptor Subunit Coding Sequences

	α1	α2	α3	α4	α5	α6	β1	β2	β3	γ1	γ2	γ3	δ
α1		70	67	59	66	60	49	47	50	54	55	52	47
α2			67	57	68	62	49	46	48	54	55	53	48
α3				60	68	60	47	49	47	52	55	56	48
α4					58	65	48	48	47	54	53	53	49
α5						61	48	48	48	53	55	54	48
α6							48	51	49	55	56	56	48
β1								71	71	47	51	50	52
β2									72	48	49	51	53
β3										47	48	49	53
γ1											70	66	46
γ2												68	48
γ3													50

acids, have mRNAs between 2.5 and 12 kb, and contain three or four N-linked glycosylation sites (Burt and Kamatchi, 1991).

Each β-subunit contains a consensus sequence for cAMP-dependent phosphorylation (Lolait et al., 1989; Ymer et al., 1989b; Schofield et al., 1987; Moss et al., 1992), which has been demonstrated to occur on the β3-subunit (Browning et al., 1993). Phosphorylation of $GABA_A$ receptors by protein kinase A may play a role in the regulation of receptor activity and may explain why β-subunits are expressed in most brain regions that contain $GABA_A$ receptors (Leidenheimer et al., 1991; Kirkness et al., 1991). The pattern of expression varies among β-subunits, providing each with a unique tissue distribution (Ymer et al., 1989b; Persohn et al., 1992; Laurie et al.,

Table 2. Percent Amino Acid Identity (Upper Half) and Similarity (Lower Half) of Rat $GABA_A$ Receptor Subunits

	α1	α2	α3	α4	α5	α6	β1	β2	β3	γ1	γ2	γ3	δ
α1		76	71	60	72	59	37	37	38	46	46	46	33
α2	84		72	59	71	59	37	36	36	46	45	47	33
α3	82	81		60	70	59	36	36	36	45	45	44	32
α4	73	71	74		59	72	38	37	37	49	47	46	36
α5	81	82	80	75		59	37	38	37	46	45	44	33
α6	74	71	73	83	72		37	36	37	49	48	46	37
β1	62	60	58	60	59	58		80	78	36	39	36	43
β2	62	58	59	59	60	57	90		79	37	37	36	43
β3	62	59	58	58	61	59	88	89		38	37	36	43
γ1	65	64	65	67	65	68	58	61	63		74	69	36
γ2	64	63	65	66	64	68	62	60	61	85		69	38
γ3	67	65	64	63	63	64	58	59	58	80	82		38
δ	58	56	56	58	57	59	66	67	66	57	60	59	

1992a). These features of the β-subunits may prove to be important for differential regulation and pharmacology of the GABA$_A$ receptors.

2.7.2.3 The GABA$_A$ Receptor γ-Subunits

The first γ GABA$_A$ receptor subunit cDNA was cloned in 1989 (Shivers et al., 1989; Pritchett et al., 1989). γ-Subunits range in length from 428 to 450 amino acids, have mRNAs between 2.8 and 4.2 kb, and contain two or three N-linked glycosylation sites (Burt and Kamatchi, 1991).

The type of γ-subunit appears to influence the affinities and efficacies of antagonists and inverse agonists and facilitate benzodiazepine binding (Ymer et al., 1989b, 1990; Herb et al., 1992; Pritchett et al., 1989). Pharmacological differences attributed to individual γ-subunits can be further complicated by splice variants of the individual γ-subunits (Whiting et al., 1990). For example, some studies suggests that the long splice variant of the γ2-subunit (γ2L) is required in the GABA$_A$ receptor combination for ethanol to exert an action on the channel function (Wafford et al., 1991; Wafford and Whiting, 1992).

2.7.2.4 The GABA$_A$ Receptor δ-Subunit

The δ GABA$_A$ receptor subunit was first cloned in 1989 (Shivers et al., 1989). The δ-subunit is 433 amino acids in length, has an mRNA which is 2.0 kb long, and contains two N-linked glycosylation sites (Burt and Kamatchi, 1991). To date no other members of the δ subfamily have been identified.

In situ studies of the δ- and γ2-subunits have shown that these subunits have almost no overlap in their distributions, suggesting that they rarely exist in the same oligomeric GABA$_A$ receptor (Shivers et al., 1989). There is some evidence that the δ-subunit is found in benzodiazepine-insensitive receptors (Shivers et al., 1989; Sommer et al., 1990).

2.7.2.5 The GABA$_A$ Receptor ρ-Subunits

The ρ GABA$_A$ receptor subunit was first cloned in 1991 from a human retinal cDNA library (Cutting et al., 1991). A second member of the ρ subfamily, human ρ2, was later isolated (Cutting et al., 1992). ρ-Subunits are 465 and 473 amino acids long, have mRNAs from 3.1 to 4.8 kb, and contain two or three N-linked glycosylation sites (Cutting et al., 1991, 1992).

The ρ-subunits appear to correspond to the pharmacologically defined GABA$_C$ receptors (Shimada et al., 1992), but share sequence identity to the other members of the GABA$_A$ receptor family (Cutting et al., 1992). ρ Polypeptides are, however, a unique subunit subfamily based on their restricted tissue distribution [predominantly in the retina, some in the cerebellum (Cutting et al., 1991)] and their unique pharmacology (bicuculline, barbiturate, and benzodiazepine insensitivity) (Shimada et al., 1992).

2.7.3 Heterogeneity of the $GABA_A$ Receptor

2.7.3.1 $GABA_A$ Receptor Subunit Polymorphisms

Alterations in $GABA_A$ receptors may contribute to the neurological symptoms of such disorders as epilepsy or Angelman's syndrome. Polymorphic dinucleotide repeats found in many of the $GABA_A$ receptor subunits (Mutirangura et al., 1992; Johnson et al., 1992; Hinks et al., 1911; Dean et al., 1991) may provide a novel molecular tool for establishing this linkage.

2.7.3.2 $GABA_A$ Receptor Subunit Combinations

Based on findings from studies of the nicotinic acetylcholine receptor and from $GABA_A$ receptor protein purification studies, it seems likely that the $GABA_A$ receptor chloride channel pore is formed from multiple subunits in a quasisymmetric structure (Olsen et al., 1990; Vernallis et al., 1993; DeLorey and Olsen, 1992). This assumption, combined with the multiplicity of identified $GABA_A$ receptor subunit genes, allows for a vast array of potential receptors to be created. Each possible combination might have its own individual characteristics and pharmacology, but this remains to be determined. Alternatively, several members could make equivalent contributions to molecular function, but the corresponding genes might respond differently to environmental or experiential cues.

Immunohistochemical (Houser, 1991; Somogyi et al., 1989) and *in situ* hybridization (Persohn et al., 1992; Wisden et al., 1993; Laurie et al., 1992a; Zhang et al., 1992) techniques allow the analysis of $GABA_A$ receptor subunit colocalization (Olsen et al., 1992). These results can be combined with data from other methods such as autoradiography (Olsen et al., 1992; Xia and Haddad, 1992), immunochemistry (Endo and Olsen, 1992, 1993), immunoprecipitation (Endo and Olsen, 1993; Duggan et al., 1991; McKernan et al., 1991; Mertens et al., 1993; Frohman et al., 1988), and ligand binding techniques (Olsen et al., 1990, 1991b, 1992; Bureau and Olsen, 1990, 1993) to assess the combinations that may occur naturally. The function of these candidate combinations can then be tested in various heterologous cellular systems such as *Xenopus* oocytes and transfected animal cell lines (Verdoorn et al., 1990).

The identification of the individual subunits that combine *in vivo* to form channels and their respective pharmacology are the subjects of a great deal of intensive research and speculation (Burt and Kamatchi, 1991; Olsen et al., 1992). For the present, however, the physiological roles and pharmacological implications of such an enormous array of possible receptor combinations remain unknown.

2.7.3.3 Alternative Mechanisms for $GABA_A$ Receptor Heterogeneity

Heterogeneity that arises from multiple subunit combinations can be augmented by variations at the level of the individual subunit. For example, a subunit can have a single amino acid substitution (from genetic polymorphism or RNA editing) that leads to changes in $GABA_A$ receptor function (Weiland et al., 1992; Pritchett and Seeburg, 1991; Korpi et al., 1993; Angelotti et al., 1992; Sigel et al., 1992). Further diversity may originate from alternative assembly and targeting of $GABA_A$ receptors within cells (Persohn et al., 1992; Perez-Velazquez and Angelides, 1993). Alternate splicing of subunit mRNAs has been shown in various species and $GABA_A$ receptor subfamilies (Kofuji et al., 1991; Whiting et al., 1990; Bateson et al., 1991; Wafford et al., 1991). This alternative mechanism for creating diversity has been shown, in the case of the γ2 receptor subunit, to lead to variation in ligand-mediated $GABA_A$ receptor pharmacology (Wafford et al., 1991). There is also evidence for heterogeneity of receptor function due to modulation by phosphorylation (Moss et al., 1992a,b; Tehrani et al., 1989) and glycosylation (Sieghart et al., 1993). Mechanisms found in other ligand-gated ion channels such as RNA editing may also operate in the $GABA_A$ receptor system (Sommer et al., 1991; Sommer and Seeburg, 1992), conceivably increasing the diversity of the $GABA_A$ receptor response.

2.7.4 Gene Structure of the $GABA_A$ Receptor

Three genes [murine δ (Sommer et al., 1990), human β1 (Kirkness et al., 1991), and chicken β4 (Lasham et al., 1991)] and the 5′ exons of a fourth gene [human β3 (Kirkness and Fraser, 1993)] have been isolated and their exon–intron boundaries mapped (Table 3). Each of the genes consists of nine exons and span from about 13 kb (murine δ) to about 65 kb (chicken β4 and human β1). This suggests that both the exon number and exon–intron boundaries have been conserved between subunits in different $GABA_A$ receptor subfamilies and between different species. Thus the genomic structure appears to have been established prior to the divergence of the genes that encode the $GABA_A$ receptor subunits. Intron sizes vary considerably among these subunits, however. The largest intron found in the murine δ-subunit gene is between exon 1 and 2, while the chicken β4-subunit gene has the largest intron (>22 kb) between exons 3 and 4, suggesting that selective pressures were exerted for maintaining the exon–intron boundaries and exon sizes but not for maintaining intron sizes.

Examination of the highly conserved transmembrane domains (M) reveals that M1, M3, and M4 are coded for on single exons. In contrast, M2 is derived from two adjacent coding exons (Kirkness et al., 1991). It is in this M2 domain that amino acids crucial for the GluR1–4 and the neuronal nicotinic acetylcholine channel properties lie (Sommer and Seeburg, 1992; Galzi et al., 1992). For example, GluR1, GluR3, and GluR4 all contain a Gln residue, and the channels assembled from these subunits are permeable to divalent cations. Conversely, those assembled (heteromeric or homomeric assemblies) with GluR2, which contains an Arg in the equivalent position, have greatly reduced divalent ion permeability (Sommer and Seeburg, 1992).

Table 3. GABA$_A$ Receptor Gene Structure

Species	Subunit
Mouse	δ[a]
Human	β1[b]
Human	β3 (partial)[c]
Chicken	β4[d]
Drosophila	*Rdl*[e]
Lymnaea stagnalis	β[f]

[a] Sommer et al. (1990).
[b] Kirkness et al. (1991).
[c] Kirkness and Fraser (1993).
[d] Sommer and Seeburg (1992).
[e] French-Constant and Rocheleau (1992).
[f] Harvey et al. (1991).

This suggests that the ion specificity of ligand-gated ion channels may lie in the transmembrane domain (M2) and that the creation or removal of the exon–intron boundary in M2 was important to the divergent evolution of the channel properties between the superfamily members. In addition, Figure 2 clearly illustrates that the M2 is the most conserved of the GABA$_A$ receptor transmembrane domains.

Like the vertebrate GABA$_A$ receptor genes, the *Drosophila* GABA$_A$ receptor gene *(Rdl)* is comprised of nine exons and spans more than 25 kb of genomic DNA (French-Constant and Rocheleau, 1992). The first six exon–intron boundaries for the *Drosophila* GABA$_A$ receptor gene correspond to those of the chicken β4-subunit gene (Lasham et al., 1991) and murine δ-subunit gene (Sommer et al., 1990), while the last two exon–intron boundaries do not. Together with strong similarity in predicted amino acid sequence, the matching exon–intron boundaries suggest a close relationship between *Drosophila* and vertebrate GABA$_A$ receptors.

The most 3′ exon–intron boundary of the *Drosophila* GABA$_A$ receptor gene *(Rdl)* differs from those found in the vertebrate GABA$_A$ receptor genes, but is in a position similar to that in vertebrate nicotinic acetylcholine (nACh) receptors, another member of the ligand-gated ion channel superfamily (French-Constant and Rocheleau, 1992). In addition, in the *Drosophila* GABA$_A$ receptor *(Rdl)* gene (French-Constant and Rocheleau, 1992), the M2 is unbroken by an intron, suggesting that it is more closely related to the nicotinic acetylcholine or glycine receptors. One explanation may be that the *Drosophila (Rdl)* gene may in fact represent an additional GABA$_A$ receptor class that may or may not be present in vertebrates (French-Constant and Rocheleau, 1992). Of the six intron positions that have been identified in the *Lymnaea stagnalis* GABA$_A$ receptor gene, the first five introns are located at identical or similar positions to those found in the vertebrate GABA$_A$ receptor genes (Harvey et al., 1991).

The high degree of conservation in the exon–intron boundaries for the GABA$_A$ receptor genes from *Drosophila,* mollusc, chicken, mouse, and human suggests that the organization of the GABA$_A$ receptor genes was established prior to the divergence of molluscs and chordates, which occurred at least 530 million years ago and before the presumed duplications that led to the different subfamily genes (Sommer et al., 1990; Harvey et al., 1991; Kirkness et al., 1991, 1993; Lasham et al., 1991). While the exon sizes and exon–intron boundaries are highly conserved within the GABA$_A$ receptor genes, there is no evidence for conservation between members of the superfamily of ligand-gated ion channels, namely between the nACh genes and the GABA$_A$ receptor genes (Sommer et al., 1990; Kirkness et al., 1991).

The strict conservation of gene structure is, however, also observed within the family of neuronal nACh receptor subunit genes (Boulter et al., 1990). In addition, none of the nACh receptors demonstrates intronic interruptions in any of their transmembrane domains, unlike the GABA$_A$ receptor genes (Kirkness et al., 1991; Lasham et al., 1991). Exon–intron boundaries in the extracellular loop are conserved in the glycine and GABA$_A$ receptors. Together with the greater sequence similarity, this correspondence suggests that GABA$_A$ receptor and glycine receptors are more closely related to each other than to the nACh receptors (Barnard et al., 1987; Sommer et al., 1990; Lasham et al., 1991).

Based on polypeptide sequence similarity and secondary structures, Barnard et al. (1987) suggested that members of the ligand-gated ion channel superfamily evolved from a common ancestor (Barnard et al., 1987). It seems likely that the movement of exon–intron boundaries or the insertion or deletion of introns from these genes during evolution is a dynamic process. Since the structures of the GABA$_A$ receptor subunit genes and the polypeptide sequences are highly conserved across diverse species, either these genes may have arisen relatively late in evolution, or, more likely, there is strong selective pressure for the maintenance of these sequences and gene structures (Lasham et al., 1991). A better evaluation of the time course and mechanisms through which the diversity of the receptor subunits arose during evolution will be possible as the gene structures and sequence similarity become available from other members of the superfamily of ligand-gated ion channels, particularly from invertebrates (Darlison, 1992; Kirkness et al., 1991). Molecular analysis of GABA$_A$ receptor genes from nematodes, fish, amphibians, and reptiles will complement the studies described above.

2.7.5 Chromosomal Localizations of GABA$_A$ Receptor Subunits

Some of the GABA$_A$ receptor genes have been mapped in human and mouse chromosomes (Table 4). These data suggest that the GABA$_A$ receptor genes lie in clusters. Especially striking is the occurrence of individual α-, β- and γ-subunits together on single chromosomes. For example, human α2, β1, and γ1 are found together on the short arm of chromosome 4 (Kirkness et al., 1991; Dean et al., 1991; Buckle et al., 1989; Wilcox et al., 1992). Likewise α5-, β3-, and γ3-subunit receptor genes are close to one another on mouse

Table 4. Chromosomal Localization of $GABA_A$ Receptor Subunit Genes

Species	Subunit	Chromosome location
Human	α2	4p12–p13[a]
	γ1	4p14–q21.1[b]
	β1	4p12–p13[c]
Mouse	α2	5[d]
	β1	5[d]
Human	α5	15q11–q13[e]
	β3	15q11–q13[f]
Mouse	α5	7[g]
	β3	7[h]
	γ3	7[i]
Human	α1	5q34–q35,[a] 5q31.1–q33.2,[b] 5q[j]
	γ2	5q31.1–q33.2[b]
Mouse	α1	11[k]
Human	α3	Xq28[l]
Mouse	α3	X[m]
Human	δ	1p[n]
Human	ρ1	6q14–q21[o]
	ρ2	6q14–q21[o]
Mouse	ρ1	4[o]
	ρ2	4[o]
Drosophila	*Rdl*	3[p]

[a] Buckle et al. (1989).
[b] Wilcox et al. (1992).
[c] Kirkness et al. (1991), Dean et al. (1991), Buckle et al. (1989).
[d] Danciger et al. (1993).
[e] Knoll et al. (1993).
[f] Wagstaff et al. (1991b).
[g] Nakatsu et al. (1993) and Danciger et al. (1993).
[h] Wagstaff et al. (1991a).
[i] Nakatsu et al. (1993).
[j] Johnson et al. (1992).
[k] Keir et al. (1991).
[l] Hinks et al. (1991), Buckle et al. (1989), Derry and Barnard (1991), Bell et al. (1989), and Faust et al. (1993).
[m] Buckle et al. (1989) and Derry and Barnard (1991).
[n] Sommer et al. (1990).
[o] Cutting et al. (1992).
[p] French-Constant et al. (1991).

chromosome 7 (Wagstaff et al., 1991; Nakatsu et al., 1993; Danciger et al., 1993), a region syntenic to the region of human chromosome 15 where α5 and β3 have already been mapped (Knoll et al., 1993; Wagstaff et al., 1991b). One would expect this synteny to extend to all three genes, predicting that the human γ3 will map to the $GABA_A$ receptor cluster of α5 and β3 on human chromosome 15. The α1- and γ2-subunit genes map to human chromosome 5 (Johnson et al., 1992; Buckle et al., 1989; Wilcox et al., 1992), and one might predict that the remaining unmapped mammalian β-subunit, β2, would map to chromosome 5, in keeping with the other β-subunits mapping into GABA$_A$ receptor gene clusters.

The arrangement of members of different subfamilies of the GABA$_A$ receptor family in chromosomal clusters, rather than members of the same subfamily, might not have been predicted from other superfamily gene localizations. For example, in the hemoglobin family, the α-like hemoglobin genes are found together on human chromosome 16, while the β-like genes are found on chromosome 11 (Trembley et al., 1988). It seems likely that the γ1–α2–β1 and γ3–α5–β3 gene clusters, in both humans and mice, have derived from a duplicated γ–α–β prototype prior to the separation of rodents and primates. The ρ-subunits, which are clustered together on human and mouse chromosomes [6 and 4 (Cutting et al., 1992), respectively], may have arisen by a later duplication.

The close proximity of GABA$_A$ receptor genes on individual chromosomes may be related to their coordinate temporal and spatial regulation. The chromosomal locations may thus provide insight into the regulation of the genes, and possibly provide assistance in understanding which receptor subunits combine together *in vivo* to form functional channels. For example, α1, β2, and γ2 mRNAs are the most widespread and abundant subunits in the rat brain (Persohn et al., 1992), and can be found colocalized in specific brain regions, including the midbrain red nucleus and inferior colliculi (Shivers et al., 1989; Wisden et al., 1992). Both α1 and γ2 have already been identified as part of a chromosomal cluster (Johnson et al., 1992; Buckle et al., 1989; Wilcox et al., 1992) that is also likely to contain the β2 gene.

One might have expected, in light of the chromosomal clustering of the GABA$_A$ receptor subunit genes, that those subunits clustered together (i.e., α2–β1–γ1; α5–β3–γ3 and α1–γ2) would share greater sequence similarity to one another than to members of alternative chromosome clusters. For example, α2 might demonstrate a higher sequence similarity to γ1 than to γ2 or γ3, as it is clustered with γ1. This, however, does not appear to be the case, as similar sequence similarity is found between α2 and all three γ-subunits (nucleic acid identity: 54, 55, and 53%; amino acid identity 46, 45, and 47%; amino acid similarity 64, 63, and 65% for α2 versus γ1, γ2, and γ3, respectively).

The α4- and α6-subunits share high amino acid identity to each other (>72% identity, >83% similarity; Table 2), and lower identity to the other α-subunits (<60% identity, <75% similarity). These subunits remain unmapped, and it is possible that they will be found together in a cluster, much like the ρ-subunits.

Other genes from the superfamily of ligand-gated ion channels are also clustered. For example, the α3, α5, and β4 genes of the human neuronal nACh receptor (Boulter et al., 1990), δ and γ of the chicken nACh receptor (Nef et al., 1984), and α1 and α2 nACh receptor genes in *Drosophila* (Sawruk et al., 1990), all cluster together. Thus the clustering phenomenon appears to be a characteristic of the nACh and GABA$_A$ receptor gene families.

In contrast to the $GABA_A$ and nACh receptor genes, the GluR1–5 all appear to exist on separate chromosomes [GluR1, 5q31.3–33.3 (Sun et al., 1992; McNamara et al., 1992); GluR2, 4q32–33 (Sun et al., 1992; McNamara et al., 1992); GluR3, Xq25–26 (McNamara et al., 1992); GluR4, 11q22–23 (McNamara et al., 1992); GluR5, 21q22 (Patier et al., 1993)]. However, some of the GluR genes are found near the $GABA_A$ receptor genes. For example, the GluR3 gene maps to human Xq25–q26 near the location of the human $GABA_A$ receptor α3 gene, and the GluR1 gene maps to 5q31–33 very near to the $GABA_A$ receptor α1 and γ2 gene locations.

2.7.6 Evolution of the $GABA_A$ Receptor Gene Family

Multigene families almost certainly arise by gene duplication, unequal crossing over, and sequence divergence. In many cases, the different polypeptides appear to be susceptible to intense natural selection, so that cognate genes in different species are far more alike than different members of a multigene family (Tobin et al., 1991). Certainly, this appears to be the case for the $GABA_A$ receptor genes, as demonstrated by the conservation of α1 sequence among species (Table 5). In many cases the persistence of multiple gene families reflects differing physiological needs in different cells and at different developmental stages (Tobin et al., 1991). This may be true for the $GABA_A$ receptor as developmental and CNS regional studies indicate that unique $GABA_A$ receptor subunits are expressed at different points in an organism's development, and in discrete cell types (Shivers et al., 1989; Wisden et al., 1992; Laurie et al., 1992a,b).

Multigene families encode polypeptides with parallel, but often distinctive functional properties. Even in cases where there are no apparent functional differences among related polypeptides, however, multigene families

Table 5. Comparison of Rat, Mouse, Human, Bovine, and Chicken α1 $GABA_A$ Receptor Subunits[a]

	R	M	H	B	C
A. Nucleotide Sequence Identity for Coding Region					
R		96	91	90	84
M			92	91	83
H				94	84
B					83
B. Amino Acid Identity (Upper) and Similarity (Lower)					
R		100	98	99	96
M	100		98	99	96
H	99	99		99	96
B	100	100	100		96
C	98	98	98	98	

[a] Rat (R), mouse (M), human (H), bovine (B), chicken (C).

allow organisms to regulate the total amounts and the distributions of functional molecules. Such regulatory flexibility appears to be used heavily in the brain and may underlie the well-known plasticity of the mammalian central nervous system (CNS) (Tobin et al., 1991). Receptor diversity in the CNS may have evolved to increase the variety of responses to GABA, resulting in increased plasticity (Persohn et al., 1992; Schofield, 1989).

To examine the phylogenetic relationship among the GABA$_A$ receptors we have aligned the amino acid sequences (Figure 3) of the members of the four cloned rat GABA$_A$ subfamilies and constructed a dendrogram (Figure 4). The length of each branch is proportional to the inferred evolutionary distance between receptor subpopulations. The data presented in this form suggest that the β/δ-subunit subfamilies diverged initially from the α/γ-subunit subfamilies prior to the divergence of this family into individual

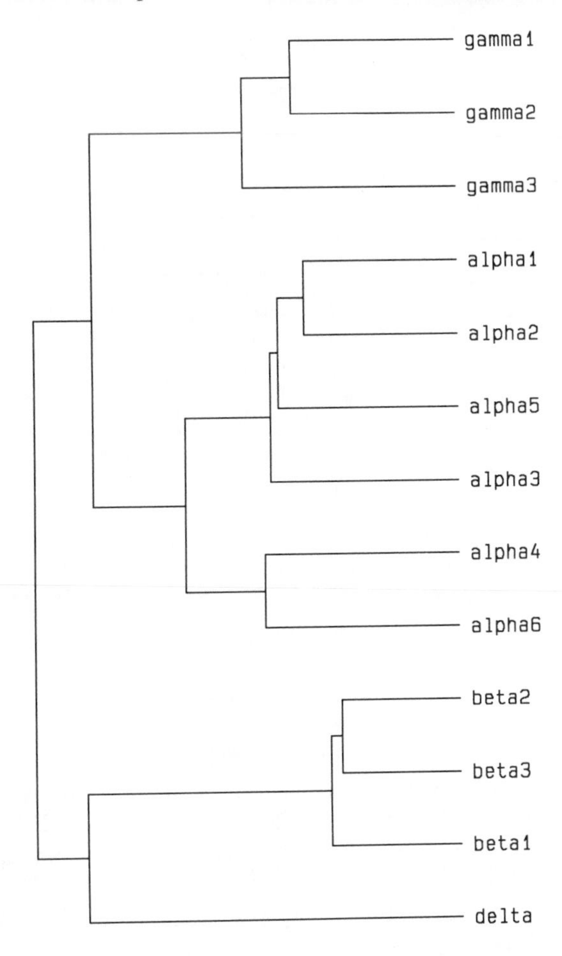

FIGURE 4. A dendrogram of the deduced amino acid sequences for the 13 rat GABA$_A$ receptor subunits. The dendrogram was performed using the Pileup program described by Feng and Doolittle (1987).

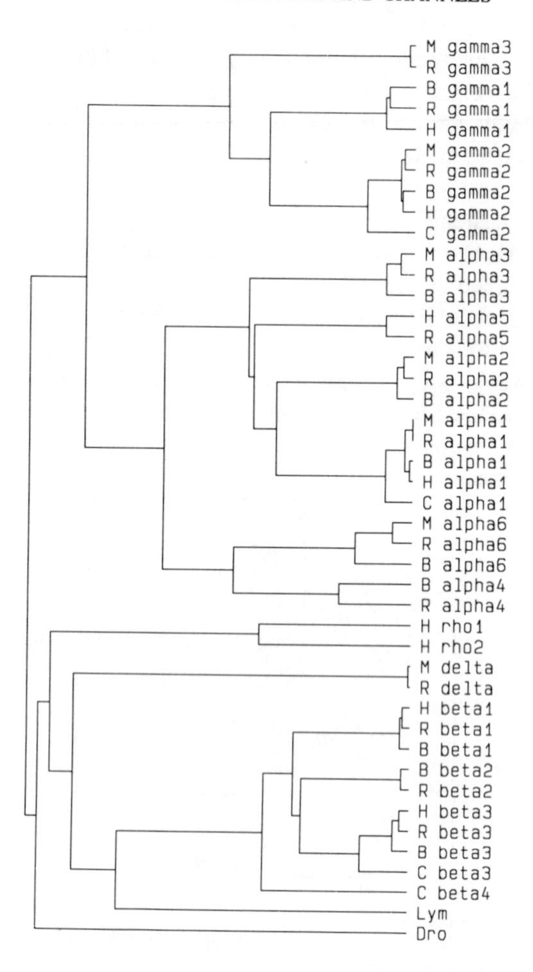

FIGURE 5. A dendrogram of the deduced amino acid sequences for the cloned rat (R), human (H), mouse (M), chicken (C), bovine (B), *Drosophila* (Dro), and *Lymnaea stagnalis* (Lym) $GABA_A$ receptor subunits. The dendrogram was performed using the Pileup program.

subunit subfamilies. The α-subunit subfamily contains members with the most diverse amino acid sequences and the separation/duplication of these members can be observed in Figure 4. Individual $GABA_A$ receptor subfamilies or "branches" include members that have common pharmacology, as would be expected from their evolutionary similarity. Even within these subfamilies, however, significant differences in ligand binding and channel modulation exist.

Alignments were also made for all of the available $GABA_A$ receptor amino acid sequences from different species and a dendrogram from these data was constructed (Figure 5). From this figure, it is clear that the ρ subfamily is more closely related to the β/δ subfamilies than to the α/γ subfamilies.

The dendrogram (Figure 5) also indicates the position of the snail *(Lymnaea stagnalis)* and fruit fly *(Drosophila)* $GABA_A$ receptor genes. The snail $GABA_A$ receptor subunit most resembles the vertebrate β-subunits and lies within the β-subunit branch. This is in contrast to the *Drosophila* cDNA, which predicts a subunit that demonstrates little similarity to the β/ρ/δ or α/γ subfamilies and clearly branches before the separation of the five established $GABA_A$ receptor subunit subfamilies. These data, indicating a closer relationship between the snail $GABA_A$ receptor subunit and the vertebrate β subfamily, suggest that the selective forces for this receptor family were more alike for snails and vertebrates than for the *Drosophila* and vertebrates.

Studies of subunits from different species strongly suggest that these sequences have been tightly conserved during evolution. For example, α1 sequences have greater than 85% identity between species (Table 5, Figure 5). This conservation of subunits between species suggests that the homologue to the chicken β4 and human ρ genes may be found in other species. Likewise it is possible, although less likely, that the vertebrate homologue to the *Drosophila* gene may be discovered and that it may represent an additional $GABA_A$ receptor subfamily. It is also possible, however, that the homologues for these genes do not exist in every species, a common finding among members of multigene superfamilies, whose numbers often differ from species to species.

2.7.7 Conclusions

From our emerging cognizance of the individual native $GABA_A$ receptor composition, CNS distribution, function, regulation, and evolution, we may be able to design specific drugs that can target discrete subpopulations of $GABA_A$ receptors. This knowledge, combined with information about the development of $GABA_A$ receptor combinations, may eventually allow comprehension of the progress of $GABA_A$ receptor drug dependence and tolerance, as well as the role of the $GABA_A$ receptor plasticity in neurological disease.

Acknowledgments

We thank George Lawless, Bob Weatherwax, and Dimitri Maslov for their help with the computer programs used in this chapter, Tim Hales, Geoff Smith, and Schoichi Endo for helpful discussions, and Sharon Belkin for artistic illustrations. Supported by the Medical Research Council of Canada Fellowship (to RFT), NIH Grant NS28772 (to RWO), and NIH Grant NS22256 (to AJT).

REFERENCES

Anand, R., Conroy, W. G., Schoepfer, R., Whiting, P., and Lindstrom, J. (1991). Neuronal nicotinic acetylcholine receptors expressed in *Xenopus* oocytes have a pentomeric quaternary structure. *J. Biol. Chem.* 266, 11192.

Angelotti, T. P., Tan, F., Chahine, K. G., and Macdonald, R. L. (1992). Molecular and electophysiological characterization of a allelic variant of the rat alpha-6 GABA-A receptor subunit. *Mol. Brain Res.* 16, 173.

Barnard, E. A., Darlison, M. G., and Seeburg, P. (1987). Molecular biology of the GABA-A receptor: The receptor/channel superfamily. *Trends Neurosci.* 10, 502.

Bateson, A. N., Harvey, R. J., Wisden, W., Glencourse, T. A., Hicks, A. A., Hunt, S. P., Bernard, E. A., and Darlison, M. G. (1990a). The chicken GABA-A receptor alpha-1 subunit: cDNA sequence and localization of the corresponding mRNA, found in Genbank.

Bateson, A. N., Harvey, R. J., Blocks, C. C., and Darlison, M. G. (1990b). Sequence of the chicken GABA-A receptor beta-3 subunit cDNA. *Nucleic Acids Res.* 18, 5557.

Bateson, A. N., Lasham, A., and Darlison, M. G. (1991). Gamma-aminobutyric acid-A receptor heterogeneity is increased by alternative splicing of a novel beta-subunit gene transcript. *J. Neurochem.* 56, 1437.

Bell, M. V., Bloomfield, J., McKinley, M., Patterson, M. N., Darlison, M. G., Barnard, E. A., and Davies, K. E. (1989). Physical linkage of a GABAA receptor subunit gene to the DXS374 locus in human Xq28. *Am. J. Hum. Genet.* 45, 883.

Boulter, J., O'Shea-Greenfield, A., Duvoisin, R. M., Connolly, J. G., Wada, E., Jensen, A., Gardner, P. D., Ballivet, M., Deneris, E. S., McKinnon, D., Heinemann, S., and Patrick, J. (1990). Alpha3, alpha5 and beta4: Three members of the rat neuronal nicotinic acetylcholine receptor-related gene family form a gene cluster. *J. Biol. Chem.* 265, 4472.

Browning, M. D., Endo, S., Smith, G. B., Dudek, E. M., and Olsen, R. W. (1993). Phosphorylation of the GABA-A receptor by cAMP-dependent protein kinase and by protein kinase C: Analysis of the substrate domain. *Neurochem. Res.* 18, 95.

Buckle, V. J., Fujita, N., Ryder-Cook, A. S., Derry, J. M. J., Barnard, P. J., Lebo, R. V., Schofield, P. R., Seeburg, P. H., Bateson, A. N., Darlison, M. G., and Barnard, E. A. (1989). Chromosomal localization of GABA-A receptor subunit genes: Relationship to human genetic diseases. *Neuron* 3, 647.

Bureau, M. and Olsen, R. W. (1990). Multiple distinct subunits of the gamma-aminobutyric acid-A receptor protein show different ligand-binding affinities. *Mol. Pharmacol.* 37, 497.

Bureau, M. H. and Olsen, R. W. (1993). GABA-A receptor subtypes: Ligand binding heterogeneity demonstrated by photoaffinity labeling and autoradiography. *J. Neurochem.* 61, 1479.

Burt, D. B. and Kamatchi, G. L. (1991). GABA-A receptor subtypes: From pharmacology to molecular biology. *FASEB J.* 5, 2916.

Cherubini, E., Gaiarsa, J., and Ben-Ari, Y. (1991). GABA: An excitatory transmitter is early postnatal life. *Trends Neurosci.* 14, 515.

Cutting, G. R., Lu, L., O'Hara, B. F., Kasch, L. M., Montrose-Fafizadeh, C., Donovan, D. M., Shimada, S., Antonarakis, S. E., Guggino, W. B., Uhl, G. R., and Kazazian, H. H. (1991). Cloning of the gamma-aminobutyric acid (GABA) P1 cDNA: A GABA receptor subunit highly expressed in the retina. *Proc. Natl. Acad. Sci. U.S.A.* 88, 2673.

Cutting, G. R., Curristin, S., Zoghbi, H., O'Hara, B. F., Seldin, M. F., and Uhl, G. R. (1992). Identification of a putative gamma-aminobutyric acid (GABA) receptor subunit rho-2 cDNA and colocalization of the genes encoding rho-2 (GABRR2) and rho-1 (GABRR1) to human chromosome 6q14-q21 and mouse chromosome 4 *Genomics* 12, 801.

Danciger, M., Farber, D. D., and Kozak, C. A. (1993). Genetic mapping of three GABA-A receptor-subunit genes in the mouse. *Genomics* 16, 361.

Darlison, M. G. (1992). Invertebrate GABA and glutamate receptors: Molecular biology reveals predictable structures but some unusual pharmacologies. *Trends Neurosci.* 15, 469.

Dean, M., Lucas-Derse, S., Bolos, A., O'Brien, S. J., Kirkness, E. F., Fraser, C. M., and Goldman, D. (1991). Genetic mapping of the beta 1 GABA receptor gene to human chromosome 4, using a tetranucleotide repeat polymorphism. *Am. J. Hum. Genet.* 49, 621.

DeLorey, T. M. and Olsen, R. W. (1992). Gamma-aminobutyric acid-A receptor structure and function. *J. Biol. Chem.* 267, 16747.

Deneris, E. S., Connolly, J., Rogers, S. W., and Duvoisin, R. (1991). Pharmacological and functional diversity of neuronal nicotinic acetylcholine receptors. *Trends Pharmacol. Sci.* 12, 34.

Deng, L., Nielsen, M., and Olsen, R. W. (1991). Pharmacological and biochemical properties of the gamma-aminobutyric acid-benzodiazepine receptor protein from codfish brain. *J. Neurochem.* 56, 968.

Derry, J. M. J. and Barnard, P. J. (1991). Mapping of the glycine receptor alpha-2-subunit gene and the GABA-A alpha-3-subunit gene on the mouse X chromosome. *Genomics* 10, 593.

Duggan, M. J., Pollard, S., and Stephenson, F. A. (1991). Immunoaffinity purification of GABA-A receptor alpha-subunit iso-oligomers. *J. Biol. Chem.* 266, 24778.

Endo, S. and Olsen, R. W. (1992). Preparation of antibodies to beta subunits of GABAA receptors. *J. Neurochem.* 59, 1444.

Endo, S. and Olsen, R. W. (1993). Antibodies specific for alpha-subunit subtypes of GABA-A receptors reveal brain regional heterogeneity. *J. Neurochem.* 60, 1388.

Faust, C. J., Gonzales, J. C., Siebold, A., Birnbaumer, M., and Herman, G. E. (1993). Comparative mapping on the mouse and human X chromosomes of a human cDNA clone encoding the vasopressin renal-type receptor (AVP2R). *Genomics* 15, 439.

Feng, D.-F. and Doolittle, R. F. (1987). Progressive sequence alignment as a prerequisite to correct phylogenetic trees. *J. Mol. Evol.* 25, 351.

French-Constant, R. H. and Rocheleau, T. (1992). *Drosophila* cyclodiene resistance gene shows conserved genomic organization with vertebrate gamma-aminobutyric acid-A receptors. *J. Neurochem.* 59, 1562.

French-Constant, R. H., Mortlock, D. P., Shaffer, C. D., MacIntyre, R. J., and Roush, R. T. (1991). Molecular cloning and transformation of cyclodiene resistance in *Drosophila:* An invertebrate gamma-aminobutyric acid subtype A receptor locus. *Proc. Natl. Acad. Sci. U.S.A.* 88, 7209.

Frohman, M. A., Dush, M. K., and Martin, G. R. (1988). Rapid production of full-length cDNAs from rare transcripts: Amplification using a single gene-specific oligonucleotide primer, *Proc. Natl. Acad. Sci. U.S.A.* 85, 8998.

Galzi, J. L., Devillers-Thiery, A., Hussy, N., Bertrand, S., Changeux, J.-P., and Bertrand, D. (1992). Mutations in the channel domain of a neuronal nicotinic receptor convert ion selectivity from cationic to anionic. *Nature* 359, 500.

Glencourse, T. A., Bateson, A. N., and Darlison, M. G. (1990). Sequence of the chicken GABA-A receptor gamma2-subunit cDNA. *Nucl. Acids Res.*, 18, 7157.

Harvey, R. J., Vreugdenhil, E., Zaman, S. H., Bhandal, N. S., Usherwood, P. N., Barnard, E. A., and Darlison, M. G. (1991). Sequence of a functional invertebrate GABA-A receptor subunit which can form a chimeric receptor with vertebrate alpha subunit. *EMBO J.* 10, 3239.

Hebebrand, J., Friedl, W., Breidenbach, B., and Propping, P. (1987). Phylogenetic comparison of the photoaffinity-labeled benzodiazepine receptor subunits. *J. Neurochem.* 48, 1103.

Herb, A., Wisden, W., Lueddens, H., Puia, G., Vicini, S., and Seeburg, P. H. (1992). The third gamma subunit of the gamma-aminobutyric acid type A receptor family. *Proc. Natl. Acad. Sci. U.S.A.* 89, 1433.

Higgins, D. G. and Sharp, P. M. (1989). Fast and sensitive multiple sequence alignment on a microcomputer. *Cabios* 5, 151.

Hinks, A. A., Johnson, K. J., Barnard, E. A., and Darlison, M. G. (1991). Dinucleotide repeat polymorphism in the human X-linked GABA-A alpha 3-subunit gene. *Nucl. Acids Res.* 19, 4016.

Houser, C. R. (1991). GABA neurons in seizure disorders: A review of immunocytochemical studies. *Neurochem. Res.* 16, 295.

Johnson, K. J., Sander, T., Hicks, A. A., van Marle, A., Janz, D., Mullan, M. J., Riley, B. P., and Darlison, M. G. (1992). Confirmation of the localization of the human GABA-A receptor alpha-1 subunit gebe (GABRA1) to distal 5q by linkage analysis. *Genomics* 14, 745.

Kato, K. (1990). Novel GABA-A receptor alpha subunit is expressed only in cerebellar granule cells. *J. Mol. Biol.* 214, 619.

Keir, W. J., Kozak, C. A., Chakraborti, A., Deitrich, R. A., and Sikela, J. M. (1991). The cDNA sequence and chromosomal localization of the murine GABA-A alpha-1 receptor gene. *Genomics* 9, 360.

Khrestchatisky, M., MacLennan, A. J., Chiang, M. Y., Xu, W., Jackson, M., Brecha, N., Sternini, C., Olsen, R. W., and Tobin, A. J. (1989). A novel alpha-subunit in rat brain GABA-A receptors. *Neuron* 3, 745.

Khrestchatisky, M., MacLennan, A. J., Tillakaratne, N. J. K., Chiang, M. Y., and Tobin, A. J. (1991). Sequence and regional distribution of the mRNA encoding the alpha2 polypeptide of rat gamma-aminobutyric acidA receptors. *J. Neurochem.* 56, 1717.

Kirkness, E. F. and Fraser, C. M. (1993). A strong promoter element is located between alternative exons of a gene encoding the human gamma-aminobutyric acid-type A receptor beta3 subunit (GABRB3). *J. Biol. Chem.* 268, 4420.

Kirkness, E. F., Kusiak, J. W., Fleming, J. T., Menninger, J., Gocayne, J. D., Ward, D. C., and Venter, J. C. (1991). Isolation, characterization, and localization of human genomic DNA encoding the beta-1 subunit of the GABA-A receptor (GABRB1). *Genomics* 10, 985.

Knoflach, F., Rhyner, T., Villa, M., Kellenberger, S., Drescher, U., Malherbe, P., Sigel, E., and Mohler, H. (1991). The gamma-3-subunit of the GABA-A-receptor confers sensitivity to benzodiazepine receptor ligands. *FEBS Lett.* 293, 191.

Knoll, J. H., Sinnett, D., Wagstaff, J., Glatt, K. A., Wilcox, A. S., Whiting, P. J., Wingrove, P., Sikela, J. M., and Lalande, M. (1993). FISH ordering of reference markers and of the gene for the alpha 5 subunit of the gamma-aminobutyric acid receptor within the Angelman and Prader–Willi syndrome chromosomal regions. *Hum. Mol. Genet.* 2, 183.

Kofuji, P., Wand, J. B., Moss, S. J., Huganir, R. L., and Burt, D. R. (1991). Generation of two forms of the gamma-aminobutyric acid-A receptor gamma-2-subunit in mice by alternative splicing. *J. Neurochem.* 56, 713.

Korpi, E. R., Kleingoor, C., Kettenmann, H., and Seeburg, P. H. (1993). Benzodiazepine-induced motor impairment linked to point mutation in cerebellar GABA-A receptor. *Nature* 361, 356.

Lasham, A., Vreugdenhil, E., Bateson, A. N., Barnard, E. A., and Darlison, M. G. (1991). Conserved organization of gamma-aminobutyric acid-A receptor genes: Cloning and analysis of the chicken beta4-subunit gene. *J. Neurochem.* 57, 352.

Laurie, D. J., Seeburg, P. H., and Wisden, W. (1992a). The distribution of 13 GABAA receptor subunit mRNAs in the rat brain. II. Olfactory bulb and cerebellum. *J. Neurosci.* 12, 1063.

Laurie, D., Wisden, W., and Seeburg, P. H. (1992b). The distribution of thirteen GABA-A receptor subunit mRNA in the rat brain. III. Embryonic and postnatal development. *J. Neurosci.* 12, 4151.

Leidenheimer, N. J., Browning, M. D., and Harris, R. A. (1991). GABA-A receptor phosphorylation: Multiple sites, actions and artifacts. *Trends Pharmacol. Sci.* 12, 84.

Levitan, E. S., Schofield, P. R., Burt, D. R., Rhee, L. M., Wisden, W., Kohler, M., Fujita, N., Rodriguez, H., Stephenson, A., Darlison, M. G., Barnard, E. A., and Seeburg, P. H. (1988). Structural and functional basis for $GABA_A$ receptor heterogeneity. *Nature* 335, 76.

Lolait, S. J., O'Carroll, A.-M., Kusano, K., Muller, J.-M., Brownstein, M. J., and Mahan, L. C. (1989). Cloning and expression of a novel rat GABA-A receptor. *FEBS Lett.* 246, 145.

Lueddens, H., Pritchett, D., Khler, M., Killisch, I., Kein nen, K., Monyer, H., Sprengel, R., and Seeburg, P. H. (1990). A cerebellar GABA-A receptor selective for a behavioural alcohol antagonist. *Nature* 346, 648.

Lunt, G. G. (1991). GABA and GABA receptors in invertebrates. *Sem. Neurosci.* 3, 251.

Macdonald, R. L. and Olsen, R. W. (1994). GABA-A receptor channels. *Annu. Rev. Neurosci.* 17, 569.

Macdonald, R. L. and Twyman, R. E. (1991). Biophysical properties and regulation of GABA-A receptor channels. *Sem. Neurosci.* 3, 219.

Malherbe, P., Sigel, E., Baur, R., Persohn, E., Richards, J. G., and Mohler, H. (1990). Functional expression and sites of gene transcription of a novel alpha subunit of the GABAA receptor in rat brain. *FEBS Lett.* 260, 261.

McKernan, R. M., Quirk, K., Prince, R., Cox, P. A., Gillard, N. P., Ragan, C. I., and Whiting, P. (1991). GABAA receptor subtypes immunopurified from rat brain with alpha subunit-specific antibodies have unique pharmacological properties. *Neuron* 7, 667.

McNamara, J. O., Eubanks, J. H., McPherson, J. D., Wasmuth, J. J., Evans, G. A., and Heinemann, S. F. (1992). Chromosomal localization of human glutamate receptor genes. *J. Neurosci.* 12, 2555.

Mertens, S., Benke, D., and Mohler, H. (1993). GABA-A receptor populations with novel subunit combinations and drug binding profiles identified in brain by alpha5- and delta-subunit-specific immunopurification. *J. Biol. Chem.* 268, 5965.

Moss, S. J., Doherty, C. A., and Huganir, R. L. (1992a). Identification of the cAMP-dependent protein kinase and protein kinase C phosphorylation sites within the major intracellular domains of the beta-1, gamma-2s, and gamma-2l subunits of the gamma-aminobutyric acid type A receptor. *J. Biol. Chem.* 267, 14470.

Moss, S. J., Smart, T. G., Blackstone, C. D., and Huganir, R. L. (1992b). Functional modulation of GABA-A receptors by cAMP-dependent protein phosphorylation. *Science* 257, 661.

Mutirangura, A., Ledbetter, S. A., Kuwano, A., Chinault, A. C., and Ledbetter, D. H. (1992). Dinucleotide repest polymorphism at the GABA-A receptor beta 3 (GABRB3) locus in the Angelman/Prader–Willi region (AS/PWS) of chromosome 15. *Hum. Mol. Genet.* 1, 67.

Nakatsu, Y., Tyndale, R. F., DeLorey, T. M., Durham-Pierre, D., Gardner, J. M., McDanel, H. J., Nguyen, Q., Wagstaff, J., Lalande, M., Sikela, J. M., Olsen, R. W., Tobin, A. J., and Brilliant, M. H. (1993). A cluster of three GABA-A receptor subunit genes is deleted in a neurological mutant of the mouse *p* locus. *Nature* 364, 448.

Nef, P., Mauron, A., Stalder, R., Alliod, C., and Ballivet, M. (1984). Structure, linkage, and sequence of the two genes encoding the delta and gamma subunits of the nicotinic acetylcholine receptor. *Proc. Natl. Acad. Sci. U.S.A.* 81, 7975.

Olsen, R. W. (1991). GABA and inhibitory synaptic transmission in the brain. *Sem. Neurosci.* 3, 175.

Olsen, R. W. and Tobin, A. J. (1990). Molecular biology of GABA-A receptors. *FASEB J.* 4, 1469.

Olsen, R. W., Bureau, M. H., Endo, S., and Smith, G. (1991a). The GABA-A receptor family in the mammalian brain. *Neurochem. Res.* 16, 317.

Olsen, R. W., Bureau, M. H., Endo, S., Smith, G. B., Brecha, N., Sternini, C., and Tobin, A. J. (1992). GABA-A receptor subtypes identified by molecular biology, protein chemistry and binding. *Mol. Neuropharmacol.* 2, 129.

Olsen, R. W., McCabe, R. T., and Wamsley, J. K. (1990). GABAA receptor subtypes: Autoradiographic comparison of GABA, benzodiazepine, and convulsant binding sites in the rat central nervous system. *J. Chem. Neuroanat.* 3, 59.

Olsen, R. W., Sapp, D. W., Bureau, M. H., Turner, D. M., and Kokka, N. (1991). Allosteric actions of central nervous system depressants including anesthetics on subtypes of the inhibitory gamma-aminobutyric acid-A receptor-chloride channel complex. *Ann. N.Y. Acad. Sci.* 625, 145.

Otero de Bengtsson, M. S., Lacorazza, H. D., Biscoglio de Jimenez Bonino, M. J., and Medina, J. H. (1993). Involvement of a disulfide bond in the binding of flunitrazepam to the benzodiazepine receptor from bovine cerebral cortex. *J. Neurochem.* 60, 536.

Patier, M.-C., Dutiaux, A., Lambolez, B., Bochet, P., and Rossier, J. (1993). Assignment of the human glutamate receptor gene GLUR5 to 21q22 by screening a chromosome 21 YAC library. *Genomics* 15, 696.

Perez-Velazquez, L. and Angelides, K. J. (1993). Assembly of GABA-A receptor subunits determines sorting and localization in polarized cells. *Nature* 361, 457.

Persohn, E., Malherbe, P., and Richards, J. G. (1992). Comparative molecular neuroanatomy of cloned GABA-A receptor subunits in the rat CNS. *J. Comp. Neurol.* 326, 193.

Pritchett, D. B. and Seeburg, P. H. (1990). Gamma-aminobutyric acidA receptor alpha 5-subunit creates novel type II benzodiazepine receptor pharmacology. *J. Neurochem.* 54, 1802.

Pritchett, D. B. and Seeburg, P. H. (1991). Gamma-aminobutyric acid type A receptor point mutation increases affinity of compounds for the benzodiazepam site. *Proc. Natl. Acad. Sci. U.S.A.* 88, 1421.

Pritchett, D. B., Sontheimer, H., Shivers, B. D., Ymer, S., Kettenmann, H., Schofield, P. R., and Seeburg, P. H. (1989). Importance of a novel GABAA receptor subunit for benzodiazepine pharmacology. *Nature* 338, 582.

Salpeter, M. M. and Loring, R. H. (1985). Nicotinic acetylcholine receptors in vertebrate muscle: Properties, distribution and neural control. *Prog. Neurobiol.* 25, 297.

Sargent, P. B. (1993). The diversity of neuronal nicotinic acetylcholine receptors. *Annu. Rev. Neurosci.* 16, 403.

Sawruk, E., Udri, C., Betz, H., and Schmitt, B. (1990). SBD, a novel structural subunit of the *Drosophila* nicotinic acetylcholine receptor, shares its genomic localization with two alpha-subunits. *FEBS Lett.* 273, 177.

Schofield, P. R. (1989). $GABA_A$ receptor complexity. *Trends Pharmacol. Sci.* 10, 476.

Schofield, P. R., Darlison, M. G., Fujita, N., Burt, D. R., Stephenson, F. A., Rodriguez, H., Rhee, L. M., Ramachandran, J., Reale, V., Glencourse, T. A., Seeburg, P. G., and Barnard, E. A. (1987). Sequence and functional expression of the GABAA receptor shows a ligand-gated receptor superfamily. *Nature* 328, 221.

Schofield, P. R., Pritchett, D. B., Sontheimer, H., Kettenmann, H., and Seeburg, P. H. (1989). Sequence and expression of human GABA(A) receptor alpha 1 and beta 1 subunits. *FEBS Lett.* 244, 361.

Shimada, S., Cutting, G., and Uhl, G. R. (1992). Gamma-aminobutyric acid A or C receptor? Gamma-aminobutyric acid rho 1 receptor RNA induces bicuculline-, barbiturate-, and benzodiazepine-insensitive gamma-aminobutyric acid responses in *Xenopus* oocytes. *Mol. Pharmacol.* 41, 683.

Shivers, B. D., Killisch, I., Sprengel, R., Sontheimer, H., and Kohler, M. (1989). Two novel GABAA receptor subunits exist in distinct neuronal subpopulations. *Neuron* 3, 327.

Sieghart, W., Item, C., Buchstaller, A., Fuchs, K., Hoger, H., and Adamiker, D. (1993). Evidence for the existence of differential O-glycosylated alpha-5 subunits of the gamma-aminobutyric acid-A receptor in the rat brain. *J. Neurochem.* 60, 93.

Sigel, E., Baur, R., Kellenberger, S., and Malherbe, P. (1992). Point mutations affecting antagonist affinity and agonist dependent gating of GABA-A receptor channels. *EMBO J.* 11, 2017.

Snodgrass, S. R. (1992). GABA and epilepsy: Their complex relationship and evolution of our understanding. *J. Child Neurol.* 7, 77.

Sommer, B. and Seeburg, P. H. (1992). Glutamate receptor channels: Novel properties and new clones. *Trends Pharmacol. Sci.* 13, 291.

Sommer, B., Kohler, M., Sprengel, R., and Seeburg, P. H. (1991). RNA editing in brain controls a determinant of ion flow in glutamate-gated channels. *Cell* 67, 11.

Sommer, B., Poustka, A., Spurr, N. K., and Seeburg, P. H. (1990). The murine GABA-A receptor delta gene: Structure and assignment to human chromosome 1. *DNA Cell Biol.* 9, 561.

Somogyi, P., Takagi, H., Richards, J. G., and Möhler, H. (1989). Subcellular localization of benzodiazepine/GABAA receptors in the cerebellum of rat, cat, and monkey using monoclonal antibodies. *J. Neurosci.* 9, 2197.

Sun, W., Ferrer-Montiel, A. V., Schinder, A. F., McPherson, J. P., Evans, G. A., and Montal, M. (1992). Molecular cloning, chromosomal mapping, and functional expression of human brain glutamate receptors. *Proc. Natl. Acad. Sci. U.S.A.* 89, 1443.

Tehrani, M. H. J., Hablitz, J. J., and Barnes, E. M. (1989). cAMP increases the rate of GABA-A receptor desensitization in chick cortical neurons. *Synapse* 4, 126.

Ticku, M. K. (1991). Drug modulation of GABA-A-mediated transmission. *Sem. Neurosci.* 3, 211.

Tobin, A. J., Khrestchatisky, M., MacLennan, A. J., Chiang, M. Y., Tillakaratne, N. J. K., Xu, W., Jackson, M. B., Brecha, N., Sternini, C., and Olsen, R. W. (1991). Structural, developmental, and functional heterogeneity of rat GABA-A receptors. In *Neuroreceptor Mechanisms in Brain.* Kito, S., Segawa, T., and Olsen, R. W., Eds., Plenum, New York, 365–374.

Trembley, J. P., Robitaille, R., and Atwood, H. L. (1988). GABA and benzodiazepine receptors in invertebrate species. In *GABA and Benzodiazepine Receptors,* Vol. 1. Squires, R. F., Ed., CRC Press, Inc., Boca Raton, FL, chap. 23.

Verdoorn, T. A., Draguhn, A., Ymer, S., Seeburg, P. H., and Sakmann, B. (1990). Functional properties of recombinant rat GABA-A receptors depend upon subunit composition. *Neuron* 4, 919.

Vernallis, A. B., Conroy, W. G., and Berg, D. K. (1993). Neurons assemble acetylcholine receptors with as many as three kinds of subunits while maintaining subunit segregation among receptor subtypes. *Neuron* 10, 451.

Wafford, K. A. and Whiting, P. J. (1992). Ethanol potentiation of GABA-A receptors requires phosphorylation of the alternatively spiced variant of the gamma2 subunit. *FEBS Lett.* 313, 113.

Wafford, K. A., Burnett, D. M., Leidenheimer, N. J., Burt, D. R., Wang, J. B., Kofuji, P., Dunwiddie, T. V., Harris, R. A., and Sikela, M. S. (1991). Ethanol sensitivity of the GABA-A receptor expressed in *Xenopus* oocytes requires 8 amino acids contained in the gamma-2L subunit. *Neuron* 7, 27.

Wagstaff, J., Chaillet, J. R., and Lalande, M. (1991a). The GABA-A receptor beta-3 subunit gene: Characterization of a cDNA from chromosome 15q11-q13 and mapping to a region of conserved synteny on mouse chromosome 7. *Genomics* 11, 1071.

Wagstaff, J., Knoll, J. H. M., Fleming, J., Kirkness, E. F., Martin-Gallardo, A., Greenberg, F., Graham, J. M., Menninger, J., Ward, D., Venter, J. C., and Lalande, M. (1991b). Localization of the gene encoding the GABA-A receptor beta 3 subunit to the Angelman/Prader–Willi region of human chromosome 15. *Am. J. Hum. Genet.* 49, 330.

Wang, J. B., Kofuji, P., Fernando, J. C., Moss, S. J., Huganir, R. L., and Burt, D. R. (1992). Alpha-1, alpha-2 and alpha-3 subunits of GABA-A receptors: Comparison in seizure-prone and -resistant mice and during development. *J. Mol. Neurosci.* 3, 177.

Weiland, H. A., Luddens, H., and Seeburg, P. H. (1992). A single histidine in GABA-A receptors in essential for benzodiazepine agonist binding. *J. Biol. Chem.* 267, 1426.

Whiting, P., McKernan, R. M., and Iversen, L. (1990). Another mechanism for creating diversity in gamma-aminobutyrate type A receptors: RNA splicing directs expression of two forms of gamma2 subunit, one of which contains a protein kinase C phosphorylation site. *Proc. Natl. Acad. Sci. U.S.A.* 87, 9966.

Wilcox, A. S., Warrington, J. A., Gardiner, K., Berger, R., Whiting, P., Altherr, M. R., Wasmuth, J. J., Patterson, D., and Sikela, J. M. (1992). Human chromosomal localization of genes encoding the gamma1 and gamma2 subunits of the gamma-aminobutyric acid receptor indicates that members of this gene family are often clustered in the genome. *Proc. Natl. Acad. Sci. U.S.A.* 89, 5857.

Wilson-Shaw, D., Robinson, M., Gambarana, C., Siegel, R. E., and Sikela, J. M. (1991). A novel gamma subunit of the GABA-A receptor identified using the polymerase chain reaction. *FEBS Lett.* 284, 211.

Wisden, W., Herb, A., Weiland, H., Keinanen, K., Luddens, H., and Seeburg, P. H. (1991). Cloning, pharmacological characteristics and expression pattern of the rat GABA-A receptor alpha 4 subunit. *FEBS Lett.* 289, 227.

Wisden, W., Laurie, D. J., Monyer, H., and Seeburg, P. H. (1992). The distribution of 13 GABAA receptor subunit mRNAs in the rat brain. I. telencephalon, diencephalon, mesencephalon. *J. Neurosci.* 12, 1040.

Xia, Y. and Haddad, G. G. (1992). Ontogeny and distribution of GABA-A receptors in rat brainstem and rostral brain regions. *Neuroscience* 49, 973.

Ymer, S., Draguhn, A., Kohler, M., Schofield, P. R., and Seeburg, P. H. (1989). Sequence and expression of a novel GABAA receptor alpha subunit. *FEBS Lett.* 258, 119.

Ymer, S., Draguhn, A., Wisden, W., Werner, P., Keinanen, K., Schofield, P. R., Sprengel, R., Pritchett, D. B., and Seeburg, P. H. (1990). Structural and functional characterization of the gamma1 subunit of GABA-A/benzodiazepine receptors. *EMBO J.* 9, 3261.

Ymer, S., Schofield, R., Draguhn, A., Werner, P., Kohler, M., and Seeburg, P. H. (1989b). GABA-A receptor beta subunit heterogeneity: Functional expression of cloned cDNAs. *EMBO J.* 8, 1665.

Zhang, J. H., Sato, M., and Tohyama, M. (1992). Co-expression of the alpha 1 and beta 2 subunits of the GABA-A receptor in the magnocellular preoptic nucleus. *Mol. Brain Res.* 15, 171.

Zhao, Z. Y. and Joho, R. H. (1990). Isolation of distantly related members in a multigene family using the polymerase chain reaction technique. *Biochem. Biophys. Res. Commun.* 167, 174.

8

Inhibitory Glycine Receptors

Dieter Langosch

2.8.0 Introduction

Glycine and γ-aminobutyric acid (GABA) are the major inhibitory neurotransmitters in the CNS. Glycine predominates in the spinal cord and brainstem, whereas GABAergic synapses are abundant in cortex and cerebellum. Both transmitters activate chloride-selective channel proteins in the postsynaptic membrane. Thereby, membrane depolarization of the cell in response to excitatory signals is efficiently antagonized. Affinity purification of the glycine receptor (GlyR) protein and cloning of corresponding cDNAs have shown that the GlyR is a member of the ligand-gated ion channel superfamily. Functional domains were mapped by electrophysiological analysis of mutated receptors expressed in heterologous cell systems (Langosch et al., 1990a; Becker, 1992; Betz, 1992).

2.8.1 Functional Properties

The single channel characteristics underlying glycine-gated whole-cell currents have been investigated by patch-clamp analysis of cultured mouse spinal neurons and spinal slices (Hamill et al., 1983; Smith et al., 1989; Bormann et al., 1987; Takahashi and Momiyama, 1991; Takahashi et al., 1992; Twyman and MacDonald, 1991). GlyR channels are exquisitely anion selective, and the rank order of anion permeability is $SCN^- > I^- > Br^- > Cl^- > F^-$. These relative permeabilities are different from the respective ion mobilities in water and inversely correlated to the elementary conductances observed with each anion. Quantitative analysis of this property indicated that the channel contains at least two binding sites for permeating anions. The permeability sequence of organic anions is bicarbonate > formate > acetate,

0-8493-8322-6/95/$0.00+$.50

indicating a pore diameter of 0.52 nm for the open channel. Glycine-gated whole-cell currents are sensitive to transmembrane voltage as they exhibit pronounced outward rectification, whereas the single channel current–voltage relationships are linear. This apparent discrepancy was resolved by voltage jump experiments that revealed that the outward rectification of whole-cell currents is caused by a lower probability of channel activation at negative voltages. The GlyR channel adopts a variety of elementary conductance states that appear to depend on the developmental stage of the preparation. For embryonic (E20), neonatal (P2-P3), and adult (P20) spinal slices (Takahashi and Momiyama, 1991; Takahashi et al., 1992), as well as in spinal neurons cultured for 3–4 weeks (Hamill et al., 1983; Smith et al., 1989; Bormann et al., 1987; Twyman and MacDonald, 1991), a main-state of 42–48 pS plus a range of additional conductances (10–12, 18–19, 27–32, and 79 pS) were observed. A large conductance of 93–94 pS was seen at high and low frequency in E20 slices (Takahashi et al., 1992) and cultured neurons (Smith et al., 1989), respectively, but not in adult (P20) slices (Takahashi et al., 1992). This is consistent with a developmental regulation of different GlyR isoforms. Multiple conductance states similar to those described above have also been recorded from postsynaptic neurons in the falling phase of glycinergic inhibitory synaptic currents; this indicates that GlyRs in cultured cells are equivalent to the receptors mediating synaptic inhibition (Takahashi and Momiyama, 1991). GlyR activity appears to be modulated by posttranslational modification. In neurons isolated from spinal trigeminal nucleus, phosphorylation by protein kinase A appears to dramatically increase the glycine-induced Cl^- currents by increasing the frequency of the channel openings (Song and Huang, 1990).

2.8.2 Pharmacology

The agonist profile of the GlyR is limited to only a few ligands. Iontophoretic application of a series of α- and β-amino acids mimics the depressant effect of glycine on motoneurons in the spinal cord. The following rank order of potency was obtained: glycine > β-alanine > taurine > L- and D-α-alanine > L-serine >> D-serine (Werman et al., 1968). These inhibitory efficacies parallel the ability of these amino acids to displace the high-affinity antagonist [^{3}H]strychnine from the receptor, indicating a largely competitive interaction between glycinergic agonists and strychnine. Besides strychnine and its derivatives, there are only a few ligands binding to the GlyR with affinities in the low nanomolar range; these include the steroid RU 5135 (3α-hydroxy-16-imino-5β-17-aza-androstan-11-one) and 1,5-diphenyl-3,7-diazaadamantan-9-ol. The glycinergic ligands as reviewed by Becker (1992) are summarized in Table 1.

Picrotoxin and picrotoxinin, which is the biologically active component of the picrotoxin, are noncompetitive inhibitors and therefore thought to interact with the anion pore of the GlyR. Different affinities for this alkaloid (in the micromolar range) have been reported for glycine responses recorded

Table 1. Ligand Affinities at the Inhibitory Glycine Receptor[a]

Ligand	K_D or K_i[b]	Species
Amino acid agonists		
Glycine	2–40 μ*M*	Rat, mouse,
β-Alanine	4–57 μ*M*	Pig, pigeon
Taurine	29–135 μ*M*	
β-Aminoisobutyric acid	200 μ*M*	Rat
Strychnine and derivatives		
Strychnine	2–14 n*M*[c]	Rat, mouse,
Pseudostrychnine	8–10 n*M*	Pig, pigeon
α-Colubrine	12–18 n*M*	
2-Aminostrychnine	8–23 n*M*	
Brucine	15–180 n*M*	
2-Nitrostrychnine	53–500 n*M*	
Cacotheline	240 n*M*	Rat
N-Methylstrychnine	500 n*M*	Rat
	31 n*M*	Pigeon
Isostrychnine	37–200 n*M*	Rat
	>700 n*M*	Pigeon
THIP and analogues		
Iso-THAZ	1.7–6.6 μ*M*	Rat
4,5-TAZA	2.8–67 μ*M*	Rat
THAZ	20–48 μ*M*	Rat
THIP	39–120 μ*M*	Rat
β-Spiro[pyrrolidinoindolines]		
	>20 μ*M*	Rat
Steroids		
RU 5135	3.6–9.8 n*M*	Rat
$GABA_A$ receptor ligands		
Pitrazepin	70–580 n*M*	Rat
(+)Bicuculline	2–7 μ*M*	Rat
Bicuculline MeCl/MeI	55–77 μ*M*	Rat, mouse
Flunitrazepam	9 μ*M*	Rat
Bromazepam, nitrazepam	10 μ*M*	Rat
Diazepam	13 μ*M*	Rat
PK 8165	4–14 μ*M*	Rat
Norhamane	52 μ*M*	Rat
Avermectin B_{1a}	0.8–1.3 μ*M*	Rat
Opioids		
Thebaine	900 n*M*	Rat
Sinomenine	23 μ*M*	Rat
L-Methadone	30 μ*M*	Rat
Alkaloids		
Boldine	67 μ*M*	Rat
Gelsemine	18 μ*M*	
Hydrastine	13 μ*M*	
Laudanosine	16 μ*M*	
Vincamine	31 μ*M*	
Miscellaneous agents		
1,5-Diphenyl-3,7,-diazaadamantan-9-ol	27 n*M*	Rat
Imipramine	17 μ*M*	Pigeon

Table 1. (Continued).

[a] THIP, 4,5,6,7-tetrahydroisoxazolo[5,4-*c*]pyridin-3-ol; THAZ, 5,6,7,8-tetrahydro-4*H*-isoxazolo-[4,5-d]azepin-3-ol; 4,5-TAZA, 2,3,6,7-tetrahydro-1*H*-azepine-4-carboxylic acid; RU 5135, 3α-hydroxy-16-imino-5β-17-aza-androstan-11-one; PK 8165, 2-phenyl-4-(2-(4-piperidinyl)ethyl)quinoline.

[b] Ligand affinities were derived from published displacement data of [^{3}H]strychnine binding. The table is modified from Becker (1992) in which the complete bibliography is given. (With kind permission from Springer-Verlag, Heidelberg.)

[c] A neonatal glycine receptor isoform containing the α2*-subunit is characterized by low affinity strychnine binding *($K_i \approx 10$ μM)* (Becker et al., 1988).

from a variety of CNS regions at different developmental stages (Davidoff and Aprison, 1969; Evans, 1978; Barker et al., 1983) pointing to heterogeneity of the respective GlyRs.

2.8.3 Receptor Structure and Subtypes

2.8.3.1 Purification and Reconstitution

Upon solubilization from rat spinal cord membranes with Triton X-100 or Na-cholate, picomole quantities of GlyR protein can be affinity purified in a one-step procedure using 2-aminostrychnine coupled to an agarose matrix (Pfeiffer et al., 1982; Graham et al., 1985). The purified receptor is a large (250-kDa) glycoprotein that contains membrane-spanning α-(48 kDa) and β-subunits (58 kDa) arranged as a pentamer (Langosch et al., 1988; Schmitt et al., 1987). *In vitro,* phosphorylation by protein kinase C specifically labeled the α-subunit on serine residues; the functional significance of this modification, however, is unclear (Ruiz-Gomez et al., 1991). Upon reconstitution in liposomes, the purified receptor mediates glycine-activated and strychnine-sensitive chloride currents (Garcia-Calvo et al., 1992; Riquelme et al., 1990).

Gephyrin (93 kDa) is a copurifying, peripheral polypeptide associated with cytoplasmic domains of the GlyR (Schmitt et al., 1987; Altschuler et al., 1986). Its ability to bind cooperatively and with high affinity to tubulin suggested that gephyrin links GlyRs to subsynaptic microtubules (Kirsch et al., 1991).

2.8.3.2 Primary Structure of Subunits and Subunit I Isoforms

Based on the sequences of tryptic peptides obtained from purified GlyR, cDNAs corresponding to its α1- (Grenningloh et al., 1987) and β-subunit (Grenningloh et al., 1990a) were cloned from a rat library. Homology screening subsequently identified additional α-subunit cDNAs, rat α2 (Kuhse et al., 1991) and α2* (Kuhse et al., 1990a), human α1 and α2 (Grenningloh et al., 1990b), and rat α3 (Kuhse et al., 1990b). Partial genomic sequences encoding an α4-subunit have been identified (B. Matzenbach, Y. Maulet, and H. Betz, unpublished). Homologs to the β-subunit could, however, not yet be isolated.

The deduced amino acid sequences are homologous to each other; an identity of ≥82% is shared by the different a subunits whereas the identity between the α1- and β-subunit amounts to only 47% (Becker, 1992). An alignment of the deduced amino acid sequences from rat is shown in Figure 1 and their evolutionary distances depicted by the phylogenetic tree in Figure 2.

GlyR subunits are homologous to γ-aminobutyric acid and acetylcholine receptors; together, these proteins constitute the large superfamily of ligand-gated ion channels (Langosch et al., 1990b). GlyR polypeptides share a number of structural principles common to all members of this superfamily (Figure 3). A cleaved signal sequence is followed by a large extracellular N-terminal domain containing potential N-glycosylation sites and a typical pair of cystein residues. Four hydrophobic segments (M1–M4) are predicted to span the lipid bilayer. A large intracellular loop connecting M3 to M4 contains consensus sites for phosphorylation (Ruiz-Gomez et al., 1991) and constitutes the most divergent region of these polypeptides (Grenningloh et al., 1987, 1990a; Langosch et al., 1990b).

Alternative splicing contributes to the diversity of GlyR α polypeptides. An α1-subunit variant, termed $\alpha 1_{ins}$, contains an eight amino acid insertion between M3 and M4 and is created by alternative splice acceptor site selection (Malosio et al., 1991a). Alternative usage of two versions of exon 3 gives rise to two α2-subunit variants, α2A and α2B, of which the former is more abundantly expressed in the CNS (Kuhse et al., 1991).

The α2-subunit gene is located on the short arm (Sp21.2–p22.1) of the human X chromosome, whereas the α1-subunit gene is autosomally located (Grenningloh et al., 1990b).

2.8.4 Localization

2.8.4.1 Distribution of GlyR Polypeptides

For immunocytochemical analyses, monoclonal antibodies have been generated that recognize epitopes within the extracellular domains of the GlyR subunits: mAb2b and mAb4a (Pfeiffer et al., 1984; Schröder et al., 1991). Staining with these antibodies indicated GlyR immunoreactivity in rat spinal cord and brainstem (van den Pol and Gorcs, 1988; Basbaum, 1988; Triller et al., 1985). This pattern coincides with the distribution of GlyRs as defined by $[^3H]$strychnine binding (Zarbin et al., 1981; Probst et al., 1988). In addition to that, GlyR-specific immunolabeling was observed in higher brain regions (olfactory bulb, cerebellum, and cerebral cortex) (van den Pol and Gorcs, 1988; Triller et al., 1987; Becker et al., 1993; Kirsch and Betz, 1993) where no high affinity $[^3H]$strychnine binding had been reported (Zarbin et al., 1981; Probst et al., 1988).

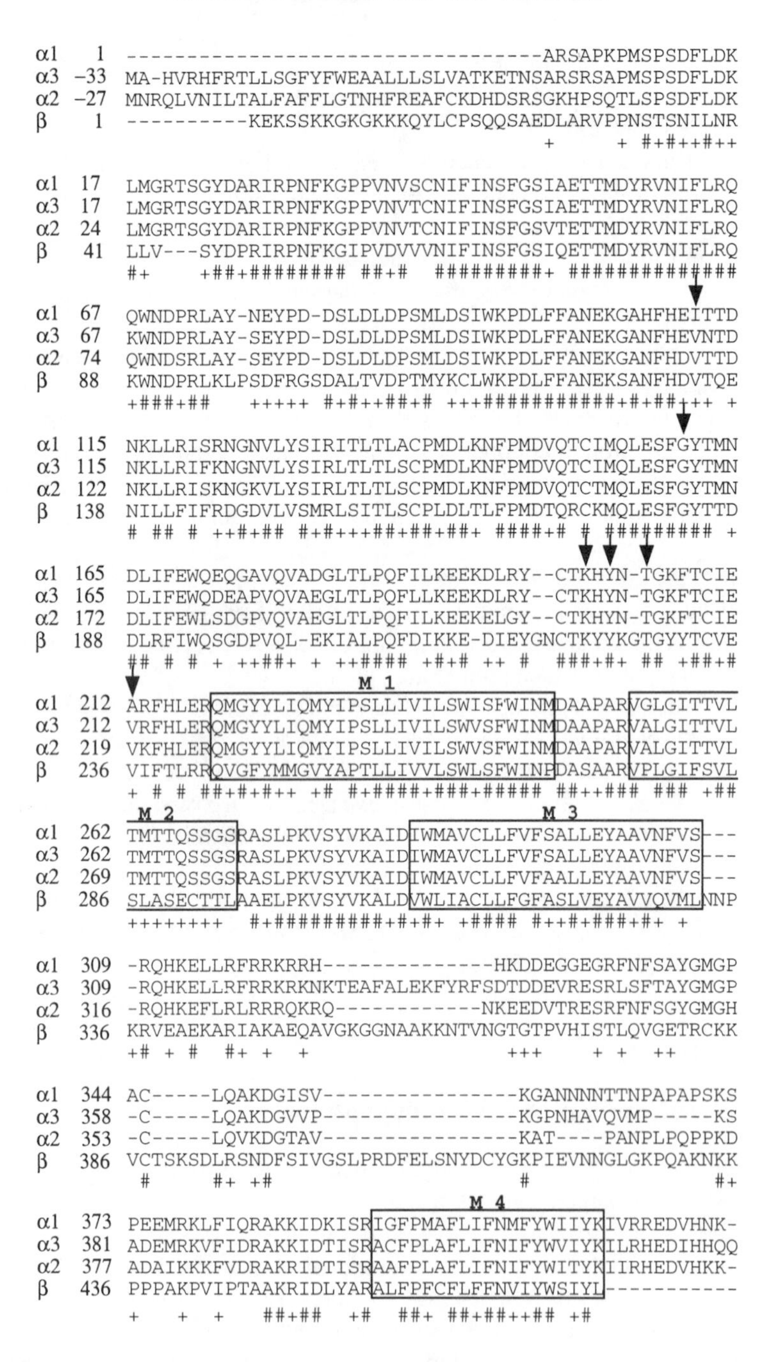

FIGURE 1. Alignment of rat GlyR α- and β-subunit primary structures predicted from cDNA sequences. Conserved and conservatively exchanged positions are marked by # and +, respectively. Predicted transmembrane segments are boxed and amino acid stretches implicated in ligand binding to the α-subunit are marked by arrows.

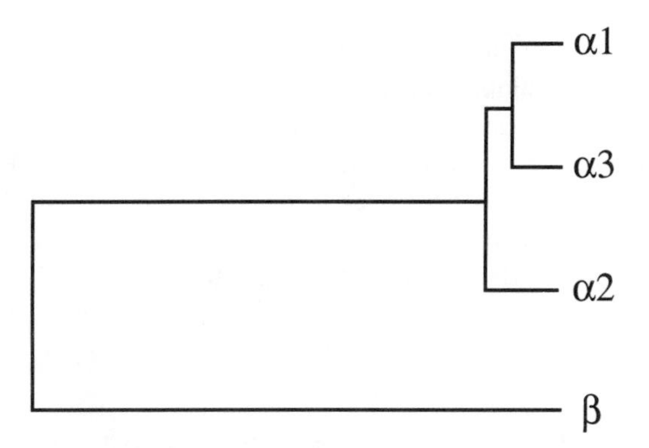

FIGURE 2. Phylogenetic tree (Feng and Doolittle, 1990) of GlyR subunits. The length of each branch of the tree correlates with the evolutionary distance between the different polypeptides.

2.8.4.2 Localization of GlyR Transcripts

GlyR subunit mRNAs have been localized not only in spinal cord and brainstem but also in various higher brain regions by Northern blot analysis (Grenningloh et al., 1987, 1990a), polymerase chain amplification (Kuhse et al., 1990a,b), and *in situ* hybridization (Malosio et al., 1991b; Sato et al., 1991; Fujita et al., 1991). Transcripts of the α1-subunit type predominated in spinal cord and brainstem but were also found in regions of the diencephalon and mesencephalon; α2 mRNA was expressed in cerebral cortex and the

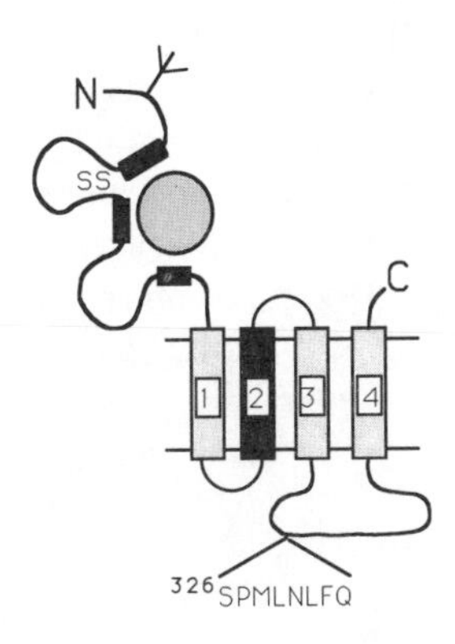

FIGURE 3. Model of the GlyR α1-subunit. The extracellular domain comprises three segments (symbolized by shaded boxes) contributing to the ligand-binding sites. It is followed by four predicted transmembrane segments (M1–M4). The bound ligand is indicated by a shaded sphere and the channel-lining M2 segment is highlighted by darker shading. Y Denotes a potential glycosylation site and SS a predicted disulfide bridge between cysteines 138 and 154. The eight amino acid insertion starting with serine 326 in the cytoplasmic loop separating M3 and M4 is characteristic of $\alpha 1_{ins.}$

hippocampal complex but only weakly in spinal cord; α3 mRNA was found mainly in cerebellum but also in the olfactory bulb and hippocampus. Transcripts encoding the β-subunit are strongly expressed in most CNS regions examined and thus exceed α-subunit-specific mRNAs. Distribution of the α1-subunit transcript correlates well with that of [^{3}H]-strychnine binding; therefore, it was concluded, that the α1 subunit represents the major component of adult GlyRs (Malosio et al., 1991b; Sato et al., 1991).

2.8.5 Heterologous Expression

GlyR α- and β-subunits were functionally expressed in *Xenopus* oocytes and HEK-293 cells. Expression of homooligomeric receptors containing only α-subunits was sufficient to generate large chloride currents cooperatively gated by glycine (Hill coefficients ≈3). These currents were inhibited by nanomolar concentrations of strychnine and micromolar concentrations of picrotoxinin (Kuhse et al., 1990a,b; Grenningloh et al., 1990b; Schmieden et al., 1989, 1992; Sontheimer et al., 1989; Pribilla et al., 1992; Takagi et al., 1992). When α- and β-subunits were coexpressed in HEK-293 cells, glycine application elicited whole cell currents that where resistant to block by picrotoxinin. At the same time, an antiserum specific for the β-subunit precipitated [^{3}H]strychnine binding sites from the solubilized membranes of α/β coexpressing cells (Pribilla et al., 1992). Upon nuclear injection of a β-subunit cDNA into oocyte nuclei, low level chloride currents could be activated by high glycine concentrations (EC_{50} = 24 m*M)* but not blocked by strychnine (Grenningloh et al., 1990a). In conclusion, it appears that α- but not β-subunits efficiently assemble to form functional homooligomeric GlyRs; upon coexpression, α/β heterooligomers are formed with pharmacological properties different from that of α homooligomers.

Single channels recorded upon expression of α1- or α2-subunits in *Xenopus* oocytes had elementary conductances of 75 pS and 88 pS, respectively (Takahashi et al., 1992). A similar value (86 pS) was obtained when recording from HEK-293 cells transfected with α1-subunit cDNA (Langosch et al., 1992).

2.8.6 Functional Domains

2.8.6.1 Ligand-Binding Sites

2.8.6.1.1 Interaction of Ligand Binding Domains

Glycine responses recorded from spinal neurons (Werman et al., 1968; Lewis et al., 1989; Tokutomi et al., 1989) or cells heterologously expressing the GlyR (Schmieden et al., 1989, 1992; Sontheimer et al., 1989) show positive cooperativity characterized by Hill coefficients of 1.7–3.0. This suggests that at least three agonist molecules must bind to activate the receptor. Several lines of evidence suggest that the binding domains for amino acid agonists and the inhibitor strychnine are distinct but overlap to different degrees with individual α-subunit isoforms (O'Connor, 1989; Aprison, 1990; Young and Snyder, 1974). (1) Channel activation by glycine is competitively antagonized by strychnine albeit in a noncooperative way (Schmieden et al., 1989; Sontheimer et al., 1989; Aprison, 1990). (2) Gating of the GlyR by the agonists β-alanine and taurine lacks cooperativity (Schmieden et al., 1989, 1992). (3) The β-alanine and taurine responses of α1 but not of α2 GlyRs are highly sensitive to strychnine (Schmieden et al., 1992).

2.8.6.1.2 Structure of Ligand-Binding Domains

Biochemical and molecular biology approaches have allowed a more detailed analysis of the GlyR's ligand binding site. [^{3}H]Strychnine is covalently incorporated into the α-subunit of the GlyR in a glycine-displaceable way upon UV illumination (Pfeiffer et al., 1982; Graham et al., 1983, 1985). Further, glycine activation of homooligomeric recombinant α-subunit GlyRs is strychnine-sensitive (Kuhse et al., 1990b; Schmieden et al., 1989, 1992; Sontheimer et al., 1989). Together, these results qualify the a subunit as ligand-binding polypeptide. Analysis of proteolytic fragments of native GlyR labeled with [^{3}H]strychnine limited the site of incorporation to amino acids 171–220 of the rat α1-subunit (Graham et al., 1983; Ruiz-Gomez et al., 1990). Additional information about the position of ligand-binding sites arose from the functional comparison of α-subunit variants and from site-directed mutagenesis. A rat α2-subunit variant, termed α2*, differs from α2-subunits in carrying a glutamate residue at position 167 in place of a glycine. Expressed in *Xenopus* oocytes, this variant displayed a 30- to 40-fold lowered affinity for glycine and about a 500-fold reduced sensitivity toward strychnine. Mutagenesis confirmed that residue 167 is an important determinant of agonist and antagonist binding to the α2-subunit (Kuhse et al., 1990a). An equivalent domain of the human α1-subunit including Gly-160 and Tyr-161 plus another site around Lys-200 and Tyr-202 was identified as contributing to the binding site for strychnine by analyzing systematically mutated receptors (Vandenberg et al., 1992a). By a similar approach, Thr-204 of the human α1-subunit, which is close to the second strychnine-binding domain, was found central to its ability to bind glycine, but not strychnine (Vandenberg et al., 1992b).

The agonists β-alanine and taurine activated recombinant GlyR isoforms with different efficacies relative to glycine (Kuhse et al., 1990b; Schmieden et al., 1992). These agonists elicited maximal whole-cell responses that, when related to the glycinergic currents, where significantly weaker with rat α2 or α3 GlyRs as compared to those seen with rat α1 receptors. A systematic interchange of residues different between α1 and α2/α3 subunits revealed that Ile-111 and Ala-212 of the rat α1-subunit determine its high relative responses to β-alanine and taurine (Schmieden et al., 1992).

Taken together, these findings imply that three different polypeptide segments of the extracellular N-terminal domain of the α-subunit contribute to the formation of the ligand-binding pockets. A different degree of allosteric coupling between these domains may be responsible for the different pharmacological properties observed with the various α-subunit GlyRs.

2.8.6.2 Channel-Forming Domain

Recombinant α/β heterooligomers are virtually resistant to block by the noncompetitive antagonist picrotoxinin whereas α homooligomers are sensitive to micromolar concentrations (Pribilla et al., 1992). When a β chimera, in which the second transmembrane segment M2 was replaced by the corresponding α sequence, was coexpressed with α-subunits, picrotoxinin-sensitive channels were produced. As picrotoxinin is generally regarded as a channel blocker, this indicates that the M2 segment is a channel-forming domain.

A synthetic peptide corresponding to the M2 segment of the α1-subunit produced channel activity in liposomes and planar lipid bilayers (Langosch et al., 1991). The ionic selectivity of these peptide channels was modulated by charged amino acids bordering the M2 segment. Thus, these residues may be the structural correlate of the anion-binding sites determined earlier for native GlyRs (Bormann et al., 1987).

2.8.7 Developmental Regulation

2.8.7.1 Neonatal GlyR

In the spinal cord of newborn rodents, a distinct neonatal GlyR isoform is expressed (Becker et al., 1988). This protein exhibits strychnine binding of low affinity and contains an α-subunit different from the adult isoform in molecular weight (49 vs. 48 kDa) and antigenic properties (Schröder et al., 1990; Becker et al., 1988; Hoch et al., 1990). The neonatal isoform of rat spinal cord is fully replaced by the adult GlyR within two weeks after birth (Becker et al., 1988); cortical GlyRs that correspond to the neonatal isoform, however, persist at low levels in adulthood (Becker et al., 1993). Primary cultures of mouse spinal cord predominantly express the neonatal form (Hoch et al., 1990) but are stimulated to express the adult GlyR in the presence of the *N*-methyl-D-aspartate receptor antagonists 2-amino-5-phosphonovalerate or MK-801 (Hoch et al., 1992).

2.8.7.2 Regulation of Transcript Synthesis

In situ hybridization and polymerase chain amplification have shown that the expression level of transcripts corresponding to the α1-subunit increases during development whereas α2 transcript expression decreases from high prenatal to low adult levels. Transcripts of the α3- and β-subunits are expressed at all stages of development (Kuhse et al., 1991; Malosio et al., 1991b).

A differential regulation of GlyR transcripts was also shown by recording glycine-activated currents in *Xenopus* oocytes expressing poly(A)$^+$ mRNA obtained from different developmental stages. Sucrose density gradient fractionation separated two classes of GlyR transcripts. A rapidly sedimenting GlyR mRNA prevailed in adult spinal cord and a light mRNA in neonatal spinal cord and adult cortex (Akagi and Miledi, 1988). Translation of the high-mol wt adult mRNA species was suppressed in the presence of antisense oligonucleotides specific for the α1-subunit whereas the light fraction from neonatal spinal cord was not affected (Akagi and Miledi, 1989). It was concluded that neonatal GlyR contains α2- and/or α2*-subunits, a notion supported by patch clamp recordings of neonatal spinal slices (see Section 2.8.1). During development, this receptor apparently is replaced by the adult isoform containing α1- plus β-subunits (Becker, 1992).

2.8.8 Pathology

A role of GlyR deficiencies in the pathogenesis of neurological disorders has been demonstrated for two hereditary diseases, the "spastic" mutation of the mouse, which has been assigned to chromosome 3 (Eicher and Lane, 1980; Becker, 1990) and inherited myoclonus, occurring in Hereford cattle (Gundlach, 1990). Animals homozygotic for these mutations exhibit disturbed motor functions that closely resemble the symptoms of sublethal strychnine poisoning. This physiological parameter is paralleled by a severe reduction of [^{3}H]strychnine-binding sites in membranes derived from spinal cord and other areas of the affected animals (Gundlach, 1990; White, 1985; Becker et al., 1986). More recent data indicate that the "spastic" mutation of *spa/spa* mice selectively interferes with the postnatal accumulation of the adult GlyR isoform, whereas perinatal expression of the neonatal receptor is not detectably affected. These data suggest that the mutant phenotype is caused by a selective deficit in developmental regulation and/or structure of adult GlyR (Becker et al., 1992).

Acknowledgments

I thank Drs. H. Betz and J. Bormann for support and critical reading of the manuscript. This work was supported by the Deutsche Forschungsgemeinschaft (SFB 169) and the Fonds der Chemischen Industrie.

REFERENCES

Akagi, H. and Miledi, R. (1988). Heterogeneity of glycine receptors and their messenger RNAs in rat brain and spinal cord. *Science* 242, 270.

Akagi, H. and Miledi, R. (1989). Discrimination of heterogeneous mRNAs encoding strychnine-sensitive glycine receptors in *Xenopus* oocytes by antisense oligonucleotides. *Proc. Natl. Acad. Sci. U.S.A.* 86, 8103.

Altschuler, R. A., Betz, H., Parakkal, M., Reeks, K., and Wenthold, R. (1986). Identification of glycinergic synapses in the cochlear nucleus through immunocytochemical localization of the postsynaptic receptor. *Brain Res.*, 369, 316.

Aprison, M. H. (1990). The discovery of the neurotransmitter role of glycine. In *Glycine Neurotransmission*. Ottersen, O. and Storm-Mathisen, V., Eds., Wiley, Chichester, England, pp. 1–23.

Barker, J. L., McBurney, R. N., and Mathers, D. A. (1983). Convulsant-induced depression of amino acid responses in cultured mouse spinal neurones studied under voltage clamp. *Br. J. Pharmacol.* 80, 619.

Basbaum, A. I. (1988). Distribution of glycine receptor immunoreactivity in the spinal cord of the rat: Cytochemical evidence for a differential glycinergic control of lamina I and V nociciptive neurons. *J. Comp. Neurol.* 6, 1358.

Becker, C.-M. (1992). Convulsants acting on the inhibitory glycine receptor. In *Handbook of Experimental Pharmacology*. Herken, H. and Hucho, F., Eds., Springer, Heidelberg, Germany, pp. 539–575.

Becker, C.-M., (1990). Disorders of the inhibitory glycine receptor: The spastic mouse. *FASEB J.* 4, 2767.

Becker, C.-M., Betz, H., and Schröder, H. (1993). Expression of inhibitory glycine receptors in postnatal rat cerebral cortex. *Brain Res.* 606, 220.

Becker, C.-M., Herrmans-Borgmeyer, I., Schmitt, B., and Betz, H. (1986). The glycine receptor deficiency of the mutant mouse spastic: Evidence for normal glycine receptor structure and localization. *J. Neurosci.* 6, 1358.

Becker, C.-M., Hoch, W., and Betz, H. (1988). Glycine receptor heterogeneity in rat spinal cord during postnatal development. *EMBO J.* 7, 3717.

Becker, C.-M., Schmieden, V., Tarroni, P., Strasser, U., and Betz, H. (1992). Isoform-selective deficit of glycine receptors in the mouse mutant spastic. *Neuron* 8, 283.

Betz, H. (1992). Structure and function of inhibitory glycine receptors. *Q. Rev. Biophys.* 25, 381.

Bormann, J., Hamill, O. P., and Sakman, B. (1987). Mechanism of anion permeation through channels gated by glycine and γ-aminobutyric acid in mouse cultured spinal cord neurons. *J. Physiol.* 385, 243.

Davidoff, R. A. and Aprison, M. H. (1969). Picrotoxin antagonism of the inhibition of interneurons by glycine. *Life Sci.* 8, 107.

Eicher, E. M. and Lane, P. (1980). Assignment of LG XVI to chromosome 3 in the mouse. *J. Hered.* 71, 315.

Evans, R. H. (1978). The effect of amino acids and antagonists on the isolated hemisected spinal cord of the immature rat. *Br. J. Pharmacol.* 62, 171.

Feng, D. F. and Doolittle, R. F. (1990). Progressive alignment and phylogenetic tree construction of protein sequences. *Methods Enzymol.* 183, 375.

Fujita, M., Sato, K., Sato, M., Inoue, T., Kozuka, T., and Tohyama, M. (1991). Regional distribution of the cells expressing glycine receptor β subunit mRNA in the rat brain. *Brain Res.* 560, 23.

Garcia-Calvo, M., Valdivieso, F., Mayor, F., Jr., and Vazques, J. (1992). Sensitive procedures for measuring chloride fluxes mediated by the purified glycine receptor incorporated into phospholipid vesicles. *Neurosci. Lett.* 136, 102.

Graham, D., Pfeiffer, F., and Betz, H. (1983). Photoaffinity-labelling of the glycine receptor of rat spinal cord. *Eur. J. Biochem.* 131, 519.

Graham, D., Pfeiffer, F., Simler, R., and Betz, H. (1985). Purification and characterization of the glycine receptor of pig spinal cord. *Biochemistry* 24, 990.

Grenningloh, G., Pribilla, I., Prior, P., Multhaup, G., Beyreuther, K., Taleb, O., and Betz, H. (1990a). Cloning and expression of the 58 kd β subunit of the inhibitory glycine receptor. *Neuron* 4, 963.

Grenningloh, G., Rienitz, A., Schmitt, B., Methfessel, C., Zensen, M., Beyreuther, K., Gundelfinger, E. D., and Betz, H. (1987). The strychnine-binding subunit of the glycine receptor shows homology with nicotinic acetylcholine receptors. *Nature* 328, 215.

Grenningloh, G., Schmieden, V., Schofield, P. R., Seeburg, P. H., Siddique, T., Mohandas, T. K., Becker, C.-M., and Betz, H. (1990b). Alpha subunit variants of the human glycine receptor: Primary structures, functional expression and chromosomal localization of the corresponding gene. *EMBO J.* 9, 771.

Gundlach, A. L. (1990). Disorder of the inhibitory glycine receptor: Inherited myoclonus in Hereford calves. *FASEB J.* 4, 2761.

Hamill, O. P., Bormann, J., and Sakmann, B. (1983). Activation of multiple-conductance state chloride channels in spinal neurones by glycine and GABA. *Nature* 305, 805.

Hoch, W., Betz, H., and Becker, C.-M. (1990). Primary cultures of mouse spinal cord express the neonatal isoform of the inhibitory glycine receptor. *Neuron* 3, 339.

Hoch, W., Betz, H., Schramm, M., Wolters, I., and Becker, C.-M. (1992). Modulation by NMDA receptor antagonists of glycine receptor isoform expression in cultured spinal cord neurons. *Eur. J. Neurosci.* 4, 389.

Kirsch, A. and Betz, H. (1993). Widespread expression of gephyrin, a putative receptor-tubulin linker protein in rat brain. *Brain Res.* 621, 301.

Kirsch, J., Langosch, D., Prior, P., Littauer, U. Z., Schmitt, B., and Betz, H. (1991). The 93-kDa glycine receptor-associated protein binds to tubulin. *J. Biol. Chem.* 266, 22242.

Kuhse, J., Kuryatov, A., Maulet, Y., Malosio, M. L., Schmieden, V., and Betz, H. (1991). Alternative splicing generates two isoforms of the α2 subunit of the inhibitory glycine receptor. *FEBS Lett.* 283, 73.

Kuhse, J., Schmieden, V., and Betz, H. (1990a). A single amino acid exchange alters the pharmacology of neonatal rat glycine receptor subunit. *Neuron* 5, 867.

Kuhse, J., Schmieden, V., and Betz, H. (1990b). Identification and functional expression of a novel ligand binding subunit of the inhibitory glycine receptor. *J. Biol. Chem.* 265, 22317.

Langosch, D., Becker, C.-M. and Betz, H. (1990a). The inhibitory glycine receptor: A ligand gated chloride channel of the central nervous system. *Eur. J. Biochem.* 194, 1.

Langosch, D., Becker, C.-M., and Betz, H. (1990b). The inhibitory glycine receptor: Ligand-gated chloride channel of the central nervous system. *Eur. J. Biochem.* 194, 1.

Langosch, D., Hartung, K., Grell, E., Bamberg, E., and Betz, H. (1991). Ion channel formation by synthetic transmembrane segments of the inhibitory glycine receptor — a model study. *Biochim. Biophys. Acta* 1063, 36.

Langosch, D., Pribilla, I., Takagi, T., Hartung, K., and Bormann, J. (1992). The inhibitory glycine receptor: Structure-function studies on a neuronal chloride channel. In *Membrane Proteins: Structures, Interactions and Models*. Pullman, A., Jortner, J., and Pullman, B., Eds., Kluwer, Dordrecht, The Netherlands, pp. 233-241.

Langosch, D., Thomas, L., and Betz, H. (1988). Conserved structure of ligand gated ion channels: The inhibitory glycine receptor is a pentamer. *Proc. Natl. Acad. Sci. U.S.A.* 85, 7394.

Lewis, C. A., Ahmed, Z., and Faber, D. S. (1989). Characteristics of glycine-activated conductances in cultured medullary neurons from embryonic rat. *Neurosci. Lett.* 96, 185.

Malosio, M. L., Grenningloh, G., Kuhse, J., Schmieden, V., Schmitt, B., Prior, P., and Betz, H. (1991a). Alternative splicing generates two variants of the α1 subunit of the inhibitory glycine receptor. *J. Biol. Chem.* 266, 2048.

Malosio, M.-L., Marqueze-Pouey, B., Kuhse, J., and Betz, H. (1991b). Widespread expression of glycine receptor subunit mRNAs in the adult and developing brain. *EMBO J.* 10, 2401.

O'Connor, V. M. (1989). Chemical modification of overlapping but conformationally distinct recognition sites for glycine and strychnine in isolated spinal cord membranes. *J. Physiol.* 415, 49.

Pfeiffer, F., Graham, D., and Betz, H. (1982). Purification by affinity chromatography of the glycine receptor of rat spinal cord. *J. Biol. Chem.* 257, 9389.

Pfeiffer, F., Simler, R., Grenningloh, G., and Betz, H. (1984). Monoclonal antibodies and peptide mapping reveal structural similarities between the subunits of the glycine receptor of rat spinal cord. *Proc. Natl. Acad. Sci. U.S.A.* 81, 7224.

Pribilla, I., Takagi, T., Langosch, D., Bormann, J., and Betz, H. (1992). Atypical M2 segment of the β subunit confers picrotoxin resistance to inhibitory glycine receptor channels. *EMBO J.* 11, 4305.

Probst, A., Cortes, R., and Palacios, J. M. (1988). The distribution of glycine receptors in the human brain: A light microscopic autoradiographic study using [3H]-Strychnine. *Neuroscience* 17, 11.

Riquelme, G., Morato, E., Lopez, E., Ruiz-Gomez, A., Ferragut, J. A., Gonzalez-Ros, J. M., and Mayor, F., Jr. (1990). Agonist binding to purified glycine receptor reconstituted into giant liposomes elicits two types of chloride currents. *FEBS Lett.* 276, 54.

Ruiz-Gomez, A., Morato, E., Garcia-Calvo, M., Valdivieso, F., and Mayor, F., Jr. (1990). Localization of the strychnine binding site on the 48-kilodalton subunit of the glycine receptor. *Biochemistry* 29, 7033.

Ruiz-Gomez, A., Vaello, M. L., Valdivieso, F., and Mayor, F. (1991). Phosphorylation of the 48-kDa subunit of the glycine receptor by protein kinase. C, *J. Biol. Chem.* 266, 559.

Sato, K., Zhang, J. H., Saika, T., Sato, M., Tada, K., and Tohyama, M. (1991). Localization of glycine receptor α1 subunit mRNA-containing neurons in the rat brain: An analysis using in situ hybridization histochemistry. *Neuroscience* 43, 381.

Schmieden, V., Grenningloh, G., Schofield, P. R., and Betz, H. (1989). Functional expression in *Xenopus* oocytes of the strychnine binding 48 kd subunit of the glycine receptor. *EMBO J.* 8, 695.

Schmieden, V., Kuhse, J., and Betz, H. (1992). Agonist pharmacology of neonatal and adult glycine receptor α subunits: Identification of amino acid residues involved in taurine activation. *EMBO J.* 11, 2025.

Schmitt, B., Knaus, P., Becker, C.-M., and Betz, H. (1987). The Mr 93,000 polypeptide of the postsynaptic glycine receptor is a peripheral membrane protein. *Biochemistry* 26, 805.

Schröder, S., Hoch, W., Becker, C.-M., Grenningloh, G., and Betz, H. (1991). Mapping of antigenic epitopes on the α1 subunit of the inhibitory glycine receptor. *Biochemistry* 30, 42.

Smith, S. M., Zorec, R., and McBurney, R. N. (1989). Conductance states activated by glycine and GABA in rat cultured spinal neurones. *J. Membr. Biol.* 108, 45.

Song, Y. and Huang, L. M. (1990). Modulation of glycine receptor chloride channels by cAMP-dependent protein kinase in spinal trigeminal neurons. *Nature* 348, 242.

Sontheimer, H., Becker, C.-M., Pritchett, D. B., Schofield, P. R., Grenningloh, G., Kettenmann, H., Betz, H., and Seeburg, P. H. (1989). Functional chloride channels by mammalian cell expression of rat glycine receptor subunit. *Neuron* 2, 1491.

Takagi, T., Pribilla, I., Kirsch, J., and Betz, H. (1992). Coexpression of the receptor-associated protein gephyrin changes the ligand binding affinities of α2 glycine receptors. *FEBS Lett.* 303, 178.

Takahashi, T. and Momiyama, A. (1991). Single-channel currents underlying glycinergic inhibitory postsynaptic responses in spinal neurons. *Neuron* 7, 965.

Takahashi, T., Momiyama, A., Hirai, K., Hishinuma, A., and Akagi, H. (1992). Functional correlation of fetal and adult forms of glycine receptors with developmental changes in inhibitory synaptic receptor channels. *Neuron* 9, 1155.

Tokutomi, N., Kaneda, M., and Akaike, N. (1989). What confers specificity on glycine for its receptor site. *Br. J. Pharmacol.* 97, 353.

Triller, A., Cluzeaud, F., and Korn, H. (1987). Gamma-aminobutyric acid-containing terminals can be apposed to glycine receptors at central synapses. *J. Cell Biol.* 104, 947.

Triller, A., Cluzeaud, F., Pfeiffer, F., Betz, H., and Korn, H. (1985). Distribution of glycine receptors at central synapses: An immunoelectron microscopicy study. *J. Cell Biol.* 101, 683.

Twyman, R. E. and MacDonald, R. L. (1991). Kinetic properties of the glycine receptor main-and sub-conductance states of mouse spinal cord neurones in culture. *J. Physiol.* 435, 303.

van den Pol, A. N. and Gorcs, T. (1988). Glycine and glycine receptor immunoreactivity in brain and spinal chord. *J. Neurosci.* 8, 472.

Vandenberg, R. J., French, C. R., Barry, P. H., Shine, J., and Schofield, P. R. (1992a). Antagonism of ligand-gated ion channel receptors: Two domains of the glycine receptor α subunit form the strychnine-binding site. *Proc. Natl. Acad. Sci. U.S.A.* 89, 1765.

Vandenberg, R. J., Handford, D. A., and Schofield, P. R. (1992b). Distinct agonist- and antagonist-binding sites on the glycine receptors. *Neuron* 9, 491.

Werman, R., Davidoff, R. A., and Aprison, M. A. (1968). Inhibitory action of glycine on spinal neurons in the cat. *Nature* 214, 681.

White, W. F. (1985). The glycine receptor in the mutant mouse spastic (spa): Strychnine binding characteristics and pharmacology. *Brain Res.* 329, 1.

Young, A. and Snyder, S. H. (1974). The glycine synaptic receptor: Evidence that strychnine binding is associated with the ionic conductance mechanism. *Proc. Natl. Acad. Sci. U.S.A.* 71, 4002.

Zarbin, M. A., Wamsley, J. K., and Kuhar, M. J. (1981). Glycine receptor: Light microscopic autoradiographic localization with [3H]-strychnine. *J. Neurosci.* 1, 532.

9

Cyclic Nucleotide-Gated Channels

King-Wai Yau and Tsung-Yu Chen

2.9.0 Introduction

Cyclic nucleotide-gated ion channels are relatively new entries in the world of ion channels. They began with the surprising discovery in the mid-1980s that the cation channel mediating phototransduction in retinal photoreceptors is directly activated by intracellular cGMP (Fesenko et al., 1985; Yau and Nakatani, 1985). Until that time, it was known only that certain ligand-gated and voltage-gated channels can be phosphorylated by cyclic nucleotide-dependent protein kinases, with resulting modulations of channel function (for review, see Swope et al., 1992; Levitan, 1985, 1994). The direct gating of ion channels by cyclic nucleotides represents a departure from the conventional course of action of these second messengers, which invariably involves kinases and phosphorylations of effector proteins. The advantage of direct gating by cyclic nucleotides is obviously one of speed and efficiency, which may be particularly relevant to, for example, sensory transduction and control of heart rate. Over the past several years, the number of ion channels shown to be affected by direct cyclic nucleotide binding has rapidly multiplied. For the sake of clarity, it may be useful here to divide this collection of channels into two categories: (1) cyclic nucleotide-activated channels, which require cyclic nucleotide binding in order to open at all, and (2) cyclic nucleotide-modulated channels, the openings of which do not obligatorily require cyclic nucleotide binding, but are affected by it. At present, the better understood channels belong to the first category.

2.9.1 Cyclic Nucleotide-Activated Channels

These channels all share the feature that cyclic nucleotide binding is required before channel opening can occur. In this sense, they are similar to conventional

0-8493-8322-6/95/$0.00+$.50

ligand-gated channels like the acetylcholine, GABA, glycine, and glutamate channels, except that the cyclic nucleotide binding site is intracellular rather than extracellular as in the other channels. At present there are five established members of this family: (1) the cGMP-activated channel in retinal rod cells, (2) a similar but distinct channel in retinal cone cells, (3) the cGMP/cAMP-activated channel in olfactory receptor neurons, (4) the cGMP-activated channel in bird pinealocytes, and (5) the cGMP-activated channel in invertebrate photoreceptors. The first three have been molecularly cloned, and they show strong structural homology to each other. The last two also have physiological properties similar to the others. In addition, there is some evidence for a putative cGMP-activated channel in certain bipolar cells of the retina. Finally, there are suggestions of a cAMP-activated cation channel in molluscan neurons (Sudlow et al., 1994) and a possible cGMP-activated cation channel in mammalian skeletal muscle (McGeoch and Guidotti, 1992).

2.9.1.1 Retinal Rod cGMP-Activated Channel

This channel plays a central role in phototransduction in rod photoreceptor neurons of the vertebrate retina (for review, see Yau and Baylor, 1989; McNaughton, 1990), and was the first cyclic nucleotide-gated channel to be discovered (Fesenko et al., 1985). In phototransduction, which takes place in the outer segment of the retinal sensory neurons, absorption of light by the visual pigment (called rhodopsin in rods) triggers a G protein-mediated signalling cascade that leads to the activation of a cGMP-specific phosphodiesterase and the hydrolysis of cytoplasmic cGMP (for review, see Pugh and Lamb, 1990; Stryer, 1991; Kaupp and Koch, 1992; Lagnado and Baylor, 1992; Koutalos and Yau, 1993). In darkness, the cytoplasmic cGMP level is relatively high, causing the steady opening of a cGMP-activated channel on the plasma membrane of the outer segment. This open channel sustains an influx of Na^+, Ca^{2+}, and Mg^{2+} [collectively called the "dark current" (Hagins et al., 1970)] into the cell, which keeps the cell in a partially depolarized state and maintains a steady release of the neurotransmitter glutamate from the cell's synaptic terminal. The hydrolysis of cGMP in the light leads to the closure of these channels, hence a stoppage of the dark current and the generation of an electrical hyperpolarization. This hyperpolarizing response to light reduces or stops the glutamate release from the photoreceptor synaptic terminal, which constitutes the signal passed on to second-order visual neurons in the retina.

The biochemical properties of this channel have been derived exclusively from the bovine protein, because large amounts of the protein can be obtained from bovine eye. On the other hand, the physiological properties of the channel, to be described below, have come mostly from experiments on amphibians (for review, see Yau and Baylor, 1989).

Unlike practically all other ligand-gated channels, this channel does not appear to show any desensitization to prolonged presence of ligand (see Yau and Baylor, 1989 for review; also Watanabe and Matthews, 1990). Earlier indications to the contrary (Puckett and Goldin, 1986; Koch et al., 1987) might be an artifact from membrane vesicle preparations. While unusual, the lack of desensitization of this channel to ligand is essential for phototransduction, by allowing the channels to stay open in the steady presence of cGMP in darkness and to be closed only by light. The activation of the channel shows a steep, sigmoidal dependence on cGMP concentration, with a Hill coefficient that can be higher than three, suggesting at least four binding sites on the channel complex for cGMP. The half-activating cGMP concentration $(K_{1/2})$ is somewhat variable in different studies, but it is probably around 50 μM in physiological conditions (Nakatani and Yau, 1988). The preference of the channel for cGMP as ligand over cAMP is high, but not perfect. Thus, although it is not the physiological ligand, cAMP can activate the channel, with a $K_{1/2}$ about 30-fold higher than for cGMP (Tanaka et al., 1989; Altenhofen et al., 1991; Ildefonse et al., 1992). The macroscopic current elicited by a saturating concentration of cAMP is also considerably smaller than that elicited by a saturated concentration of cGMP, with the underlying mechanism still not entirely clear. Certain analogs of cGMP, such as 8-bromo-cGMP and the fluorescent derivative 8-(5-thioacetamidofluorescein)-cGMP, are more effective, by approximately an order of magnitude, than cGMP in activating the channel (Caretta et al., 1985; Koch and Kaupp, 1985). The phosphorothioate analogs (Sp)-cGMP[S] and (Rp)-cGMP[S], on the other hand, are less effective than cGMP (Zimmerman et al., 1985). Finally, the channel can be permanently locked in the activated state by the photoaffinity analog 8-*p*-azidophenacylthio-cGMP (Brown et al., 1993). Certain transition metal divalent cations, such as Ni^{2+}, also increase the affinity of the channel for cGMP when applied to the cytoplasmic side of the membrane (Ildefonse et al, 1992; Karpen et al., 1993).

As pointed out above, this channel is permeable to both monovalent and divalent cations. It actually prefers divalent over monovalent cations. Thus, in physiological conditions, about 15 to 20% of the inward dark current is carried by divalent cations (ca. 15% by Ca^{2+} and 5% by Mg^{2+}), even though the concentration of extracellular Na^+ is more than 30 times higher by comparison. The Ca^{2+} influx through this channel has the important role of mediating a negative feedback regulation on the gain of the phototransduction process (for review, see Pugh and Lamb, 1990; Kaupp and Koch, 1992; Lagnado and Baylor, 1992; Koutalos and Yau, 1993; Yau, 1991). Not only do divalent cations permeate through this channel, but in the process they also partially block the channel (Haynes and Yau, 1985; Matthews, 1986; Yau et al., 1986), most likely by interacting with negatively charged binding sites within the channel (Menini, 1990; Zimmerman and Baylor, 1992). In the absence of divalent cations, the current–voltage relation of the channel shows only a mild outward rectification, reflecting a slightly larger open probability of the channel at positive voltages (Karpen et al., 1988). In the presence of extracellular divalent cations (as would be under physiological conditions), however, the current–voltage relation is strongly outward-rectifying, reflecting the voltage-dependent blockage by divalent cations. This blockage is quite substantial, with the effective single-channel conductance reduced from approximately 25 pS (see below) to approximately 100 fS, which can be measured only with noise analysis (Bodoia and Detwiler, 1985; Gray and Attwell, 1985). The blockage may serve to reduce quantal noise associated with the openings of these channels in darkness, thus providing a higher signal-to-noise ratio required for detecting dim light (see Yau and Baylor, 1989).

The single-channel openings in native membrane, as observed with excised-patch recordings from rod cells in the absence of divalent cations, typically last less than 1 msec. At low cGMP concentrations, the prominent openings show a conductance of approximately 25 pS, and they occur either singly or in flickery bursts of a few milliseconds or longer. In addition, smaller channel openings (subopen states) are also evident. The fast gating of this channel has made the kinetics difficult to analyze (Matthews and Watanabe, 1988; Torre et al., 1992). One proposed idea is that the sequential binding of cGMP molecules to the multiple binding sites on the channel complex may lead to progressively larger conductance states (Ildefonse and Bennett, 1991). Even at saturating cGMP concentrations, the channel still shows rapid transitions between the open and the closed states; that is, the open probability is less than one.

So far, only a couple of chemicals have been reported to block this channel with much effectiveness. The first is L-*cis*-diltiazem, which blocks the channel at micromolar concentrations in both rod membrane preparations and excised patches of plasma membrane (Koch and Kaupp, 1985; Stern et al., 1986; Haynes, 1992; McLatchie and Matthews, 1992). The other chemical is 3′,4′-dichlorobenzamil, which also blocks at micromolar concentrations, apparently from the cytoplasmic side (Nicol et al., 1987). In addition, the channel is also moderately sensitive to tetracaine blockage (Schnetkamp, 1987; Quandt et al., 1991).

Very little is known about possible modulation of this channel. There is some evidence that phosphorylation of this channel may alter the affinity of the channel for cGMP (Gordon et al., 1992). The ligand affinity is also mildly affected by Ca^{2+}-calmodulin, though perhaps indirectly through a tightly associated cytoskeleton (spectrin)-like protein (Hsu and Molday, 1992).

2.9.1.1.1 Bovine Rod cGMP-Activated Channel

This is the first rod cGMP-activated channel to be purified and molecularly cloned (Cook et al., 1986, 1987; Kaupp et al., 1989). When expressed in the *Xenopus* oocyte, the cloned cDNA gives functional cGMP-activated channels that broadly resembles the native channel in properties (but see below). The apparent molecular weight of the purified protein is around 63 kDa, while the calculated molecular weight based on the cloned cDNA is close to 80 kDa (690 amino acids, see Figure 1). A good part of this discrepancy appears to arise from post- or cotranslational removal of the first 92 amino acids on the N-terminus of the protein in the native rod cell (Molday et al., 1991), possibly having to do with proper protein targeting. The primary structure of the channel protein has an approximate 80-amino acid domain near the C-terminus that shows significant homology to either of the two cGMP-binding

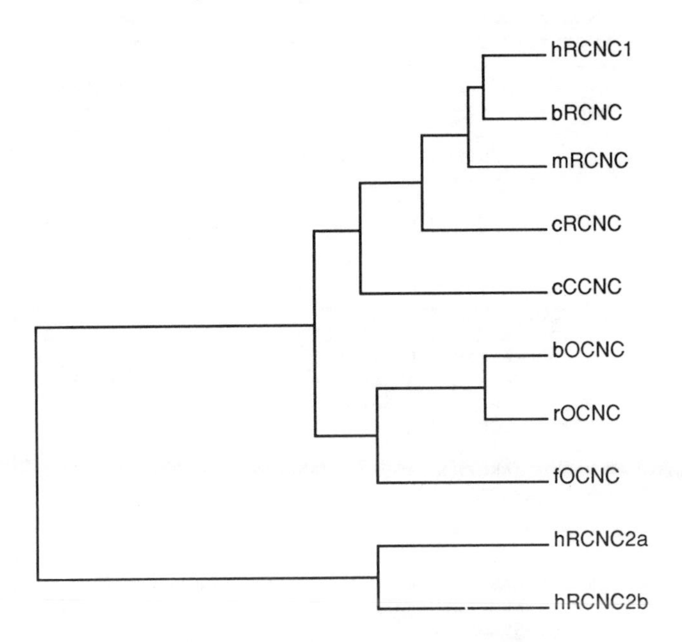

FIGURE 1. Primary structures of cloned cyclic nucleotide-activated channels. (A) Phylogenetic tree relating the different channel members. (B) Amino acid sequence comparisons between them. hRCNC1, human rod cyclic nucleotide-activated channel subunit 1. bRCNC, mRCNC, cRCNC, bovine, mouse, and chicken homologs of the rod cyclic nucleotide-activated channel. cCCNC, chicken cone cyclic nucleotide-activated channel. bOCNC, rOCNC, fOCNC, bovine, rat, and fish olfactory cyclic nucleotide-activated channel. hRCNC2a, hRCNC2b, human rod cyclic nucleotide-activated channel subunit 2, with alternative spliced forms a and b. Shadings indicate identical amino acid residues across all lower eight sequences; upper two sequences are also shaded if amino acid is likewise identical in corresponding positions.

domains in cGMP-dependent protein kinase (PKG), and to a lesser degree to the regulatory subunits of cAMP-dependent protein kinase (PKA) and the catabolite gene activator protein, CAP, in *Escherichia coli* (Kaupp et al., 1989). The crystal structure of the CAP dimer with two bound cAMP molecules has been determined (McKay and Steitz, 1981; McKay et al., 1982). By analogy to CAP, key arginine, glutamate, and glycine residues in PKA have been identified that interact with the cyclic nucleotide ribose phosphate moiety (Weber et al., 1987); the same residues are present in PKG (Weber et al., 1989). The cGMP-activated channel also has these residues (Arg-559, Glu-544, and Gly-543, see Figure 2). In addition, a threonine residue has been identified in PKG that likely forms a hydrogen bond with the 2-NH_2 group of cGMP. The corresponding residue in PKA is alanine, which cannot form the same hydrogen bond. This threonine/alanine difference is thought to be involved in the discrimination between cGMP and cAMP at the binding pocket (Weber et al., 1989); this has been verified by site-directed mutagenesis experiments showing

```
hRCNC2a    --------------------------------------------------
hRCNC2b    MPRELSRIEEEKEDEEEEEEEEEEEEEEEVTEVLLDSCVVSQVGVGQSEE    50
hRCNC1     M--------------------------KNNIINTQQSFVTMPNV------    18
bRCNC      M--------------------------KKVIINTWHSFVNIPNV------    18
mRCNC      M--------------------------KTNIINTWHSFVNIPNV------    18
cRCNC      M--------------------------KVGVIETHHSHPIIPSV------    18
cCCNC      MAK------INTQHSYPGMHGLSVRTTDEDIERIENGFIRTHSLCEDTSS    44
bOCNC      --------------------------MTEKANGVKSSPANNHNH------    18
rOCNC      -------------------------MMTEKSNGVKSSPANNHNH------    19
fOCNC      M----------------------------TGQAALERSVSSHRL------    16

hRCNC2a    --------------------------------------------------
hRCNC2b    DGTRPQSTSDQKLWEEVGEEAKKEAEEKAKEEAEEVAEEEAEKEPQDWAE   100
hRCNC1     -IVPDIEKEIRRMENGACSSFSEDDD-SASTS-EES--------ENE---    54
bRCNC      -VGPDVEKEITRMENGACSSFSGDDD-DSASMFEES--------ETE---    55
mRCNC      -IVPAIEKEIRRMENGACSSFSDDDN-GSLS--EES--------ENE---    53
cRCNC      -VVQDTSEDPGLIEKGE-NRFA----------------------------    38
cCCNC      ELQRVISMEGRHLSGSQTSPFTGRGAMARLSRFVVSLRSWATRHLHH---    91
bOCNC      -HAPPAIKASGKDDHRASSRPQSAAA------------------DDT---    46
rOCNC      -HPPPSIKANGKDDHRAGSRPQSVAA-----------------DDDT---    48
fOCNC      -SVRSRLEGEAERAESAISRTDGDDD------------------TCS---    44

hRCNC2a    --------------------------------------------------
hRCNC2b    TKEEPEAEAEAASSGVPATKQHPEVQVEDTDADSCPLMAEENPPSTVLPP   150
hRCNC1     -------NPHARGSF--SYKSLRKGGPSQRE----QYL-----PGAI---    83
bRCNC      -------NPHARDSF--RSNTHGSGQPSQRE----QYL-----PGAI---    84
mRCNC      -------DSF----F--RSNSYKRRGPSQRE----QHL-----PGTM---    78
cRCNC      -----------------------------RQ----WYL-----PGAF---    47
cCCNC      -------EDQRPDSFLERIRGPELVEVSSRQSNIRSFLGIREQPGGV---   131
bOCNC      -------SSE--------LQQLAEMDAPQQR------------RGGF---    66
rOCNC      -------SPE--------LQRLAEMDTPRRG------------RGGF---    68
fOCNC      -------ELQ-------RVTALELPSAEMLE----AFT----QRRPL---    69

hRCNC2a    --------------------------------------------------
hRCNC2b    PSPAKSDTLIVPSSASGTHRKKLPSEDDEAEELKALSPAESPVVAWSDPT   200
hRCNC1     --------------------------------------------------    83
bRCNC      --------------------------------------------------    84
mRCNC      --------------------------------------------------    78
cRCNC      --------------------------------------------------    47
cCCNC      --------------------------------------------------   131
bOCNC      --------------------------------------------------    66
rOCNC      --------------------------------------------------    68
fOCNC      --------------------------------------------------    69

hRCNC2a    --------------------------------------------------
hRCNC2b    TPKDTDGQDRAASTASTNSAIINDRLQELVKLFKERTEKVKEKLIDPDVT   250
hRCNC1     -----------------------ALFNVNNSSNKDQEPEEKKK------K   104
bRCNC      -----------------------ALFNVNNSSNKEQEPKEKKK------K   105
mRCNC      -----------------------ALFNVNNSSNKDQEPKEKKK------K    99
cRCNC      -----------------------AQYNINNNSNKDEE----KK------K    64
cCCNC      -----------------------NGPWPLARFNVNFSNNTNED------K   152
bOCNC      -----------------------RRIARLVGVLREWA----YR------N    83
rOCNC      -----------------------QRIVRLVGVIRDWA----NK------N    85
fOCNC      -----------------------ARLVNLVLSLREWA----HK------S    86
```

FIGURE 1b(1).

that converting the alanine residue to threonine in PKA markedly increases the cGMP affinity of the enzyme (Shabb et al., 1990). In the cGMP-activated channel, the threonine residue (Thr-560) is also present (Figure 2). When this residue is mutated into alanine, the expressed channel shows a decrease in

```
hRCNC2a   ------------------------------------MLCCKFKHRPWKKY   14
hRCNC2b   SDEESPKPSPAKKAPEPAPDTKPAEAEPVEEEHYCDMLCCKFKHRPWKKY  300
hRCNC1    KKEKKSKSDNKNENKNDPE-KKKKKKDKEK-----KKKEEKSKDKKEEEK  148
bRCNC     KKEKKSKPDDKNENKKDPEKKKKKKEKDKDK-----KKKEEKGKDKKEEEK 150
mRCNC     KKEKKSKADDKNEIKKDPEKKKKKKEKEKEK-----KKKEEKTKEKKEEEK 144
cRCNC     KKEKKSKSENKKDGERQKNKEKKEKH-KNK-----DKKKGKEEEK----K  104
cCCNC     KEEKKEVKEEKKEEKKEEKKEEKKDDKKDD-----KKDDKKDDKKKEEQK  197
bOCNC     FREEEPRPDSFLERFRGPELHTVTTQ-QGD-----GKGDKDGEGKGTKKK  127
rOCNC     FREEEPRPDSFLERFRGPELQTVTTH-QGD-----DKGGKDGEGKGTKKK  129
fOCNC     LVETEQRPDSFLERFRGPQAA------NDQ-----SAAPADAPKKTFKER  125

hRCNC2a   QFPQSIDPLTNLMYVLWLFFVVMAWNWNCWLIPVRWAFPYQTPDNIHHWL   64
hRCNC2b   QFPQSIDPLTNLMYVLWLFFVVMAWNWNCWLIPVRWAFPYQTPDNIHHWL  350
hRCNC1    KEVVVIDPSGN-TYYNWLFCITLPVMYNWTMVIARACFDELQSDYLEYWL  197
bRCNC     KEVVVIDPSGN-TYYNWLFCITLPVMYNWTMIIARACFDELQSDYLEYWL  199
mRCNC     KEVVVIDPSGN-TYYNWLFCITLPVMYNWTMIIARACFDELQSDYLEYWL  193
cRCNC     KDIFIIDPAGN-MYYNWLFCITMPVMYNWTMIIARACFDELQNDYLAVWF  153
cCCNC     KEVFVIDPSSN-MYYNWLTIIAAPVFYNWCMLICRACFDELQIDHIKLWL  246
bOCNC     FELFVLDPAGD-WYYRWLFLIALPVLYNWCLLVARACFSDLQKGYYIVWL  176
rOCNC     FELFVLDPAGD-WYYRWLFVIAMPVLYNWCLLVARACFSDLQRNYFVVWL  178
fOCNC     WEGFVVSQSDD-IYYYWLFFIALASLYNWIMLVARACFDQLQDENFFLWV  174

hRCNC2a   LMDYLCDLIYFLDITVFQTRLQFVRGGDIITDKKDMRNNYLKSRRFKMDL  114
hRCNC2b   LMDYLCDLIYFLDITVFQTRLQFVRGGDIITDKKDMRNNYLKSRRFKMDL  400
hRCNC1    ILDYVSDIVYLIDMFV-RTRTGYLEQGLLVKEELKLINKYKSNLQFKLDV  246
bRCNC     AFDYLSDVVYLLDMFV-RTRTGYLEQGLLVKEERKLIDKYKSTFQFKLDV  248
mRCNC     IFDYVSDVVYLADMFV-RTRTGYLEQGLLVKDRMKLIEKYKANLQFKLDV  242
cRCNC     IVDYVSDVIYIADMFV-RTRTGYLEQGLLVKEEQKLKEKYKSSLQFKLDF  202
cCCNC     FLDYCSDIIYVFDMFV-RFRTGFLEQGLLVKDEKKLRDHYTQTVQFKLDV  295
bOCNC     VLDYVSDVVYIADLFI-RLRTGFLEQGLLVKDTKKLRDNYIHTMQFKLDV  225
rOCNC     VLDYFSDTVYIADLII-RLRTGFLEQGLLVKDPKKLRDNYIHTLQFKLDV  227
fOCNC     GLDYLCDVIYILDTCI-RLRTGYLEQGLLVKDLAKLRDNYIRTLQFKLDF  223

hRCNC2a   LSLLPLDFLYLKVGVN-PLLRLPRCLKYMAFFEFNSRLESILSKAYVYRV  163
hRCNC2b   LSLLPLDFLYLKVGVN-PLLRLPRCLKYMAFFEFNSRLESILSKAYVYRV  449
hRCNC1    LSLIPTDLLYFKLGWNYPEIRLNRLLRFSRMFEFFQRTETRTNYPNIFRI  296
bRCNC     LSVIPTDLLYIKFGWNYPEIRLNRLLRISRMFEFFQRTETRTNYPNIFRI  298
mRCNC     LSVIPTDLLYIKFGWNYPEIRLNRLLRISRMFEFFQRTETRTNYPNIFRI  292
cRCNC     LSIIPTDLLYFKLGLNYPELRINRLLRVARMFEFFQRTETRTNYPNIFRI  252
cCCNC     LSLLPTDLAYLKLGLNYPELRFNRLLRIARLFEFFDRTETRTNYPNMFRI  345
bOCNC     ASIIPTDLIYFAVGIHNPEVRFNRLLHFARMFEFFDRTETRTSYPNIFRI  275
rOCNC     ASIIPTDLIYFAVGIHSPEVRFNRLLHFARMFEFFDRTETRTSYPNIFRI  277
fOCNC     LSILPTELLFFVTGY-VPQLRFNRLLRFSRMFEFFDRTETRTNYPNAFRI  272

hRCNC2a   IRTTAYLLYSLHLNSCLYYWASAYQGLGSTHWVYDGVGNSYIRCYYFAVK  213
hRCNC2b   IRTTAYLLYSLHLNSCLYYWASAYQGLGSTHWVYDGVGNSYIRCYYFAVK  499
hRCNC1    SNLVMYIVIIIHWNACVFYSISKAIGFGNDTWVYPDINDPEFGRLARKYV  346
bRCNC     SNLVMYIIIIIHWNACVYFSISKAIGFGNDTWVYPDVNDPDFGRLARKYV  348
mRCNC     SNLVMYIVIIIHWNACVYYSISKAIGFGNDTWVYPDVNDPEFGRLARKYV  342
cRCNC     SNLVMYIVIIIHWNACVYYSISKAIGFGADTWVYPNTSHPEFARLTRKYV  302
cCCNC     GNLVLYILIIIHWNACIYFAISKVIGFGTDSWVYPNVSIPEYGRLSRKYI  395
bOCNC     SNLILYILIIIHWNACIYYAISKSIGFGVDTWVYPNITDPEYGYLSREYI  325
rOCNC     SNLVLYILVIIHWNACIYYVISKSIGFGVDTWVYPNITDPEYGYLAREYI  327
fOCNC     CNLILYILVIIHWNACIYYAISKALGLSSDTWVYSGQNKT----LSFCYV  318
```

FIGURE 1b(2).

sensitivity to cGMP (by about 30-fold, becoming close to the cAMP sensitivity) without any change in sensitivity to cAMP (Altenhofen et al., 1991). Thus, the idea involving a hydrogen bond between a threonine residue and cGMP also

```
hRCNC2a   TLITIGGLPDPKTLFEIVFQLLNYFTGVFAFSVMIGQM--RDVVGAA---   258
hRCNC2b   TLITIGGLPDPKTLFEIVFQLLNYFTGVFAFSVMIGQM--RDVVGAA---   544
hRCNC1    YSLYWSTL-TLTTIGETPPPVRDSEYVFVVVDFLIGVLIFATIVGNIGSM   395
bRCNC     YSLYWSTL-TLTTIGETPPPVRDSEYFFVVADFLIGVLIFATIVGNIGSM   397
mRCNC     YSLYWSTL-TLTTIGETPPPVLDSEYIFVVVDFLIGVLIFATIVGNIGSM   391
cRCNC     YSLYWSTL-TLTTIGETPPPVRDSEYFFVVVDFLVGVLIFATIVGNVGSM   351
cCCNC     YSLYWSTL-TLTTIGETPPPVKDEEYLFVVIDFLVGVLIFATIVGNVGSM   444
bOCNC     YCLYWSTL-TLTTIGETPPPVKDEEYLFVIFDFLIGVLIFATIVGNVGSM   374
rOCNC     YCLYWSTL-TLTTIGETPPPVKDEEYLFVIFDFLIGVLIFATIVGNVGSM   376
fOCNC     YCFYWSTL-TLTTIGEMPPPVKDEEYVFVVFDFLVGVLIFATIVGNVGSM   367

hRCNC2a   ----TAGQTYYRSCMDSTVKYMNFYKIPKSVQNRVKTWYEYTWHSQGMLD   304
hRCNC2b   ----TAGQTYYRSCMDSTVKYMNFYKIPKSVQNRVKTWYEYTWHSQGMLD   590
hRCNC1    ISNMNAARAEFQARIDAIKQYMHFRNVSKDMEKRVIKWFDYLWTNKKTVD   445
bRCNC     ISNMNAARAEFQARIDAIKQYMHFRNVSKDMEKRVIKWFDYLWTNKKTVD   447
mRCNC     ISNMNAARAEFQSRVDAIKQYMHFRNVSKDMEKRVIKWFDYLWTNKKTVD   441
cRCNC     ISNMNAARAEFQAKIDAIKQYMHFRNVSKDMEKRVIKWFDYLWTNKKAVD   401
cCCNC     ISNMNASRAEFQAKVDSIKQYMHFRKVTKDLEARVIKWFDYLWTNKKTVD   494
bOCNC     ISNMNATRAEFQAKIDAVKHYMQFRKVSKEMEAKVIRWFDYLWTNKKSVD   424
rOCNC     ISNMNATRAEFQAKIDAVKHYMQFRKVSKDMEAKVIKWFDYLWTNKKTVD   426
fOCNC     IANMNATRAEFQTRIDAIKHYMHFRKVNRTLETRVIKWFDYLWTNKKTVD   417

hRCNC2a   ESELMVQLPDKMRLDLAIDVNYNIVSKVALFQGCDRQMIFDMLKRLRSVV   354
hRCNC2b   ESELMVQLPDKMRLDLAIDVNYNIVSKVALFQGCDRQMIFDMLKRLRSVV   640
hRCNC1    EKEVLKYLPDKLRAEIAINVHLDTLKKVRIFADCEAGLLVELVLKLQPQV   495
bRCNC     EREVLKYLPDKLRAEIAINVHLDTLKKVRIFADCEAGLLVELVLKLQPQV   497
mRCNC     EREVLRYLPDKLRAEIAINVHLDTLKKVRIFADCEAGLLVELVLKLQPQV   491
cRCNC     EREVLKYLPDKLRAEIAINVHLETLKKVRIFADCEAGLLVELVLKLQPQV   451
cCCNC     EKEVLKNLPDKLKAEIAINVHLDTLKKVRIFQDCEAGLLIELVLKLKPTV   544
bOCNC     EREVLKNLPAKLRAEIAINVHLSTLKKVRIFQDCEAGLLVELVLKLRPQV   474
rOCNC     EREVLKNLPAKLRAEIAINVHLSTLKKVRIFQDCEAGLLVELVLKLRPQV   476
fOCNC     EQEVLKNLPDKLRAEIAINVHLDTLKKVRIFQDCEAGLLVELVLKLRPQV   467

hRCNC2a   YLPNDYVCKKGEIGREMYIIQAGQVQVLGGPDGKSVLVTLKAGSVFGEIS   404
hRCNC2b   YLPNDYVCKKGEIGREMYIIQAGQVQVLGGPDGKSVLVTLKAGSVFGEIS   690
hRCNC1    YSPGDYICKKGDIGREMYIIKEGKLAVVAD-DGVTQFVVLSDGSYFGEIS   544
bRCNC     YSPGDYICKKGDIGREMYIIKEGKLAVVAD-DGITQFVVLSDGSYFGEIS   546
mRCNC     YSPGDYICKKGDIGREMYIIKEGKLAVVAD-DGITQFVVLSDGSYFGEIS   540
cRCNC     YSPGDYICRKGDIGREMYIIKEGKLAVVAD-DGVTQFVVLSDGSYFGEIS   500
cCCNC     FSPGDYICKKGDIGREMYIIKEGKLAVVAD-DGITQFVVLSDGSYFGEIS   593
bOCNC     FSPGDYICRKGDIGKEMYIIKEGKLAVVAD-DGVTQYALLSAGSCFGEIS   523
rOCNC     FSPGDYICRKGDIGKEMYIIKEGKLAVVAD-DGVTQYALLSAGSCFGEIS   525
fOCNC     YSPGDYICRKGDIGKEMYIIKEGQLAVVAD-DGVTQFALLTAGGCFGEIS   516

hRCNC2a   LLAVGG---GNRRTANVVAHGFTNLFILDKKDLNEILVHYPESQKLLRKK   451
hRCNC2b   LLAVGG---GNRRTANVVAHGFTNLFILDKKDLNEILVHYPESQKLLRKK   737
hRCNC1    ILNIKGSKAGNRRTANIKSIGYSDLFCLSKDDLMEALTEYPDAKTMLEEK   594
bRCNC     ILNIKGSKAGNRRTANIKSIGYSDLFCLSKDDLMEALTEYPDAKGMLEEK   596
mRCNC     ILNIKGSKAGNRRTANIKSIGYSDLFCLSKDDLMEVLTEYPDAKTMLEEK   590
cRCNC     ILNIKGSKAGNRRTANIRSIGYSDLFCLSKDDLMEALTEYPDAKAMLEEK   550
cCCNC     ILNIKGSKSGNRRTANIRSIGYSDLFCLSKDDLMEALTEYPEAKKALEEK   643
bOCNC     ILNIKGSKMGNRRTANIRSLGYSDLFCLSKDDLMEAVTEYPDAKRVLEER   573
rOCNC     ILNIKGSKMGNRRTANIRSLGYSDLFCLSKDDLMEAVTEYPDAKKVLEER   575
fOCNC     ILNIQGSKMGNRRTANIRSIGYSDLFCLSKDDLMEAVAEYPDAQKVLEER   566
```

FIGURE 1b(3).

applies to the channel. Finally, since the cloned protein contains a single cGMP-binding site, and since the Hill coefficient of channel activation described earlier can be higher than 3, the channel complex must be at least a

```
hRCNC2a   ARRMLRSNNKPKEEKSVLILPPRAGTPKLFNAALAMTGKMGGKGAKGGKL   501
hRCNC2b   ARRMLRSNNKPKEEKSVLILPPRAGTPKLFNAALAMTGKMGGKGAKGGKL   787
hRCNC1    GKQILMKDGLLDLNIANAGSDPKDLEEKV----TRMEGSV---DLLQTRF   637
bRCNC     GKQILMKDGLLDINIANAGSDPKDLEEKV----TRMESSV---DLLQTRF   639
mRCNC     GKQILMKDGLLDINIANMGSDPKDLEEKV----TRMEGSV---DLLQTRF   633
cRCNC     GKQILMKDGLLDIEVANLGSDPKDLEEKV----AYMEGSM---DRLQTKF   593
cCCNC     GRQILMKDNLIDEEAAKAGADPKDLEEKI----DRLETAL---DTLQTRF   686
bOCNC     GREILMKEGLLDENEVAASMEV-DVQEKL----EQLETNM---DTLYTRF   615
rOCNC     GREILMKEGLLDENEVAASMEV-DVQEKL----EQLETNM---DTLYTRF   617
fOCNC     GREILRKQGLLDESVAAGGLGVIDTEEKV----ERLDASL---DILQTRF   609

hRCNC2a   AHLRARLKELAALEAAAKHEELVEQAKSSQDVKGEEGSAAPDQHTHPKEA   551
hRCNC2b   AHLRARLKELAALEAAAKHEELVEQAKSSQDVKGEEGSAAPDQHTHPKEA   837
hRCNC1    ARILAEYESM--------------QQKLKQRLTKVEKFLKP---------   664
bRCNC     ARILAEYESM--------------QQKLKQRLTKVEKFLKP---------   666
mRCNC     ACILAEYESM--------------QQKLKQRLSKVEKFLKP---------   660
cRCNC     ARLLAEYDAA--------------QQKLKKRLTQIEKILKP---------   620
cCCNC     ARLLAEYSSS--------------QQKVKQRLARVETRVKK---------   713
bOCNC     ARLLAEYTGA--------------QQKLKQRITVLETKMKQ---------   642
rOCNC     ARLLAEYTGA--------------QQKLKQRITVLETKMKQ---------   644
fOCNC     ARLLGEFTST--------------QRRLKQRITALERQLCHTGLGLLSDN   645

hRCNC2a   ATDPPAPRTPPEPPGSPPSSPPPASLGSCEGEEEGPAEPEEHSVRICMSP   601
hRCNC2b   ATDPPAPRTPPEPPGSPPSSPPPASLGRPEGEEEGPAEPEEHSVRICMSP   887
hRCNC1    LID-------------------TEFSSI------------E-------GP   676
bRCNC     LID-------------------TEFSAI------------E-------GS   678
mRCNC     LIE-------------------TEFSAL------------E-------EP   672
cRCNC     VME-------------------QEFLDF------------E-------EA   632
cCCNC     Y-G-------------------SGSLSV------------G-------EP   724
bOCNC     NNE-------------------DDSLSD------------G-------MN   654
rOCNC     NHE-------------------DDYLSD------------G-------IN   656
fOCNC     EAE-------------------GEHAGVPTHTHADIHAQPE-------TH   669

hRCNC2a   GPEPGEQILSVKMPEEREEKAE   623
hRCNC2b   GPEPGEQILSVKMPEEREEKAE   909
hRCNC1    GAES---------GPIDST---   686
bRCNC     GTES---------GPTDSTQD-   690
mRCNC     GGES---------EPTESLQD-   684
cRCNC     DPPT---------DKPGVTKTE   645
cCCNC     EPEK---------PEEQKKD--   735
bOCNC     SPEP---------PAEKP----   663
rOCNC     TPEP---------TAAE-----   664
fOCNC     TRTS---------AETNSEEET   682
```

FIGURE 1b(4).

tetramer. From what follows, it now appears that the cGMP-activated channel also bears homology in structural motif to voltage-gated K^+ channels, which has been shown to have four subunits (MacKinnon, 1991). Thus, the same is likely true for the cGMP-activated channel as well.

When the amino acid sequence of this channel was first derived, no homology to other ion channels was noted (Kaupp et al., 1989). Subsequently, however, a region similar to the S4 domain in voltage-gated channels has been recognized (Jan and Jan, 1990). In voltage-gated channels, the S4 domain is transmembranous and serves as the voltage sensor for the channel. The cGMP-activated channel, however, cannot be activated by voltage alone, even though in the presence of cGMP it shows a slight voltage dependence

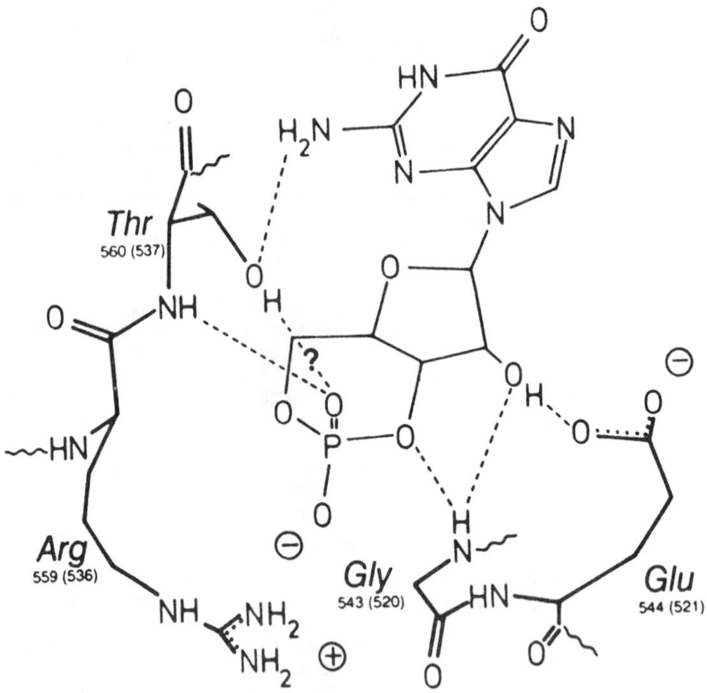

FIGURE 2. Predicted sites of contact between cGMP and the binding site in cGMP-activated channels. Hypothetical hydrogen bonds (see Shabb et al., 1990) are indicated by interrupted lines. The numbers refer to positions of amino acid residues in the sequence of the bovine rod (olfactory) channel proteins. The hydrogen bond marked by a question mark may not be involved in cAMP or cGMP binding. (From Altenhofen et al., 1991, with permission.)

in its open probability (see above). Thus it is not clear at present what function, if any, the S4-like domain serves in this channel, nor is it even clear whether it is transmembranous. It is suggested that the S4 region may be part of the core structure of a common ancester of both voltage-gated and cyclic nucleotide-activated channels, which has evolved with time to take on the important function of being the voltage sensor in voltage-gated channels (Jan and Jan, 1990).

Besides the S4 domain, a region of this channel resembling the pore-forming region of voltage-gated K^+ channels is now also recognized (Guy et al., 1991; Goulding et al., 1993; Kramer et al., 1994). Unlike K^+ channels, the cGMP-activated channel does not discriminate between Na^+ and K^+; at the same time, it is permeant to and also blocked by divalent cations (see above). Using mutagenesis, Heginbotham et al. (1992) discovered that these differences in conduction properties can be explained by the presence of two amino acids residues, tyrosine and glycine, in the pore region of K^+ channels but not in that of the cyclic nucleotide-activated channel (Figure 3a). Interestingly, the pore region of voltage-gated Ca^{2+} channels also lack these two amino acid residues (Figure 3a). Ca^{2+} channels are nonselectively permeable to monova-

A

```
Shaker       F W W A V V T M T T V G Y G D M T
drk1         F W W A T I T M T T V G Y G D I Y
slo          V Y F L I V T M S T V G Y G D V Y

AKT1         M Y W S I T T L T T V G Y G D L H
KAT1         L Y W S I T T L T T T G Y G D F H
eag          L Y F T M T C M T S V G F G N Y A

bRCNC        L Y W S T L T L T T I G     E T P
Deletion     F W W A V V T M T T V G     D M T

CA B1   I    F A V L T V F Q C I T M     E G W
CA B1  II    A A I M T V F Q I L T G     E D W
CA B1 III    W A L L T L F T V S T G     E G W
CA B1  IV    Q A L M L L F R S A T G     E A W
```

B

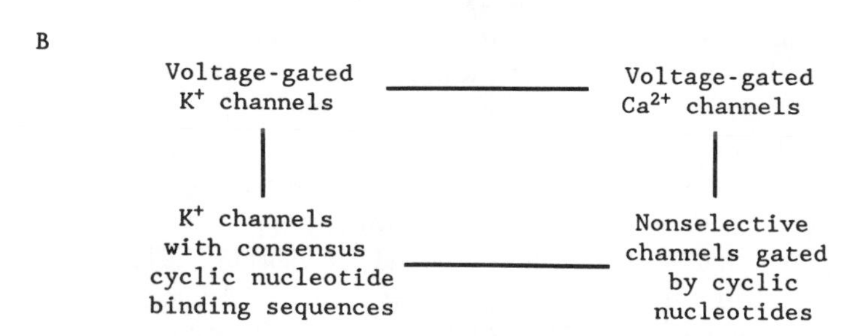

FIGURE 3. (A) Alignment of the deduced amino acid sequences of the pore regions of several members of the S4-containing ion channel superfamily. *Shaker* and *drk1* are voltage-activated K^+ channels (Tempel et al., 1987; Frech et al., 1989). *slo* is a Ca^{2+}-activated K^+ channel (Atkinson et al., 1991; Adelman et al., 1992). *AKT1* and *KAT1* encode K^+ channels from the plant *Arabidopsis thaliana* (see text). *eag* is a K^+ channel from *Drosophila* (see text). *AKT1, KAT1,* and *eag* contain sequences that are consensus sites for cyclic nucleotide-binding domains. *bRCNC* is the bovine rod cGMP-activated channel. *Deletion* is the *Shaker* mutant missing the tyrosine-glycine pair (see text). CA B1 I–IV are the homologous regions from the four domains of the brain Ca^{2+} channel (Mori et al., 1991). (B) Diagram illustrating the possible evolutionary link between several members of the S4 ion channel superfamily. (Modified from Heginbotham et al., 1992, with permission.)

lent cations when divalent cations are absent, but this conductivity to monovalent cations becomes blocked in the presence of divalent cations. The property of divalent cation block on monovalent cation flux, as exhibited by both the cGMP-activated channel and the Ca^{2+} channel, appears to be due to a glutamate residue corresponding in position to an aspartate next to the glycine mentioned above in the pore of K^+ channels; this acidic residue is present in the cGMP-activated channel as well as in the corresponding positions in all of the four repeats of the Ca^{2+} channel (Figure 3a). When this acidic residue in the cGMP-activated channel is mutated into a neutral one such as asparagine, glutamine, or glycine, the sensitivity of the channel to divalent cations is reduced by two orders of magnitude (Root and MacKinnon, 1993). Replacement of the corresponding residue in Na^+ channel by glutamate likewise results in Ca^{2+} channel-like ion conduction properties (Heinemann et al., 1992). Finally, as mentioned above, the deletion mutant of the *Shaker* K^+ channel (i.e., with the tyrosine–glycine pair described above being removed) has an aspartate residue in the corresponding position (Figure 3a). When this aspartate is mutated into glutamate, the Ca^{2+} block on the *Shaker* deletion mutant is also increased (Heginbotham et al., 1992). Thus, this acidic residue appears to influence the way Ca^{2+} interacts with the pore in a broad spectrum of channels. Figure 3b shows the possible evolutionary link between several members of the superfamily of ion channels bearing the S4 domain as proposed by Heginbotham et al. (1992).

Figure 4 shows the putative topology of the rod cGMP-activated channel in the membrane (see also Wohlfart et al., 1992; Chen et al., 1993; Goulding et al., 1992). The transmembrane domains are primarily based on a hydrophobicity plot of the protein assuming α-helical structure in the transmembranous regions. The S4 domain is also assumed to be in a transmembrane domain even though this is pure speculation at present. What is certain from immunochemical labelings is that the N-terminus is cytoplasmic (Molday et al., 1991), and that the amino-acid domain 321–339 just C-terminal of H4 is extracellular, with Asn-327 being an N-glycosylation site (Wohlfart et al., 1992). At the same time, the cGMP-binding site has to be cytoplasmic, based on electrophysiological experiments. The channel also appears to be tightly associated with a 240-kDa protein that exhibits immunochemical cross-reactivity with spectrin (Molday et al., 1990). An observed influence by Ca^{2+}-calmodulin on the cGMP affinity of the channel appears to be conferred through this latter protein (Hsu and Molday, 1993).

Based on Northern blot hybridizations and PCR amplifications, there is some suggestion that the same channel may be present in heart and kidney as well (Dhallan et al., 1992).

2.9.1.1.2 Human Rod cGMP-Activated Channel

A cDNA (hRCNC1) was cloned from a human retinal cDNA library based on homology to the bovine channel described above (Dhallan et al., 1992; see also Pittler et al., 1992). It encodes a protein sequence of 686 amino acids in length that is 91% identical to the bovine channel (Figure 1). The gene for this protein is located on human chromosome 4, and contains at least 10 exons (Dhallan et al., 1992). One large exon encodes the C-terminal two-thirds of the protein, whereas seven small exons encode the N-terminal one-third of the protein. It is also found that alternative splicing removes one of the small exons in a subset of transcripts in the human retina, producing an internal in-frame deletion of 36 codons. When expressed in a human embryonic kidney cell line (HEK 293), the full-length cDNA clone produces functional ion channels, but not the differentially spliced variant. Thus, the significance of the latter transcript remains unclear.

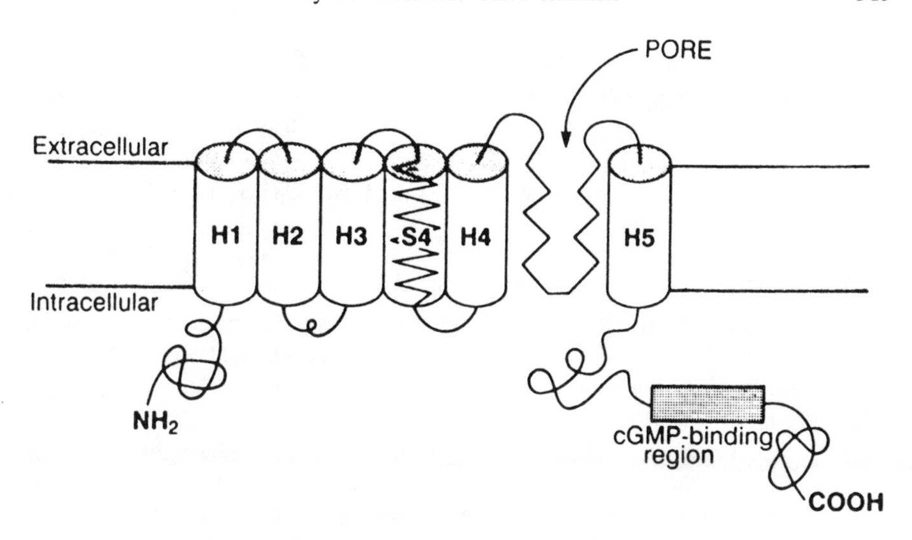

FIGURE 4. Putative folding pattern of the cyclic nucleotide-activated channels in the membrane. The H1–H5 numbering is according to Kaupp et al. (1989). (Modified from Bönigk et al., 1993, with permission).

Although the cloned bovine and human rod channels, when expressed, both show properties that are broadly similar to those of the native channel in either amphibians or mammals, there are two prominent differences. The first is that the cloned channels exhibit predominantly prolonged and stable (i.e., rectangular-looking) openings of 25–30 pS conductance that can last up to tens of milliseconds, which are not observed in the native channel. Second, the cloned channel is 50 to 100 times less sensitive to the drug L-*cis*-diltiazem than the native channel. Recently, another cDNA has been cloned from human retina that represents a second subunit (hRCNC2) of the native rod cGMP-activated channel (Chen et al., 1993). hRCNC2 has many of the structural features of hRCNC1, including the S4 region, the pore region, the cGMP-binding site, and an overall similar hydrophobicity profile, but it cannot form functional cGMP-activated channels by itself. When coexpressed with hRCNC1, however, hRCNC2 gives cGMP-activated channels with flickery openings (i.e., bursts of brief openings and closings) that are characteristic of the native channel. At the same time, these channels have a high sensitivity to diltiazem (K_i ~1 μM at +60 mV) like the native bovine channel (Quandt et al., 1991). Finally, immunocytochemistry has also located hRCNC2 to rod outer segments (Chen et al., 1993). Thus, it appears that hRCNC1 and hRCNC2 are both subunits of the rod cGMP-activated channel mediating phototransduction in human retina. The homologs of hRCNC2 in other animal species, such as bovine, remain to be identified. Most recently, subunit 2 of the retinal rod cGMP-activated channel has been shown to be a component of the 240-kDa protein mentioned above (Chen et al., 1994). Physiological experiments in the same study also demonstrated that subunit 2 mediates the Ca^{2+} calmodulin modulation of the rod cGMP-activated channel.

2.9.1.1.3 Mouse Rod cGMP-Activated Channel

The mouse homolog of the bovine and human (hRCNC1) cGMP-activated channel has also been cloned (Pittler et al., 1992). It encodes a protein of 684 amino acids showing about 90% identity to the bovine homolog (Figure 1). The mouse gene is located on chromosome 5. No functional expression of this clone has yet been reported.

2.9.1.1.4 Chicken Rod cGMP-Activated Channel

Most recently, the chicken homolog of this channel has been cloned (Bönigk et al., 1993). The encoded protein has 645 amino acids and bears 76% identity to the bovine homolog (Figure 1). As found for the bovine protein, there appears to be post- or cotranslational processing that results in a smaller protein in the native cell. Interestingly, this channel shows a highly nonlinear current–voltage relation even in the absence of divalent cations, which is quite different from the cloned homologs from other animal species. No physiological studies on the native chicken rod channel, however, have yet been carried out to allow proper comparison.

2.9.1.2 Retinal Cone cGMP-Activated Channel

Retinal cones differ from rods in that they are 25 to 100 times less sensitive to light than rods. The kinetics of cone responses to light are also several fold faster (see, for example, Nakatani and Yau, 1989). These physiological differences, however, are likely to reside primarily in those stages of the phototransduction cascade upstream of the cGMP-activated channel. In terms of the cGMP dependence and the gating kinetics, the rod and cone channels are quite similar to each other, at least based on electrical recordings from the fish cone channel (Haynes and Yau, 1985, 1990; Watanabe and Murakami, 1991; Picones and Korenbrot, 1992). One difference between the two is that the current–voltage relation of the cone channel under physiological conditions (i.e., in the presence of extracellular divalent cations) shows exponential increases in current at both increasingly positive and negative voltages. In contrast, the rod channel shows an exponential increase in current at only increasingly positive voltages, and the current is relatively constant at negative voltages. The channel also appears to have an even stronger preference for divalent over monovalent cations compared to the rod channel (Perry and McNaughton, 1991); this property is likely connected with its characteristic current–voltage relation, but its functional significance is still unclear at present.

2.9.1.2.1 Chicken Cone cGMP-Activated Channel

So far, only the cone channel in chicken has been cloned (Bönigk et al., 1993). The cone protein encoded by the cloned cDNA has 735 amino residues, versus 645 residues for the chicken rod channel. The sequence identity between this channel and the bovine rod channel is 65%. Interestingly, the similarity between the chicken cone and the bovine olfactory channels (64%) is greater than that between the bovine rod and olfactory channels (57%). The key features of the rod cGMP-activated channel, including the S4 domain, the pore region and the cGMP-binding domain, can all be recognized from the cone channel sequence. Furthermore, as in the case of the rod channel, the cone channel in native tissue appears to be proteolytically processed. A structural comparison between the chicken rod and cone channels indicates that their pore regions are very similar, so the basis for the difference in electrical properties between them remains to be examined.

It is also an open question whether the native cone channel, like at least the human rod channel, is composed of more than one subunit species.

2.9.1.3 cAMP/cGMP-Activated Channel in Olfactory Receptor Neurons

Olfactory transduction is the signaling process by which odorants lead to electrical excitation of olfactory receptor neurons in the vertebrate olfactory epithelium. One well-established mechanism for this process turns out to be remarkably similar to visual transduction in basic design. Thus, odorant stimulation of specific receptor proteins on the cilia of these sensory neurons activates, via a G-protein (G_{olf}), an adenylate cyclase (type III); this leads to an increase in cytoplasmic cAMP level, which in turn opens a cAMP/cGMP-activated cation channel on the plasma membrane and produces an electrical depolarization (for review, see Lancet, 1986; Reed, 1992). The two main differences between visual and olfactory transductions are (1) the second messenger in olfaction is cAMP, versus cGMP in vision, and (2) there is an increase in cyclic nucleotide concentration and hence opening of channels in olfaction, versus a decrease in cyclic nucleotide concentration and a closing of channels in vision. The cyclic nucleotide-activated channel mediating olfaction is broadly similar to the photoreceptor channel in activation and conduction properties (Nakamura and Gold, 1987; Kurahashi, 1989; Zufall et al., 1991; Frings et al., 1992; Kurahashi and Kaneko, 1993). One interesting difference between the two channels, however, is that the olfactory channel has a much higher affinity to both cGMP and cAMP compared to the rod channel, with a $K_{1/2}$ for cGMP of 1–4 μM and for cAMP of 2–20 μM (Nakamura and Gold, 1987; Kurahashi, 1989; Zufall et al., 1991; Frings et al., 1992; Kurahashi and Kaneko, 1993). The maximum macroscopic currents induced by the two ligands are also very similar (Nakamura and Gold, 1987), a property again different from the rod channel. At this point it remains a mystery why the channel has a slightly higher affinity for cGMP, even though the second messenger in olfactory transduction is cAMP.

Like the rod cGMP-activated channel, the olfactory cAMP/cGMP activated channel is also modulated by Ca^{2+} (Kramer and Siegelbaum, 1992), apparently through calmodulin as well (Chen and Yau, 1994).

2.9.1.3.1 Rat Olfactory cAMP/cGMP-Activated Channel

This is the first olfactory cAMP/cGMP-activated channel to be cloned and functionally expressed (Dhallan et al., 1990). The encoded protein has 664 amino acids (Figure 1). The channel shows remarkable homology to, for example, the bovine rod channel, with an overall 57% identity in amino acid sequence. The hydrophobicity plots of the two proteins are also very similar, suggesting similar folding patterns in the membrane. Compared with the native rat olfactory channel (Frings et al., 1992), it has a similar sensitivity to cGMP (1–2 μ*M*), but is considerably less sensitive to cAMP, with a $K_{1/2}$ of 40 to 70 μ*M* depending on voltage. Another difference is the long stable openings shown by the cloned channel (expressed in HEK 293 cells), much like those shown by the cloned rod channel mentioned earlier, a feature that is not generally observed in the native channel. These differences suggest that, like the rod channel, the native olfactory channel may be composed of more than one protein subunit species. Some most recent cloning work have confirmed this idea (Bradley et al., 1994; Liman and Buck, 1994).

2.9.1.3.2 Bovine Olfactory cAMP/cGMP-Activated Channel

This channel has also been cloned (Ludwig et al., 1990). The encoded protein has 663 amino acids (Figure 1). When expressed in *Xenopus* oocytes (Altenhofen et al., 1991), its properties are very similar to those of the cloned rat olfactory channel, including the peculiar feature that cAMP is much less effective than cGMP in activating the channel (see above). Like the rod channel, there is a threonine residue (Thr-537) in the cyclic nucleotide binding domain that is likely to be involved in hydrogen bonding with cGMP. As expected from the experiment already described above for the bovine rod channel, when this threonine is mutated into alanine the affinity of the channel for cGMP becomes reduced and approaches that for cAMP, which remains unchanged (Altenhofen et al., 1991). On the other hand, mutating the threonine into serine, which can also form a hydrogen bond with cGMP through its hydroxyl group, actually enhances the affinity of the channel for cGMP compared to wild-type.

2.9.1.3.3 Catfish Olfactory cAMP/cGMP-Activated Channel

This channel is also very similar to that in mammals. The cloned cDNA encodes a protein of 682 amino acids (Figure 1) that shows 54% identity with the bovine rod channel and 63% identity with the rat and bovine olfactory channels (Goulding et al., 1992). A surprising finding is worth noting here, however. The expressed channel (in *Xenopus* oocyte) shows roughly equal sensitivities to cAMP and cGMP ($K_{1/2}$ of 42 and 40 μ*M,* respectively), a feature broadly characteristic of the native olfactory channel of different animal species, including fish. On the other hand, these $K_{1/2}$ values are also 15-fold too high compared to the native channel. Another peculiarity is that, while the cloned channel shows roughly equal affinities for cAMP and cGMP, its sequence actually has a threonine residue (Thr-530) appropriately positioned in the cyclic nucleotide-binding domain for hydrogen bonding with cGMP. Thus, based on what is described earlier, this channel should have a higher affinity for cGMP than for cAMP. Why this is not the case remains a puzzle.

2.9.1.4 cGMP-Activated Channel in Pineal Gland

In fish and bird, the neurons in the pineal gland are light-sensitive (Pu and Dowling, 1981; Tamotsu and Morita, 1986; Robertson and Takahashi, 1988; Zatz and Mullen, 1988). In lamprey, for example, pineal photoreceptor cells give an electrical hyperpolarizing response just like retinal rods and cones (Pu and Dowling, 1981; Tamotsu and Morita, 1986). It is therefore likely that pineal and retinal photoreceptor neurons have a common phototransduction pathway, including the presence of a cGMP-activated channel. This has been confirmed by excised-patch recordings from dissociated pineal photoreceptor cells (Dryer and Henderson, 1991). The activation of the pineal channel by cGMP shows a similar $K_{1/2}$ value as the retinal channels; at the same time, it prefers cGMP over cAMP. Molecular cloning of the channel, however, remains to be done in order to know the degree of homology between the pineal and the retinal channels. It would also be interesting to find out whether the same channel is present in mammalian pinealocytes.

2.9.1.5 cGMP-Activated Channel in Invertebrate Photoreceptors

Unlike vertebrate retinal photoreceptors, invertebrate photoreceptors generally give a depolarizing response to light. The phototransduction process in invertebrate sensory neurons, however, appears to be more complex, with the possibility of parallel pathways (Anderson and Brown, 1988; Payne et al., 1988; Ranganathan et al., 1991; Lisman et al., 1992; Minke and Selinger, 1992). Currently, two second messengers, Ca^{2+} and cGMP, are thought to be important in this process. The evidence supporting a role of Ca^{2+}, involving both an IP_3-induced intracellular release and possibly also a Ca^{2+} influx from the extracellular space, is very strong (Anderson and Brown, 1988; Payne et al., 1988; Ranganathan et al., 1991; Lisman et al., 1992; Minke and Selinger, 1992). Its exact role in the transduction process, nonetheless, remains somewhat mirky, particularly in relation to the light-activated channels. As for cGMP, there is so far little firm evidence for a light-induced change in cGMP content in invertebrate photoreceptor cells (Brown et al., 1992). Its main supporting evidence has come from some experiments indicating the presence of a cGMP-activated channel in the light-sensitive lobe of *Limulus* ventral photoreceptors (Lisman et al., 1992; Bacigalupo et al., 1991; Johnson and Bacigalupo, 1992). In these experiments, involving excised, inside-out membrane patches, cGMP or the hydrolysis-resistant analog 8-bromo-cGMP evoked sustained channel openings that rapidly disappeared when the agonist was removed. The properties of these open channels are quite similar to those of the light-activated channels recorded from cell-attached patches. Thus, both show large (40–43 pS) and small (15–18 pS) single-channel conductances with linear current–voltage relations, reversal potentials near 0 mV, and open times of ca. 1 msec time constant. The open probability is nearly voltage independent at negative voltages, but increases dramatically at positive voltages, again like the light-activated channel. So far there is limited data on the activation dependence on cGMP concentration; the $K_{1/2}$ of the channel is possibly in the range of 10 to 30 μM cGMP. No information is available on whether the channel is sensitive to cAMP as well. Ca^{2+} does not activate the channel in the absence of cGMP (Bacigalupo et al., 1991). The above properties are broadly similar to those of the cGMP-activated channel in retinal rod photoreceptors, except that the invertebrate channel does not appear to be blocked by external divalent cations. Recent results using bilayer reconstitution methods indicate that squid and scallop photoreceptors may also have a cGMP-activated channel (Nasi and Gomez, 1990, 1991).

The structures of these invertebrate cGMP-activated channels remain to be elucidated by cloning studies. At the same time, it is not clear how light leads to the openings of these channels, and how these channels are linked to the IP_3-mediated Ca^{2+} pathway.

2.9.1.6 Putative cGMP-Activated Channel of Retinal On-Bipolar Cells

Retinal bipolar cells are postsynaptic to photoreceptor cells. They are of two types: one type depolarizes (on-type), and the other hyperpolarizes (off-type), in response to illumination. Off-bipolar cells have conventional glutamate receptor channels on them so that they are depolarized in darkness by the steady release of glutamate from the photoreceptors, and give a hyperpolarizing response to light when the reduced glutamate release leads to closure of these channels. On-bipolar cells are unusual in that they typically (for certain exceptions, see Saito et al., 1979; Hirano and MacLeish, 1991) have a nonselective cation channel that is closed in darkness by glutamate and opens when glutamate release is reduced in the light, thus giving a depolarizing response. There is now evidence that this cation channel is opened by the presence of cytoplasmic cGMP (Nawy and Jahr, 1990; Shiells and Falk, 1990, 1992a). The signal transduction pathway operating at this synapse appears to involve a special metabotropic glutamate receptor (which is sensitive to the agonist 2-amino-4-phosphonobutyrate, or APB) that via a G protein stimulates a cGMP phosphodiesterase to remove cGMP, hence causing channel closure (Shiells and Falk, 1992a; Nawy and Jahr, 1991). This signaling pathway is strongly reminiscent of the phototransduction pathway, though different proteins may be involved in the two processes (Shiells and Falk, 1992b; Vardi et al., 1993). It also remains to be definitively shown (such as by excised-patch recording) that the cation channel involved in this pathway is indeed directly gated by cGMP, instead of, say, controlled through phosphorylation by a cGMP-dependent protein kinase. Molecular cloning of this channel would ultimately help to answer this question as well. Incidentally, even if this channel is directly activated by cGMP, there is a good possibility that it is distinct from the photoreceptor channels, because it is unaffected by divalent cations (Shiells and Falk, 1992c).

2.9.2 Cyclic Nucleotide-Modulated Channels

This category includes several ion channels that do not require cyclic nucleotide binding for their opening, but their open probability is affected (increased or decreased) by cyclic nucleotide binding. By extrapolation from existing molecular information on a few of them, it is possible that all of them have a cyclic nucleotide-binding domain highly homologous to that in cyclic nucleotide-activated channels.

2.9.2.1 Renal cGMP-Modulated Channel

In the apical membrane of renal inner medullary collecting duct cells, there is an amiloride-sensitive cation channel that is inhibited by a rise in intracellular cGMP produced by the circulating hormone atrial natriuretic peptide (ANP) (Maack and Kleinert, 1986; Zeidel et al., 1987). This action serves to suppress Na^+ absorption across the collecting duct and therefore to increase urinary Na^+ excretion. This channel is not voltage dependent, is selective for monovalent cations (P_{Na}:P_{Cl} = 13:1), has a single-channel conductance of 28 pS, and is inhibited by micromolar concentrations of amiloride (Light et al., 1988). From excised, inside-out membrane patch experiments, it is now known that this channel is inhibited by direct cGMP binding (Light et al., 1989). The open probability of active channels in excised patches decreases from approximately 0.72 to 0.46 in the presence of 100 μ*M* cGMP, which produces maximal inhibition. Half-maximal inhibition occurs at ca. 7 μ*M* cGMP. The hydrolysis-resistant analog 8-bromo-cGMP has the same effect, indicating that cGMP hydrolysis is not required for the action. This is the first example of a channel that has its open probability reduced when cyclic nucleotide is bound to it. This channel is also selective for cGMP, because cAMP at up to 1 m*M* has no effect. What is particularly surprising about this channel is that, in addition to being directly inhibited by cGMP, it is also inhibited through a cGMP-dependent protein kinase (Light et al., 1990). This kinase does not appear to act on the channel directly by phosphorylation, but instead it appears to act through a G_i protein, which may be an agent that activates this channel (Light et al., 1990). The dual action of cGMP perhaps serves to provide a channel modulation that has one component with rapid onset and offset (through direct binding) and another component with slow recovery (through the kinase) (Light et al., 1990). This complex action observed in the renal channel raises the question whether an analogous control may apply to one or more of the cyclic nucleotide-activated channels described above.

2.9.2.2 cAMP-Modulated K^+ Channel in Larval *Drosophila* Muscle

In the longitudinal ventrolateral muscle of larval *Drosophila* there is a tetraethylammonium (TEA)-sensitive, K^+-selective (P_K/P_{Na} = 10) channel that is directly activated by cAMP (Delgado et al., 1991). This channel has a very low, basal probability of opening that is voltage independent. cAMP increases this open probability, with a $K_{1/2}$ of 50 μM and a Hill coefficient of near 3. cGMP is ineffective up to 80 μM, nor is AMP (80 μM), IP_3 (50 μM), ATP (1 mM), or Ca^{2+} (100 μM). There is no information on the physiological function of this channel, except that it contributes to the total resting conductance, and hence the resting membrane potential, of the muscle fiber. It is also not clear whether this channel is present in adult muscle or other excitable tissues in *Drosophila*. Since the channel has a nonzero open probability even in the absence of cAMP, we have classified it as a cyclic nucleotide-modulated channel by our criterion in Section 2.9.0.

2.9.2.3 cAMP-Modulated Pacemaker (I_f) Channel in Cardiac Pacemaker Cells

In cardiac pacemaker cells of the sinoatrial node, there is a hyperpolarization-activated cationic current (I_f) that is involved in the generation of the slow diastolic depolarization underlying spontaneous activity of the cells (DiFrancesco, 1993). This current is modulated by adrenaline and acetylcholine in opposite ways, leading to acceleration and deceleration of the pacemaker activity, respectively. Both neurotransmitters act by modulating adenylate cyclase activity and hence the level of intracellular cAMP, with a rise in cAMP produced by adrenaline leading to an increase of the I_f current. The modulation of voltage-activated channels by cAMP through phosphorylation has been well documented (see Levitan, 1985, 1994 for review), but in this case the effect appears to be a direct result of cAMP binding to the I_f channel. cAMP activates I_f by shifting its activation curve to more positive voltages, making the current more readily activated upon hyperpolarization of the cell; the fully activated current is not affected by cAMP (DiFrancesco and Tortora, 1991). The maximum shift of the activation curve is approximately 10 mV. This action of cAMP does not appear to be mediated by a G protein because it is not affected by the absence of GTP or the presence of the G protein inhibitor GDP-βS. The cAMP effect is half-maximal at 0.2 μM cAMP, with a Hill coefficient of 0.85. The Hill coefficient of near unity is interesting because it suggests that there may be only one cyclic nucleotide-binding site on the channel, unlike the situation with the cyclic nucleotide-activated channels in sensory neurons described above. cGMP and cCMP also have the same action, but are less effective, with $K_{1/2}$s of 8 and 12 μM, respectively. This channel is the first example of an ion channel the gating of which is dually controlled by voltage and direct cAMP binding.

2.9.2.4 *eag* K^+ Channel in *Drosophila*

The *eag (ether a go-go)* mutants in *Drosophila*, with a leg-shaking phenotype, show spontaneous repetitive firing of their motor neurons and elevated transmitter release at the larval neuromuscular junction (Kaplan and Trout, 1969; Ganetzky and Wu, 1985). Voltage-clamp experiments have shown that several K^+ currents are affected by *eag* mutations, including a transient K^+ current, a delayed noninactivating K^+ current, and a fast and a slow Ca^{2+}-activated K^+ currents (Wu et al., 1983; Zhong and Wu, 1991). The gene underlying the *eag* is now cloned (Warmke et al., 1991). The encoded protein has 1174 amino acids, with hydropathic analysis suggestive of a folding pattern similar to those of K^+ channels. Furthermore, the transmembrane domains in the *eag* protein show homology (approximately 25% identity) to the corresponding sequences in the *Shaker* family of K^+ channels. Particularly striking are the S4 region and the pore region. These findings together suggest that the *eag* protein constitutes a subunit common to different K^+ channels, though its size is about twice as large as the proteins of the *Shaker* family, with the nonoverlapping part being at the C-terminus. Subsequently, it was recognized that the *eag* protein bears even closer homology to the cyclic nucleotide-activated channels, including having a putative cyclic nucleotide-binding domain at the appropriate location (Guy et al., 1991). Recent electrophysiological evidence has confirmed that the *eag* protein indeed confers to K^+ channels the property of being modulated by cGMP (Zhong and Wu, 1993). Furthermore, the *eag* protein can also form a functional homomeric channel by itself (Robertson et al., 1993). Other recent experiments have indicated that this channel is dually controlled by both voltage and cAMP, and is permeable to both K^+ and Ca^{2+} (Brüggeman et al., 1993). Finally, it now appears that there is a family of *eag*-related genes in both *Drosophila* and mammals (Warmke and Ganetzky, 1993).

2.9.2.5 *KAT1* and *AKT1* Channels in Plant

Unlike animal cells, plant cells are generally exposed to low K^+ concentrations. For various cellular functions, these cells have to accumulate K^+ by uptake from the external environment, either by high-affinity K^+ transport systems or by K^+ channels (Gustin et al., 1986; Schroeder and Hedrich, 1989), helped by a generally very negative membrane potential of –120 to –250 mV maintained by an electrogenic H^+-ATPase. Using suppression of the mutant phenotype of a K^+ transport-deficient strain of yeast as an assay, two cDNAs have been independently isolated from the plant *Arabidopsis thaliana* (Sentenac et al., 1992; Anderson et al., 1992). These cDNAs, designated *AKT1* and *KAT1*, are derived from different genes but are highly homologous to each other. They encode proteins of 838 and 677 amino acids, respectively. Very surprisingly, they also show homology to cyclic nucleotide-activated channels and *Shaker* K^+ channels, including a similar predicted folding pattern in the membrane and the presence of a S4 domain, a homologous pore region, and a cyclic nucleotide-binding domain. The presence of a cyclic nucleotide-binding domain is interesting because there is yet no conclusive evidence for the presence of cyclic nucleotides in higher plants (Spiteri et al., 1989). In addition, at least *AKT1* also has an ankyrin-like region near the C-terminus, possibly for anchoring the channel protein to the cytoskeleton (Sentenac et al., 1992). The *KAT1* cDNA has been expressed in the *Xenopus* oocyte

(Schachtman et al., 1992). The resulting channel is K^+-selective, and is activated by hyperpolarizations more negative than –100 mV. It has a single-channel conductance of 28 pS and is blocked by TEA and barium on the extracellular side. This inward-rectifying K^+ channel, however, is different from those in animal cells, which based on recent cloning studies apparently belong to a new family of K^+ channels with a different structural motif (Ho et al., 1993; Kubo et al., 1993). No detailed studies have yet been carried out on the *AKT1* and *KAT1* proteins to see, for example, whether they require cyclic nucleotides in order to open, or whether their openings are simply modulated by cyclic nucleotides. Accordingly, for the time being they are put under the category of cyclic nucleotide-modulated channels. In any case, these proteins in plant suggest that there is indeed an ancient common ancester of both voltage-gated and cyclic nucleotide-gated channels (Jan and Jan, 1990).

2.9.3 Conclusion

In a span of less than 10 years, the number of cyclic nucleotide-gated channels identified has grown rapidly, with the different members involved in a variety of cellular functions. Undoubtedly, many other members with interesting properties and functions remain to be discovered.

Acknowledgments

This work is partially supported by a grant from the National Eye Institute (EY 06837) to King-Wai Yau.

REFERENCES

Adelman, J. P., Shen, K. E., Kavanaugh, M. P., Warren, R. A., Wu, Y. N., Lagrutta, A., Bond, C. T., and North, R. A. (1992). Calcium-activated potassium channels expressed from cloned complementary DNAs, *Neuron* 9, 209.

Ahmad, I., Redmond, L. J., and Barnstable, C. J. (1990). Developmental and tissue-specific expression of the rod photoreceptor cGMP-gated ion channel gene. *Biochem. Biophys. Res. Commun.* 173, 463.

Altenhofen, W., Ludwig, J., Eismann, E., Kraus, W., Bönigk, W., and Kaupp, U. B. (1991). Control of ligand specificity in cyclic nucleotide-gated channels from rod photoreceptors and olfactory epithelium. *Proc. Natl. Acad. Sci. U.S.A.* 88, 9868.

Anderson, J. A., Huprikar, S. S., Kochian, L. V., Lucas, W. J., and Gaber, R. R. (1992). Functional expression of a probable *Arabidopsis thaliana* potassium channel in *Saccharomyces cerevisiae*. *Proc. Natl. Acad. Sci. U.S.A.* 89, 3736.

Anderson, R. E. and Brown, J. E. (1988). Phosphoinositides in the retina. *Prog. Retinal Res.* 8, 211.

Atkinson, N. S., Robertson, G. A., and Ganetzky, B. (1991). A component of calcium-activated potassium channels encoded by the *Drosophila slo* locus. *Science* 253, 551.

Bacigalupo, J., Johnson, E. C., Vergara, C., and Lisman, J. E. (1991). Light-dependent channels from excised patches of *Limulus* ventral photoreceptors are opened by cGMP. *Proc. Natl. Acad. Sci. U.S.A.* 88, 7938.

Bodoia, R. D. and Detwiler, P. B. (1985). Patch-clamp recordings of the light-sensitive dark noise in retinal rods from the lizard and frog. *J. Physiol. (London)* 367, 183.

Bönigk, W., Altenhofen, W., Müller, F., Dose, A., Illing, M., Molday, R. S., and Kaupp, U. B. (1993). Rod and cone photoreceptor cells express distinct genes for cGMP-gated channels. *Neuron* 10, 865.

Bradley, J., Li, J., Davidson, N., Lester, H. A., and Zinn, K. (1994). Heteromeric olfactory cyclic nucleotide-gated channels: A new subunit that confers increased sensitivity to cAMP. *Proc. Natl. Acad. Sci. U.S.A.* in press.

Brown, J. E., Faddis, M., and Combs, A. (1992). Light does not induce an increase in cyclic-GMP content of squid or *Limulus* photoreceptors. *Exp. Eye Res.* 54, 403.

Brown, R. L., Gerber, W. V., and Karpen, J. W. (1993). Specific labeling and permanent activation of the retinal rod cGMP-activated channel by the photoaffinity analog 8-*p*-azidophenacylthio-cGMP. *Proc. Natl. Acad. Sci. U.S.A.* 90, 5369.

Brüggeman, A., Pardo, L. A., Stühmer, W., and Pongs, O. (1993). *Ether-a-go-go* encodes a voltage-gated channel permeable to K^+ and Ca^{2+} and modulated by cAMP. *Nature* 365, 445.

Caretta, A., Cavaggioni, A., and Sorbi, R. T. (1985). Binding stoichiometry of a fluorescent cGMP analogue to membranes of retinal rod outer segments. *Eur. J. Biochem.* 153, 49.

Chen, T.-Y., Illing, M., Molday, L. L., Hsu, Y.-T., Yau, K.-W., and Molday, R. S. (1994). Subunit 2 (or β) of retinal rod cGMP-gated cation channel is a component of the 240 kd channel-associated protein and mediates Ca^{2+} calmodulin modulation. *Proc. Natl. Acad. Sci. U.S.A.* in press.

Chen, T.-Y., Peng, Y.-W., Dhallan, R. S., Ahamed, B., Reed, R. R., and Yau, K.-W. (1993). A new subunit of the cyclic nucleotide-gated cation channel in retinal rods. *Nature* 362, 764.

Chen, T.-Y. and Yau, K.-W. (1994). Direct modulation by Ca^{2+} calmodulin of cyclic nucleotide-activated channel rat olfactory receptor neurons. *Nature* 368, 545.

Cook, N. J., Hanke, W., and Kaupp, U. B. (1987). Identification, purification, and functional reconstitution of the cyclic GMP-dependent channel from rod photoreceptors. *Proc. Natl. Acad. Sci. U.S.A.* 84, 585.

Cook, N. J., Zeilinger, C., Koch, K.-W., and Kaupp, U. B. (1986). Solubilization and functional reconstitution of the cGMP-dependent cation channel from bovine rod outer segments. *J. Biol. Chem.* 261, 17033.

Delgado, R., Hidalgo, P., Diaz, F., Latorre, R., and Labarca, P. (1991). A cyclic AMP-activated K^+ channel in *Drosophila* larval muscle is persistently activated in dunce. *Proc. Natl. Acad. Sci. U.S.A.* 88, 557.

Dhallan, R. S., Macke, J. P., Eddy, R. L., Shows, T. B., Reed, R. R., Yau, K.-W., and Nathans, J. (1992). Human rod photoreceptor cGMP-gated channel: Amino acid sequence, gene structure, and functional expression. *J. Neurosci.* 12, 3248.

Dhallan, R. S., Yau, K.-W., Schrader, K. A., and Reed, R. R. (1990). Primary structure and functional expression of a cyclic nucleotide-activated channel from olfactory neurons. *Nature* 347, 184.

DiFrancesco, D. (1993). Pacemaker mechanisms in cardiac tissue. *Annu. Rev. Physiol.* 55, 451.

DiFrancesco, D. and Tortora, P. (1991). Direct activation of cardiac pacemaker channels by intracellular cyclic AMP. *Nature* 351, 145.

Dryer, S. E. and Henderson, D. (1991). A cyclic GMP-activated channel in dissociated cells of the chick pineal gland. *Nature* 353, 756.

Fesenko, E. E., Kolesnikov, S. S., and Lyubarsky, A. L. (1985). Induction by cyclic GMP of cationic conductance in plasma membrane of retinal rod outer segment. *Nature* 313, 310.

Frech, G. C., VanDongen, M. J., Schuster, G., Brown, A. M., and Joho, R. H. (1989). A novel potassium channel with delayed rectifier properties isolated from rat brain by expression cloning. *Nature* 340, 642.

Frings, S., Lynch, J. W., and Lindemann, B. (1992). Properties of cyclic nucleotide-gated channels mediating olfactory transduction: Activation, selectivity and blockage. *J. Gen. Physiol.* 100, 45.

Ganetzky, B. and Wu, C.-F. (1985). Genes and membrane excitability in *Drosophila*. *Trends Neurosci.* 8, 322.

Gordon, S. E., Brautigan, D. L., and Zimmerman, A. L. (1992). Protein phosphatases modulate the apparent agonist affinity of the light-regulated ion channel in retinal rods. *Neuron* 9, 739.

Goulding, E. H., Ngai, J., Kramer, R. H., Colicos, S., Axel, R., Siegelbaum, S. A., and Chess, A. (1992). Molecular cloning and single-channel properties of the cyclic nucleotide-gated channel from catfish olfactory neurons. *Neuron* 8, 45.

Goulding, E. H., Tibbs, G. R., Liu, D., and Siegelbaum, S. A. (1993). Role of H5 domain in determining pore diameter and ion permeation through cyclic nucleotide-gated channels. *Nature* 364, 61.

Gray, P. and Attwell, D. (1985). Kinetics of light-sensitive channels in vertebrate photoreceptors. *Proc. R. Soc. London* B223, 379.

Gustin, M. C., Martinac, B., Saimi, Y., Culbertson, M. R., and Kung, C. (1986). Ion channels in yeast. *Science* 233, 1195.

Guy, H. R., Durell, S. R., Warmke, J., Drysdale, R., and Ganetzky, B. (1991). Similarities in amino acid sequences of Drosophila *eag* and cyclic nucleotide-gated channels. *Science* 254, 730.

Hagins, W. A., Penn, R. D., and Yoshikami, S. (1970). Dark current and photocurrent in retinal rods. *Biophys. J.* 10, 380.

Haynes, L. W. (1992). Block of the cyclic GMP-gated channel of vertebrate rod and cone photoreceptors by *L-cis*-diltiazem. *J. Gen. Physiol.* 100, 783.

Haynes, L. W. and Yau, K.-W. (1985). Cyclic GMP-sensitive conductance in outer segment membrane of catfish cones. *Nature* 317, 61.

Haynes, L. W. and Yau, K.-W. (1990). Single-channel measurement from the cyclic GMP-activated conductance of catfish retinal cones. *J. Physiol.* 429, 451.

Heginbotham, L., Abramson, T., and MacKinnon, R. (1992). Functional connection between the pores of distantly related ion channels as revealed by mutant K^+ channels. *Science* 258, 1152.

Heinemann, S. H., Terlau, H., Stühmer, W., Imoto, K., and Numa, S. (1992). Calcium channel characteristics conferred on the sodium channel by single mutations. *Nature* 356, 441.

Hirano, A. A. and MacLeish, P. R. (1991). Glutamate and 2-amino-4-phosphonobutyrate evoke an increase in potassium conductance in retinal bipolar cells. *Proc. Natl. Acad. Sci. U.S.A.* 88, 805.

Ho, K., Nichols, C. G., Lederer, W. J., Lytton, J., Vassilev, P. M., Kanazirska, M. V., and Hebert, S. C. (1993). Cloning and expression of an inwardly rectifying ATP-regulated potassium channel. *Nature* 362, 31.

Hsu, Y.-T. and Molday, R. S. (1993). Modulation of the cGMP-gated channel of rod photoreceptor cells by calmodulin. *Nature* 361, 76.

Ildefonse, M. and Bennett, N. (1991). Single-channel study of the cGMP-dependent conductance of retinal rods from incorporation of native vesicles into planar lipid bilayers. *J. Memb. Biol.* 123, 133.

Ildefonse, M., Crouzy, S., and Bennett, N. (1992). Gating of retinal rod cation channel by different nucleotides: Comparative study of unitary currents. *J. Memb. Biol.* 130, 91.

Jan, L.-Y. and Jan, Y. N. (1990). A superfamily of ion channels. *Nature* 345, 672.

Johnson, E. C. and Bacigalupo, J. (1992). Spontaneous activity of the light-dependent channel irreversibly induced in excised patches from *Limulus* ventral photoreceptors. *J. Memb. Biol.* 130, 33.

Kaplan, W. D. and Trout, W. E. (1969). The behavior of four neurological mutants of *Drosophila*. *Genetics* 61, 399.

Karpen, J. W., Brown, R. L., Stryer, L., and Baylor, D. A. (1993). Interactions between divalent cations and the gating machinery of cyclic GMP-activated channels in salamander retinal rods. *J. Gen. Physiol.* 101, 1.

Karpen, J. W., Zimmerman, A. L., Stryer, L., and Baylor, D. A. (1988). Gating kinetics of the cyclic GMP-activated channel of retinal rods: Flash photolysis and voltage-jump studies. *Proc. Natl. Acad. Sci. U.S.A.* 85, 1287.

Kaupp, U. B. and Koch, K.-W. (1992). Role of cGMP and Ca^{2+} in vertebrate photoreceptor excitation and adaptation. *Annu. Rev. Physiol.* 54, 153.

Kaupp, U. B., Niidome, T., Tanabe, T., Terada, S., Bönigk, W., Stühmer, W., Cook, N. J., Kangawa, K., Matsuo, H., Hirose, T., Miyata, T., and Numa, S. (1989). Primary structure and functional expression from complementary DNA of the rod photoreceptor cyclic GMP-gated channel. *Nature* 342, 762.

Koch, K.-W. and Kaupp, U. B. (1985). Cyclic GMP directly regulates a cationic conductance in membranes of bovine rods by a cooperative mechanism. *J. Biol. Chem.* 260, 6788.

Koch, K.-W., Cook, N. J., and Kaupp, U. B. (1987). The cGMP-dependent channel of vertebrate rod photoreceptors exists in two forms of different cGMP sensitivity and pharmacological behavior. *J. Biol. Chem.* 262, 14415.

Koutalos, Y. and Yau, K.-W. (1993). A rich complexity emerges in phototransduction. *Curr. Opinion Neurobiol.*, 3, 513.

Kramer, R. H. and Siegelbaum, S. A. (1992). Intracellular Ca^{2+} regulates the sensitivity of cyclic nucleotide-gated channels in olfactory receptor neurons. *Neuron* 9, 897.

Kramer, R. H., Goulding, E., and Siegelbaum, S. A. (1994). Potassium channel inactivation peptide blocks cyclic nucleotide-gated channels by binding to the conserved pore domain. *Neuron* 12, 655.

Kubo, Y., Baldwin, T. J., Jan, Y. N., and Jan, L. Y. (1993). Primary structure and functional expression of a mouse inward rectifier potassium channel. *Nature* 362, 127.

Kurahashi, T. (1989). Activation by odorants of cation-selective conductance in the olfactory receptor cell isolated from the newt. *J. Physiol.* 419, 177.

Kurahashi, T. and Kaneko, A. (1993). Gating properties of the cAMP-gated channel in toad olfactory receptor cells. *J. Physiol. (London)* 466, 287.

Lagnado, L. and Baylor, D. A. (1992). Signal Flow in visual transduction. *Neuron* 8, 995.

Lancet, D. A. (1986). Vertebrate olfactory reception. *Annu. Rev. Neurosci.* 9, 329.

Levitan, I. B. (1985). Phosphorylation of ion channels. *J. Memb. Biol.* 87, 177.

Levitan, I. B. (1994). Modulation of ion channels by protein phosphorylation and dephosphorylation. *Annu. Rev. Physiol.* 56, 193.

Light, D. B., Corbin, J. D., and Stanton, B. A. (1990). Dual ion-channel regulation by cyclic GMP and cyclic GMP-dependent protein kinase. *Nature* 344, 336.

Light, D. B., McCann, F. V., Keller, T. M., and Stanton, B. A. (1988). Amiloride-sensitive cation channel in apical membrane of inner medullary collecting duct. *Am. J. Physiol.* 255, F278.

Light, D. B., Schwiebert, E. M., Karlson, K. H., and Stanton, B. A. (1989). Atrial natriuretic peptide inhibits a cation channel in renal inner medullary collecting duct cells. *Science* 243, 383.

Liman, E. R. and Buck, L. B. (1994). A second subunit of the olfactory cyclic nucleotide-gated channel confers high sensitivity to cAMP. *Neuron* in press.

Lisman, J., Erickson, M. A., Richard, E. A., Cote, R. H., Bacigalupo, J., Johnson, E., and Kirkwood, A. (1992). Mechanisms of amplification, deactivation, and noise reduction in invertebrate photoreceptors. In *Sensory Transduction.* Corey, D. P. and Roper, S. D., Eds., The Rockefeller University Press, New York.

Ludwig, J., Margalit, T., Eismann, E., Lancet, D., and Kaupp, U. B. (1990). Primary structure of cAMP-gated channel from bovine olfactory epithelium. *FEBS Lett.* 270, 24.

Maack, T. and Kleinert, H. D. (1986). Renal and cardiovascular effects of atrial natriuretic factor. *Biochem. Pharmacol.* 35, 2057.

MacKinnon, R. (1991). Determination of the subunit stoichiometry of a voltage-activated potassium channel. *Nature* 350, 232.

Matthews, G. (1986). Comparison of the light-sensitive and cyclic GMP-sensitive conductances of the rod photoreceptor: Noise characteristics. *J. Neurosci.* 6(9), 2521.

Matthews, G. and Watanabe, S.-I. (1988). Activation of single ion channels from toad retinal rod inner segments by cyclic GMP: Concentration dependence. *J. Physiol. (London)* 403, 389.

McGeoch, J. E. M. and Guidotti, G. (1992). An insulin-stimulated cation channel in skeletal muscle: Inhibition by calcium causes oscillation. *J. Biol. Chem.* 267, 832.

McKay, D. B. and Steitz, T. A. (1981). Structure of catabolite gene activator protein at 2.9 Å resolution suggests binding to left-handed B-DNA. *Nature (London)* 290, 744.

McKay, D. B., Weber, I. T., and Steitz, T. A. (1982). Structure of catabolite gene activator protein at 2.9-Å resolution. *J. Biol. Chem.* 257, 9518.

McLatchie, I. M. and Matthews, H. R. (1992). Voltage-dependent block by *L-cis*-diltiazem of the cyclic GMP-activated conductance of salamander rods. *Proc. R. Soc. London* B247, 113.

McNaughton, P. A. (1990). Light response of vertebrate photoreceptors. *Physiol. Rev.* 70, 847.

Menini, A. (1990). Currents carried by monovalent cations through cyclic GMP-activated channels in excised patches from salamander rods. *J. Physiol. (London)* 424, 167.

Minke, B. and Selinger, Z. (1992). Inositol lipid pathway in fly photoreceptors: Excitation, calcium mobilization and retinal degeneration. *Prog. Retinal Res.* 11, 99.

Molday, L. L., Cook, N. J., Kaupp, U. B., and Molday, R. S. (1990). The cGMP-gated cation channel of bovine rod photoreceptor cells is associated with a 240-kDa protein exhibiting immunochemical cross-reactivity with spectrin. *J. Biol. Chem.* 265, 18690.

Molday, R. S., Molday, L. L., Dosé, A., Clark-Lewis, I., Illing, M., Cook, N. J., Eismann, E., and Kaupp, U. B. (1991). The cGMP-gated channel of the rod photoreceptor cell: Characterization and orientation of the amino terminus. *J. Biol. Chem.* 266, 21917.

Mori, Y., Friedrich, T., Kim, M.-S., Mikami, A., Nakai, J., Ruth, P., Bosse, E., Hofman, F., Flockerzi, V., Furuichi, T., Mikoshiba, K., Imoto, K., Tanabe, T., and Numa, S. (1991). Primary structure and functional expression from complementary DNA of a brain calcium channel. *Nature* 350, 398.

Nakamura, T. and Gold, G. H. (1987). A cyclic nucleotide-gated conductance in olfactory receptor cilia. *Nature* 325, 442.

Nakatani, K. and Yau, K.-W. (1988). Guanosine 3′,5′-cyclic monophosphate-activated conductance studied in a truncated rod outer segment of the toad. *J. Physiol. (London)* 395, 731.

Nakatani, K. and Yau, K.-W. (1989). Sodium-dependent calcium extrusion and sensitivity regulation in retinal cones of the salamander. *J. Physiol.* 409, 525.

Nasi, E. and Gomez, M. (1990). Recording from solitary photoreceptors and reconstituted rhadbomeric membranes of the squid. *Biophys. J.* 57(2), 368 (abstr.).

Nasi, E. and Gomez, M. (1991). Light-activated channels in scallop photoreceptors: Recordings from cell-attached and perfused excised patches. *Biophys. J.* 59(2), 540 (abstr.).

Nawy, S. and Jahr, C. E. (1990). Suppression by glutamate of cGMP-activated conductance in retinal bipolar cells. *Nature* 346, 269.

Nawy, S. and Jahr, C. E. (1991). cGMP-gated conductance in retinal bipolar cells is suppressed by the photoreceptor transmitter. *Neuron* 7, 677.

Nicol, G. D., Schnetkamp, P. P. M., Saimi, Y., Cragoe, E. J., and Bownds, M. D. (1987). A derivative of amiloride blocks both the light-and cyclic GMP-regulated conductances in rod photoreceptors. *J. Gen. Physiol.* 90, 651.

Payne, R., Walz, B., Levy, S., and Fein, A. (1988). The localization of calcium release in *Limulus* ventral photoreceptors and its control by negative feedback. *Phil. Trans. Roy. Soc.* 320, 359.

Perry, R. J. and McNaughton, P. A. (1991). Response properties of cones from the retina of the tiger salamander. *J. Physiol.* 433, 561.

Picones, A. and Korenbrot, J. I. (1992). Permeation and interaction of monovalent cations with the cGMP-gated channel of cone photoreceptors. *J. Gen. Physiol.* 100, 647.

Pittler, S. J., Lee, A. K., Altherr, M. R., Howard, T. A., Seldin, M. F., Hurwitz, R. L., Wasmuth, J. J., and Baehr, W. (1992). Primary structure and chromosomal localization of human and mouse rod photoreceptor cGMP-gated cation channel. *J. Biol. Chem.* 267, 6257.

Pu, G. A. and Dowling, J. E. (1981). Anatomical and physiological characteristics of pineal photoreceptor cell in the larval lamprey, *Petromyzon marinus. J. Neurophysiol.* 46, 1018.

Puckett, K. L. and Goldin, S. M. (1986). Guanosine 3′,5′-cyclic monophosphate stimulates release of actively accumulated calcium in purified disks from rod outer segments of bovine retina. *Biochemistry* 25, 1739.

Pugh, E. N. and Lamb, T. D. (1990). Cyclic GMP and calcium: The internal messengers of excitation and adaptation in vertebrate photoreceptors. *Vision Res.* 30, 1923.

Quandt, F. N., Nicol, G. D., and Schnetkamp, P. P. M. (1991). Voltage-dependent gating and block of the cyclic-GMP-dependent current in bovine rod outer segments. *Neuroscience* 42, 629.

Ranganathan, R., Harris, W. A., and Zuker, C. S. (1991). The molecular genetics of invertebrate phototransduction. *Trends Neurosci.* 14, 486.

Reed, R. R. (1992). Signaling Pathways in Odorant Detection. *Neuron* 8, 205.

Robertson, G. A., Warmke, J. W., and Ganetzky, B. (1993). Functional expression of the *Drosophila EAG* K^+ channel gene. *Biophys. J.* 64(2), 340 (abstr.).

Robertson, L. M. and Takahashi, J. S. (1988). Circadian clock in cell culture. II. In vitro photic entrainment of melatonin oscillation from dissociated chick pineal cells. *J. Neurosci.* 8, 22.

Root, M. J. and MacKinnon, R. (1993). Identification of an external divalent cation-binding site in the pore of a cGMP-activated channel. *Neuron* 11, 459.

Saito, T., Kondo, H., and Toyoda, J.-I. (1979). Ionic mechanisms of two types of on-center bipolar cells in the carp retina. *J. Gen. Physiol.* 73, 73.

Schachtman, D. P., Schroeder, J. I., Lucas, W. J., Anderson, J. A., and Gaber, R. F. (1992). Expression of an inward-rectifying potassium channel by the *Arabidopsis* KAT1 cDNA. *Science* 258, 1654.

Schnetkamp, P. P. M. (1987). Sodium ions selectively eliminate the fast component of guanosine cyclic 3′5′-phosphate induced Ca^{2+} release from bovine rod outer segment disks. *Biochemistry* 26, 3249.

Schroeder, J. I. and Hedrich, R. (1989). Involvement of ion channels and active transport in osmoregulation and signaling of higher plant cells. *Trends Biochem. Sci.* 14, 187.

Sentenac, H., Bonneaud, N., Minet, M., Lacroute, F., Salmon, J.-M., Gaymard, F., and Grignon, C. (1992). Cloning and expression in yeast of a plant potassium ion transport system. *Science* 256, 663.

Shabb, J. B., Ng, L., and Corbin, J. D. (1990). One amino acid change produces a high affinity cGMP-binding site in cAMP-dependent protein kinase. *J. Biol. Chem.* 265, 16031.

Shiells, R. A. and Falk, G. (1990). Glutamate receptors of rod bipolar cells are linked to a cyclic GMP cascade via a G-protein. *Proc. R. Soc. London B* 242, 91.

Shiells, R. A. and Falk, G. (1992a). The glutamate-receptor linked cGMP cascade of retinal on-bipolar cells is pertussis and cholera toxin-sensitive. *Proc. R. Soc. London B* 247, 17.

Shiells, R. A. and Falk, G. (1992b). Retinal on-bipolar cells contain a nitric oxide-sensitive guanylate cyclase. *NeuroReport* 3, 845.

Shiells, R. A. and Falk, G. (1992c). Properties of the cGMP-activated channel of retinal on-bipolar cells. *Proc. R. Soc. London B* 247, 21.

Spiteri, A., Viratelle, O. M., Raymond, P., Roncillac, M., Labouesse, J., and Pradet, A. (1989). Artefactual origin of cyclic AMP in higher plant tissues. *Plant Physiol.* 91, 624.

Stern, J. H., Kaupp, U. B., and MacLeish, P. R. (1986). Control of the light-regulated current in rod photoreceptors by cyclic GMP, calcium and *l-cis*-diltiazem. *Proc. Natl. Acad. Sci. U.S.A.* 83, 1163.

Stryer, L. (1991). Visual excitation and recovery. *J. Biol. Chem.* 266, 10711.

Sudlow, L. C., Huang, R. C., Green, D. J., and Gillette, R. (1993). cAMP-activated Na^+ current of molluscan neurons is resistant to kinase inhibitors and is gated by cAMP in the isolated patch. *J. Neurosci.* 13, 5188.

Swope, S. L., Moss, S. J., Blackstone, C. D., and Huganir, R. L. (1992). Phosphorylation of ligand-gated ion channels: A possible role of synaptic plasticity. *FASEB J.* 6, 2514.

Tamotsu, S. and Morita, Y. (1986). Photoreception in pineal organs of larval and adult lampreys, *Lampetra japonica*. *J. Comp. Physiol.* A159, 1.

Tanaka, J. C., Eccleston, J. F., and Furman, R. E. (1989). Photoreceptor channel activation by nucleotide derivatives. *Biochemistry* 28, 2776.

Tempel, B. L., Papazian, D. M., Schwarz, T. L., Jan, Y. N., and Jan, L. Y. (1987). Sequence of a probable potassium channel component encoded at *Shaker* locus of *Drosophila*. *Science* 237, 770.

Torre, V., Straforini, M., Sesti, F., and Lamb, T. D. (1992). Different channel-gating properties of two classes of cyclic GMP-activated channel in vertebrate photoreceptors. *Proc. R. Soc. London* B250, 209.

Vardi, N., Matesic, D. F., Manning, D. R., Liebman, P. A., and Sterling, P. (1993). Identification of a G-protein in depolarizing rod bipolar cells. *Vis. Neurosci.* 10, 473.

Warmke, J. W. and Ganetzky, B. (1993). A novel potassium channel gene family: *EAG* homologs in *Drosophila*, mouse and human. *Biophys. J.* 64(2), 340 (abstr.).

Warmke, J., Drysdale, R., and Ganetzky, B. (1991). A distinct potassium channel polypeptide encoded by the *Drosophila eag* locus. *Science* 252, 1560.

Watanabe, S.-I. and Matthews, G. (1990). Cyclic GMP-activated channels of rod photoreceptors show neither fast nor slow desensitization. *Vis. Neurosci.* 4, 481.

Watanabe, S.-I. and Murakami, M. (1991). Similar properties of cGMP-activated channels between cones and rods in the carp retina. *Vis. Neurosci.* 6, 563.

Weber, I. T., Shabb, J. B., and Corbin, J. D. (1989). Predicted structure of the cGMP binding domains of the cGMP-dependent protein kinase: A key alanine/threonine difference in evolutionary divergence of cAMP and cGMP binding sites. *Biochemistry* 28, 6122.

Weber, I. T., Steitz, T. A., Bubis, J., and Taylor, S. S. (1987). Predicted structures of cAMP binding domains of type I and II regulatory subunits of cAMP-dependent protein kinase. *Biochemistry* 26, 343.

Wohlfart, P., Haase, W., Molday, R. S., and Cook, N. J. (1992). Antibodies against synthetic peptides used to determine the topology and site of glycosylation of the cGMP-gated channel from bovine rod photoreceptors. *J. Biol. Chem.* 267, 644.

Wu, C.-F., Ganetzky, B., Haugland, F. N., and Liu, A. X. (1983). Potassium currents in *Drosophila:* Different components affected by mutations of two genes. *Science* 220, 1076.

Yau, K.-W. (1991). Calcium and light adaptation in retinal photoreceptors. *Curr. Opinion Neurobiol.* 1, 252.

Yau, K.-W. and Baylor, D. A. (1989). Cyclic GMP-activated conductance of retinal photoreceptor cells. *Annu. Rev. Neurosci.* 12, 289.

Yau, K.-W. and Nakatani, K. (1985). Light-suppressible, cyclic GMP-sensitive conductance in the plasma membrane of a truncated rod outer segment. *Nature* 317, 252.

Yau, K.-W., Haynes, L. W., and Nakatani, K. (1986). Roles of calcium and cyclic GMP in visual transduction. In *Membrane Control of Cellular Activity*, Lüttgau, H. C., Ed., Gustav Fischer, Stuttgart, Germany, 343.

Zatz, M. and Mullen, D. A. (1988). Two mechanisms of photoendocrine transduction in cultured chick pineal cells: Pertussis toxin blocks the acute but not the phase-shifting effects of light on the melatonin rhythm. *Brain Res.* 453, 63.

Zeidel, M. L., Silva, P., Brenner, B. M., and Seifter, J. L. (1987). cGMP mediates effects of atrial peptides on medullary collecting duct cells. *Am. J. Physiol.* 252, F551.

Zhong, Y. and Wu, C.-F. (1991). Alteration of four identified K^+ currents in *Drosophila* muscle by mutations in *eag*. *Science* 252, 1562.

Zhong, Y. and Wu, C.-F. (1993). Modulation of different K^+ currents in *Drosophila:* A hypothetical role for the eag subunit in multimeric K^+ channels. *J. Neurosci.* 13, 4669.

Zimmerman, A. L. and Baylor, D. A. (1992). Cation interactions within the cyclic GMP-activated channel of retinal rods from the tiger salamander. *J. Physiol. (London)* 449, 759.

Zimmerman, A. L., Yamanaka, G., Eckstein, F., Baylor, D. A., and Stryer, L. (1985). Interaction of hydrolysis-resistant analogs of cyclic GMP with the phosphodiesterase and light-sensitive channel of retinal rod outer segments. *Proc. Natl. Acad. Sci. U.S.A.* 82, 8813.

Zufall, F., Firestein, S., and Shepherd, G. M. (1991). Analysis of single cyclic nucleotide-gated channels in olfactory receptor cells. *J. Neurosci.* 11, 3573.

Index

B

C

D

H

I

J

K

L

M

N

Q

R

S